T0293799

Circulating Tumor Cells in Cancer

Circulating Tumor Cells in Cancer

Edited by Jonah Randall

New York

Hayle Medical,
750 Third Avenue, 9th Floor,
New York, NY 10017, USA

Visit us on the World Wide Web at:
www.haylemedical.com

ISBN: 978-1-64647-530-8

Cataloging-in-Publication Data

Circulating tumor cells in cancer / edited by Jonah Randall.
p. cm.
Includes bibliographical references and index.
ISBN 978-1-64647-530-8
1. Cancer cells. 2. Cancer cells--Growth. 3. Tumor markers.
4. Metastasis. I. Randall, Jonah.

RC269.7 .C57 2023
616.994 07--dc23

Table of Contents

Preface

This book was inspired by the evolution of our times; to answer the curiosity of inquisitive minds. Many developments have occurred across the globe in the recent past which has transformed the progress in the field.

A circulating tumor cell (CTC) refers to a type of a cell that has been shed from a primary tumor into the lymphatic system or the vascular system, and is carried all over the body through blood circulation. CTCs have the ability to extravasate and become a source for the creation of new tumors in distant organs. The identification and analysis of CTC can aid in early patient prognoses and deciding suitable personalized therapy. The detection of CTCs using liquid biopsy has a number of advantages over conventional tissue biopsies. Liquid biopsies are non-invasive and can be used multiple times. They also provide valuable information about metastatic risk, disease progression and treatment success. Analysis of blood samples from cancer patients has an increased possibility of CTC detection as the disease progresses. This book unravels the recent studies on circulating tumor cells in cancer. It presents researches and studies performed by experts across the globe. This book is a vital tool for all researching and studying circulating tumor cells.

This book was developed from a mere concept to drafts to chapters and finally compiled together as a complete text to benefit the readers across all nations. To ensure the quality of the content we instilled two significant steps in our procedure. The first was to appoint an editorial team that would verify the data and statistics provided in the book and also select the most appropriate and valuable contributions from the plentiful contributions we received from authors worldwide. The next step was to appoint an expert of the topic as the Editor-in-Chief, who would head the project and finally make the necessary amendments and modifications to make the text reader-friendly. I was then commissioned to examine all the material to present the topics in the most comprehensible and productive format.

I would like to take this opportunity to thank all the contributing authors who were supportive enough to contribute their time and knowledge to this project. I also wish to convey my regards to my family who have been extremely supportive during the entire project.

Editor

Analysis of Circulating Tumor Cells in Patients with Non-Metastatic High-Risk Prostate Cancer before and after Radiotherapy Using Three Different Enumeration Assays

Joanna Budna-Tukan [1,*,†], Monika Świerczewska [1,†], Martine Mazel [2], Wojciech A. Cieślikowski [3], Agnieszka Ida [3], Agnieszka Jankowiak [1], Andrzej Antczak [3], Michał Nowicki [1], Klaus Pantel [4], David Azria [5], Maciej Zabel [6,7] and Catherine Alix-Panabières [2,*]

1 Department of Histology and Embryology, Poznan University of Medical Sciences, 60-781 Poznan, Poland; mswierczewska@ump.edu.pl (M.Ś.); agajankowiak@hotmail.com (A.J.); mnowicki@ump.edu.pl (M.N.)
2 Laboratory of Rare Human Circulating Cells (LCCRH), University Medical Centre of Montpellier, 34093 Montpellier, France; martine-mazel@chu-montpellier.fr
3 Department of Urology, Poznan University of Medical Sciences, 61-285 Poznan, Poland; w.cieslikowski@gmail.com (W.A.C.); agnieszka.ida@gmail.com (A.I.); aa26@poczta.onet.pl (A.A.)
4 Department of Tumor Biology, University Medical Centre Hamburg-Eppendorf, 20246 Hamburg, Germany; pantel@uke.de
5 Radiation Oncology Department, Montpellier Cancer Institute, 34298 Montpellier, France; david.azria@icm.unicancer.fr
6 Division of Histology and Embryology, Department of Human Morphology and Embryology, Wroclaw Medical University, 50-368 Wroclaw, Poland; mazab@ump.edu.pl
7 Division of Anatomy and Histology, University of Zielona Góra, 65-046 Zielona Góra, Poland
* Correspondence: jbudna@ump.edu,pl (J.B.-T.); c-panabieres@chu-montpellier.fr (C.A.-P.);

† These authors contributed equally to this work.

Abstract: The characterization of circulating tumor cells (CTCs) can lead to a promising strategy for monitoring residual or relapsing prostate cancer (PCa) after local therapy. The aim of this study was to compare three innovative technologies for CTC enumeration in 131 high-risk patients with PCa, before and after radiotherapy, combined with androgen deprivation. The CTC number was tested using the FDA-cleared CellSearch® system, the dual fluoro-EPISPOT assay that only detects functional CTCs, and the in vivo CellCollector® technology. The highest percentage of CTC-positive patients was detected with the CellCollector® (48%) and dual fluoro-EPISPOT (42%) assays, while the CellSearch® system presented the lowest rate (14%). Although the concordance among methods was only 23%, the cumulative positivity rate was 79%. A matched-pair analysis of the samples before, and after, treatment suggested a trend toward a decrease in CTC count after treatment with all methods. CTC tended to be positivity correlated with age for the fluoro-EPISPOT assay and with PSA level from the data of three assays. Combining different CTC assays improved CTC detection rates in patients with non-metastatic high-risk PCa before and after treatment. Our findings do not support the hypothesis that radiotherapy leads to cancer cell release in the circulation.

Keywords: liquid biopsy; CTC; prostate cancer; early diagnostic; PSA; radiotherapy

---

1. Introduction

The management of patients with prostate cancer (PCa) is challenging because more than 40% of surgically treated patients will experience a biochemical disease recurrence [1]. Moreover, according to D'Amico criteria, 15% of PCa patients are already high risk at diagnosis, and are defined as clinical T stage ≥cT2c, a Gleason score ≥8, or a prostate specific antigen (PSA) >20 ng/mL [2]. Local therapy, such as radical prostatectomy and/or radiation, is complex because PCa heterogeneous phenotypes lead to different clinical outcomes, ranging from indolent to lethal metastatic diseases [3]. Thus, PCa diagnosis is a crucial issue as early disease detection significantly increases recovery rate and overall survival. The most commonly used diagnostic tools, such as serum PSA level, transrectal ultrasonography (TRUS), Gleason histologic grading of biopsy specimens, and the clinical tumor, node, metastasis (TNM) staging, do not allow for the precise determination of each patient's current tumor status [1]. Indeed, serum PSA level does not unequivocally differentiate between aggressive and indolent cancer, and can be affected by several factors, such as age or urinary tract infections [4]. Moreover, the reliability of PSA testing for PCa screening is insufficient because patients with PSA lower than 4.0 ng/mL can have PCa [5]. Biopsy analysis can give false-negative results, caused by small tumor size or uneven distribution of cancer cells within the tumor [6]. Hence, novel, specific, and improved prognostic markers are needed to determine clinically significant tumors causing progressive disease, and this information can then be implemented into personalized treatment of cancer patients at early disease stages [7,8].

The characterization of circulating tumor cells (CTCs) is a new promising method to understand the biology of residual or relapsing PCa after local therapy. As CTCs are shed into the bloodstream during primary tumor growth and progression, they can be used to predict disease progression, select therapeutic targets, and monitor therapy outcome [9]. It has been shown that CTC count is proportional to survival in metastatic breast, colorectal, and prostate cancer [10]. In patients with non-metastatic cancers (TNM-stage M0), including PCa, the clinical relevance of CTC counts is still under investigation. The strongest data sets exist for breast cancer, demonstrating independent prognostic relevance in patients receiving (neo-)adjuvant chemotherapy [11–14].

The major challenge in CTC research is the isolation of these very rare cells (typically one single CTC in 10^6–10^7 leukocytes from peripheral blood of patients with cancer) [15]. A number of new innovative technologies, that improve CTC detection, have recently been described. The most widely used approach is based on immunoaffinity, where cells are captured by targeting tumor-associated antigens (positive enrichment). They include immunomagnetic assays, like CellSearch® [16], Adna Test [17], or MACS [18], microfluidic devices like CTC-Chip [19], Herringbone Chip [20], and spiral ClearCell® FX chip [21] or finally combination of mentioned above, like Ephesia [22], LiquidBiopsy [23] and IsoFlux [24]. On the contrary, some technologies use antibodies against leukocyte-associated antigens (negative enrichment), i.e., EasyStep Human CD45 Depletion Kit [25] or QMS [26]. Among them, the EpCAM-dependent CellSearch® system, which is the only method currently approved by the Food and Drug Administration (FDA) for CTC selection and enumeration, is regarded as the "gold standard" and used for monitoring metastatic breast, colon, and prostate cancer [10]. Studies in patients with PCa showed a correlation between CTC number and clinical outcome [27].

However, CTC detection still faces many problems. Cancer heterogeneity seems to be the major obstacle because even cancer cells derived from the same tumor can present various immunophenotypes. Epithelial to mesenchymal transition (EMT) is considered a crucial reason for the change. During this process the expression of epithelial markers, such as EpCAM and/or cytokeratins can be significantly downregulated and may limit CTC detection by EpCAM-dependent technologies, like CellSearch® [28].

This is the rationale to combine EpCAM-dependent with EpCAM-independent technologies. Another limitation of the "gold standard" system is the fact that the targeted epithelial phenotype (EpCAM- and cytokeratin-positive) does not mirror the tumor-specific phenotype. Thus the cytokeratin-negative, non-epithelial sub-population of CTCs, is lost during the analysis, which is restricted to the initial biomarker selection [29,30]. Moreover, CellSearch® presents quite a high rate of apoptotic cells captured [31], and which cannot determine whether the detected CTCs are viable and able to form metastases. The fluoro-EPISPOT assay (Epithelial ImmunoSPOT, CHU Montpellier patent 2002), based on the detection of CTC-secreted proteins, is a promising alternative for the detection of functional CTCs because only viable cells can secrete, shed or release proteins [32]. Furthermore, in vivo CTC isolation techniques, which significantly increase the volume of blood tested, have gained significant attention. Among them, the in vivo CellCollector® has been used in clinical studies on PCa and other solid tumors [33,34].

In this study, we compared these three techniques for CTC detection in patients with high-risk PCa treated by radiotherapy (RT). Among them only the automated CellSearch® system has already been studied by many different groups all over the world. For prostate cancer, worldwide there are only five reports of the EPISPOT assay being used [32,34–37] and three reports using the CellCollector® [34,37,38], with two in each group being a component of our multicenter ERA-NET TRANSCAN study. To the best of our knowledge, there are no reports using these two assays on patients with non-metastatic prostate cancer receiving radiotherapy. We can demonstrate that the combined use of these three assays increased the detection rate of CTCs in patients with non-metastatic PCa, that are known to have very low CTC concentrations [15]. To evaluate the potential clinical relevance of CTC detection, we analyzed the correlation between the CTC counts obtained with each method and conventional risk factors.

2. Results

2.1. Patient Characteristics

Between 2013 and 2016, 131 patients with non-metastatic high-risk PCa were recruited at diagnosis before any therapy (baseline), and 68/131 could be seen again after 256 (±126) days after the initiation of therapy (summary of the baseline clinical characteristics in Table 1). At baseline, their median age was 68.5 years, the median PSA concentration was 27.85 ng/mL, and the median Gleason score on biopsy was 7. During treatment, metastases were detected in 19 patients who were excluded from the statistical analysis.

Table 1. Clinical characteristics of the patients.

Parameter	Overall	CellSearch® System			EPISPOT			CellCollector®			Combined		
		CTC Negative	CTC Positive	p-Value	CTC Negative	CTC Positive	p-Value	CTC Negative	CTC Positive	p-Value	CTC Negative	CTC Positive	p-Value
Patients, n (%)	131	94 (82)	20 (18)		67 (54)	57 (46)		60 (48)	66 (52)		16 (12)	115 (88)	
Age, Median (IQR), Min–Max	68.5, (64.25–72), 51–89	68 (64–72) 51–80	70.5 (66.75–72.75) 60–89	0.15	69 (66–72) 53–89	66 (63–72) 51–79	0.053	69 (64–72) 51–89	67 (64.25–72) 53–78	0.50	69, (67.5–72.25), 59–80	68, (64–72), 51–89	0.22
BMI, Median (IQR), Min–Max	27.61, (25.6–29.8), 20.45–46.17	27.4, (25.5–29.7), 20.8–46.17	28.01, (26.8–29.4), 23.2–41.9	0.37	27.13, (25.5–29.5), 20.8–41.91	27.7, (26.03–30.67), 20.45–37.18	0.33	27.46, (25.24–29.11), 20.45–37.18	27.68, (26.17–31.7), 21.51–46.17	0.19	25.59, (23.5–26.9), 20.8–29.4	27.68, (26.12–30.66), 20.45–46.17	0.0036
PSA, Median (IQR), Min–Max	27.85, (15–35–40.75), 0.5–191	28.2 (20–42) 3.89–191	25.6 (7–09–40.25) 0.5–172	0.26	26.795 (10.365–41.15) 0.5–172	28.2 (17.6–40) 5–191	0.23	25 (10.5–41.1) 2.5–191	28.5 (22.5–37) 0.5–136.9	0.23	24.83, (11.78–38.95), 3.9–66	28.05, (15.625–40.75), 0.5–191	0.49
Biopsy Gleason score; n (%)													
3+3	30 (23)	22 (23.9)	4 (20)	0.77	13 (19.7)	15 (26.8)	0.069	13 (22)	17 (25.8)	0.14	3 (18.8)	27 (23.9)	0.35
3+4	40 (31)	30 (32.6)	4 (20)		18 (27.3)	20 (35.7)		19 (32.2)	21 (31.8)		4 (25)	36 (31.8)	
4+3	24 (19)	19 (20.7)	2 (10)		10 (15.2)	12 (21.4)		10 (16.9)	12 (18.2)		4 (25)	20 (17.7)	
≥4+4	35 (28)	21 (22.8)	10 (50)		25 (37.8)	9 (16.1)		17 (28.9)	16 (24.2)		5 (31.2)	30 (26.6)	
Clinical T stage; n (%)													
T1a	1 (0.8)	0 (0)	0 (0)	0.64	0 (0)	1 (1.8)	0.078	0 (0)	1 (1.6)	0.24	0 (0)	1 (0.9)	0.95
T1c	76 (60.3)	56 (62.2)	9 (47.4)		36 (55.4)	35 (63.6)		31 (53.4)	45 (70.3)		10 (62.5)	66 (60)	
T2a	10 (7.9)	6 (6.7)	3 (15.8)		4 (6.2)	6 (10.9)		7 (12.1)	3 (4.7)		1 (6.2)	9 (8.2)	
T2b	3 (2.4)	3 (3.3)	0 (0)		2 (3.1)	0 (0)		1 (1.7)	2 (3.1)		0 (0)	3 (2.7)	
T2c	23 (18.3)	16 (17.8)	4 (21.1)		13 (20)	10 (18.2)		12 (20.7)	9 (14.1)		3 (18.8)	20 (18.2)	
T3a	12 (9.5)	8 (8.9)	3 (15.8)		10 (15.4)	2 (3.6)		6 (10.3)	4 (6.2)		2 (12.5)	10 (9.1)	
T3b	1 (0.8)	1 (1.1)	0 (0)		0 (0)	1 (1.8)		1 (1.7)	0 (0)		0 (0)	1 (0.9)	

BMI, body mass index; PSA, prostate specific antigen.

The number of samples tested with each method at baseline and/or after treatment are listed in Table 2.

Table 2. Number of blood samples tested with each circulating tumor cells (CTC) enumeration method.

Samples Tested	CellSearch®	EPISPOT	CellCollector®
No. of samples tested at baseline	114	124	126
No. of samples tested after radiotherapy	63	68	64
No. of samples tested at both time points	53	62	64

2.2. *Comparison of CTC Detection Rate Obtained with Three Assays*

First, the total number of CTC-positive samples (baseline and after the treatment) obtained using each method was compared (Figure 1).

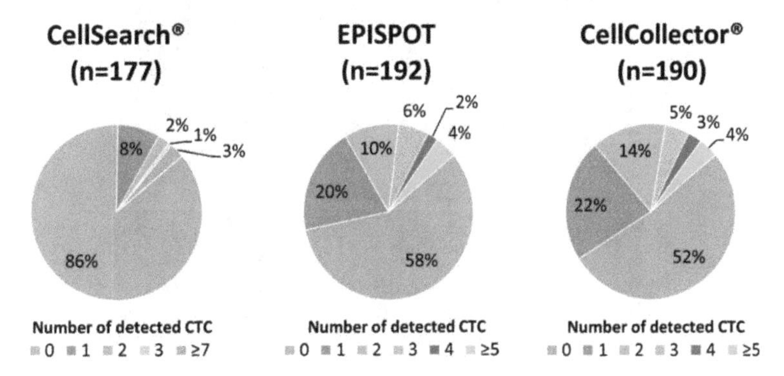

Figure 1. Distribution of detected circulating tumor cells (CTC) using the CellSearch® system, dual fluoro-EPISPOT assay, and CellCollector®. The results of CTC quantification in all samples (i.e., samples collected one day before and at least three months after radiotherapy) tested with the indicated assay.

With the CellSearch® system, 24/177 (14%) patient samples had ≥1 CTCs (range 1–54, median 1) (Figure 2 for representative images of PanCK-positive and CD45-negative CTCs).

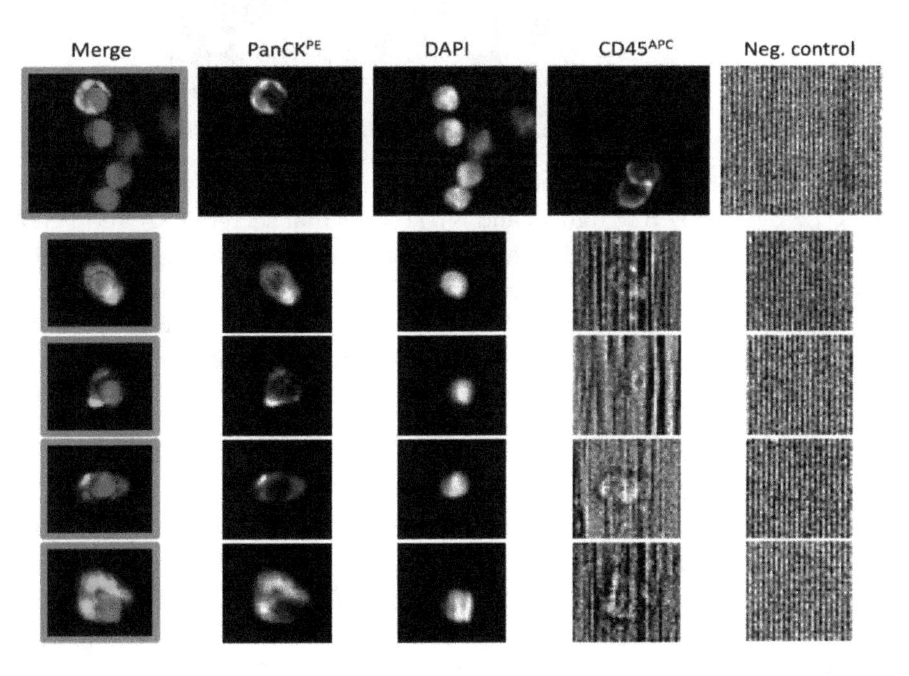

Figure 2. Representative images of CTCs from patients with high-risk PCa, that was detected using the CellSearch® system. Cells were identified as tumor cells according to the following criteria: EpCAM-positive, panCK-positive, DAPI-positive, CD45-negative and negative for the last channel. CK: Cytokeratin; PanCK: anti-CK8, -18, -19 antibodies; PE: Phycoerythrin; APC: Allophycocyanin.

Using the dual fluoro-EPISPOT assay, which detects only viable PSA/FGF2-secreting CTCs, 81/192 (42%) patient samples had ≥1 CTCs (range 1-25, median 2) (Figure 1). Very few PSA-secreting CTCs also secreted FGF2 (9 out of 81) (Figure 3). Concerning the control groups for the EPISPOT assay in prostate cancer, we already reported data on the detection of PSA-secreting cells (defined as CTCs in prostate cancer) in patients with benign diseases (e.g., acute prostatitis, benign hyperplasia of prostate) as well as in healthy donors [39,40]; we have never observed any positive events in these two control groups. We also reported that PBMC from healthy donors do not secreted PSA nor FGF2 (Figure 3).

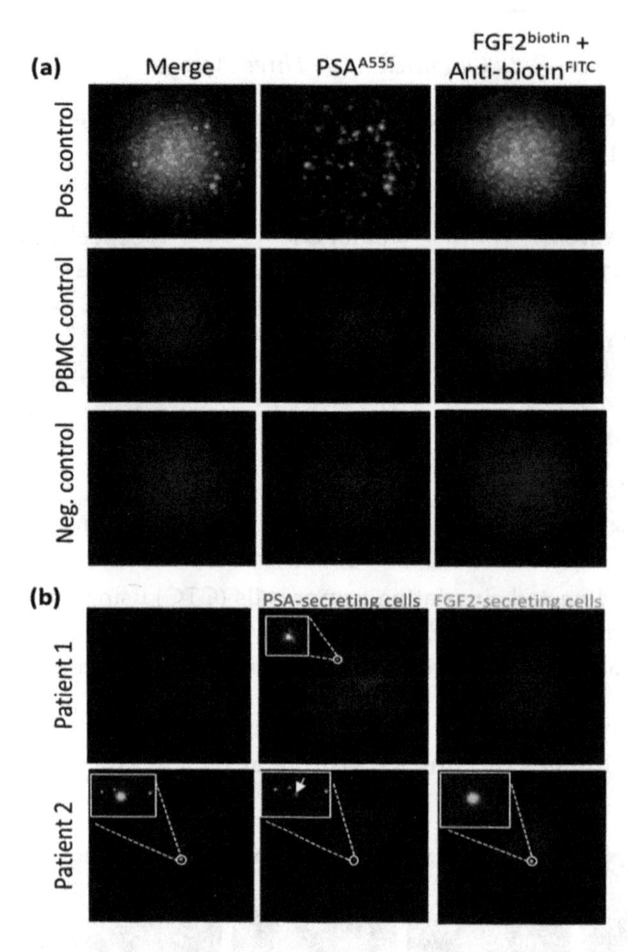

Figure 3. Detection of viable CTCs using the dual fluorescent EPISPOT$^{PSA/FGF2}$ assay. (**a**) Positive and negative controls. LNCaP (shown in figure) and NBTII cells that secrete PSA and FGF2, respectively, were used as positive controls (2000 cells/well), whereas wells with peripheral blood mononuclear cells (PBMC) and without cells were used as negative controls. Each immuno-spot corresponds to the protein "fingerprint" of one viable cell. (**b**) Patient samples. PSA-secreting cells are considered as CTCs in blood samples from patients with PCa (Patient 1). A subset of CTCs can secrete FGF-2 in addition to PSA (Patient 2). Representative images of PSA-positive, FGF2-positive and double PSA/FGF2 immuno-spots (merge) corresponding to viable CTCs. Immuno-spots were detected and observed using the C.T.L. Elispot Reader, 50× magnification.

Finally, 91/190 (48%) patient samples had ≥1 CTCs (range 1–16, median 2) using CellCollector® (Figure 1). Among the 197 PanCK-positive CTCs, 184 also expressed PSA (93%) (Figure 4).

By combining all the samples (baseline and after 256 (±126) days of treatment) assessed with the three methods (n = 161 samples), the number of positive samples (i.e., CTCs detected at least by one of the three assays) increased to 79% (127/161), with 8.7% (14/161) of blood samples harboring ≥ 5 CTCs. The concordance among the three assays (all positive or all negative) was 23%, while the concordance between two assays was approximately 50% for each comparison. The positive concordance was observed in samples with low number of CTCs, harboring up to 3 CTCs (Figure 5).

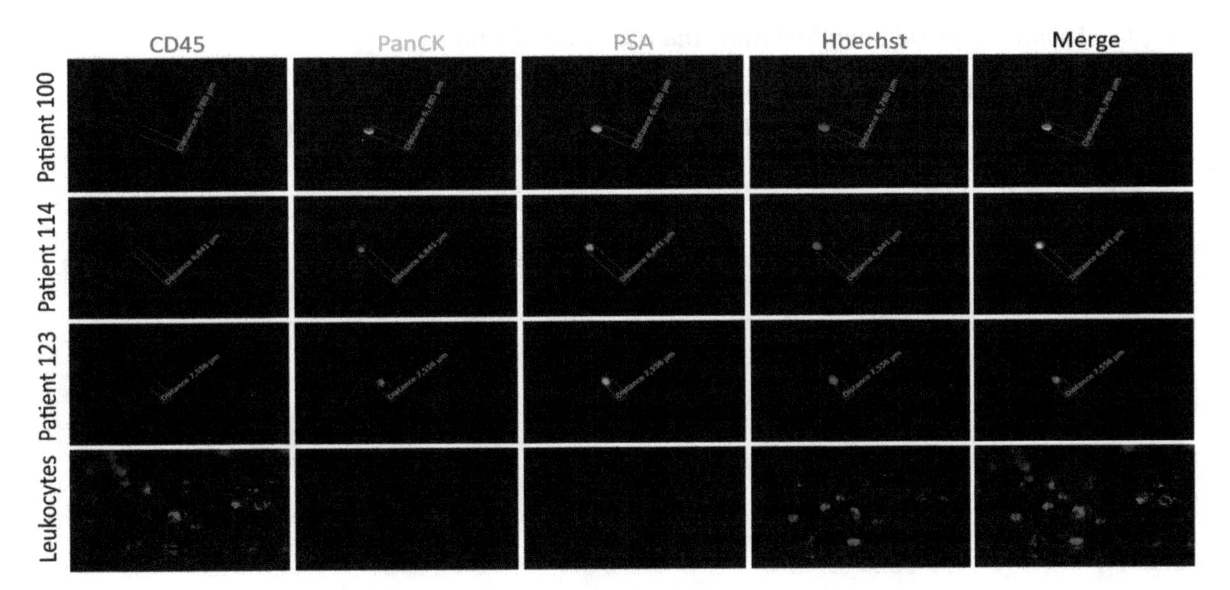

Figure 4. Detection of CTCs captured by the CellCollector® in vivo system. Representative images of CTCs isolated in vivo and non-specifically bound leukocytes (CD45-positive). The cells were identified as tumor cells according to the following criteria: panCK-positive (green), CD45-negative (red) and optionally, PSA-positive (orange). The nucleus was stained with Hoechst 33258 (blue). The images were obtained using a fluorescent microscope (Carl Zeiss, Axio Imager 2, 20×) and analyzed with the Carl Zeiss, Axio Vision 4.8 software.

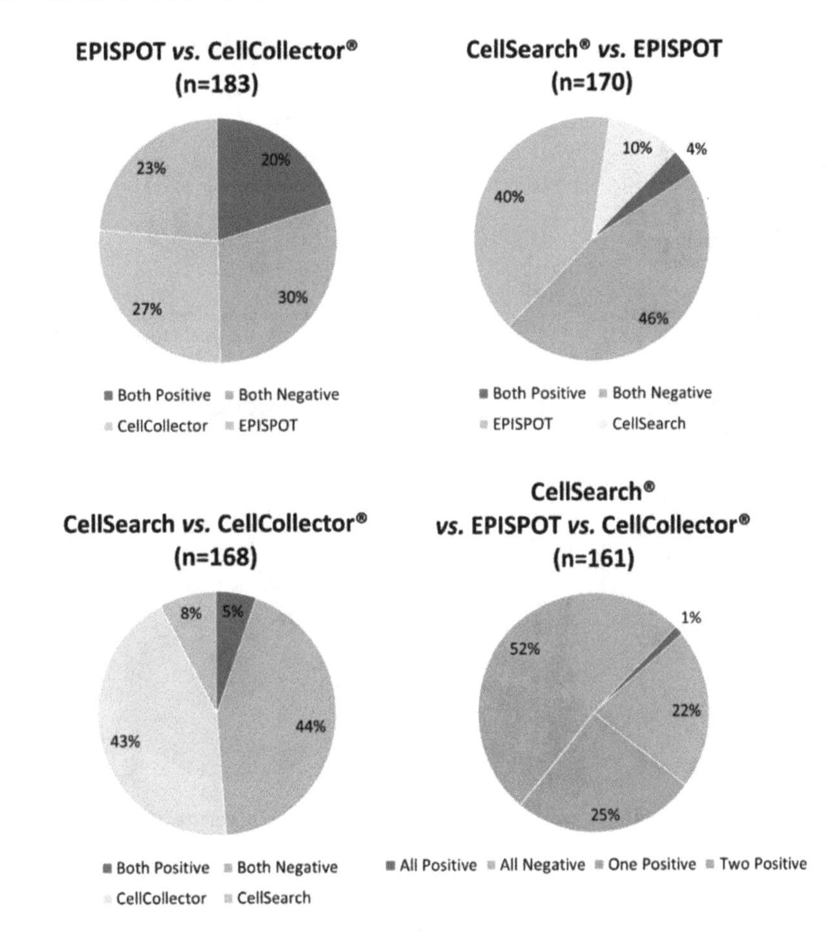

Figure 5. Concordance between two or three CTC detection assays. The chart shows the number of concordant positive (red) or negative (blue) results for the same sample obtained with two or three detection assays, as indicated.

2.3. *Analysis of Matched Blood Samples before and after Radiotherapy*

Matched-pair analysis of CTC counts before and after radiotherapy for the patients seen at both occasions ($n = 68$) did not find any significant difference for all three assays (CellSearch® $p = 0.28$, dual fluoro-EPISPOT $p = 0.27$, CellCollector® $p = 0.36$) (Figure 6).

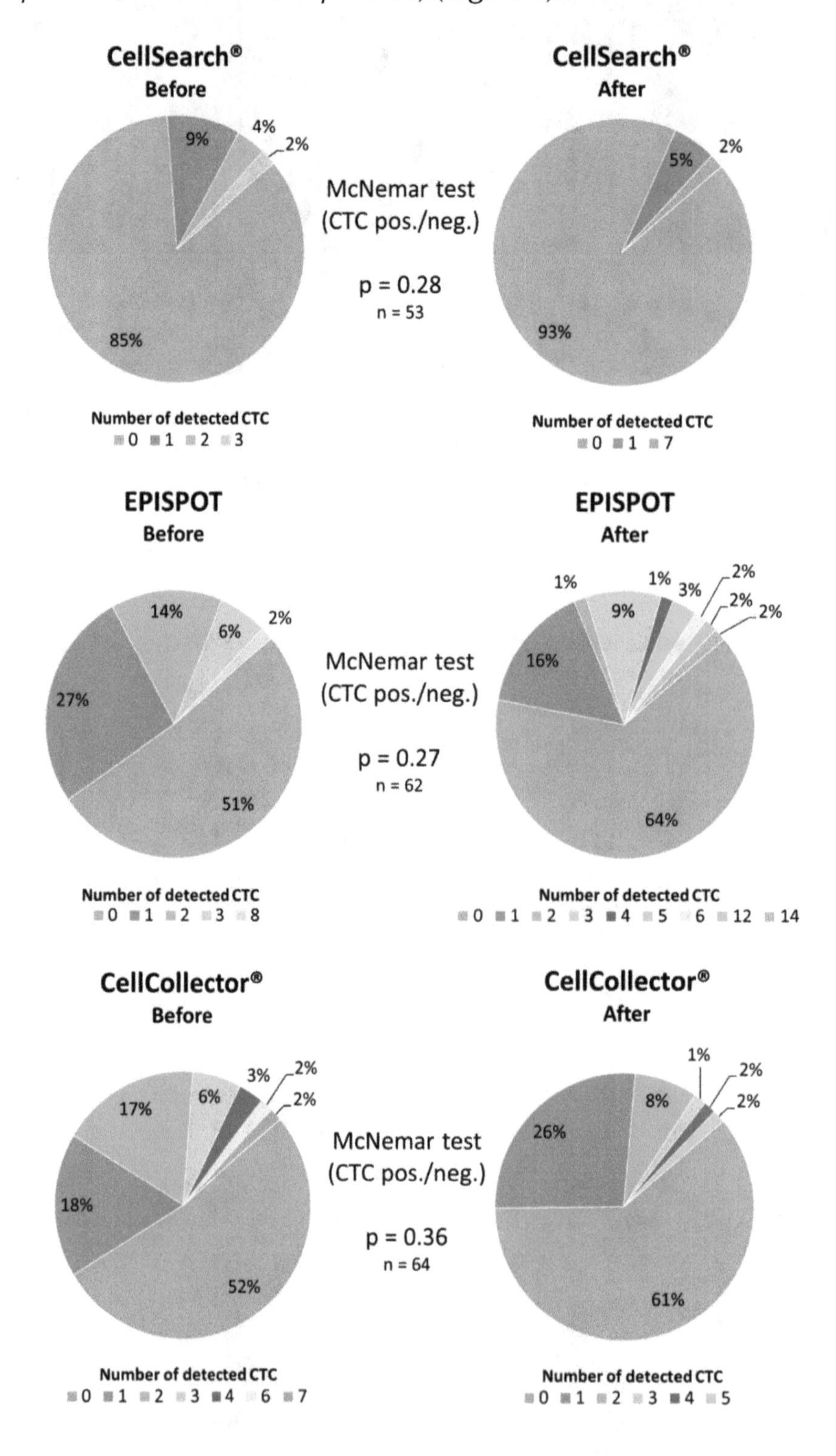

Figure 6. Matched-pair analysis of CTCs quantified with the CellSearch®, dual fluoro-EPISPOT$^{PSA/FGF2}$ or CellCollector® assay before and at least three months after radiotherapy start.

Nevertheless, CTC positivity rate tended to decrease in the samples collected after treatment. Specifically, using the CellSearch® system, 15% of patients had a positive CTC count before, and 7%

after treatment. Most of them (83%; 44/53) presented no change, whereas in 13% (7/53) CTC number was decreased and in 4% (2/53) increased at the second analysis. With the dual fluoro-EPISPOT assay, CTC positivity rate decreased from 49% before to 36% after the initiation of radiotherapy. Moreover, in 39% (24/62) of patients CTC number did not change between assays, whereas in 32% (20/62) decreased and in 29% (18/62) increased. With the CellCollector® test, CTC positivity decreased from 48% before to 39% after radiotherapy; CTC number did not change between assays in 39% (25/64), decreased in 36% (23/64) and increased in 25% of patients.

2.4. Correlation Between CTC Count and Clinical Parameters

Analysis of the correlations between baseline CTC counts (each method separately) and clinical parameters (age, body mass index, PSA median level, Gleason score and clinical tumor stage) only found that patients with ≥1 CTCs (obtained with the dual fluoro-EPISPOT$^{\mathbf{PSA/FGF2}}$) tended to be younger than patients with negative CTC counts ($p = 0.053$) (Table 1). Finally, an analysis of the combined CTC counts (three methods together) showed a significant correlation between CTC counts and patients' body mass index ($p = 0.0036$) (Table 1).

3. Discussion

In this study, we compared three different CTC identification/enumeration methods (CellSearch®, dual fluoro-EPISPOT$^{\mathbf{PSA/FGF2}}$ assay and CellCollector®) in a cohort of patients with localized high-risk PCa, treated with radiotherapy, combined with androgen deprivation. A comparison of the results obtained with each assay showed that the CellSearch® system presented the lowest CTC detection rate (only 14% of samples compared with 42% and 48%, with dual fluoro-EPISPOT$^{\mathbf{PSA/FGF2}}$ and CellCollector®, respectively). Previous studies using CellSearch® reported CTC positivity rates in patients with PCa that varied from 14% [41] to 37% [34], in non-metastatic prostate cancer patients. This observation, although unexpected, can be explained. CellSearch® technology regarded as "gold standard" for CTC detection was cleared by FDA for metastatic cancers [10]. We applied CellSearch® in our study on non-metastatic PCa patients, based on the available data describing its use in non-metastatic breast [42], colorectal [43], esophagus [44], and bladder [45] cancers. However, the concentration of CTCs in non-metastatic cancer patients is very low, and increasing the blood volume has helped to increase the rate of CTC detection in early stage breast cancer [11]. Interestingly, the EpCAM-based IsoFlux System, with only slightly higher blood volume suitable for testing (up to 10 mL), presented a very high CTC recovery rate in prostate and hepatocellular cancer patients. The IsoFlux System was designed to maximize the recovery of low EpCAM-expressing cells, specifically, those undergoing EMT or expressing stem cell markers through the application of another antibody cocktail. Additionally, the modification of magnetic bead coupling, flow rate, and microfluidic chamber dimension resulted in higher system sensitivity, compared to CellSearch® [24,46]. Further studies, with increased number of patients, are required to fully evaluate IsoFlux System diagnostic usefulness.

The positivity rates of the other two tests were comparable. Importantly, the EPISPOT assay uses only 10 mL of blood, which is comparable to the volume required by CellSearch®. Thus, the higher number of CTCs can be attributed to different principles of detection. Since the EPISPOT is an EpCAM-independent assay, the putative EMT process does not influence the detection outcome, which leads to a higher yield. However, the major novelty of the EPISPOT assay is the detection of proteins that are shed, actively released, or secreted by viable CTCs, which can contribute to tumor cell dissemination towards more distant sites [36]. Indeed, as not all circulating cells detached from a solid tumor are viable, the real risk of metastasis is hard to estimate on the basis of the count of all (dead and viable) CTCs. Our previous studies on breast cancer patients clearly indicated that patients with CK19-releasing viable CTCs showed significantly reduced overall survival rates compared to patients that do not harbor them. The stratification of breast cancer patients to low- and high-risk groups was improved when EPISPOT and CellSearch® assays were combined [47]. The current EPISPOT technology is oriented against PSA protein product, being the biomarker of prostate-derived CTCs, and additionally the stem cell growth factor, FGF2, which was found to mediate the in vitro growth of micro-metastatic cells [48]. Corresponding analysis of prognosis prediction will be conducted with prospective clinical data of recruited PCa patients. The EPISPOT assay has been validated in patients with breast [49], colon [50], and prostate cancer [32]. Previous studies using this assay, in patients with localized PCa, reported CTC-positivity rates that ranged from 58.7% [34] to 71% [51].

The positivity rate obtained from the CellCollector® in vivo capture system was slightly higher than the one from the EPISPOT assay, and the values were in the range of those reported by Kuske et al. [34] in patients with high-risk PCa, and by Gorges et al. [33], in patients with lung cancer. The main advantage of this tool is its ability to screen higher blood volumes by catching the CTCs directly from the blood. Clinical studies, with localized PCa patients, showed higher sensitivity and increased CTCs capturing rates, as compared to the CellSearch® system [33]. Importantly, control tests on blood from healthy volunteers showed that no "CTC-like" events were found [33,52], proving the high specificity of the device. Additionally, molecular analysis on in vivo isolated CTCs is feasible, enabling in depth analysis of the mechanisms governing cancer progression and the optimization of treatment profiles [53]. However, the distinctive feature of in vivo CTC isolation brings some concerns as well. Compared to ex vivo CellSearch® and EPISPOT technologies, the Cell-Collector requires 30 min of incubation in the patient's arm vein, thus causing greater discomfort comparing to regular blood collection. Although no evident side effects of this device have been noted [52], the prolonged venipuncture is burdened with a minor risk of bruises, diaphoresis with hypotension, syncope and injury of muscles and nerves [54]. An alternative to in vivo CTC capture is the use of diagnostic leukapheresis [55], which enables higher CTC yields than conventional enrichment strategies using 10–20 mL of blood.

The CTCs are a highly heterogeneous population of tumor cells. Thus, EpCAM-based enrichment technologies cannot identify all CTC sub-populations. Recently, we applied novel combinations of in vivo CTC, captured by CellCollector® and downstream RT-qPCR multiplex analysis, directed to mRNA of epithelial, EMT, and stem cell markers. We found that EMT markers were frequently expressed on CTCs, which would otherwise have remained undetected by enrichment based solely on epithelial markers. This observation elucidates the need for antibody cocktails against various CTC antigens in order to capture different CTCs phenotypes [37]. On the other hand, there is a strong need to efficiently exclude cells other than CTCs from the analysis. The leukocyte common antigen—CD45 is widely used for this purpose in various CTC detection assays, including CellSearch® and CellCollector. The effectiveness of non-CTCs demarcation can be strengthened by combining targeting of CD45 with other biomarkers, like those specific for myeloid cells (CD11b), tumor-associated macrophages (CD68), erythrocytes (CD235a), endothelial cells (CD146), and hematopoietic stem cells (CD34) [56,57].

Nowadays, CTC detection technologies enable the investigation of other biomarkers expressed by CTCs. Recently, attention was focused on programmed cell death ligand 1 (PD-L1) expression, given it prevents the immune system response and blocks anti-cancer immunity. In metastatic breast cancer [58], head and neck squamous cell carcinoma [59,60], and lung cancer [61] patients it has been show that combined approach of CTC detection with parallel PD-L1 immunostaining may predict the efficacy of immune checkpoint blockade therapies and the likelihood of resistance development.

When radiotherapy efficiently reduces tumor size in patients with PCa, the number of cancer cells in the circulation should also partially or totally decrease. However, non-effective RT leads to a lack of tumor response, and possibly the novel presence/increase of CTC number. Specifically, Martin et al., basing his findings on the results obtained in patients with non-small cell lung cancer (NSCLC), suggested that fractionated radiotherapy disrupts the tumor mass, thus promoting the passage of tumor cells into the circulation [62]. At an early phase of radiotherapy, when cancer cells suffer only from sub-lethal damage, the sudden increase of CTCs could, in principle, contribute to the formation of distant metastases. The paired analysis of our samples did not find any significant change in CTC counts per patient (obtained with any of the three assays) before, and after, treatment. Only a moderate tendency to decrease CTC-positive patient numbers was observed (CellSearch®:

15% versus 7%, EPISPOT: 49% versus 36%, CellCollector®: 48% versus 39%). These data do not support the hypothesis that radiotherapy in PCa leads to CTC release into the circulation. However, the time interval between the start of the RT and the second visit was approximately 256 days and detected CTCs were more likely the result of disease progression than tumor disruption following RT. For stronger evidence, immunofluorescent evaluation of phosphorylated histone H2AX (γ-H2AX) in CTCs would be helpful. This biomarker reveals the degree of double-stranded DNA breaks caused by RT. The presence of viable, γ-H2AX-positive CTCs in the peripheral circulation is the evidence of their RT-induced origin [62]. However, consecutive molecular analysis of radiotherapy-mobilized CTCs during the treatment course seems relevant because it has been suggested that such cells could return and colonize the tumor of origin. This process was described by Vilalta et al. in pre-clinical models of breast cancer as "self-seeding" [63], and could be responsible for primary tumor re-growth. It has been hypothesized that chemoattractant factors, released during radiotherapy by tumor cells, including granulocyte-macrophage colony stimulating factor (GM-CSF) [63], IL-6, or IL-8 [64], promote CTC recruitment back to the primary tumor site.

Other groups also investigated the effect of radiotherapy on CTC count. Lowes et al., using the CellSearch® system in patients with PCa, observed a trend toward CTC reduction after radiotherapy, but the results were not significant, which is consistent with our results. However, the percentage of CTC-positive patients before radiotherapy was very high (73% compared with 14% in our study). Like us, they did not find any correlation between CTC numbers before, and after, therapy and PSA level, clinical tumor stage, and Gleason score [65]. Tombal et al. used nested RT-PCR for CTC detection and found a significant correlation between CTC-positivity and biochemical recurrence, and mean PSA doubling time [66]. Moreover, the response to salvage radiotherapy was observed mainly in patients without CTCs. On the other hand, a study in patients with esophageal squamous cell carcinoma found that CTC-positivity by nested RT-PCR, before therapy, is not a predictive factor for the response to radiotherapy. However, based on the observation that patients who become CTC-negative after radiotherapy had a response rate of 86%, it was concluded that CTC-positivity after treatment is a poor outcome indicator. In this study, CTC-positivity was correlated with lymph node metastases and adverse 2-year progression-free survival, but not with age, tumor size, and clinical tumor stage [67]. Similarly, Dorsey et al. found a significant decrease in CTC number (based on their elevated telomerase activity) after radiotherapy in patients with localized NSCLC. The CTC count was correlated with the clinical course and response to treatment [68]. Discrepancies between studies could be explained by the differences in patients' status, sample size, type of assay used for the CTC count, and sampling time points.

In our study, we did not find any correlation between the CTC count (each method separately) before radiotherapy, and any of the tested clinical parameters (age, body mass index, PSA median level, Gleason score and clinical tumor stage). Theil et al. [38] tested the CellCollector® in patients with PCa and detected a correlation between CTC number and tumor stage, suggesting some clinical relevance as a marker of poor prognosis. However, clinical follow-up studies are required to validate the CellCollector® prognostic relevance. Moreover, CTC identification (i.e., CTC release from the nanowire) must be rapidly improved for its use in clinical trials. Surprisingly, patients with CTC-positive samples (obtained with the dual fluoro-EPISPOT assay that detect PSA-secreting cells) tended to be younger ($p = 0.053$), an observation that is inconsistent with the finding that PSA level increases with age [69]. The prognostic significance of patients' age as an independent predictor of biochemical recurrence is still controversial. Some reports recognized advanced age as an indicator of poorer biochemical outcome [70], while others did not find any association in patients undergoing radical prostatectomy [71] or radiotherapy [72]. However, another study suggested that in high-risk younger patients, biochemical recurrence-free survival is poorer than in older patients [73]. In addition, PSA serum level was not significantly different in patients with CTC-negative and positive samples detected with the dual fluoro-EPISPOT assay. These findings suggest that the major source of serum PSA was not PSA-secreting CTCs, but rather the tumor itself. Finally, while combining three methods together, we found a significant correlation between CTC counts and patients' BMI. The data indicate subtle [74] to strong [75,76] BMI's effect on PCa's incidence, aggressiveness, and mortality. The possible cause is the high concentration of leptin, insulin, insulin-like growth factor-I and low of adiponectin, which together, are likely to promote tumor growth and progression [77]. Additionally, obese men tend to have lower level of testosterone, which assures the normal prostate condition [78]. On the contrary, Schiffmann et al. described the obesity paradox in surgically-treated PCa patients, presenting a decreased risk of metastases after the treatment, possibly caused by the diabetes treatment, common for obese men [79]. The association between BMI and CTC level is not widely described, however studies on breast cancer patients show inconsistent results. Ortmann et al. found no significant relation, neither before, nor after, chemotherapy, between BMI and CTC number, suggesting they are independent prognostic factors [80], while Fayanju et al. indicated negative prognostic significance of their combination for non-obese patients exclusively [81].

4. Materials and Methods

4.1. Study Design

This study was performed between 2013 and 2016, within the framework of the international multicenter project, "Circulating Tumor Cells as Biomarkers for Minimal Residual Disease in Prostate Cancer" (CTC-SCAN, an ERA-NET TRANSCAN project coordinated by Klaus Pantel, Hamburg). During this time, 131 patients with high-risk PCa, according to the D'Amico criteria (PSA $\geq$ 20 ng/mL AND/OR Gleason score on biopsy $\geq$ 8 AND/OR clinical tumor stage $\geq$ 2c), were enrolled at the Department and Clinic of Urology and Urologic Oncology at Poznan University of Medical Sciences, Poland, and at the Cancer Institute of Montpellier (ICM), France. Patients with metastases were not included in the study. The patients were examined during the first visit, before starting radiotherapy combined with hormone therapy (all 131), and at least three months after the beginning of treatment (68/131). The mean number of days ($\pm$ standard deviation) between two blood samples was 256 days ($\pm$126). The experimental protocol of this study was approved by the Local Bioethics Committees at Poznan University of Medical Sciences (registered under the number 28/13), at the University Medical Centre of Montpellier and ICM (registered under the number 2013-A00523-42). All patients were informed about the study design/procedures and signed a written informed consent. Blood sampling dates, endocrine therapy, and treatment duration are presented in Table S1.

4.2. CTC Isolation and Detection

For CTC detection, using the CellSearch® system (Silicon Biosystem, Menarini, Florence, Italy) and the dual fluoro-EPISPOT$^{PSA/FGF2}$ assay, 10mL of blood was collected in CellSave® tubes, and EDTA tubes, respectively. For CTC detection with CellCollector® CANCER01 (Gilupi GmbH, Potsdam, Germany), the device was placed directly in the arm vein for 30 min.

Within 24 h after blood draw, the samples, collected in CellSave® tubes in Poznan, were shipped at room temperature to the Laboratory of Rare Human Circulating Cells (LCCRH) at the University Medical Center of Montpellier, France, where the samples were analyzed on arrival day.

4.2.1. CellSearch® System

The FDA-cleared CellSearch® system has been used as a "gold standard" for CTC detection in patients with metastatic PCa for almost one decade [82,83]; however, in this project, CellSearch® was used to detect CTCs in patients without metastases. The CellSearch® Circulating Epithelial Cell Kit (Silicon Biosystems, Menarini) uses 7.5 mL of blood for CTC enrichment and enumeration. The CellSearch® system is based on magnetic beads (ferrofluid) coated with antibodies against EpCAM for CTC enrichment. EpCAM-positive cells are captured and then immunostained with antibodies against cytokeratins (CK 8, 18 and 19) and with DAPI (nuclear staining). EpCAM- and CK-positive and CD45 (common leukocyte antigen)-negative cells are identified as CTCs [41,82].

4.2.2. Dual Fluoro-EPISPOT$^{PSA/FGF2}$ Assay

CTCs from 10 mL of blood samples were first enriched by negative selection using 20 µL of RosetteSep™ Human Circulating Epithelial Tumor Cell Enrichment Cocktail (STEMCELL Technologies, Vancouver, Canada) per 1mL of blood. After incubation and phase separation with 1.073 density gradient, CTCs were collected from the interphase and washed twice.

For this study, the dual fluoro-EPISPOT$^{\mathbf{PSA/FGF2}}$ assay was used to detect CTCs that secrete PSA and fibroblast growth factor 2 (FGF2, an important self-renewal regulator of normal and cancer stem cells), as previously described [39,40]. We previously showed that this assay can detect CTCs that secrete PSA and FGF2 from blood samples of patients with PCa [36]. Briefly, after negative enrichment, CTCs were cultured at 37 °C and 5% CO_2 on nitrocellulose membranes pre-coated with 1.04 µg/µL anti-PSA H50 antibody (provided by Prof Hans Lilja, Department of Biotechnology, University of Turku, Finland) and 0.5 µg/µL anti-FGF2 500-M38 antibody (Peprotech), diluted in PBS and blocked with 5% BSA/PBS. After 48 h, PSA and FGF2, produced by viable cells and captured by the primary antibodies, were detected by incubation with fluorochrome-conjugated anti-PSA-H117-A555 antibody (1.0 µg/µL; provided by Prof. Hans Lilja, Department of Biotechnology, University of Turku, Finland) and anti-FGF2 500-P18Bt antibody labeled with biotin (0.5 µg/µL; Peprotech) and then with 1:20 anti-biotin-FITC antibody (Miltenyi Biotec) diluted in 0.5% bovine serum albumin (BSA)/PBS. After washing, PSA and FGF2 immuno-spots were imaged and counted using a fluorescent microscope, and a video-camera imaging and computer-assisted analysis system (KS ELISPOT, Carl Zeiss Vision) and also the C.T.L. ELISPOT reader [84,85].

The LNCap (PCa) and NBTII cell lines from the American Type Culture Collection (ATCC) were used as positive control for PSA, and FGF2 expression/secretion, respectively. LNCaP cells (ATCC® CRL-1740™) were cultured in RPMI 1640 medium (Gibco), 10% fetal bovine serum (FBS) (Sigma Aldrich) and antibiotics (Sigma Aldrich). NBTII cells (ATCC® CRL-1655™) were grown in DMEM medium with GlutaMAX (Gibco), 10% FBS (Sigma Aldrich), and antibiotics (Sigma Aldrich). At 80% confluence in 75 cm^2 flasks, the cells were trypsinized (0.25% trypsin/EDTA, Gibco), washed, and counted in order to seed 2000 cells/well (two wells) in each plate. For the dual fluoro-EPISPOT assay, LNCap, NBTII and enriched CTCs were cultured in RPMI 1640 medium (Sigma Aldrich) with 10% FBS (Sigma Aldrich), antibiotics (Sigma Aldrich), 1% L-glutamine (Sigma Aldrich), and 1% Insulin-Transferrin-Selenium (Gibco).

4.2.3. CellCollector®

The CellCollector® CANCER01 (Gilupi GmbH) allows the in vivo capture and detection of CTCs [52]. The device is composed of medical stainless-steel wire, coated with polycarboxylate hydrogel and functionalized with anti-EpCAM antibodies. The CellCollector® was placed in the vein of the arm using a standard cannula (32 mm) for 30 min. After its removal, the cells captured on the wire were fixed in cold acetone for 10 min, permeabilized in 0.1% Triton X-100/PBS for 10 min, and blocked in 3% BSA/PBS for 30 min. Then, the cells were incubated for 1h with fluorochrome-conjugated antibodies against PanCK (A488 antibody cocktail, 1:50; eBioscience, Exbio) and PSA (H117-A555 antibody, 1:80; provided by Prof Hans Lilja, Department of Biotechnology, University of Turku, Finland), all diluted in 3% BSA/PBS. To confirm the absence of leukocytes, the cells were incubated with the anti-CD45-A647 antibody (1:25; Exbio), diluted in 3% BSA/PBS. The nuclei were stained with Hoechst 33258 (Sigma Aldrich) for 10 min. The images were then obtained using a fluorescent microscope (Axio Imager 2, Carl Zeiss) and a 20× objective and were analyzed with AxioVision 4.8 (Carl Zeiss). The cells were identified as CTCs using the following criteria: Intact morphology, polymorphic, diameter ≥4 µm, CK- and/or PSA-positivity, Hoechst 33258-positivity, and CD-45 negativity.

4.3. Statistical Analysis

For analysis with clinical-pathological variables, CTC counts were stratified as negative (no CTC detected) or positive (at least one CTC detected). To test for correlations between CTC positivity and clinical-pathological variables, the Wilcoxon-test was used for continuous variables, and the chi-square (likelihood) test for categorical variables. To compare the CTC counts before, and at least three months after, the initiation of radiotherapy, the McNemar's test was used. To assess the clinical risk of biochemical progression after radiotherapy, patients with negative and positive CTC counts were compared using the chi-square (likelihood) test. All tests were two-sided and a p-value < 0.05 was considered to be significant. Data were analyzed using RStudio (version 1.1.383), an integrated development environment for R (version 3.4.2). The readxl package (version 1.0.0) was used to load the Excel file containing the experimental data into R, and the dplyr package (version 0.7.4) for data manipulation, the MASS package for the chi-square likelihood test, and the Stats package (version 3.4.2) for the Wilcoxon test.

5. Conclusions

Combining different CTC assays improved CTC detection rates in patients with non-metastatic, high-risk PCa before, and after, treatment. Our findings do not support the hypothesis that radiotherapy leads to cancer cell release in the circulation. Additional clinical trials are required before the implementation of CTC detection in the standard clinical practice of patients with PCa receiving radiotherapy. Moreover, a better molecular characterization of CTCs might provide additional information on the survival mechanisms implemented during their rough passage in the bloodstream and on the biology behind extravasation in distant organs [37,86].

Author Contributions: Conceptualization, J.B.-T. and C.A.-P.; methodology, M.Ś., M.M.; software, D.A.; validation, A.A., M.N.; formal analysis, K.P.; investigation, A.J., M.Ś., M.M.; resources, W.A.C., A.I., A.A., and D.A.; data curation, M.M.; writing—original draft preparation, J.B-T.; writing—review and editing, C.A.-P.; visualization, M.M., M.Ś.; supervision, C.A.-P., M.Z., M.N.; project administration, C.A.-P., K.P., and M.Z.; funding acquisition, K.P.

References

1. Scher, H.I.; Morris, M.J.; Larson, S.; Heller, G. Validation and clinical utility of prostate cancer biomarkers. *Nat. Rev. Clin. Oncol.* **2013**, *10*, 225–234. [CrossRef] [PubMed]
2. Chang, A.J.; Autio, K.A.; Roach, M.; Scher, H.I. High-risk prostate cancer-classification and therapy. *Nat. Rev. Clin. Oncol.* **2014**, *11*, 308–323. [CrossRef] [PubMed]
3. Carter, H.B. Differentiation of lethal and non lethal prostate cancer: PSA and PSA isoforms and kinetics. *Asian J. Androl.* **2012**, *14*, 355–360. [CrossRef] [PubMed]
4. Andriole, G.L., Jr. PSA screening and prostate cancer risk reduction. *Urol. Oncol.* **2012**, *30*, 936–937. [CrossRef] [PubMed]
5. Dimakakos, A.; Armakolas, A.; Koutsilieris, M. Novel tools for prostate cancer prognosis, diagnosis, and follow-up. *Biomed Res. Int.* **2014**, *2014*, 890697. [CrossRef] [PubMed]
6. Wolters, T.; van der Kwast, T.H.; Vissers, C.J.; Bangma, C.H.; Roobol, M.; Schroder, F.H.; van Leenders, G.J.L.H. False-negative prostate needle biopsies: Frequency, histopathologic features, and follow-up. *Am. J. Surg. Pathol.* **2010**, *34*, 35–43. [CrossRef] [PubMed]
7. Thalgott, M.; Rack, B.; Horn, T.; Heck, M.M.; Eiber, M.; Kubler, H.; Retz, M.; Gschwend, J.E.; Andergassen, U.; Nawroth, R. Detection of circulating tumor cells in locally advanced high-risk prostate cancer during neoadjuvant chemotherapy and radical prostatectomy. *Anticancer Res.* **2015**, *35*, 5679–5685.
8. Thalgott, M.; Rack, B.; Maurer, T.; Souvatzoglou, M.; Eiber, M.; Kress, V.; Heck, M.M.; Andergassen, U.; Nawroth, R.; Gschwend, J.E.; et al. Detection of circulating tumor cells in different stages of prostate cancer. *J. Cancer Res. Clin. Oncol.* **2013**, *139*, 755–763. [CrossRef] [PubMed]

9. Alix-Panabieres, C.; Pantel, K. Clinical applications of circulating tumor cells and circulating tumor DNA as liquid biopsy. *Cancer Discov.* **2016**, *6*, 479–491. [CrossRef] [PubMed]
10. Coumans, F.A.W.; Ligthart, S.T.; Uhr, J.W.; Terstappen, L.W.M.M. Challenges in the enumeration and phenotyping of CTC. *Clin. Cancer Res.* **2012**, *18*, 5711–5718. [CrossRef] [PubMed]
11. Rack, B.; Schindlbeck, C.; Juckstock, J.; Andergassen, U.; Hepp, P.; Zwingers, T.; Friedl, T.W.P.; Lorenz, R.; Tesch, H.; Fasching, P.A.; et al. Circulating tumor cells predict survival in early average-to-high risk breast cancer patients. *J. Natl. Cancer Inst.* **2014**, *106*. [CrossRef] [PubMed]
12. Janni, W.J.; Rack, B.; Terstappen, L.W.M.M.; Pierga, J.-Y.; Taran, F.-A.; Fehm, T.; Hall, C.; de Groot, M.R.; Bidard, F.-C.; Friedl, T.W.P.; et al. Pooled analysis of the prognostic relevance of circulating tumor cells in primary breast cancer. *Clin. Cancer Res.* **2016**, *22*, 2583–2593. [CrossRef] [PubMed]
13. Riethdorf, S.; Muller, V.; Loibl, S.; Nekljudova, V.; Weber, K.; Huober, J.; Fehm, T.; Schrader, I.; Hilfrich, J.; Holms, F.; et al. Prognostic impact of circulating tumor cells for breast cancer patients treated in the neoadjuvant "Geparquattro" trial. *Clin. Cancer Res.* **2017**, *23*, 5384–5393. [CrossRef] [PubMed]
14. Bidard, F.-C.; Michiels, S.; Riethdorf, S.; Mueller, V.; Esserman, L.J.; Lucci, A.; Naume, B.; Horiguchi, J.; Gisbert-Criado, R.; Sleijfer, S.; et al. Circulating tumor cells in breast cancer patients treated by neoadjuvant chemotherapy: A meta-analysis. *J. Natl. Cancer Inst.* **2018**, *110*, 560–567. [CrossRef] [PubMed]
15. Hong, B.; Zu, Y. Detecting circulating tumor cells: Current challenges and new trends. *Theranostics* **2013**, *3*, 377–394. [CrossRef] [PubMed]
16. Cristofanilli, M.; Hayes, D.F.; Budd, G.T.; Ellis, M.J.; Stopeck, A.; Reuben, J.M.; Doyle, G.V.; Matera, J.; Allard, W.J.; Miller, M.C.; et al. Circulating tumor cells: A novel prognostic factor for newly diagnosed metastatic breast cancer. *J. Clin. Oncol.* **2005**, *23*, 1420–1430. [CrossRef] [PubMed]
17. Andreopoulou, E.; Yang, L.-Y.; Rangel, K.M.; Reuben, J.M.; Hsu, L.; Krishnamurthy, S.; Valero, V.; Fritsche, H.A.; Cristofanilli, M. Comparison of assay methods for detection of circulating tumor cells in metastatic breast cancer: AdnaGen AdnaTest BreastCancer select/detect versus Veridex CellSearch system. *Int. J. Cancer* **2012**, *130*, 1590–1597. [CrossRef] [PubMed]
18. Pluim, D.; Devriese, L.A.; Beijnen, J.H.; Schellens, J.H.M. Validation of a multiparameter flow cytometry method for the determination of phosphorylated extracellular-signal-regulated kinase and DNA in circulating tumor cells. *Cytometry A* **2012**, *81*, 664–671. [CrossRef] [PubMed]
19. Nagrath, S.; Sequist, L.V.; Maheswaran, S.; Bell, D.W.; Irimia, D.; Ulkus, L.; Smith, M.R.; Kwak, E.L.; Digumarthy, S.; Muzikansky, A.; et al. Isolation of rare circulating tumour cells in cancer patients by microchip technology. *Nature* **2007**, *450*, 1235–1239. [CrossRef] [PubMed]
20. Stott, S.L.; Hsu, C.-H.; Tsukrov, D.I.; Yu, M.; Miyamoto, D.T.; Waltman, B.A.; Rothenberg, S.M.; Shah, A.M.; Smas, M.E.; Korir, G.K.; et al. Isolation of circulating tumor cells using a microvortex-generating herringbone-chip. *Proc. Natl. Acad. Sci. USA* **2010**, *107*, 18392–18397. [CrossRef] [PubMed]
21. Khoo, B.L.; Warkiani, M.E.; Tan, D.S.-W.; Bhagat, A.A.S.; Irwin, D.; Lau, D.P.; Lim, A.S.T.; Lim, K.H.; Krisna, S.S.; Lim, W.-T.; et al. Clinical validation of an ultra high-throughput spiral microfluidics for the detection and enrichment of viable circulating tumor cells. *PLoS ONE* **2014**, *9*, e99409. [CrossRef] [PubMed]
22. Autebert, J.; Coudert, B.; Champ, J.; Saias, L.; Guneri, E.T.; Lebofsky, R.; Bidard, F.-C.; Pierga, J.-Y.; Farace, F.; Descroix, S.; et al. High purity microfluidic sorting and analysis of circulating tumor cells: Towards routine mutation detection. *Lab Chip* **2015**, *15*, 2090–2101. [CrossRef] [PubMed]
23. Winer-Jones, J.P.; Vahidi, B.; Arquilevich, N.; Fang, C.; Ferguson, S.; Harkins, D.; Hill, C.; Klem, E.; Pagano, P.C.; Peasley, C.; et al. Circulating tumor cells: Clinically relevant molecular access based on a novel CTC flow cell. *PLoS ONE* **2014**, *9*, e86717. [CrossRef] [PubMed]
24. Harb, W.; Fan, A.; Tran, T.; Danila, D.C.; Keys, D.; Schwartz, M.; Ionescu-Zanetti, C. Mutational analysis of circulating tumor cells using a novel microfluidic collection device and qPCR Assay. *Transl. Oncol.* **2013**, *6*, 528–538. [CrossRef] [PubMed]
25. Liu, Z.; Fusi, A.; Klopocki, E.; Schmittel, A.; Tinhofer, I.; Nonnenmacher, A.; Keilholz, U. Negative enrichment by immunomagnetic nanobeads for unbiased characterization of circulating tumor cells from peripheral blood of cancer patients. *J. Transl. Med.* **2011**, *9*, 70. [CrossRef]
26. Lara, O.; Tong, X.; Zborowski, M.; Farag, S.S.; Chalmers, J.J. Comparison of two immunomagnetic separation technologies to deplete T cells from human blood samples. *Biotechnol. Bioeng.* **2006**, *94*, 66–80. [CrossRef]
27. Scher, H.I.; Jia, X.; de Bono, J.S.; Fleisher, M.; Pienta, K.J.; Raghavan, D.; Heller, G. Circulating tumour cells as

prognostic markers in progressive, castration-resistant prostate cancer: A reanalysis of IMMC38 trial data. *Lancet Oncol.* **2009**, *10*, 233–239. [CrossRef]

28. Scheel, C.; Weinberg, R.A. Cancer stem cells and epithelial-mesenchymal transition: Concepts and molecular links. *Semin. Cancer Biol.* **2012**, *22*, 396–403. [CrossRef]
29. Wang, L.; Balasubramanian, P.; Chen, A.P.; Kummar, S.; Evrard, Y.A.; Kinders, R.J. Promise and limits of the CellSearch platform for evaluating pharmacodynamics in circulating tumor cells. *Semin. Oncol.* **2016**, *43*, 464–475. [CrossRef]
30. Andree, K.C.; van Dalum, G.; Terstappen, L.W.M.M. Challenges in circulating tumor cell detection by the CellSearch system. *Mol. Oncol.* **2016**, *10*, 395–407. [CrossRef]
31. Sharma, S.; Zhuang, R.; Long, M.; Pavlovic, M.; Kang, Y.; Ilyas, A.; Asghar, W. Circulating tumor cell isolation, culture, and downstream molecular analysis. *Biotechnol. Adv.* **2018**, *36*, 1063–1078. [CrossRef] [PubMed]
32. Alix-Panabieres, C.; Pantel, K. Liquid biopsy in cancer patients: Advances in capturing viable CTCs for functional studies using the EPISPOT assay. *Expert Rev. Mol. Diagn.* **2015**, *15*, 1411–1417. [CrossRef] [PubMed]
33. Gorges, T.M.; Penkalla, N.; Schalk, T.; Joosse, S.A.; Riethdorf, S.; Tucholski, J.; Lucke, K.; Wikman, H.; Jackson, S.; Brychta, N.; et al. Enumeration and molecular characterization of tumor cells in lung cancer patients using a novel in vivo device for capturing circulating tumor cells. *Clin. Cancer Res.* **2016**, *22*, 2197–2206. [CrossRef] [PubMed]
34. Kuske, A.; Gorges, T.M.; Tennstedt, P.; Tiebel, A.-K.; Pompe, R.; Preisser, F.; Prues, S.; Mazel, M.; Markou, A.; Lianidou, E.; et al. Improved detection of circulating tumor cells in non-metastatic high-risk prostate cancer patients. *Sci. Rep.* **2016**, *6*, 39736. [CrossRef] [PubMed]
35. Kruck, S.; Gakis, G.; Stenzl, A. Disseminated and circulating tumor cells for monitoring chemotherapy in urological tumors. *Anticancer Res.* **2011**, *31*, 2053–2057.
36. Alix-Panabieres, C. EPISPOT assay: Detection of viable DTCs/CTCs in solid tumor patients. *Recent Results Cancer Res.* **2012**, *195*, 69–76. [CrossRef]
37. Markou, A.; Lazaridou, M.; Paraskevopoulos, P.; Chen, S.; Swierczewska, M.; Budna, J.; Kuske, A.; Gorges, T.M.; Joosse, S.A.; Kroneis, T.; et al. Multiplex gene expression profiling of in vivo isolated circulating tumor cells in high-risk prostate cancer patients. *Clin. Chem.* **2018**, *64*, 297–306. [CrossRef]
38. Theil, G.; Fischer, K.; Weber, E.; Medek, R.; Hoda, R.; Lucke, K.; Fornara, P. The use of a new cellcollector to isolate circulating tumor cells from the blood of patients with different stages of prostate cancer and clinical outcomes—A proof-of-concept study. *PLoS ONE* **2016**, *11*, e0158354. [CrossRef]
39. Alix-Panabieres, C.; Rebillard, X.; Brouillet, J.-P.; Barbotte, E.; Iborra, F.; Segui, B.; Maudelonde, T.; Jolivet-Reynaud, C.; Vendrell, J.-P. Detection of circulating prostate-specific antigen-secreting cells in prostate cancer patients. *Clin. Chem.* **2005**, *51*, 1538–1541. [CrossRef]
40. Alix-Panabieres, C.; Vendrell, J.-P.; Pelle, O.; Rebillard, X.; Riethdorf, S.; Muller, V.; Fabbro, M.; Pantel, K. Detection and characterization of putative metastatic precursor cells in cancer patients. *Clin. Chem.* **2007**, *53*, 537–539. [CrossRef]
41. Loh, J.; Jovanovic, L.; Lehman, M.; Capp, A.; Pryor, D.; Harris, M.; Nelson, C.; Martin, J. Circulating tumor cell detection in high-risk non-metastatic prostate cancer. *J. Cancer Res. Clin. Oncol.* **2014**, *140*, 2157–2162. [CrossRef] [PubMed]
42. Lucci, A.; Hall, C.S.; Lodhi, A.K.; Bhattacharyya, A.; Anderson, A.E.; Xiao, L.; Bedrosian, I.; Kuerer, H.M.; Krishnamurthy, S. Circulating tumour cells in non-metastatic breast cancer: A prospective study. *Lancet Oncol.* **2012**, *13*, 688–695. [CrossRef]
43. Gazzaniga, P.; Gianni, W.; Raimondi, C.; Gradilone, A.; Lo Russo, G.; Longo, F.; Gandini, O.; Tomao, S.; Frati, L. Circulating tumor cells in high-risk nonmetastatic colorectal cancer. *Tumour Biol.* **2013**, *34*, 2507–2509. [CrossRef] [PubMed]
44. Reeh, M.; Effenberger, K.E.; Koenig, A.M.; Riethdorf, S.; Eichstadt, D.; Vettorazzi, E.; Uzunoglu, F.G.; Vashist, Y.K.; Izbicki, J.R.; Pantel, K.; et al. Circulating tumor cells as a biomarker for preoperative prognostic staging in patients with esophageal cancer. *Ann. Surg.* **2015**, *261*, 1124–1130. [CrossRef] [PubMed]
45. Karl, A.; Tritschler, S.; Hofmann, S.; Stief, C.G.; Schindlbeck, C. Perioperative search for circulating tumor cells in patients undergoing radical cystectomy for bladder cancer. *Eur. J. Med. Res.* **2009**, *14*, 487–490. [CrossRef] [PubMed]
46. Sanchez-Lorencio, M.I.; Ramirez, P.; Saenz, L.; Martinez Sanchez, M.V.; de La Orden, V.; Mediero-Valeros, B.;

Veganzones-De-Castro, S.; Baroja-Mazo, A.; Revilla Nuin, B.; Gonzalez, M.R.; et al. Comparison of two types of liquid biopsies in patients with hepatocellular carcinoma awaiting orthotopic liver transplantation. *Transplant. Proc.* **2015**, *47*, 2639–2642. [CrossRef] [PubMed]

47. Ramirez, J.-M.; Fehm, T.; Orsini, M.; Cayrefourcq, L.; Maudelonde, T.; Pantel, K.; Alix-Panabieres, C. Prognostic relevance of viable circulating tumor cells detected by EPISPOT in metastatic breast cancer patients. *Clin. Chem.* **2014**, *60*, 214–221. [CrossRef]
48. Pantel, K.; Alix-Panabieres, C.; Riethdorf, S. Cancer micrometastases. *Nat. Rev. Clin. Oncol.* **2009**, *6*, 339–351. [CrossRef]
49. Alix-Panabieres, C.; Vendrell, J.-P.; Slijper, M.; Pelle, O.; Barbotte, E.; Mercier, G.; Jacot, W.; Fabbro, M.; Pantel, K. Full-length cytokeratin-19 is released by human tumor cells: a potential role in metastatic progression of breast cancer. *Breast Cancer Res.* **2009**, *11*, R39. [CrossRef]
50. Deneve, E.; Riethdorf, S.; Ramos, J.; Nocca, D.; Coffy, A.; Daures, J.-P.; Maudelonde, T.; Fabre, J.-M.; Pantel, K.; Alix-Panabieres, C. Capture of viable circulating tumor cells in the liver of colorectal cancer patients. *Clin. Chem.* **2013**, *59*, 1384–1392. [CrossRef]
51. Schwarzenbach, H.; Alix-Panabieres, C.; Muller, I.; Letang, N.; Vendrell, J.-P.; Rebillard, X.; Pantel, K. Cell-free tumor DNA in blood plasma as a marker for circulating tumor cells in prostate cancer. *Clin. Cancer Res.* **2009**, *15*, 1032–1038. [CrossRef] [PubMed]
52. Saucedo-Zeni, N.; Mewes, S.; Niestroj, R.; Gasiorowski, L.; Murawa, D.; Nowaczyk, P.; Tomasi, T.; Weber, E.; Dworacki, G.; Morgenthaler, N.G.; et al. A novel method for the in vivo isolation of circulating tumor cells from peripheral blood of cancer patients using a functionalized and structured medical wire. *Int. J. Oncol.* **2012**, *41*, 1241–1250. [CrossRef] [PubMed]
53. Strati, A.; Markou, A.; Parisi, C.; Politaki, E.; Mavroudis, D.; Georgoulias, V.; Lianidou, E. Gene expression profile of circulating tumor cells in breast cancer by RT-qPCR. *BMC Cancer* **2011**, *11*, 422. [CrossRef] [PubMed]
54. Galena, H.J. Complications occurring from diagnostic venipuncture. *J. Fam. Pract.* **1992**, *34*, 582–584. [PubMed]
55. Fischer, J.C.; Niederacher, D.; Topp, S.A.; Honisch, E.; Schumacher, S.; Schmitz, N.; Zacarias Fohrding, L.; Vay, C.; Hoffmann, I.; Kasprowicz, N.S.; et al. Diagnostic leukapheresis enables reliable detection of circulating tumor cells of nonmetastatic cancer patients. *Proc. Natl. Acad. Sci. USA* **2013**, *110*, 16580–16585. [CrossRef] [PubMed]
56. Van der Toom, E.E.; Verdone, J.E.; Gorin, M.A.; Pienta, K.J. Technical challenges in the isolation and analysis of circulating tumor cells. *Oncotarget* **2016**, *7*, 62754–62766. [CrossRef]
57. Barriere, G.; Fici, P.; Gallerani, G.; Fabbri, F.; Zoli, W.; Rigaud, M. Circulating tumor cells and epithelial, mesenchymal and stemness markers: Characterization of cell subpopulations. *Ann. Transl. Med.* **2014**, *2*, 109. [CrossRef]
58. Mazel, M.; Jacot, W.; Pantel, K.; Bartkowiak, K.; Topart, D.; Cayrefourcq, L.; Rossille, D.; Maudelonde, T.; Fest, T.; Alix-Panabieres, C. Frequent expression of PD-L1 on circulating breast cancer cells. *Mol. Oncol.* **2015**, *9*, 1773–1782. [CrossRef]
59. Strati, A.; Koutsodontis, G.; Papaxoinis, G.; Angelidis, I.; Zavridou, M.; Economopoulou, P.; Kotsantis, I.; Avgeris, M.; Mazel, M.; Perisanidis, C.; et al. Prognostic significance of PD-L1 expression on circulating tumor cells in patients with head and neck squamous cell carcinoma. *Ann. Oncol.* **2017**, *28*, 1923–1933. [CrossRef]
60. Kulasinghe, A.; Kenny, L.; Punyadeera, C. Circulating tumour cell PD-L1 test for head and neck cancers. *Oral Oncol.* **2017**, *75*, 6–7. [CrossRef]
61. Dhar, M.; Wong, J.; Che, J.; Matsumoto, M.; Grogan, T.; Elashoff, D.; Garon, E.B.; Goldman, J.W.; Sollier Christen, E.; Di Carlo, D.; et al. Evaluation of PD-L1 expression on vortex-isolated circulating tumor cells in metastatic lung cancer. *Sci. Rep.* **2018**, *8*, 2592. [CrossRef] [PubMed]
62. Martin, O.A.; Anderson, R.L.; Russell, P.A.; Cox, R.A.; Ivashkevich, A.; Swierczak, A.; Doherty, J.P.; Jacobs, D.H.M.; Smith, J.; Siva, S.; et al. Mobilization of viable tumor cells into the circulation during radiation therapy. *Int. J. Radiat. Oncol. Biol. Phys.* **2014**, *88*, 395–403. [CrossRef] [PubMed]
63. Vilalta, M.; Rafat, M.; Giaccia, A.J.; Graves, E.E. Recruitment of circulating breast cancer cells is stimulated by radiotherapy. *Cell Rep.* **2014**, *8*, 402–409. [CrossRef] [PubMed]
64. Kim, M.-Y.; Oskarsson, T.; Acharyya, S.; Nguyen, D.X.; Zhang, X.H.-F.; Norton, L.; Massague, J. Tumor self-seeding by circulating cancer cells. *Cell* **2009**, *139*, 1315–1326. [CrossRef] [PubMed]

65. Lowes, L.E.; Lock, M.; Rodrigues, G.; D'Souza, D.; Bauman, G.; Ahmad, B.; Venkatesan, V.; Allan, A.L.; Sexton, T. Circulating tumour cells in prostate cancer patients receiving salvage radiotherapy. *Clin. Transl. Oncol.* **2012**, *14*, 150–156. [CrossRef] [PubMed]
66. Tombal, B.; van Cangh, P.J.; Loric, S.; Gala, J.-L. Prognostic value of circulating prostate cells in patients with a rising PSA after radical prostatectomy. *Prostate* **2003**, *56*, 163–170. [CrossRef]
67. Yin, X.-D.; Yuan, X.; Xue, J.-J.; Wang, R.; Zhang, Z.-R.; Tong, J.-D. Clinical significance of carcinoembryonic antigen-, cytokeratin 19-, or survivin-positive circulating tumor cells in the peripheral blood of esophageal squamous cell carcinoma patients treated with radiotherapy. *Dis. Esophagus* **2012**, *25*, 750–756. [CrossRef]
68. Dorsey, J.F.; Kao, G.D.; MacArthur, K.M.; Ju, M.; Steinmetz, D.; Wileyto, E.P.; Simone, C.B.; Hahn, S.M. Tracking viable circulating tumor cells (CTCs) in the peripheral blood of non-small cell lung cancer (NSCLC) patients undergoing definitive radiation therapy: Pilot study results. *Cancer* **2015**, *121*, 139–149. [CrossRef]
69. Prcic, A.; Begic, E.; Hiros, M. Usefulness of total PSA value in prostate diseases diagnosis. *Acta Inform. Med.* **2016**, *24*, 156–161. [CrossRef]
70. Turk, H.; Celik, O.; Un, S.; Yoldas, M.; Isoglu, C.S.; Karabicak, M.; Ergani, B.; Koc, G.; Zorlu, F.; Ilbey, Y.O. Predictive factors for biochemical recurrence in radical prostatectomy patients. *Cent. Eur. J. Urol.* **2015**, *68*, 404–409. [CrossRef]
71. Grossfeld, G.D.; Latini, D.M.; Lubeck, D.P.; Mehta, S.S.; Carroll, P.R. Predicting recurrence after radical prostatectomy for patients with high risk prostate cancer. *J. Urol.* **2003**, *169*, 157–163. [CrossRef]
72. Geara, F.B.; Bulbul, M.; Khauli, R.B.; Andraos, T.Y.; Abboud, M.; Al Mousa, A.; Sarhan, N.; Salem, A.; Ghatasheh, H.; Alnsour, A.; et al. Nadir PSA is a strong predictor of treatment outcome in intermediate and high risk localized prostate cancer patients treated by definitive external beam radiotherapy and androgen deprivation. *Radiat. Oncol.* **2017**, *12*, 149. [CrossRef] [PubMed]
73. Hong, S.K.; Nam, J.S.; Na, W.; Oh, J.J.; Yoon, C.Y.; Jeong, C.W.; Kim, H.J.; Byun, S.-S.; Lee, S.E. Younger patients have poorer biochemical outcome after radical prostatectomy in high-risk prostate cancer. *Asian J. Androl.* **2011**, *13*, 719–723. [CrossRef] [PubMed]
74. Davies, N.M.; Gaunt, T.R.; Lewis, S.J.; Holly, J.; Donovan, J.L.; Hamdy, F.C.; Kemp, J.P.; Eeles, R.; Easton, D.; Kote-Jarai, Z.; et al. The effects of height and BMI on prostate cancer incidence and mortality: A Mendelian randomization study in 20,848 cases and 20,214 controls from the PRACTICAL consortium. *Cancer Causes Control* **2015**, *26*, 1603–1616. [CrossRef] [PubMed]
75. Haque, R.; Van Den Eeden, S.K.; Wallner, L.P.; Richert-Boe, K.; Kallakury, B.; Wang, R.; Weinmann, S. Association of body mass index and prostate cancer mortality. *Obes. Res. Clin. Pract.* **2014**, *8*, e374–e381. [CrossRef]
76. Xie, B.; Zhang, G.; Wang, X.; Xu, X. Body mass index and incidence of nonaggressive and aggressive prostate cancer: A dose-response meta-analysis of cohort studies. *Oncotarget* **2017**, *8*, 97584–97592. [CrossRef] [PubMed]
77. Presti, J.C., Jr. Obesity and prostate cancer. *Curr. Opin. Urol.* **2005**, *15*, 13–16. [CrossRef]
78. Cunha, G.R.; Hayward, S.W.; Wang, Y.Z.; Ricke, W.A. Role of the stromal microenvironment in carcinogenesis of the prostate. *Int. J. Cancer* **2003**, *107*, 1–10. [CrossRef]
79. Schiffmann, J.; Karakiewicz, P.I.; Rink, M.; Manka, L.; Salomon, G.; Tilki, D.; Budaus, L.; Pompe, R.; Leyh-Bannurah, S.-R.; Haese, A.; et al. Obesity paradox in prostate cancer: Increased body mass index was associated with decreased risk of metastases after surgery in 13,667 patients. *World J. Urol.* **2018**, *36*, 1067–1072. [CrossRef]
80. Ortmann, U.; Janni, W.; Andergassen, U.; Beck, T.; Beckmann, M.W.; Lichtenegger, W.; Neugebauer, J.K.; Salmen, J.; Schindlbeck, C.; Schneeweiss, A.; et al. Correlation of high body mass index and circulating tumor cell positivity in patients with early-stage breast cancer. *J. Clin. Oncol.* **2012**, *30*, 1600.
81. Fayanju, O.M.; Hall, C.S.; Bauldry, J.B.; Karhade, M.; Valad, L.M.; Kuerer, H.M.; DeSnyder, S.M.; Barcenas, C.H.; Lucci, A. Body mass index mediates the prognostic significance of circulating tumor cells in inflammatory breast cancer. *Am. J. Surg.* **2017**, *214*, 666–671. [CrossRef] [PubMed]
82. Bono, J.S.; de Scher, H.I.; Montgomery, R.B.; Parker, C.; Miller, M.C.; Tissing, H.; Doyle, G.V.; Terstappen, L.W.W.M.; Pienta, K.J.; Raghavan, D. Circulating tumor cells predict survival benefit from treatment in metastatic castration-resistant prostate cancer. *Clin. Cancer Res.* **2008**, *14*, 6302–6309. [CrossRef] [PubMed]
83. Scher, H.I.; Heller, G.; Molina, A.; Attard, G.; Danila, D.C.; Jia, X.; Peng, W.; Sandhu, S.K.; Olmos, D.;

Riisnaes, R.; et al. Circulating tumor cell biomarker panel as an individual-level surrogate for survival in metastatic castration-resistant prostate cancer. *J. Clin. Oncol.* **2015**, *33*, 1348–1355. [CrossRef] [PubMed]
84. Pantel, K.; Alix-Panabieres, C. Functional studies on viable circulating tumor cells. *Clin. Chem.* **2016**, *62*, 328–334. [CrossRef] [PubMed]
85. Soler, A.; Cayrefourcq, L.; Mazel, M.; Alix-Panabieres, C. EpCAM-Independent enrichment and detection of viable circulating tumor cells using the EPISPOT assay. *Methods Mol. Biol.* **2017**, *1634*, 263–276. [CrossRef] [PubMed]
86. Strilic, B.; Offermanns, S. Intravascular survival and extravasation of tumor cells. *Cancer Cell* **2017**, *32*, 282–293. [CrossRef] [PubMed]

Pilot Study of Circulating Tumor Cells in Early-Stage and Metastatic Uveal Melanoma

Kartik Anand [1], Jason Roszik [2], Dan Gombos [3], Joshua Upshaw [4], Vanessa Sarli [4], Salyna Meas [4], Anthony Lucci [4], Carolyn Hall [4] and Sapna Patel [2,*]

[1] Houston Methodist Cancer Center, Houston, TX 77030, USA; kartikanand88@gmail.com
[2] Department of Melanoma Medical Oncology, UT MD Anderson Cancer Center, Houston, TX 77030, USA; jroszik@mdanderson.org
[3] Department of Head and Neck Surgery, Section of Ophthalmology, UT MD Anderson Cancer Center, Houston, TX 77030, USA; dgombos@mdanderson.org
[4] Department of Surgical Oncology, UT MD Anderson Cancer Center, Houston, TX 77030, USA; jrupshaw@mdanderson.org (J.U.); vnsarli@mdanderson.org (V.S.); smeas@mdanderson.org (S.M.); alucci@mdanderson.org (A.L.); cshall@mdanderson.org (C.H.)
* Correspondence: sppatel@mdanderson.org.

Abstract: Nearly 50% of uveal melanoma (UM) patients develop metastatic disease, and there remains no current standard assay for detection of minimal residual disease. We conducted a pilot study to check the feasibility of circulating tumor cell (CTC) detection in UM. We enrolled 40 patients with early or metastatic UM of which 20 patients had early-stage disease, 19 had metastatic disease, and one was not evaluable. At initial blood draw, 36% of patients had detectable CTCs (30% in early-stage vs. 42% in metastatic), which increased to 54% at data cutoff (40% in early-stage vs. 68% in metastatic). Five early-stage patients developed distant metastases, 60% (3/5) had detectable CTCs before radiographic detection of the metastasis. Landmark overall survival (from study enrollment) at 24 months was statistically lower in CTC-positive vs. negative early-stage UM ($p < 0.05$). Within this small dataset, the presence of CTCs in early-stage UM predicted an increased risk of metastatic disease and was associated with worse outcomes.

Keywords: uveal melanoma; liquid biopsy; circulating tumor cells; pilot study

---

1. Introduction

Uveal melanoma, the most common primary intra-ocular malignancy, is a rare neoplasm that constitutes less than 5% of total melanoma incidence [1]. The most common site involved is the choroid (90%), followed by the ciliary body (6%) and the iris (4%) [2]. In the U.S., the age-adjusted risk of uveal melanoma is 5.1 per million [3]. Caucasians are at increased risk of uveal melanoma and the mean age of diagnosis is 62 years [4]. Risk factors include fair skin, blonde hair, and light eye color, presence of choroidal nevus and presence of germline breast cancer 1-associated protein 1 (BAP1) mutation. The role of ultraviolet exposure as a risk factor remains controversial [5]. Less than 5% of the cases at initial ocular presentation have distant metastasis. Primary tumor management generally involves globe sparing local therapy such as laser or radiation; alternatively, enucleation is an option. Prognostic factors to help predict the risk of metastasis include cytogenetics, gene expression profiling by RNA-based techniques and mutational analysis [6]. Increased expression of ABCB5 protein is also linked to poor prognosis and increased risk of metastasis [7]. It is common practice to have surveillance scans with abdominal imaging (CT or MRI) every 3 to 6 months post primary tumor management [6]. As approximately half of patients with uveal melanoma develop metastatic disease, there is need to enhance our ability to detect minimal residual disease (MRD). A "liquid biopsy" approach, based on the assessment of rare circulating tumor cells (CTCs) or DNA from peripheral blood samples, provides a promising method to permit identification of early recurrence or to monitor disease status. Liquid biopsies are easily obtained, minimally-invasive, longitudinal snapshots that can be used to measure micro-metastatic disease burden, monitor disease progression, and provide real-time genomic assessments of primary tumor/metastatic lesions. As hematogenous spread remains the route of dissemination in uveal melanoma, there is particular interest in the role of CTCs. CTCs have been shown to initiate metastasis in pre-clinical mouse xenograft models [8]. The use of CTCs as a prognostic marker has been evaluated in numerous solid organ malignancies including breast, prostate, colon, and bladder and esophageal cancer where high CTC count is correlated with increased risk of metastasis [9]. We designed a study to evaluate the role of CTCs in uveal melanoma in patients presenting to the University of Texas MD Anderson Cancer Center, Melanoma and Skin Center. Our main objective was to check the feasibility of CTC detection in uveal melanoma. Our secondary objectives were to evaluate if CTCs varied with risk status in early-stage uveal melanoma and if the presence of CTCs predicted risk of distant metastasis.

2. Materials and Methods

2.1. Patients

We enrolled 40 patients who presented to the University of Texas MD Anderson Cancer Center Uveal Melanoma medical oncology clinic with early-stage or metastatic uveal melanoma from 1 December 2014 to 1 February 2018 in an IRB-approved study (LAB11-0314). All patients provided written, informed consent and were 18 years of age or older. The demographic information collected included age, sex and race, date of diagnosis of primary uveal melanoma or metastasis, and dates of CTC collection. Data from 39 patients was available for evaluation.

2.2. Risk Stratification

Primary uveal melanoma patients were risk-stratified based on the commercially available DecisionDX-UM assay, which uses RT-PCR to determine the gene expression of 15 genes in the tumor sample. The assay is validated to determine the risk of distant metastasis and stratifies early-stage uveal melanoma based on increasing risk of 5-year distant metastasis into Class 1A, 1B and 2 with Class 1 tumors considered lowest-risk and Class 2 tumors considered highest-risk.

2.3. Circulating Tumor Cell Analysis

No patients reported adverse events or complications from blood collection. Serial peripheral venous blood draws were collected; the first (baseline) sample was collected after primary tumor diagnosis or diagnosis of metastatic disease. We used one 10 mL tube of blood containing CellSaveTM preservative for the detection of CTCs using the CellSearch Circulating Melanoma Cell Assay®. Circulating tumor cell assessments were performed within 72 h of blood collection as per the manufacturer's protocol. The CellSearch® Circulating Melanoma Cell test uses ferrofluids coated with CD146 antibodies to immunomagnetically enrich melanoma cells, label the nuclei of these cells with the fluorescent dye 4,2-diamidino-2-phenylindole dihydrochloride (DAPI), and stain them using fluorescently labeled antibodies to detect the combination of high molecular weight melanoma-associated antigen (HMW-MAA; clone 9.2.27), CD45, and CD34. A semi-automated fluorescence-based microscope system was employed to identify circulating tumor cells CD146+, HMW-MAA+, CD45−, CD34−, and nucleated (DAPI+) cells.

2.4. Statistical Analysis

We used *t*-tests to compare the mean number of circulating tumor cells between early-stage and metastatic uveal melanoma. Landmark overall survival was calculated using the landest "R" package.

Kaplan–Meier curves were derived using the 'survival' package in R, for comparison of landmark overall survival between groups of patients stratified by the presence of one or more baseline circulating tumor cell. The p-values were two-tailed and values <0.05 were considered statistically significant.

3. Results

The median age of the data set was 52 years (20–83 years) and all patients were non-Hispanic Caucasians. Forty-four percent (17/39) of the patients were male and 56% (22/39) female. At the time of study enrollment, 51% (20/39) had early-stage disease and 49% (19/39) had metastatic disease. For early-stage disease, results of gene expression profiling (Decision-DX assay) were available for 75% (15/20) of the patients; 87% (13/15) were Class 2 and 13% (2/15) were Class 1. Mutational analysis data was present for 19 of 20 patients (49%) out of which 12 (63%) had a GNAQ mutation, 5 (26%) had a GNA11 mutation, one (5%) patient was wild-type for both genes and one (5%) had another mutation separate from GNAQ or GNA11. The median time from diagnosis to blood sampling to checking for CTC for the whole cohort was 20.35 months (20.10 months for early-stage vs. 24.65 months for metastatic disease). At the time of data cutoff, 1 June 2018, the median study follow up for the whole cohort was 16.4 months (16.84 months for early-stage vs. 14.56 months for metastatic disease) (Table 1).

Table 1. Patient characteristics

Total Patients (n) = 39	
Sex	
Male (n)	17 (44%)
Female (n)	22 (56%)
Median age	52 years (20–83 years)
Race	
Non-Hispanic White (n)	39 (100%)
Disease status at study enrollment	
Early-stage disease (n)	20 (51%)
Class 1 (n)	2
Class 2 (n)	13
Unknown (n)	5
Metastatic (n)	19 (49%)
Mutation analysis (n)	19
GNAQ (n)	12 (63%)
GNA11 (n)	5 (27%)
Wildtype (n)	1 (5%)
Other (n)	1 (5%)
Median time between diagnosis and blood sampling	20.35 months
Early-stage	20.10 months
Metastatic	24.65 months
Total Study follow up	16.40 months
Early-stage	16.84 months
Metastatic	14.56 months

3.1. CTCs Are More Frequently Detected in Metastatic Uveal Melanoma Compared to Early-Stage Uveal Melanoma

At initial blood draw, at least one CTC was detected in 36% (14/39) of the patients in the whole cohort, 30% (6/20) of the early-stage disease group, and 42% (8/19) of the patients in the metastatic group (Table 2). Of the eight patients for whom ≥1 CTC was detected at initial blood draw in the metastatic group, three patients had one CTC detected, two patients had two CTCs, and one patient had five, twenty-two and thirty-eight CTCs respectively. In the early-stage disease group, of the six patients who had CTCs detected at initial draw, three had one CTC, one had two CTCs and two had three CTCs. At the median follow up of 16.4 months, the rate of detection of CTCs increased to 54% (21/39), 13 (68.4%) patients in metastatic group had ≥1 CTC detected vs. eight (40%) in the early-stage disease group. Three patients (two patients in the metastatic group and one in the early-stage group) with detectable CTC at initial draw had no detectable CTC during subsequent draws in the study follow-up. The 21 patients included those three patients as well. Out of the eight patients with early-stage disease who had ≥1 CTC detected, five had Class 2 status, one had Class 1A status, and two had an unknown gene expression profile. The mean number of CTCs was higher for the metastatic group compared to the early-stage group, nine (1–38) vs. 1.83 (1–3) respectively. (p-value = 0.184) (Figure 1).

Table 2. Circulating tumor cells at initial draw

Early-Stage Uveal Melanoma (*n*)	**20**
No CTC detected (*n*)	14 (70%)
CTC detected ≥ 1 (*n*)	14 (70%)
1 CTC	3
Class 1	1
Class 2	2
Unknown	0
2 CTCs	1
Class 1	0
Class 2	0
Unknown	1
3 CTCs	2
Class 1	0
Class 2	1
Unknown	1
Metastatic Uveal Melanoma (*n*)	**19**
No CTC detected (*n*)	11 (58%)
CTC detected ≥ 1 (*n*)	8 (42%)
1 CTC	3
2 CTCs	2
5 CTCs	1
22 CTCs	1
38 CTCs	1

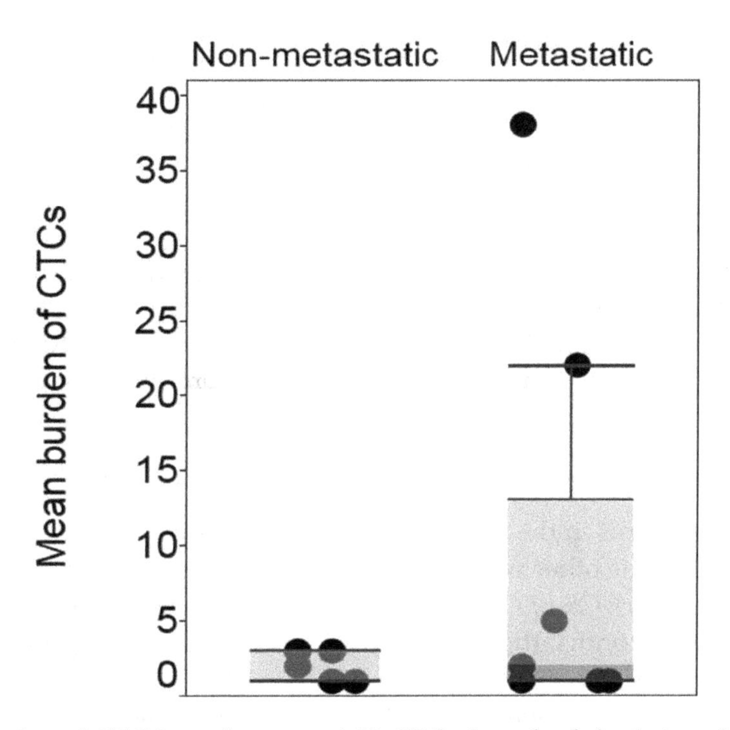

Figure 1. Mean burden of CTC in early-stage, 1.83 CTCs (standard deviation: 0.98), vs. metastatic uveal melanoma 9 CTCs (standard deviation: 13.7). $p > 0.05$.

3.2. *Presence of CTC Predicts Metastasis Risk in Early-Stage Uveal Melanoma*

Metastatic disease developed in five out of 20 (25%) patients in the early-stage group. Eighty percent of these (4/5) were Class 2 and 20% (1/5) had unknown gene expression profiles; however, this patient was known to harbor a monosomy 3 karyotype, a well-established risk factor. In three out of five patients (60%), CTCs were detected before metastatic disease was detected by radiographic imaging. Patients 001, 012, and 018 had CTCs detected 24.9 months, 9.5 months and 3.2 months, respectively, before radiographic detection of distant metastasis. (Table 3, Figure 2).

Table 3. Characteristics of patients who developed metastasis in early-stage uveal melanoma

Patient	Class by Gene Expression	Mutational Analysis	Date of Diagnosis	Date of CTC Detection	Date of Metastasis by Imaging	Vital Status at Study Cutoff
Patient 001	Class 2	GNA11	12/15/2012	12/3/2014	11/20/2017	Alive
Patient 003	Class 2	*GNAQ*	9/15/2014	CTC never detected	7/14/2015	Alive
Patient 012	Class 2	*GNAQ*	1/15/2015	10/19/2015	8/2/2016	Deceased
Patient 018	Unknown *	wild-type	10/19/2015	4/27/2016	1/31/2017	Deceased
Patient 029	Class 2	not tested	10/20/2016	CTC never detected	11/22/2017	Alive

* monosomy 3. Data: month/day/year.

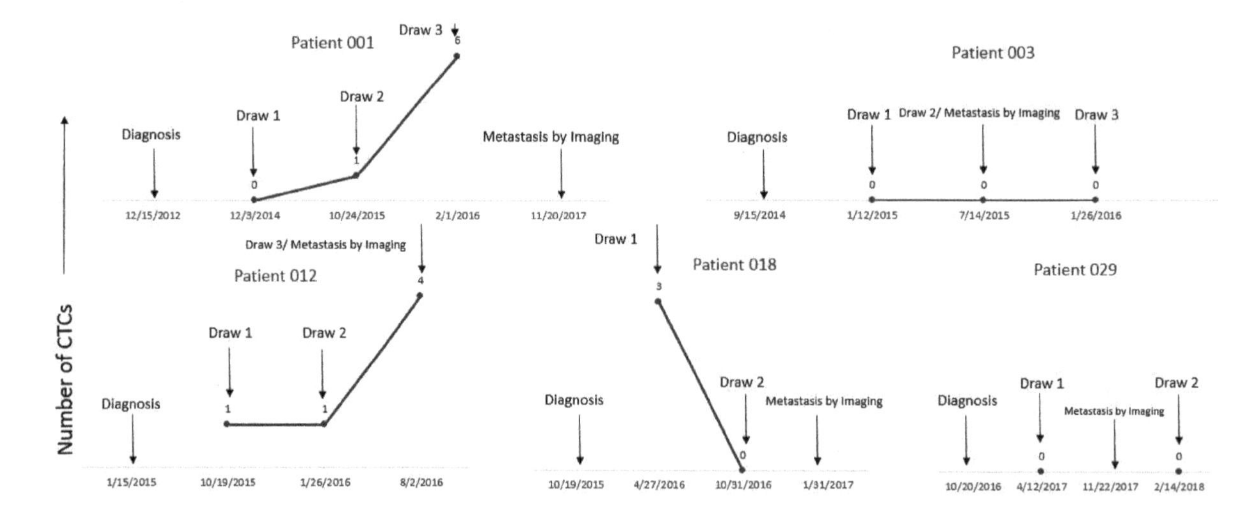

Figure 2. Three out of five patients who developed distant metastasis had CTCs detected before radiographic detection of metastasis. Patients 001, 012, and 018 had CTCs detected 24.9, 9.5 and 3.2 months prior to radiographic evidence of metastasis, respectively. Patient 003 and 029 were "non-secretors" for CTCs as they had no CTC detected pre and post metastasis detection by radiographic imaging.

3.3. *CTC Detection Is Risk Factor for Increased Mortality in Early-Stage Uveal Melanoma*

At study cutoff two patients in the early-stage group had died, secondary to disease progression. One patient was Class 2 and the other was unknown with monosomy 3. Both patients had detectable CTCs at initial draw. Landmark OS of CTC positive vs. negative was compared for early-stage uveal melanoma at 12, 24, and 36 months; the p-value at 12 months was not applicable, at 24 months was 0.047 and at 36 months was $p = 0.051$ (Figure 3). Nine of 19 (47%) patients in the metastatic group had died at study cutoff. Of the 9 patients, six patients had detectable CTC at either initial draw (4) or subsequent draw (2).

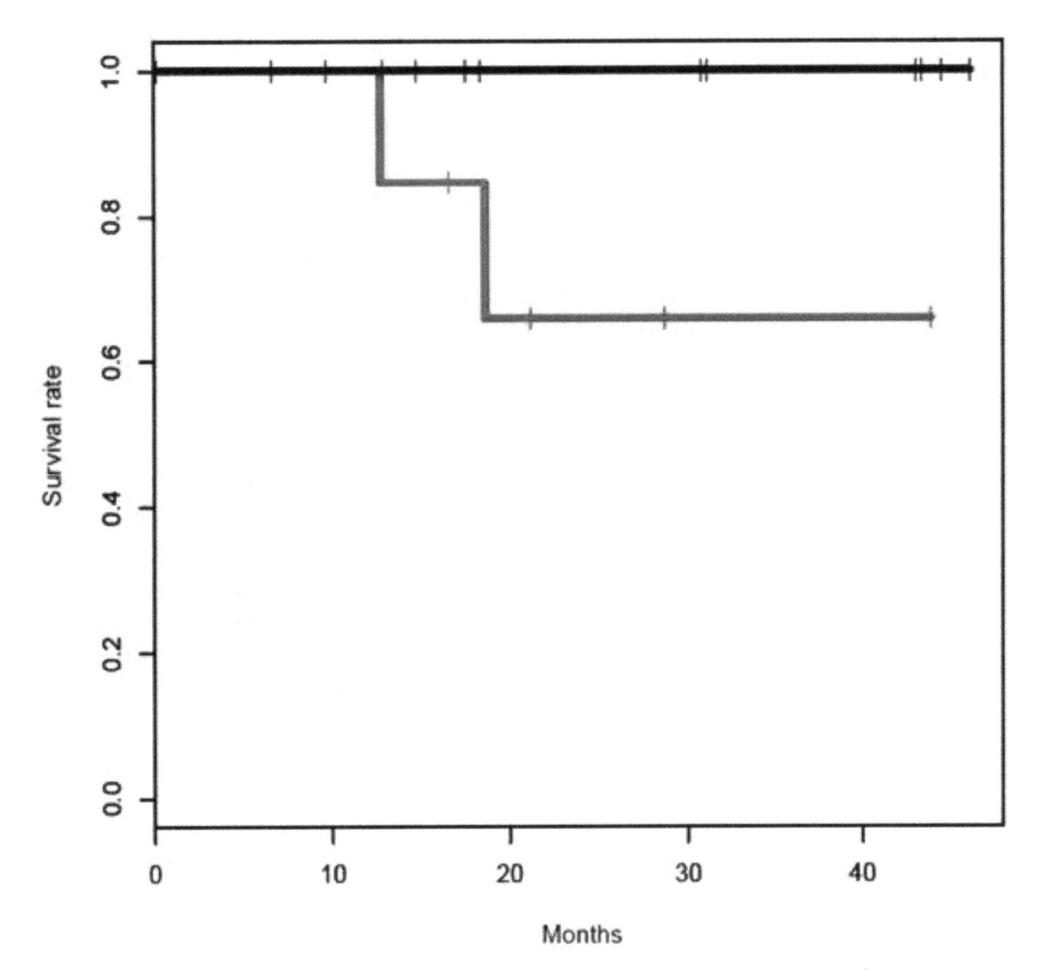

Figure 3. Landmark overall survival at 24 months was poor in CTC detected vs. CTC not-detected early-stage uveal melanoma ($p < 0.05$).

4. Discussion

This study was a pilot project conducted with the primary aim to check feasibility of CTC detection in uveal melanoma patients. At initial draw, 36% (14/39) of the patients had detectable CTCs, which increased to 54% (21/39) during course of the study follow-up. CTCs were most frequently detected amongst the metastatic group compared to the early-stage disease group. Our study, along with others [10–13], confirms that CTCs are detectable in primary uveal melanoma. In early-stage disease, the presence of CTCs correlated with gene expression profiling. Of the patients with ≥1 CTC, 75% (6/8) had gene expression profiling and 83% (5/6) had Class 2 disease. In the early-stage disease group, three patients (60%) had detectable CTCs in the blood before detection of distant metastasis by radiological imaging (MRI) and two patients who did not have any CTC present before detection of distant metastasis by imaging remained negative for CTC presence for at least one more draw. Non-detection of CTCs was limited by the timing of the blood draw in relation to CTC shedding or biological factors [14]. We speculated that there were two groups of early-stage uveal melanoma patients—"CTC secretors" and "CTC non-secretors". Nonetheless, our study along with the studies by Schuster et al. [13] and Keilholz et al. [15] show the importance of CTC detection in predicting early metastasis in uveal melanoma. In our study comparing early-stage CTC detected vs. non detected uveal melanoma, landmark overall survival at 24 months was statistically low ($p < 0.05$). The reason for not reaching statistical difference at 36 months was likely the small cohort of patients, one of the study limitations. The IRB-approved protocol is ongoing in order to increase accrual, and to permit sequential circulating tumor cell measurements during routine follow-up visits. Measurement of circulating tumor DNA (ctDNA) is another important tool in liquid biopsy to monitor disease status of uveal melanoma. A study by Beasley et al. [12] showed that ctDNA was able to detect metastatic disease earlier than imaging while Bidard et al. showed that in metastatic uveal melanoma ctDNA predicted poor outcomes [16]. Liquid biopsy remains an active area of interest in regards to uveal melanoma. CTCs can even be tested for high-risk cytogenetics. Tura et al. correlated the presence of monosomy 3 in CTCs with its presence in primary tumors [17]. Additional studies coupling CTC detection with other strategies such as ctDNA and single-cell sequencing [18] may provide better complementary diagnostic and prognostic information compared to traditional radiographic surveillance. There remains a need to incorporate liquid biopsy in clinical trials evaluating adjuvant therapy post primary tumor treatment in early-stage uveal melanoma to indicate patients at high-risk of distant metastasis.

5. Conclusions

Circulating tumor cells are more frequently detected in the metastatic stage compared to in early-stage uveal melanoma. In early-stage uveal melanoma, most patients in whom circulating tumor cells were detected had adverse prognostic risk factors (Class 2 status by gene expression profiling and monosomy 3). The presence of circulating tumor cells in early-stage disease predicted increased risk of distant metastasis and worse clinical outcomes. The clinical utility of circulating tumor cells in uveal melanoma is not yet known; however, in early-stage disease, longitudinal tracking of circulating tumor cells may serve as a surrogate or complementary tool to radiographic imaging for the detection of microscopic residual disease. In metastatic uveal melanoma, they could also be tracked alongside radiographic scans to determine disease burden and/or response to therapy, ultimately serving as a less costly tool for more efficient assessment than traditional diagnostic imaging. Further studies of early-stage and metastatic uveal melanoma are needed before recommending the routine use of circulating tumor cells in clinical practice.

Author Contributions: Conceptualization, S.P. and D.G.; methodology, J.U., C.H., V.S., S.M., and A.L.; validation, C.H., A.L.; formal analysis, J.R., C.H., V.S., and S.M.; resources, A.L.; data curation, K.A., J.U., and S.P.; writing—original draft preparation, K.A., C.H., and S.P.; writing—review and editing, all authors.

References

1. Chang, A.E.; Karnell, L.H.; Menck, H.R. The National Cancer Data Base report on cutaneous and noncutaneous melanoma: A summary of 84,836 cases from the past decade. The American College of Surgeons Commission on Cancer and the American Cancer Society. *Cancer* **1998**, *83*, 1664–1678. [CrossRef]
2. Shields, C.L.; Furuta, M.; Thangappan, A.; Nagori, S.; Mashayekhi, A.; Lally, D.R.; Kelly, C.C.; Rudich, D.S.; Nagori, A.V.; Wakade, O.A.; et al. Metastasis of uveal melanoma millimeter-by-millimeter in 8033 consecutive eyes. *Arch. Ophthalmol.* **2009**, *127*, 989–998. [CrossRef] [PubMed]
3. Singh, A.D.; Turell, M.E.; Topham, A.K. Uveal melanoma: Trends in incidence, treatment, and survival. *Ophthalmology* **2011**, *118*, 1881–1885. [CrossRef] [PubMed]
4. Andreoli, M.T.; Mieler, W.F.; Leiderman, Y.I. Epidemiological trends in uveal melanoma. *Br. J. Ophthalmol.* **2015**, *99*, 1550–1553. [CrossRef] [PubMed]
5. Kaliki, S.; Shields, C.L. Uveal melanoma: Relatively rare but deadly cancer. *Eye* **2017**, *31*, 241–257. [CrossRef] [PubMed]
6. Tsai, K.K.; Bollin, K.B.; Patel, S.P. Obstacles to improving outcomes in the treatment of uveal melanoma. *Cancer* **2018**, *124*, 2693–2703. [CrossRef] [PubMed]
7. Broggi, G.; Musumeci, G.; Puzzo, L.; Russo, A.; Reibaldi, M.; Ragusa, M.; Longo, A.; Caltabiano, R. Immunohistochemical Expression of ABCB5 as a Potential Prognostic Factor in Uveal Melanoma. *Appl. Sci.* **2019**, *9*, 1316. [CrossRef]
8. Baccelli, I.; Schneeweiss, A.; Riethdorf, S.; Stenzinger, A.; Schillert, A.; Vogel, V.; Klein, C.; Saini, M.; Bauerle, T.; Wallwiener, M.; et al. Identification of a population of blood circulating tumor cells from breast cancer patients that initiates metastasis in a xenograft assay. *Nat. Biotechnol.* **2013**, *31*, 539–544. [CrossRef] [PubMed]

9. Alix-Panabieres, C.; Pantel, K. Clinical Applications of Circulating Tumor Cells and Circulating Tumor DNA as Liquid Biopsy. *Cancer Discov.* **2016**, *6*, 479–491. [CrossRef] [PubMed]
10. Suesskind, D.; Ulmer, A.; Schiebel, U.; Fierlbeck, G.; Spitzer, B.; Spitzer, M.S.; Bartz-Schmidt, K.U.; Grisanti, S. Circulating melanoma cells in peripheral blood of patients with uveal melanoma before and after different therapies and association with prognostic parameters: A pilot study. *Acta Ophthalmol.* **2011**, *89*, 17–24. [CrossRef] [PubMed]
11. Tura, A.; Luke, J.; Merz, H.; Reinsberg, M.; Luke, M.; Jager, M.J.; Grisanti, S. Identification of circulating melanoma cells in uveal melanoma patients by dual-marker immunoenrichment. *Invest. Ophthalmol. Vis. Sci.* **2014**, *55*, 4395–4404. [CrossRef] [PubMed]
12. Beasley, A.; Isaacs, T.; Khattak, M.A.; Freeman, J.B.; Allcock, R.; Chen, F.K.; Pereira, M.R.; Yau, K.; Bentel, J.; Vermeulen, T. Clinical Application of Circulating Tumor Cells and Circulating Tumor DNA in Uveal Melanoma. *JCO Precis. Oncol.* **2018**, *2*, 1–12. [CrossRef]
13. Schuster, R.; Bechrakis, N.E.; Stroux, A.; Busse, A.; Schmittel, A.; Scheibenbogen, C.; Thiel, E.; Foerster, M.H.; Keilholz, U. Circulating tumor cells as prognostic factor for distant metastases and survival in patients with primary uveal melanoma. *Clin. Cancer Res.* **2007**, *13*, 1171–1178. [CrossRef] [PubMed]
14. Plaks, V.; Koopman, C.D.; Werb, Z. Cancer. Circulating tumor cells. *Science* **2013**, *341*, 1186–1188. [CrossRef] [PubMed]
15. Keilholz, U.; Goldin-Lang, P.; Bechrakis, N.E.; Max, N.; Letsch, A.; Schmittel, A.; Scheibenbogen, C.; Heufelder, K.; Eggermont, A.; Thiel, E. Quantitative detection of circulating tumor cells in cutaneous and ocular melanoma and quality assessment by real-time reverse transcriptase-polymerase chain reaction. *Clin. Cancer Res.* **2004**, *10*, 1605–1612. [CrossRef] [PubMed]
16. Bidard, F.C.; Madic, J.; Mariani, P.; Piperno-Neumann, S.; Rampanou, A.; Servois, V.; Cassoux, N.; Desjardins, L.; Milder, M.; Vaucher, I.; et al. Detection rate and prognostic value of circulating tumor cells and circulating tumor DNA in metastatic uveal melanoma. *Int. J. Cancer* **2014**, *134*, 1207–1213. [CrossRef] [PubMed]
17. Tura, A.; Merz, H.; Reinsberg, M.; Luke, M.; Jager, M.J.; Grisanti, S.; Luke, J. Analysis of monosomy-3 in immunomagnetically isolated circulating melanoma cells in uveal melanoma patients. *Pigment. Cell Melanoma. Res.* **2016**, *29*, 583–589. [CrossRef] [PubMed]
18. Navin, N.; Hicks, J. Future medical applications of single-cell sequencing in cancer. *Genome Med.* **2011**, *3*, 31. [CrossRef] [PubMed]

3

Circulating Tumor Cells in Right- and Left-Sided Colorectal Cancer

Chiara Nicolazzo [1], Cristina Raimondi [2], Angela Gradilone [1], Alessandra Emiliani [2], Ann Zeuner [3], Federica Francescangeli [3], Francesca Belardinilli [1], Patrizia Seminara [2], Flavia Loreni [1], Valentina Magri [4], Silverio Tomao [2] and Paola Gazzaniga [1,*]

1 Department of Molecular Medicine, Circulating tumor cells Unit, Sapienza University of Rome, 00161 Rome, Italy
2 Department of Radiological, Oncological and Pathological Sciences, Division of Medical Oncology, Sapienza University of Rome, 00161 Rome, Italy
3 Department of Hematology, Oncology and Molecular Medicine, Istituto Superiore di Sanità, 00161 Rome, Italy
4 Department of Surgical Sciences, Sapienza University of Rome, 00161 Rome, Italy
* Correspondence: paola.gazzaniga@uniroma1.it.

Abstract: Molecular alterations are not randomly distributed in colorectal cancer (CRC), but rather clustered on the basis of primary tumor location underlying the importance of colorectal cancer sidedness. We aimed to investigate whether circulating tumor cells (CTC) characterization might help clarify how different the patterns of dissemination might be relative to the behavior of left- (LCC) compared to right-sided (RCC) cancers. We retrospectively analyzed patients with metastatic CRC who had undergone standard baseline CTC evaluation before starting any first-line systemic treatment. Enumeration of CTC in left- and right-sided tumors were compared. The highest prognostic impact was exerted by CTC in left-sided primary cancer patients, even though the lowest median number of cells was detected in this subgroup of patients. CTC exhibit phenotypic heterogeneity, with a predominant mesenchymal phenotype found in CTC from distal compared to proximal primary tumors. Most CTC in RCC patients exhibited an apoptotic pattern. CTC in left-sided colon cancer patients exhibit a predominant mesenchymal phenotype. This might imply a substantial difference in the biology of proximal and distal cancers, associated with different patterns of tumor cells dissemination. The poor prognosis of right-sided CRC is not determined by the hematogenous dissemination of tumor cells, which appears to be predominantly a passive shedding of non-viable cells. Conversely, the subgroup of poor-prognosis left-sided CRC is reliably identified by the presence of mesenchymal CTC.

Keywords: colorectal cancer; sidedness; circulating tumor cells; epithelial-mesenchymal transition; prognosis; CellSearch®; ScreenCell®

---

1. Introduction

There is discrepancy between the latest advances in molecular segmentation of colorectal cancer (CRC), which recently led to the consensus molecular subtyping of the disease, and the insufficient availability of biomarkers ready for routine clinical use [1]. We certainly recognize that colorectal cancer is a heterogeneous and complex disease and we recently witnessed the elegant demonstration of its spatial and temporal dynamic nature [2]. RAS gene mutations have historically been the watershed for molecular stratification of metastatic colorectal cancer (mCRC) and, so far, RAS genes mutational analysis is the only discriminant between frontline treatment options in this setting. In the last few years, we have made progress in the search for strategies to repeatedly monitor the molecular makeup of colorectal cancer along its Darwinian temporal evolution and we gained preliminary, although promising, elements on the possibility to adapt therapies on the basis of the continuous mutability of colorectal cancer clones [3]. Evidence is emerging that molecular alterations are not randomly distributed in colorectal cancer, but rather clustered on the basis of primary tumor location and underlie the prognostic and predictive significance of colorectal cancer sidedness. Colorectal cancers indeed exhibit differences in epidemiology, clinical presentation, and outcomes depending on the location of the primary tumor. It has been reported that right-sided tumors have a lower incidence, a more advanced stage at presentation, and are associated with worse prognosis compared to left-sided colorectal tumors [4]. Further than RAS mutational status and EGFR activation, biological reasons behind the differences between left- and right-sided colorectal cancers are laid aside. Several publications have demonstrated that distinct metastatic patterns in colorectal cancer patients exist based on primary tumor location [5,6]. This might be attributable to a number of reasons, including the embryological origin of proximal and distal cancers, the anatomical location and venous drainage, and the biology of right-sided and left-sided cancers. Prominent differences have been also demonstrated in terms of tumor microenvironment, which results to be inflamed and with marked stromal infiltration in left cancers as compared to right cancers, which in turn are characterized by lower inflammatory status and higher expression of pro-angiogenic factors. This might also explain the different pattern of response to biological agents, with left-sided colorectal cancer (LCC) more prone to respond to anti-EGFR therapies and right-sided colorectal cancer (RCC) more responsive to antiangiogenic drugs. Cancer-specific molecular alterations and tumor microenvironment structure might both affect the dissemination pattern of cancer cells and ultimately determine the outcome of patients. The dissemination of cancer cells from primary tumors into distant sites represents the first event in the multistep process known as the invasion–metastasis cascade. Individual cancer cells and multi-cellular cohorts (clusters) arising from primary tumors intravasate and travel to distant tissues thus representing an intermediate between primary tumors and eventually formed metastatic colonies [7]. We have previously demonstrated that the presence of even a single circulating tumor cell in the peripheral blood of patients with mCRC significantly affects their prognosis [8]. Of additional relevance is the well-proven heterogeneity in the biological properties of either single or clustered CTC, which often exhibit various combinations of epithelial and mesenchymal traits [9]. Hence, a more in depth analysis of the biological programs activated in CTC might help clarify how different the patterns of dissemination might be relative to the behavior of primary tumors. The primary aim of the present study was to evaluate whether the presence of circulating tumor cells, as a surrogate of tumor in the bloodstream, might dichotomize according to sidedness and whether this might have a prognostic impact. The secondary aim was to investigate whether CTC isolated from proximal and distal colorectal cancers might differ in their biological features, specifically referring to epithelial-mesenchymal transition (EMT)-related phenotype.

2. Results

2.1. Enumeration of CTC According to Tumor Sidedness

Eighty-four metastatic colorectal cancer patients were included in this retrospective analysis. The whole population was divided into three subgroups according to primary tumor sidedness: 24 right-sided CRC, 31 left-sided CRC, and 29 rectal cancers. The demographic and clinicopathological characteristics of the whole population are shown in Table 1.

We report here that CTC were not uniformly distributed in the three subgroups of patients and were found in 46%, 39%, and 38% of patients with primary RCC, LCC, and rectal cancer, respectively. This difference was not found to be statistically significant ($p = 0.8$). Some degree of heterogeneity was also observed in terms of number of CTC. Indeed, the highest median number of CTC was observed in the group of patients with RCC (6.75, range 0–67), as compared to LCC (1.29, range 0–9), and rectal cancers (2.68, range 0–37). While the comparison between the three groups was found not statistically significant by ANOVA test ($p = 0.12$), the comparison between paired groups demonstrated a difference in CTC number distribution only between RCC and LCC. This difference was found to be statistically significant ($p = 0.03$).

Unexpectedly, the highest prognostic impact was exerted by CTC in left-sided primary cancer patients, even though the lowest median number of cells was detected in this subgroup of patients. Indeed, the prognostic impact of CTC in terms of time to progression (TTP) reached a robust statistical significance only in the subgroup of patients with primary left colon cancer (11.1 months in CTC positive vs. 25.6 months in CTC negative patients, $p = 0.009$), while being on the threshold of significance in rectal cancer patients (11.6 months in CTC positive vs. 18 months in CTC negative patients, $p = 0.058$) and not wholly significant in patients with primary right colon cancer (11.5 months in CTC positive vs. 15.2 months in CTC negative patients, $p = 0.5$) (Figure 1).

Table 1. Demographic and clinicopathological characteristics of patients analyzed for circulating tumor cells (CTC) enumeration (CellSearch®).

Characteristics	No. of Patients ($n = 84$)
Sex	
Male	50
Female	34
Primary tumor location	
Right	24
Left	31
Rectum	29
Stage of disease	
Metastatic	84
KRAS status (tumor tissue)	
Wild type	31
Mutant	29
Unknown	24

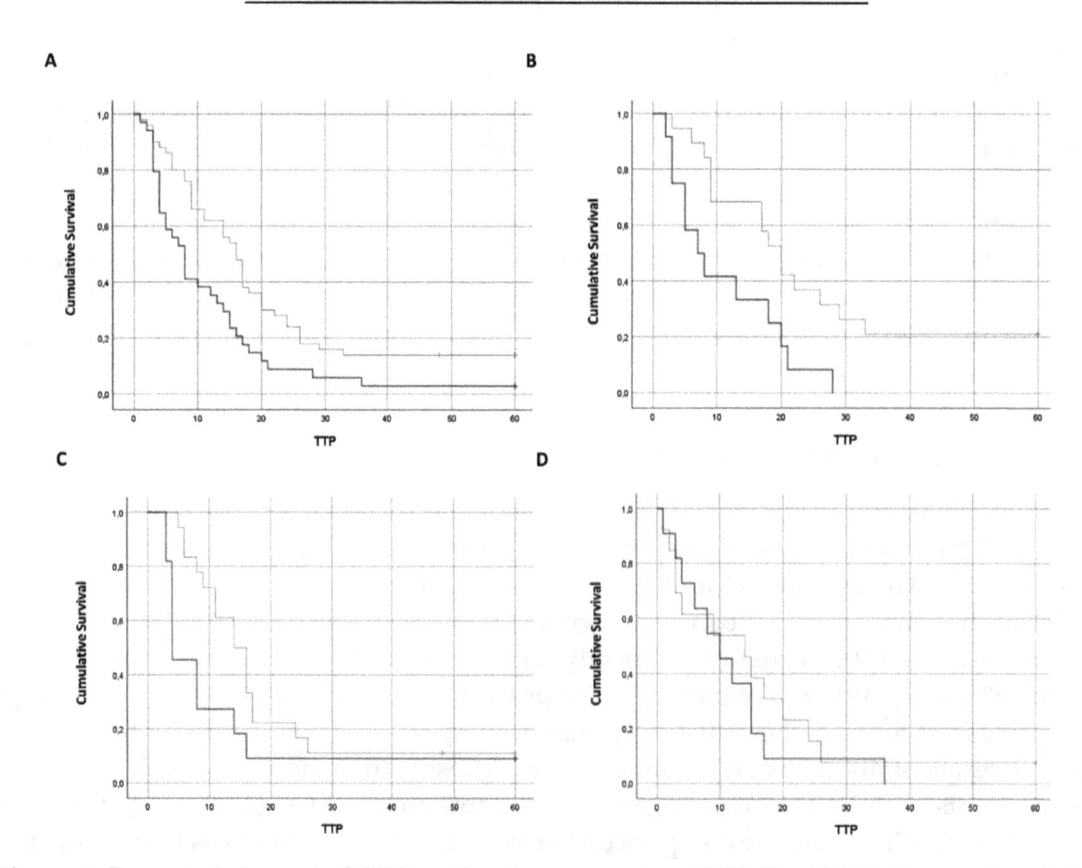

Figure 1. Prognostic impact of CTC on time to progression (TTP). Panel (**A**): Comparison in TTP

between CTC-positive vs. CTC-negative samples (all mCRC pts). Panel (**B**): Comparison in TTP between CTC-positive vs. CTC-negative samples in left colon cancer pts (p = 0.009). Panel (**C**): Comparison in TTP between CTC-positive vs. CTC-negative samples in rectal cancer pts (p = 0.058). Panel (**D**): Comparison in TTP between CTC-positive vs. CTC-negative samples in right colon cancer pts (p = 0.5).

2.2. Apoptotic Morphology of CTC

In the whole population of 84 patients, we further retrospectively review all the archived images of the CellSearch® analyses in order to investigate whether an apoptotic morphological pattern could be differentially observed in CTC from RCC patients as compared to LCC and rectal cancer patients. Apoptotic CTC were defined as all EpCAM+, CK+, DAPI +, CD45- CellSearch® events with altered morphological parameters such as speckled pattern of keratin staining and/or fragmented or disintegrated nuclei [10]. We found that 130/162 CTC (80%) visualized in RCC patients exhibited a clear apoptotic morphological pattern differently from left-sided colon cancers and rectal cancers, where only 12/118 (10%) displayed apoptotic CTC (Figure 2).

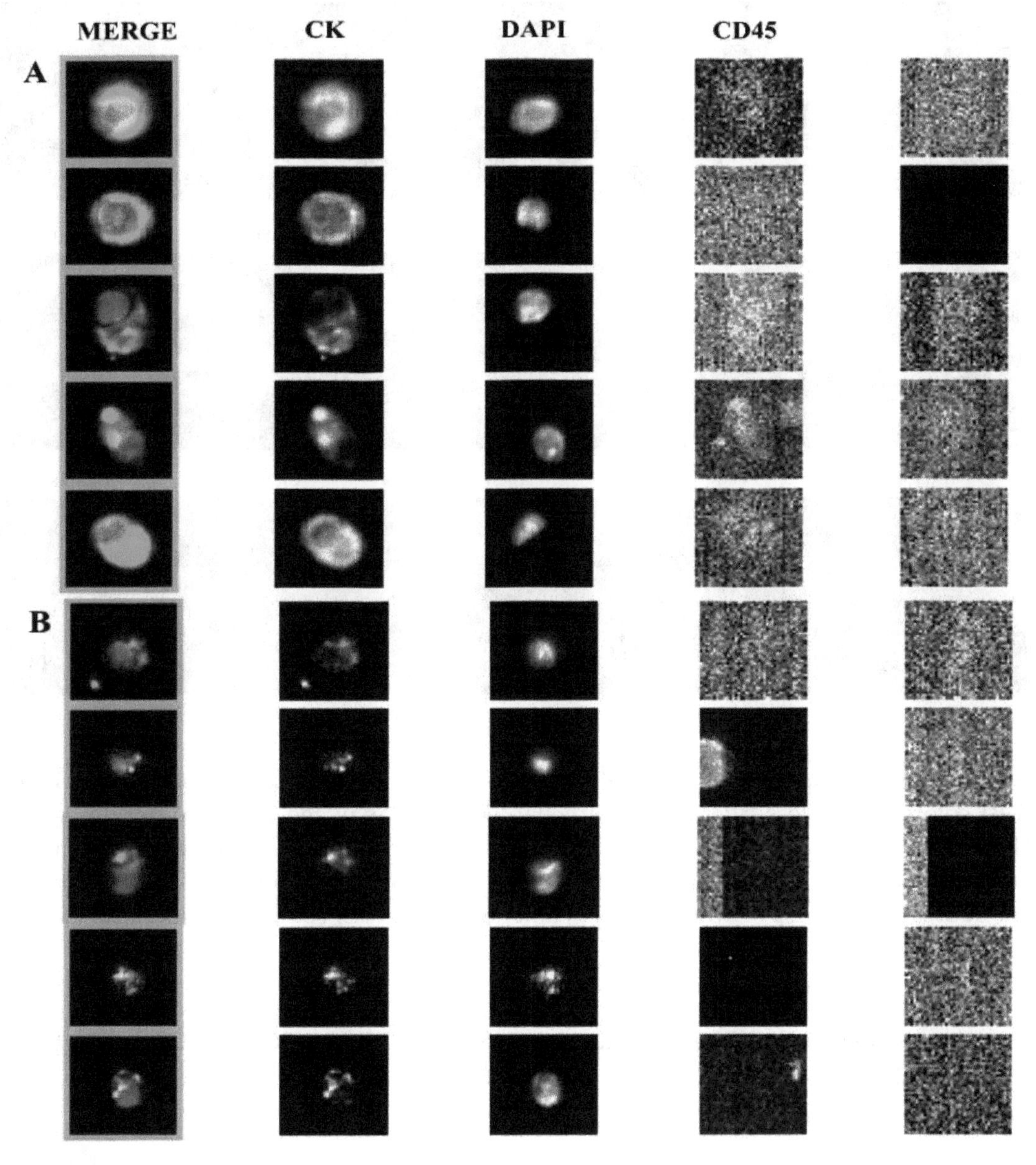

Figure 2. Images of intact CTC (panel **A**) and apoptotic CTC (panel **B**) isolated through CellSearch® from left-sided and right-sided colorectal cancers, respectively.

2.3. *Epithelial-Like and Mesenchymal-Like Features of CTC*

In order to provide a possible explanation for the unpredictable strongest prognostic significance of CTC in LCC patients, we performed a second analysis to gather data on the biology of CTC. To this purpose, we analyzed 24 patients (15 retrospectively and 9 prospectively enrolled) in order to perform the molecular characterization of CTC isolated from whole blood through a filtration device (ScreenCell® Cyto), which allows the separation of live cells for downstream cytology studies. For each filter, three microscopic fields were analyzed. Seven RCC, nine LCC, and eight rectal cancers were analyzed for epithelial-like and EMT-like markers. Vimentin and N-cadherin were chosen as mesenchymal-like markers, and CK20 was selected as colon cancer-specific epithelial marker. Despite the limited number of samples available, we found that CTC isolated from RCC patients exhibited a predominant epithelial-like phenotype (EpCAM+, CK20+, vimentin−/N-cadherin−) as compared to the mesenchymal-like traits observed in CTC from LCC cancer patients (EpCAM+/−, CK20−, vimentin+/N-cadherin+) (Figure 3).

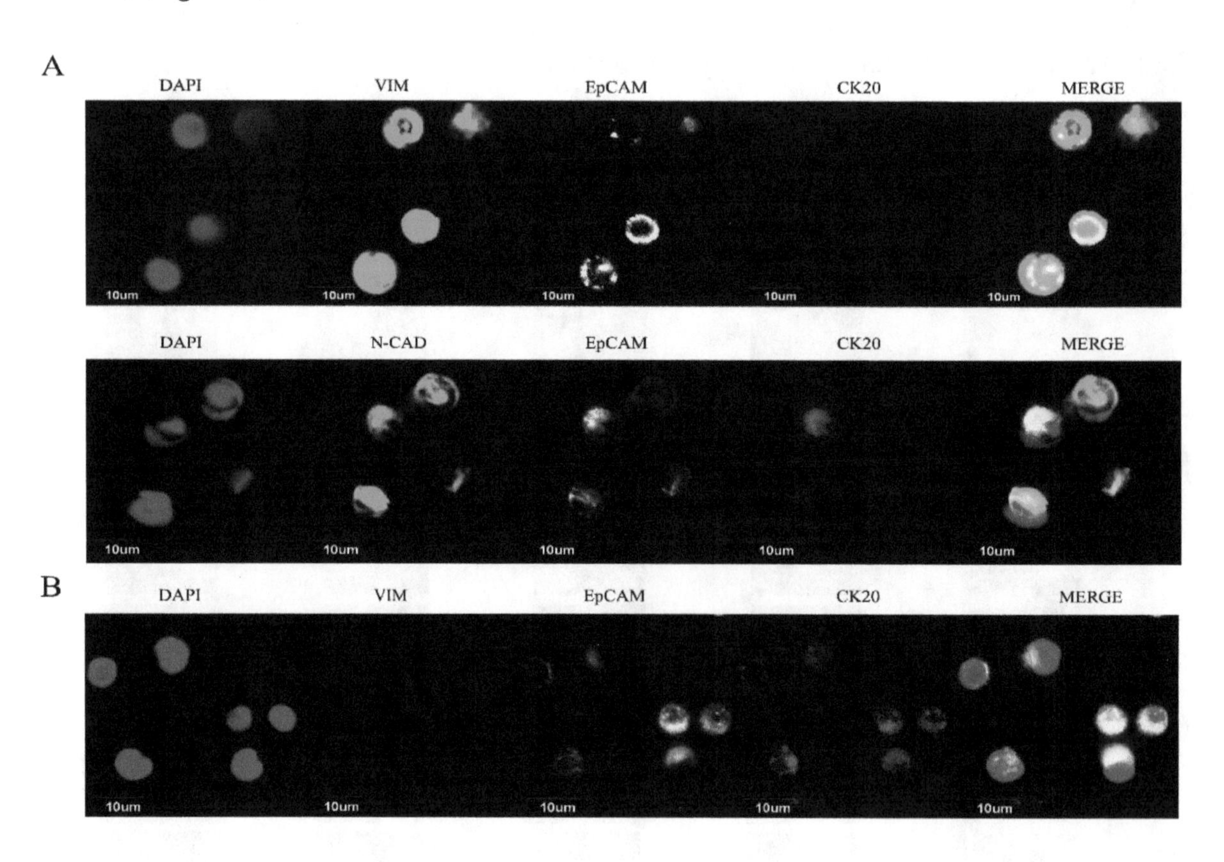

Figure 3. Epithelial-like and mesenchymal-like CTC isolated from left-sided (panel **A**) and right-sided (panel **B**) colorectal cancers. Panel **A**: Epithelial-like CTC analyzed by immunofluorescence for EpCAM (yellow), cytokeratin 20 (CK20, red), and vimentin/N-cadherin (VIM/N-CAD, green). In blue, nuclei are visualized. Original magnification 60×. Panel **B**. mesenchymal-like CTC analyzed by immunofluorescence for EpCAM (yellow), cytokeratin 20 (CK20, red), and vimentin (VIM, green). In blue, nuclei are visualized.

The percentage of epithelial-like and mesenchymal-like CTC in RCC, LCC, and rectal cancers are illustrated in Table 2.

Table 2. Percentage of epithelial-like and mesenchymal-like CTC in right-sided cancers (RCC), left-sided cancers (LCC), and rectal cancers.

Patient n.	Tumor Location	n. CTC (ScreenCell)	Epithelial-Like CTC (%)	Mesenchymal-Like CTC (%)
1	right	20	16 (80)	4 (20)
2	right	5	5 (100)	0 (0)
3	right	16	10(63)	6 (37)
4	right	25	20 (80)	5(20)
5	right	8	7 (87)	1 (13)
6	right	30	25 (83)	5 (17)
7	right	27	22 (81)	5 (19)
8	left	4	0 (0)	4 (100)
9	left	4	1 (25)	3 (75)
10	left	10	1 (10)	9 (90)
11	left	11	4(36)	7 (64)
12	left	6	2(33)	4 (67)
13	left	16	5(31)	11 (69)
14	left	7	0(0)	7 (100)
15	left	8	2(25)	6 (75)
16	left	10	2(0)	8 (100)
17	rectum	5	0(0)	5 (100)
18	rectum	10	2(20)	8 (80)
19	rectum	14	2(14)	12 (86)
20	rectum	7	2(29)	5 (71)
21	rectum	10	0 (0)	10 (100)
22	rectum	6	1 (17)	5 (83)
23	rectum	8	0 (0)	8 (100)
24	rectum	3	0 (0)	3 (100)

3. Discussion

This is the first publication reporting on the detection rate and the prognostic significance of CTC according to primary tumor sidedness in metastatic colorectal cancer patients. Although the term "colorectal cancer" is currently referred to a single tumor type, increasing evidence is emerging concerning the prognostic impact of primary tumor sidedness. A recent meta-analysis of 66 clinical studies compared the overall survival of RCC versus LCC in over 1.4 million patients and demonstrated a 20% reduced risk of death for patients whose tumors arise from the left side [11]. To date, the molecular background of proximal and distal colorectal cancers has been only preliminary unraveled and translational efforts to gain knowledge about the biological underpinnings of colorectal cancer sidedness are critically important. It has been shown that baseline detection of CTC in metastatic colorectal cancer is an independent prognostic factor for progression-free survival (PFS) and overall survival (OS) with CellSearch® [12]. We confirmed here that the presence of CTC is a significant prognostic factor in metastatic colorectal cancer patients regardless of tumor sidedness. The whole population was then sub-grouped according to primary tumor sidedness. In particular, left-sided colon cancers and rectal cancers were considered separately in this retrospective evaluation, in line with recently released data supporting the colorectal cancer "three entities" hypothesis [13–15]. In our series of patients, CTC were not uniformly distributed in the three subgroups, showing the highest prognostic impact in patients with left-sided colon cancer, while being the lowest in number in this subgroup of patients. We hypothesized that this unanticipated result could reflect a difference at the molecular level between CTC shed from left-sided as compared to right-sided primary tumors. Indeed, we found that CTC exhibit phenotypic heterogeneity, with a predominant mesenchymal phenotype found in CTCs from distal compared to proximal primary tumors. It is well recognized that the EMT program enables epithelial cancer cells to acquire properties that are critical to invasion and metastatic dissemination, such as increased motility, invasiveness, and the ability to degrade components of the extracellular matrix. The EMT program that allows cancer cells to disseminate from a primary tumor also promotes their self-renewal capability, usually depicted as the defining trait of cancer stem cells. Such EMT process seems almost invariably triggered by heterotypic signals, including TGF-β, Wnt, and interleukins, which cancer cells receive from the tumor reactive stroma, which thus plays a substantial role in dictating cancer progression [16]. Tumor cells can reach the circulation by either active invasion, which requires the acquisition of certain mesenchymal traits, or by passive shedding, due to cancer cell pushing by tumor expansive growth and facilitated by the abundance of highly abnormal blood vessels. Tumors are likely able to use both active and passive methods to enter the vasculature, depending on the site of tumor initiation, the aggressiveness of tumor cells, and the tumor micro-environmental conditions. It has been reported that most of the cancer cells that entered into the vasculature by passive shedding are non-viable cells, thus incapable of completing the efficient colonization of distant sites [17]. Distinct metastatic patterns in colorectal cancer patients based on primary tumor location have been recently demonstrated, with higher rates of liver and lung metastases in left-sided colon cancers and rectal cancers, respectively, as compared to right-sided tumors, which appear to be associated with higher rates of peritoneal metastases. We could hypothesize that proximal and distal colorectal cancers may also differ in the early steps of the metastatic cascade and that alternative modalities of tumor cells intravasation might be adopted, depending on primary tumor location. As far as we know, substantial differences exist in the tumor microenvironment of distal and proximal colorectal cancer, with the former being inflamed and with marked stromal infiltration and the latter characterized by lower inflammatory status and higher expression of pro-angiogenic factors [18]. With this in mind, we could envisage that cancer cells from primary distal cancers receive abundant signals from the surrounding reactive stroma, which are able to activate their latent EMT programs and equip them with the ability to actively intravasate into the circulation. Conversely, proximal cancers, which are frequently larger in size and plenty of disorganized and leaky blood vessels, might release into the circulation a higher number of cancer cells, which are not necessarily viable and able to sustain the following steps of the metastatic cascade. As expected, we found that the vast majority of CTC in RCC patients exhibited a clear apoptotic pattern, thus, possibly providing the rationale for the limited prognostic impact of these cells in patients with primary proximal cancers. Although we did not use any apoptosis-specific marker to identify apoptotic CTC, several reports have described apoptotic CTC as a specific CTC subtype well identifiable at CellSearch® [10,19]; they are characterized by altered morphological parameters such as speckled pattern of keratin staining and/or fragmented or disintegrated nuclei.

Our data suggest that the poor prognosis of right-sided colorectal cancer might not be determined by the hematogenous dissemination of tumor cells, which appears to predominantly be a passive shedding of non-viable cells in the blood vessels. We also demonstrate that a subgroup of poor-prognosis left-sided colon cancer exists, which is reliably identified by the presence of CTC in the blood vessels. Particularly, CTC found in left-sided colon cancer patients exhibit a phenotype with different levels of mesenchymal differentiation. This might imply a substantial difference in the biology of proximal and distal cancers, mainly related to the tumor microenvironment and strongly associated with different patterns of tumor cells dissemination from primary tumors. Although EMT is certainly triggered by stromal signals, EMT-specific traits in CTC, which are someway "stromal independent", might indicate that cancer cells in the blood are able to transcriptionally control their nature by cell-autonomous mechanisms, as recently advocated by the new CRIS-B subtype of colorectal cancer [20]. The main limitation of the study is the small population of samples available for molecular analysis, and our results need to be confirmed in a larger cohort. Investigating the intracellular properties of CTC, in their fluid microenvironment, might contribute to define how and to what extent cancer cell-specific traits contribute to the creation of accurate molecular subtypes of CRC and to the definition of reliable prognostic indicators.

4. Materials and Methods

4.1. CTC Enumeration

We retrospectively analyzed patients with mCRC who had undergone standard baseline CTC evaluation through the CellSearch® platform (Menarini Silicon Biosystems, Castel Maggiore, Bo, Italy) before starting any first-line systemic treatment between 2010 and 2015. Informed consent had been obtained in all patients. A total of 84 mCRC patients were included in this retrospective study. Information on primary tumor location was obtained from the original pathology reports. Primary tumors located in the proximal two-thirds of the transverse colon, ascending colon, and caecum were coded as right-sided. Tumors located in the distal third of the transverse colon, splenic flexure, descending colon, and sigmoid colon were categorized as left-sided. The protocol had been approved by Ethical Committee of Policlinico Umberto I (protocol n. 668/09, 9 July 2009; amended protocol 179/16, 1 March 2016).

From each patient, 7.5 mL of peripheral blood was collected in CellSave preservative tube (Menarini Silicon Biosystems) containing EDTA and a cell fixative, maintained at room temperature and processed within 72 h. The CTC enumeration was carried out through the CellSearch® system, employing CellSearch® Epithelial Cell Kit, which contains a ferrofluid-based capture reagent and immunofluorescent staining reagents. Briefly, CTC were first enriched from 7.5 mL of whole blood by anti-EpCAM-antibody-coated ferrofluid reagent and subsequently stained for cytokeratins (CK), 4′-6-Diamidino-2-phenylindole (DAPI), and CD45. An event was classified as CTC when exhibiting the phenotype EpCAM+, CK+, DAPI+, and CD45-. Apoptotic CTC were defined as all EpCAM+/CK+/CD45-cells characterized by altered morphological parameters such as speckled pattern of keratin staining and/or fragmented or disintegrated nuclei [10].

4.2. Epithelial-Like and Mesenchymal-Like CTC

To isolate CTC for cytological studies, ScreenCell® Cyto kit (ScreenCell, Sarcelles, France) was used, an EpCAM-independent device allowing size-based separation of CTC from whole blood. A total of 24 patients were analyzed. In order to fix the cells and to lyse red blood cells (RBC), 3 mL of blood was diluted in 4 mL of filtration buffer (FC). After 8 minutes of incubation at room temperature, 7 mL of diluted sample was added into device tank and filtered under a pressure gradient using a vacutainer tube. Filtration was usually completed within 3 minutes. After washing with PBS to remove RBC debris, each filter was left on absorbing paper to dry at room temperature. After hydration with Tris-buffered saline (TBS) for 10 minutes, the filters were incubated in a humid chamber overnight at 4 °C with the following primary antibodies: Goat polyclonal anti-cytokeratin (CK) 20 (N-13, 1:100; Santa Cruz Biotechnology Inc., Rockford, IL, USA), mouse monoclonal anti-epithelial cell adhesion molecule (EpCAM) (VU1D9, 1:100; Invitrogen, Rockford, IL, USA), and rabbit polyclonal anti-vimentin (H-84, 1:100; Santa Cruz Biotechnology Inc.) or rabbit polyclonal anti-N-Cadherin (D4R1H, 1:100; Cell Signaling Technology). The next day, filters were washed twice in PBS and then incubated with donkey anti-goat Alexa Fluor 647 (Molecular Probes, Eugene, OR, USA) secondary antibody for 45 minutes at room temperature in the dark. After washing in PBS, filters were incubated with goat serum (1:10) for 30 minutes to reduce cross reactivity between goat anti-rabbit, goat anti-mouse, and donkey anti-goat secondary antibodies. Serum was removed without washing and filters were incubated with goat anti-rabbit Alexa Fluor 488 and goat anti-mouse Alexa Fluor 555 (Molecular Probes) for 45 minutes at room temperature in the dark. Nuclei were stained with 4′,6-diamidino-2-phenylindole (DAPI; Invitrogen) for 15 minutes at room temperature. All antibodies were dissolved in PBS containing 3% bovine serum albumin (BSA), 3% fetal bovine serum (FBS), 0.001% NaN3 and 0.1% Triton X-100. Finally, the filters were mounted with Prolong-Gold Antifade (Invitrogen) on slides and analyzed using a FV1000 Confocal microscope (Olympus FV1000) equipped with a 60× oil immersion objective. Markers levels were evaluated based on immunofluorescence staining intensity. Results were provided as a discrete nominal (positive/negative) score.

Chi Square test was used to assess the difference of CTC positive cases according to tumor sidedness. The one-way analysis of variance (ANOVA) and T student's test were used to assess the statistical significance of CTC number distribution in RCC, LCC, and rectal cancer. Survival analysis was conducted with the Kaplan–Meier method, yielding median survival times (95% confidence intervals) and comparing survival curves with the log-rank test. Statistical significance was set at the 2-tailed 0.05 level. Computations were performed with IBM SPSS Statistics 24.0.

5. Conclusions

Circulating tumor cells from distal compared to proximal colorectal tumors display quantitative and qualitative heterogeneity, with rare cells and predominant mesenchymal phenotype found in CTC isolated from left-sided tumors. Right-sided tumors are characterized by a high percentage of apoptotic CTC. The EMT-like features of CTC in left-sided colon cancer might account for the poor prognosis observed in the subgroup of CTC positive patients.

Author Contributions: Conceptualization, C.N. and P.G.; methodology, C.N.; software, A.G.; formal analysis, C.R. and F.B.; investigation, C.N., A.G., A.Z., F.F., and F.L.; resources, A.E., P.S., S.T., and P.G.; data curation, V.M.; writing—original draft preparation, C.N., C.R. and P.G.; writing—review and editing, P.G.; visualization, C.N. and P.G.; supervision, S.T. and P.G.; project administration, P.G.; funding acquisition, P.G.

Acknowledgments: F.B. is the recipient of a fellowship from Fondazione Umberto Veronesi

References

1. Dienstmann, R.; Vermeulen, L.; Guinney, J.; Kopetz, S.; Tejpar, S.; Tabernero, J. Consensus molecular subtypes and the evolution of precision medicine in colorectal cancer. *Nat. Rev. Cancer* **2017**, *17*, 79–92. [CrossRef] [PubMed]
2. Siravegna, G.; Mussolin, B.; Buscarino, M.; Corti, G.; Cassingena, A.; Crisafulli, G.; Ponzetti, A.; Cremolini, C.; Amatu, A.; Lauricella, C.; et al. Clonal evolution and resistance to EGFR blockade in the blood of colorectal cancer patients. *Nat. Med.* **2015**, *21*, 795–801. [CrossRef] [PubMed]
3. Venesio, T.; Siravegna, G.; Bardelli, A.; Sapino, A. Liquid Biopsies for Monitoring Temporal Genomic Heterogeneity in Breast and Colon Cancers. *Pathobiology* **2018**, *8*, 5146–5154. [CrossRef] [PubMed]
4. Stintzing, S.; Tejpar, S.; Gibbs, P.; Thiebach, L.; Lenz, H.J. Understanding the role of primary tumour localisation in colorectal cancer treatment and outcomes. *Eur. J. Cancer* **2017**, *84*, 69–80. [CrossRef] [PubMed]
5. Holch, J.W.; Demmer, M.; Lamersdorf, C.; Mich, M.; Schulz, C.; von Einem, J.C.; Modest, D.P.; Heinemann, V. Pattern and Dynamics of Distant Metastases in Metastatic Colorectal Cancer. *Visc. Med.* **2017**, *33*, 70–75. [CrossRef] [PubMed]
6. Hugen, N.; Nagtegaal, I.D. Distinct metastatic patterns in colorectal cancer patients based on primary tumour location. *Eur. J. Cancer* **2017**, *75*, 3–4. [CrossRef] [PubMed]
7. Lambert, A.W.; Pattabiraman, D.R.; Weinberg, R.A. Emerging Biological Principles of Metastasis. *Cell* **2017**, *168*, 670–691. [CrossRef] [PubMed]
8. Raimondi, C.; Nicolazzo, C.; Gradilone, A.; Giannini, G.; De Falco, E.; Chimenti, I.; Varriale, E.; Hauch, S.; Plappert, L.; Cortesi, E.; et al. Circulating tumor cells: exploring intratumor heterogeneity of colorectal cancer. *Cancer Biol. Ther.* **2014**, *15*, 496–503. [CrossRef] [PubMed]
9. Micalizzi, D.S.; Haber, D.A.; Maheswaran, S. Cancer metastasis through the prism of epithelial-to-mesenchymal transition in circulating tumor cells. *Mol. Oncol.* **2017**, *11*, 770–780. [CrossRef] [PubMed]
10. Andree, K.C.; van Dalum, G.; Terstappen, L.W. Challenges in circulating tumor cell detection by the CellSearch system. *Mol. Oncol.* **2016**, *10*, 395–407. [CrossRef] [PubMed]
11. Petrelli, F.; Tomasello, G.; Borgonovo, K.; Ghidini, M.; Turati, L.; Dallera, P.; Passalacqua, R.; Sgroi, G.; Barni, S. Prognostic Survival Associated With Left-Sided vs Right-Sided Colon Cancer: A Systematic Review and Meta-analysis. *JAMA Oncol.* **2016**, *3*, 211–219. [CrossRef] [PubMed]

12. Cabel, L.; Proudhon, C.; Gortais, H.; Loirat, D.; Coussy, F.; Pierga, J.Y.; Bidard, F.C. Circulating tumor cells: clinical validity and utility. *Int. J. Clin. Oncol.* **2017**, *22*, 421–430. [CrossRef] [PubMed]
13. Marshall, J.; Lenz, H.J.; Xiu, J.; El-Deiry, W.S.; Swensen, J.; El Ghazal, H.; Gatalica, Z.; Hwang, J.J.; Philip, P.A.; Shields, A.F.; et al. Molecular variances between rectal and left-sided colon cancers. *J. Clin. Oncol.* **2017**, *35*, abstract 522. [CrossRef]
14. Salem, M.E.; Yin, J.; Renfro, L.A.; Weinberg, B.A.; Maughan, T.; Richard Adams, R.; Van Cutsem, E.; Falcone, A.; Tebbutt, N.C.; Seymour, M.T.; et al. Rectal versus left-sided colon cancers: Clinicopathological differences observed in a pooled analysis of 4182 patients enrolled to 8 clinical trials from the ARCAD database. *J. Clin. Oncol.* **2017**, *35*, abstract 675.
15. Li, F.; Lai, M. Colorectal cancer, one entity or three. *J. Zhejiang Univ. Sci. B* **2009**, *10*, 219–229. [CrossRef] [PubMed]
16. Friedl, P.; Alexander, S. Cancer invasion and the microenvironment: plasticity and reciprocity. *Cell* **2011**, *147*, 992–1009. [CrossRef] [PubMed]
17. Bockhorn, M.; Jain, R.K.; Munn, L.L. Active versus passive mechanisms in metastasis: do cancer cells crawl into vessels, or are they pushed? *Lancet Oncol.* **2007**, *8*, 444–448. [CrossRef]
18. Ulivi, P.; Scarpi, E.; Chiadini, E.; Marisi, G.; Valgiusti, M.; Capelli, L.; Casadei Gardini, A.; Monti, M.; Ruscelli, S.; Frassineti, G.L.; et al. Right- vs. Left-Sided Metastatic Colorectal Cancer: Differences in Tumor Biology and Bevacizumab Efficacy. *Int. J. Mol. Sci.* **2017**, *18*, 1240. [CrossRef]
19. Deutsch, T.M.; Riethdorf, S.; Nees, J.; Hartkopf, A.D.; Schönfisch, B.; Domschke, C.; Sprick, M.R.; Schütz, F.; Brucker, S.Y.; Stefanovic, S.; et al. Impact of apoptotic circulating tumor cells (aCTC) in metastatic breast cancer. *Breast Cancer Res. Treat.* **2016**, *160*, 277–290. [CrossRef]
20. Isella, C.; Brundu, F.; Bellomo, S.E.; Galimi, F.; Zanella, E.; Porporato, R.; Petti, C.; Fiori, A.; Orzan, F.; Senetta, R.; et al. Selective analysis of cancer-cell intrinsic transcriptional traits defines novel clinically relevant subtypes of colorectal cancer. *Nat. Commun.* **2017**, *8*, 15107. [CrossRef] [PubMed]

4

Analysis of AR/ARV7 Expression in Isolated Circulating Tumor Cells of Patients with Metastatic Castration-Resistant Prostate Cancer (SAKK 08/14 IMPROVE Trial)

Ivana Bratic Hench [1], Richard Cathomas [2], Luigi Costa [1], Natalie Fischer [3], Silke Gillessen [4,5], Jürgen Hench [1], Thomas Hermanns [6], Eloïse Kremer [7], Walter Mingrone [8], Ricardo Pereira Mestre [9], Heike Püschel [1], Christian Rothermundt [4], Christian Ruiz [1], Markus Tolnay [1], Philippe Von Burg [10], Lukas Bubendorf [1] and Tatjana Vlajnic [1,*]

1. Institute of Pathology, University Hospital Basel, Schönbeinstrasse 40, 4031 Basel, Switzerland
2. Department of Oncology/Hematology, Cantonal Hospital Graubünden, 7000 Chur, Switzerland
3. Department of Oncology, Cantonal Hospital Winterthur, 8401 Winterthur, Switzerland
4. Department of Oncology/Hematology, Cantonal Hospital St. Gallen, 9007 St. Gallen, Switzerland
5. University of Bern, 3012 Bern, Switzerland
6. Department of Urology, University Hospital Zurich, University of Zurich, 8091 Zurich, Switzerland
7. Swiss Group for Clinical Cancer Research (SAKK) Coordinating Center, 3008 Bern, Switzerland
8. Department of Medical Oncology, Cantonal Hospital Olten, 4600 Olten, Switzerland
9. Clinic of Medical Oncology, Oncology Institute of Southern Switzerland, 6500 Bellinzona, Switzerland
10. Department of Oncology/Hematology, Hospital of Solothurn, 4500 Solothurn, Switzerland

* Correspondence: tatjana.vlajnic@usb.ch

Abstract: Despite several treatment options and an initial high response rate to androgen deprivation therapy, the majority of prostate cancers will eventually become castration-resistant in the metastatic stage (mCRPC). Androgen receptor splice variant 7 (ARV7) is one of the best-characterized androgen receptor (AR) variants whose expression in circulating tumor cells (CTCs) has been associated with enzalutamide resistance. ARV7 expression analysis before and during enzalutamide treatment could identify patients requiring alternative systemic therapies. However, a robust test for the assessment of the ARV7 status in patient samples is still missing. Here, we implemented an RT-qPCR-based assay for detection of AR full length (ARFL)/ARV7 expression in CTCs for clinical use. Additionally, as a proof-of-principle, we validated a cohort of 95 mCRPC patients initiating first line treatment with enzalutamide or enzalutamide/metformin within a clinical trial. A total of 95 mCRPC patients were analyzed at baseline of whom 27.3% (26/95) had ARFL+ARV7+, 23.1% (22/95) had ARFL+ARV7−, 23.1% (22/95) had ARFL−ARV7−, and 1.1% (1/95) had ARFL−ARV7+ CTCs. In 11.6% (11/95), no CTCs could be isolated. A total of 25/95 patients had another CTC analysis at progressive disease, of whom 48% (12/25) were ARV7+. Of those, 50% (6/12) were ARV7− and 50% (6/12) were ARV7+ at baseline. Our results show that mRNA analysis of isolated CTCs in mCRPC is feasible and allows for longitudinal endocrine agent response monitoring and hence could contribute to treatment optimization in mCRPC.

Keywords: mCRPC; circulating tumor cells; liquid biopsy; androgen receptor; ARV7

1. Introduction

While most newly diagnosed prostate cancer (PC) patients have potentially curable localized disease, still a significant proportion of patients progress or present initially with locally advanced or the metastatic stage [1]. The backbone treatment of metastatic PC (mPC) is androgen deprivation therapy (ADT). While at first most mPCs are hormone-sensitive, virtually all mPC patients progress and develop resistance to ADT within a median time of approximately 12–18 months [2–4]. Therapeutic options for metastatic castration-resistant prostate cancer (mCRPC) include novel androgen receptor (AR) targeting drugs (abiraterone, enzalutamide), taxane chemotherapy, immunotherapy (sipuleucel-T), and bone tropic radioisotopes (radium-223). Current recommendations for treatment strategies mainly depend on patient characteristics, extent of metastatic disease, prior treatments, and symptoms, but there is no randomized data for the optimal therapy sequence [5]. There is a consensus that novel AR targeting drugs should be used in treatment for asymptomatic men with mCRPC progressing on or after docetaxel (without prior abiraterone or enzalutamide) [6].

Multiple mechanisms of resistance contribute to progression to mCRPC, with reactivation of the AR signaling pathway, mediated by AR amplifications, AR mutations, and expression of splice variants, being the most prominent one [7]. Androgen receptor splice variant 7 (ARV7) is one of the most abundant constitutively expressed splice variants found in PC [8]. It is a constitutively active isoform of the AR that lacks the ligand-binding domain yet retains its transcriptional activity in a ligand-independent fashion. ARV7 mRNA expression in biopsies is significantly upregulated in hormone-naïve and mCRPC patients in relation to levels found in healthy tissue [9]. More recently, Antonarakis et al. have shown that ARV7 mRNA expression in circulating tumor cells (CTCs) may predict poor response to enzalutamide or abiraterone [10,11]. Results from subsequent studies have confirmed a correlation between positive ARV7 status in CTCs and impaired clinical progression-free survival under treatment with enzalutamide or abiraterone [12]. CTCs can be tested not only for ARV7 expression, but can also be used to identify AR hotspot mutations that could further enhance the diagnostic use of liquid biopsies to optimize treatment [13]. In conclusion, liquid biopsies are likely to become a key diagnostic test for tailoring existing treatment strategies. Despite the promising data in the literature, blood-based ARV7 testing has still not entered routine clinical practice due to ongoing controversies and technical challenges [14,15]. It has been recently emphasized that robust clinical validation of such assays is required before their routine use [15].

We designed a prospective, multicenter, randomized, open-label, phase II two-arm trial with the main objective to assess the efficacy of a combination first-line treatment with enzalutamide and metformin as opposed to enzalutamide alone in mCRPC patients progressing under ADT (SAKK 08/14, IMPROVE, NTC02640534). In this part of the trial, we address the potential use of molecular CTC analysis to determine the potential impact of ARV7 expression on patient outcome. We used the AdnaTest® ProstateCancerPanel ARV7 (Qiagen) (Adna ARV7 Test), which was modified and further developed since the seminal paper of Antonarakis et al. [11]. As this test was not yet commercially available at the time of trial start, we aimed in this pilot study to examine its analytical and clinical validity by quantification of AR full length (ARFL) and ARV7 expression in isolated epithelial cell adhesion molecule-positive (EpCAM+) CTCs. Second, we aimed to assess the feasibility to detect prostate-specific markers (prostate-specific antigen (PSA), prostate-specific membrane antigen (PSMA), ARFL, and ARV7) in EpCAM+ CTCs in mCRPC patients subjected to either enzalutamide alone or to the combination enzalutamide/metformin.

2. Results

2.1. *Sensitivity and Specificity of the Adna ARV7 Test*

Analytical validity of the Adna ARV7 Test was assayed by spike-in experiments. The specificity of the test was investigated by the cell lines DU145 and PC3 and the sensitivity by LNCAP and VCAP. Gene expression of PSA, PSMA, ARFL, and ARV7 before spiking the cultured cells was individually tested on all cell lines by qRT-PCR (Figure S1). While VCAP and LNCAP featured a high expression of PSA, PSMA, ARFL, and ARV7, DU145 and PC3 were negative for PSA, ARFL, and ARV7 mRNA. We observed minute mRNA levels for PSMA in PC3, but not in DU145. These data are in line with published data [16]. We examined the Adna ARV7 Test's specificity by spiking 50×10^3 PC3 and 50×10^3 DU145 cells in 6 mL of healthy donor blood. In both cases, we detected high levels of GAPDH mRNA, but no expression of PSA, ARFL, or ARV7. In PC3-spiked samples, we observed PSMA expression while DU145-spiked samples were negative. Additionally, we ran the test directly on isolated DU145 RNA that had been processed with the AdnaTest Prostate Cancer Detect kit and obtained the same results as with DU145 cells (Table 1). These data are in concordance with expression profile data of the respective cell lines (Figure S1) and suggest a high specificity of the Adna ARV7 Test.

Table 1. The sensitivity and specificity of the Adna ARV7 Test.

Sample	GAPDH	PSA	PSMA	ARFL	ARV7
DU145 $(50 \times 10^3)^3$	+	−	−	−	−
DU145 $(\text{RNA})^2$	+	−	−	−	−
PC3 $(50 \times 10^3)^2$	+	−	+	−	−
Healthy ^{10} *	−	−	−	−/+	−
LNCaP $(200)^1$	+	+	+	+	+
LNCaP $(68)^1$	+	+	+	+	+
LNCaP $(10)^4$	+	+	+	+	++−−
LNCaP $(5)^3$	++−	+	++−	++−	++−
LNCaP95 $(5)^5$	++++−	+++−−	+	+	+
VCaP $(18 \times 10^3)^1$	+	−	+	+	+
VCaP $(10)^3$	+−+	+−+	+	+	++−
VCaP $(5)^3$	−	+−−	+	+	+

The number in superscript refers to the number of independent replicates. The number in brackets refers to the number of spiked cells. Detected gene expression level is marked with a plus, and absence with a minus. If the results between replicates differed, the result of each replicate is listed separately. * In has been detected.

The sensitivity was tested by spiking various numbers of cells from LNCaP, LNCaP95, and VCaP into healthy donor blood (Table 1). ARV7 expression was detected in samples with five LNCaP95 cells (100%, 5/5 replicates), five and ten LNCaP cells (2/3 and 2/4 replicates, respectively) or five and ten VCaP cells (3/3 and 2/3 replicates, respectively). ARFL expression was detected in samples with 10 LNCaP cells (100%, 4/4, replicates), five LNCaP cells (67%, 2/3 replicates), five LNCaP95 cells (100%, 5/5), and with 10 and five VCAP cells (3/3 replicates each) (Table 1). The varying sensitivity in the background of different cell lines likely originates from the differences in ARFL/ARV7 expression levels in these cells, since LNCaP95 and VCAP express ARFL and ARV7 at higher levels than LNCaP (Figure S1).

PSMA expression was detected in all tested blood samples that were spiked with VCaP, LNCaP, and LNCaP95 cells, respectively. PSA expression was found in samples with five LNCaP cells and in two of five blood samples with five LNCaP95 cells. We did not detect any PSA expression in the majority of VCaP-spiked samples (Table 1). This was unexpected as others and we had demonstrated (Figure S1) that the VCaP cell line expresses large quantities of PSA [17]. We analyzed PSA, PSMA, ARFL, and ARV7 expression in blood samples of 10 healthy donors. In eight healthy donors (three females, five males), we did not detect any PSA, PSMA, ARFL and ARV7 expression in blood samples. However, in the remaining two (female) blood samples, we only found expression of ARFL (Table 1).

2.2. Cross-Laboratory Variability

Cross-laboratory variability between our laboratory and Qiagen R&D Laboratory (Hilden, Germany) was objectified in two ways. First, we used five spiked-in samples with five LNCaP95 cells and five healthy donor samples (Table 2). Second, we tested 14 patient samples (Table 3). Samples were exchanged between the two laboratories and analyzed independently. Both laboratories were blinded for the results of the other. In both analyses, cDNA derived from captured EpCAM+ CTCs was exchanged between the two laboratories and all downstream analyses were performed independently.

Table 2. Adna ARV7 Test concordance between two independent laboratories. Sample set consisted of five healthy donor and five spiked-in samples containing five LNCaP 95 cells each. Numbers in brackets indicate replicates (number of samples with concordant results/total number of samples).

Laboratories	GAPDH	PSA	PSMA	ARFL	ARV7
Concordance	90% (9/10)	70% (7/10)	100% (10/10)	100% (10/10)	100% (10/10)

Table 3. Adna ARV7 Test concordance between two independent laboratories. Sample set consisted of 14 patient samples. Numbers in brackets indicate replicates (number of samples with concordant results/total number of samples).

Laboratories	GAPDH	PSA	PSMA	ARFL	ARV7
Concordance	92.86% (13/14)	78% (11/14)	100% (14/14)	100% (14/14)	100% (14/14)

Data concordance of spiked-in samples was 100% with regard to PSMA, ARFL, and ARV7 expression. GAPDH concordance was 90% and PSA concordance 70% (Table 2). Data concordance between 14 patient samples for PSMA, ARFL, and ARV7 was 100%, and for GAPDH 92.86%. Concordance for PSA was 78% (Table 3). Both concordance tests indicated lower concordance (<80%) regarding PSA expression. We tested PSA and PSMA expression in these samples also with the AdnaTest Prostate Cancer Detect assay (Adna Detect). PSA concordance between Adna ARV7 Test performed in Lab1 or Lab2 and Adna Detect was 57% and 50%, respectively (Supplementary Table S1). Hence, PSA was no longer tested in subsequent experiments. While the concordance between the two laboratories for detection of PSMA mRNA with the Adna ARV7 Test was 100%, concordance between Adna ARV7 Test and Adna Detect was only 57.1% (8/14) (Supplementary Table S1). In all six discordant PSMA cases, PSMA expression was not detected with Adna Detect. These results indicate higher sensitivity of the Adna ARV7 Test for PSMA mRNA detection in comparison to Adna Detect.

2.3. *Reproducibility of the CTC Profile Measured at Baseline by the Adna ARV7 Test*

To test the reproducibility of the Adna ARV7 Test, two subsequent blood samples were drawn from five patients. GAPDH, PSMA, ARFL, and ARV7 expression were analyzed and the results were compared (Table 4). In four patients, from whom blood samples had been taken at baseline, the data concordance was 100% for PSMA, ARFL, and ARV7. Patient 5 (P5), from whom blood samples had been taken at progression, gave different results. While the first blood sample had EpCAM+ CTCs expressing PSMA, ARFL, and ARV7, the second blood sample had no detectable EpCAM+ CTCs.

Table 4. Adna ARV7 Test concordance between two independent blood samples from five patients. Numbers in brackets refer to the replicates (number of samples with concordant results/total number of samples).

Patients	PSMA	ARFL	ARV7
P1	100% (2/2)	100% (2/2)	100% (2/2)
P2	100% (2/2)	100% (2/2)	100% (2/2)
P3	100% (2/2)	100% (2/2)	100% (2/2)
P4	100% (2/2)	100% (2/2)	100% (2/2)
P5	50% (1/2)	50% (1/2)	50% (1/2)

2.4. *Patient Cohort*

Based on our validation assay for the Adna ARV7 Test, we defined a scoring scheme for CTC analysis. Patient samples with CD45 and GAPDH expression, but negative for PC-specific markers

(PSA, PSMA, ARFL, and ARV7) were scored as "no CTCs" or "CTCs negative for PC markers". Samples with expression detection for either CD45 or GAPDH were excluded from further analysis. A total of 11 (11.6%) patients at baseline and 9 (31%) at progression had either "no CTCs" or had "CTCs negative for PC markers". A total of 13 (13.7%) patient samples at baseline and one at the end of the treatment were excluded due to failed qPCR analysis (Figure 1).

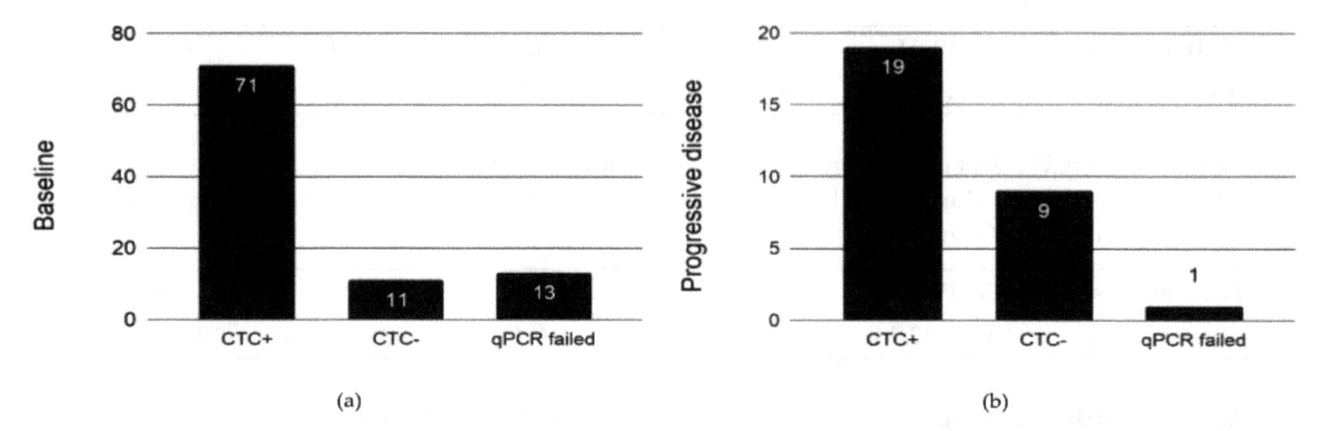

Figure 1. Analysis of circulating tumor cells (CTCs) at (**a**) baseline (n = 95) and (**b**) progressive disease (n = 29). Patient blood samples positive for CD45 and GAPDH expression, but negative for prostate cancer (PC)-specific markers (prostate-specific antigen (PSA), prostate-specific membrane antigen (PSMA), ARFL, and androgen receptor splice variant 7 (ARV7)) were scored as CTC negative. A patient blood sample is considered CTC positive if in addition to CD45 and GAPDH, at least one PC-specific marker is unambiguously detected. The number in each bar refers to the number of patients.

2.5. Expression Profile of EpCAM+ CTCs in Patients at Baseline

At baseline, three major CTC phenotypes were observed at similar proportions: 26/95 (27.4%) of patients had ARFL+ARV7+, 22/95 (23.2%) had ARFL−ARV7−, and 22/95 (23.2%) had ARFL+ARV7− CTCs (Figure 2a). Only one of the patients had ARFL−ARV7+ CTCs. About half of the patients (42/95, 44.2%) had PSMA+ARFL+ CTCs, 16/95 (16.8%) had PSMA+ARFL−, 6/95 (6.3%) had PSMA−ARFL+, and 7.4% (7/95) had PSMA−ARF− CTCs at baseline (Figure 2b).

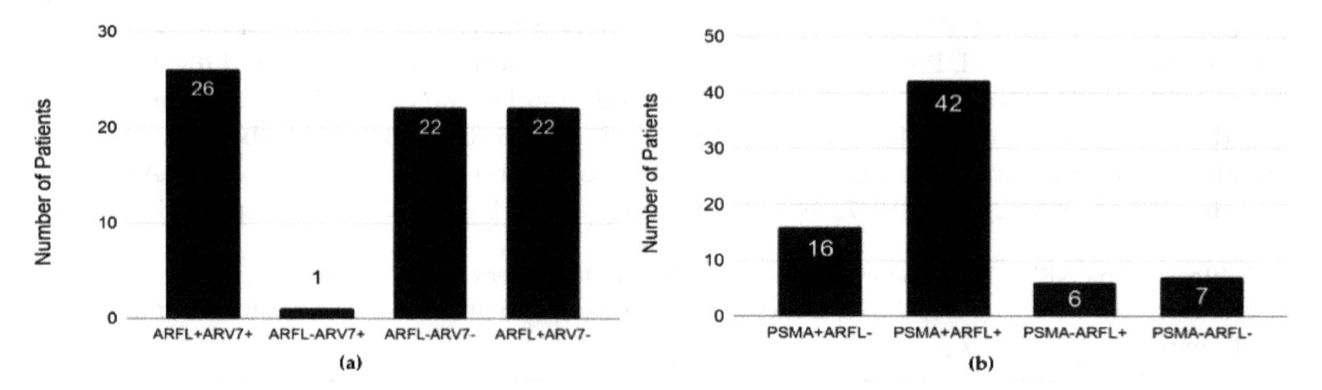

Figure 2. ARFL, ARV7, and PSMA expression in epithelial cell adhesion molecule-positive (EpCAM+) CTCs of metastatic castration-resistant prostate cancer (mCRPC) patients at baseline: (**a**) ARFL and ARV7 expression, and (**b**) PSMA and ARFL expression. The number in each bar refers to the number of patients with the indicated CTC phenotype.

2.6. Expression Profile of EpCAM+ CTCs in Patients on Enzalutamide at Progressive Disease

A total of 29/95 patients had another CTC analysis at progressive disease (PD), of whom 9 (31%) had no CTCs detected or their CTCs were negative for all tested mRNAs, 19 (65.5%) had CTCs positive for at least one PC-specific mRNA, and in one patient qPCR analysis failed. From 19 patients with evaluable CTCs data, the majority (12/19, 63.2%) had ARFL+ARV7+ and 5/19 (26.3%)

had ARFL+ARV7− CTCs. ARFL−ARV7− and ARFL-ARV7+ CTC profiles were single observations, respectively (Figure 3a). 18/19 (94.7%) patients had positive CTCs for PSMA expression. From those 19 patients, 16/19 (84.2%) had PSMA+ARFL+, two patients (2/19, 10.5%) had PSMA+ ARFL−, and one patient had PSMA−ARFL+ CTCs (Figure 3b).

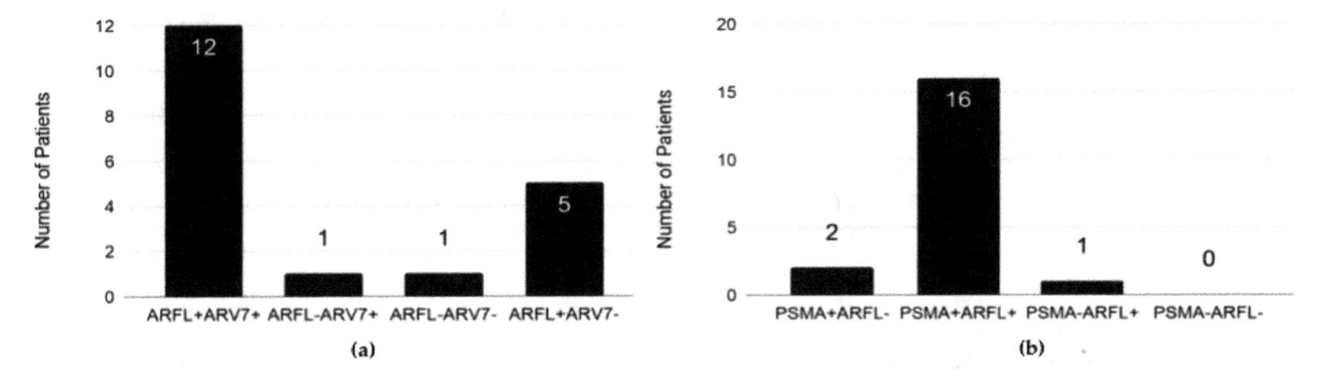

Figure 3. ARFL, ARV7 and PSMA expression in EpCAM+ CTCs of mCRPC patients at progressive disease (PD): (**a**) ARFL and ARV7 expression and (**b**) PSMA and ARFL expression. The number in each bar refers to the number of patients with the indicated CTC phenotype. The total number of analyzed patients with CTC-positive blood samples at PD was 19.

2.7. ARFL/ARV7 Conversions at Progressive Disease

To analyze conversions in ARFL and ARV7 expression, we only took patients (n = 25) into account who had evaluable data at both time points, baseline and at PD. 48% (12/25) were ARV7+ at PD. Of those, 50% (6/12) were ARV7+ and 50% (6/12) were ARV7− at baseline (Table 5). The majority of patients (13/17, 76.4%) with ARFL+ CTCs at baseline still had ARFL+ CTCs at progression. Interestingly, 5/25 patients with ARFL−ARV7− CTCs at baseline converted into all of the possible profiles: ARFL+ARV7+, ARFL−ARV7−, ARFL−ARV7+, ARFL−ARV7−, and "no CTCs". Of three patients in whom we did not detect CTCs at baseline, two patients remained negative for CTCs, and in one patient we detected ARFL+ARV7− CTCs at PD (Table 5).

Table 5. ARFL and ARV7 conversion at PD. Percentages refer to the fraction of patients within groups who at baseline shared biomarker constellation (first column). Numbers in brackets refer to the number of patients with specific ARFL/ARV7 profile analyzed at PD.

	Progressive Disease (PD)				
Baseline	**ARFL+ ARV7+**	**ARFL+ ARV7−**	**ARFL− ARV7+**	**ARFL− ARV7−**	**no CTCs**
ARFL+, ARV7+ (8)	75% (6)	12.5% (1)	0	0	12.5% (1)
ARFL+, ARV7− (9)	44.4% (4)	22.2% (2)	0	0	33.3% (3)
ARFL−, ARV7− (5)	20% (1)	20% (1)	20% (1)	20% (1)	20% (1)
no CTCs (3)	0	33.3% (1)	0	0	66.7% (2)
Total (25)	11	5	1	1	7

2.8. PSMA/ARFL Conversions at Progressive Disease

15 (out of 25) patients (60%) had PSMA+ARFL+ CTCs at PD. 13/17 (76.5%) patients who had PSMA+ CTCs at baseline retained PSMA expression at PD (Table 6). Three out of five patients who had PSMA- CTCs at baseline had PSMA+ CTCs at PD.

Table 6. ARFL and PSMA conversion at PD. Percentages refer to the fraction of patients within groups who at baseline shared biomarker constellation (first column). Numbers in brackets refer to the number of patients with specific PSMA/ARFL profile/total number of patients analyzed at PD.

	Progressive Disease (PD)				
Baseline	**PSMA+ ARFL+**	**PSMA+ ARFL−**	**PSMA− ARFL+**	**PSMA− ARFL−**	**no CTCs**
PSMA+, ARFL+ (14)	71.4% (10)	0	7.1% (1)	0	21.4% (3)
PSMA+, ARFL− (3)	33.3% (1)	66.7% (2)	0	0	0
PSMA−, ARFL+ (3)	66.7% (2)	0	0	0	33.3% (1)
PSMA−, ARFL− (2)	50% (1)	0	0	0	50% (1)
no CTCs (3)	33.3% (1)	0	0	0	66.7% (2)
Total (25)	15	2	1	0	7

3. Discussion

The potential predictive value of ARV7+ CTCs in determining resistance to enzalutamide or abiraterone has been shown in several studies [10,11,18], although there remains some controversy since some patients with ARV7-positive liquid biopsies (CTCs or RNA from whole blood) might still derive benefit from enzalutamide or abiraterone therapy [19]. A recent prospective multicenter study strongly supports the clinical significance of ARV7 detection in CTCs as a prognostic marker, which is associated with shorter progression-free survival and overall survival (OS) [20]. Moreover, nuclear-specific localization of the ARV7 protein within the CTCs might be significant in terms of guiding treatment selection in mCRPC patients [20,21]. Despite these promising data pointing towards a predictive role of ARV7, a robust method to evaluate ARV7 expression in CTCs as part of routine clinical practice has not yet been established. Therefore, we assessed the performance of the Adna ARV7 Test in a series of spike-in experiments and in mCRPC patients from a multicenter, randomized, phase 2, open-label study investigating metformin effects on mCRPC patients during enzalutamide therapy (SAKK 08/14, IMPROVE). The Adna ARV7 Test enables CTC enrichment by EpCAM-based immunoisolation. Upon CTC enrichment, mRNA transcripts of prostate-specific markers (PSA, PSMA, ARFL, and ARV7) are quantified by RT-qPCR. There have been few studies using RT-qPCR for detection of ARV7 upon EpCAM-based CTC enrichment [10,11,22,23]. It has been demonstrated that the CTC detection rate of the custom modified AdnaTest and RT-qPCR is superior to the CellSearch assay, which is currently the only FDA-approved test for isolation of CTCs [24]. In patients with mCRPC, similar CTC detection rates between the custom modified AdnaTest, where CTCs were identified based on the presence of KLK3, PSMA or EGFR transcripts, and a digital droplet PCR assay detecting prostate-specific mRNA in whole blood, have been reported [24]. Despite slight differences between these approaches, all studies agree that RT-qPCR is sensitive enough to detect RNA transcripts in CTCs. In their seminal paper, Antonarakis et al. used a modified AdnaTest Prostate Cancer CTC Panel for CTC enrichment coupled with RT-qPCR analysis using custom-made primers for ARFL and ARV7. In addition, cDNAs for PSA, PSMA, and EGFR were analyzed by multiplex PCR and reaction products were quantified with an Agilent Bioanalyzer [11].

We tested the sensitivity and specificity of the Adna ARV7 Test by a series of spike-in experiments. We demonstrated high specificity for this test as we did not detect any expression of PSA, PSMA, ARFL, and ARV7 by using pure RNA isolated from DU145 cells, which are known to lack the expression of these markers. Moreover, no PSA, ARFL, or ARV7 expression was detected in samples spiked with approximately 50,000 PC3 cells and 50,000 DU145 cells. PSMA mRNA expression was observed in PC3 spiked samples (but not in DU145 spiked samples) as expected since we observed a low, yet detectable PSMA mRNA expression in the PC3 cell line (Figure S1). In healthy male donors, we did not observe any expression for any of the tested markers. However, in two of five healthy female donors, we detected ARFL expression solely in the blood samples. Expression of ARFL in bone marrow cells, platelets, and other tissues has been reported [25–27]. Of note, we detected CD45 expression in all healthy control and spike-in samples. This result is likely explained by leukocytes binding nonspecifically to the magnetic beads in the absence of CTCs.

Several studies have demonstrated an independent prognostic value of CTC counts for OS in mCRPC, where the presence of ≥5 CTCs per 7.5 mL of blood had an adverse effect [28–32]. Therefore, this threshold of ≥5 CTCs/7.5 mL of blood has been suggested to be used as a prognostic marker for OS [28–32]. For this reason, we performed a series of spike-in experiments with at least five cells from different cell lines in healthy female donor blood to test the sensitivity of our assay. We demonstrated that the sensitivity of the Adna ARV7 Test is at least five cells/6 mL of blood. However, we observed that the reproducibility of detecting five cells in at least 6 mL of blood varied between different cell lines, where LNCaP95 and VCaP cell lines outperformed LNCaP. This is most likely due to the lower expression levels of ARFL and ARV7 in LNCaP in comparison to VCAP, where the expression levels of ARFL and ARV7 are 10× and 40× higher, respectively (Figure S1). Additionally, in our manual spiking approach, we were not able to distinguish between dead and living cells that might explain the variability between replicates.

Next, we demonstrated a high interlaboratory concordance for GAPDH, PSMA, ARFL, and ARV7. In contrast, PSA results should be interpreted with caution due to high interlaboratory variability. Analysis of biological replicates showed good reproducibility (4/5 patients, 80%) of this assay, which has not previously been investigated by others.

Interestingly, we observed a dynamic change of ARV7 mRNA expression in the clinical study cohort during treatment with enzalutamide, as previously reported by Antonarakis et al. in mCRPC patients by a modified Adna Prostate Cancer Test [10]. At baseline, 28% of our mCRPC patients were ARV7+ before the start of first-line enzalutamide therapy, which is well in line with previous reports [10,11]. At PD, six out of eight patients retained ARV7 expression in their CTCs, while two reverted from ARV7+ CTCs at baseline to either ARV7− CTCs or "no CTCs" at progression. Conversion from ARV7+ CTCs to ARV7− CTCs was also observed by Antonarakis et al. in mCRPC patients by a modified Adna Prostate Cancer Test [10]. The reversions of ARV7 expression might occur due to an inactivation of the AR signaling axis, decreasing selective pressure for ARV7 expression. Alternatively, the mRNA levels might be below the detection threshold of the Adna ARV7 Test for biological or technical reasons. However, these assumptions remain to be proven in larger patient cohorts. Currently, the clinical trial is still ongoing and no final efficacy results are available. After completion of the trial, we will be able to correlate clinical data of all patients and determine the correlation between their CTC profiles and clinical outcomes.

PSMA is known to be over-expressed in advanced PC or mCRPC [33]. While strong PSMA expression in the tumor has been associated with higher tumor stages, Gleason Scores, preoperative PSA levels, HER2 expression, and a higher risk of biochemical recurrence [34], there is no correlation between blood PSMA mRNA and tumor stage, Gleason score, or serum PSA [35]. However, it has been shown that the detection of the PSMA mRNA in blood can predict biochemical recurrence after radical prostatectomy [35]. Furthermore, PSMA expression in CTCs is characterized by a high intra-patient heterogeneity and is found in a large portion (67%) of metastatic PC patients by the CellSearch® assay [36]. Our data show PSMA expression in CTCs in the majority of the patients at baseline (58/95, 61%) and at PD (18/25, 72%). Moreover, PSMA expression in CTCs appears to be rather stable as the majority of the patients with PSMA+ CTCs at baseline retained PSMA+ CTCs (13/17, 76.4%) at progression.

The Adna ARV7 Test has certain restrictions that require consideration. First, there is no possibility to distinguish patients who do not have CTCs from patients where CTCs detection failed due to technical reasons. Second, it is not possible to evaluate whether PSMA, ARFL, and ARV7 mRNAs are expressed in the same cell. Finally, only EpCAM+ CTCs are analyzed, which might pose a selection bias as it is assumed that some CTCs decrease the expression of EpCAM as a result of epithelial–mesenchymal transition [37].

4. Material and Methods

4.1. Study Design

We aimed to prospectively evaluate the dynamics of PSA, PSMA, ARFL, and ARV7 markers on EpCAM+ CTCs from mCRPC patients during treatment with enzalutamide alone or in combination with metformin. Peripheral blood for CTC analysis was taken at baseline (before starting the therapy) and at the time of progression. The study was approved by the Swiss Group for Clinical Cancer Research (SAKK) Board (Bern, Switzerland) (2016-00127, 31 January 2016). All patients gave their signed informed consent before blood collection.

4.2. Patient Characteristics

A total of 95 asymptomatic or minimally symptomatic mCRPC patients with confirmed adenocarcinoma of the prostate progressing under ADT had been prospectively enrolled in the IMPROVE clinical trial (SAKK 08-14, Trials.gov: NCT02640534) between September 2016 and May 2019 and scheduled to undergo another line of therapy with either enzalutamide/metformin or enzalutamide alone. As patient recruitment is still ongoing, we are not allowed to present detailed information on the diagnostic characteristics of the enrolled patients at this point. Detailed patient characteristics at the time of registration and at progression are available in the Supplementary Material.

4.3. RNA Extraction from VCaP, LNCaP, PC3, and DU145 Cell Lines and Detection of PSA, PSMA, ARFL, and ARV7 Expression by RT-qPCR

Total RNA from VCaP, DU145, LNCaP, and PC3 cell lines was extracted with the RNeasy Mini Kit (Qiagen). RNA concentrations were determined using the NanoDrop ND-1000 spectrophotometer (NanoDrop Technologies). A total of 1.5 µg of total RNA was reverse transcribed into complementary DNA (cDNA) by using the SuperScript VILO cDNA Synthesis Kit (Invitrogen) according to the manufacturer's instructions. Gene expression was examined by qPCR performed on 15× diluted cDNA samples using the QuantiNova SYBR Green Kit (Qiagen). All qPCR experiments were performed on QuantStudio 3 (Thermo Fisher). The threshold cycle (Ct) of PSA; PSMA, ARFL, and ARV7 expression was determined and normalized to that of human GAPDH to obtain a ΔCt value (Ct(gene of interest) − Ct(GAPDH)) from each sample. The relative mRNA expression level was calculated by normalizing the ΔCt to DU145 cells ($2^{-\Delta\Delta Ct}$). GAPDH, PSA, PSMA, ARFL, ARV7 primers used for the qPCR analysis were obtained from the AdnaTest Prostate CancerPanel ARV7 (Qiagen) (Figure S1). Three biological replicates were analyzed for each cell line. The statistical significance in difference of expression levels was calculated with an ANOVA with a post-hoc Student's *t*-test (two-tailed with assumptions of equal variances) followed by a Bonferroni correction.

4.4. Quantitative Real-Time PCR and PSA, PSMA, ARFL and ARV7 Detection in EpCAM+ CTCs

Patient's peripheral blood samples for CTC analysis were collected in 8.5 mL ACD-A Tubes (BD Vacutainer, Becton Dickinson) before the start of the therapy and at the time of progression. All patient blood samples were stored at 4 °C immediately after withdrawal until further processing. CTC enrichment was performed within 30 h after blood draw by AdnaTest Prostate Cancer Select (Qiagen) according to the manufacturer's instructions. AdnaTest Prostate Cancer Detect kit (Qiagen) was used to isolate mRNA from isolated CTCs and Sensiscript RT kit (Qiagen) for cDNA synthesis. cDNA was stored at −20 °C. Prior to qPCR, cDNA was pre-amplified by a Multiplex PCR Plus kit (Qiagen) as follows: Initial activation at 95 °C for 5 min followed by 18 cycles (95 °C for 30 s, 60 °C for 90 s, and 72 °C for 90 s). CD45, GAPDH, PSA, PSMA, ARFL, and ARV7 expression were determined by qPCR by the help of QuantiNova SYBR Green (Qiagen) on 20× diluted pre-amplified DNA samples. qPCR primers were provided by the AdnaTest Prostate CancerPanel ARV7 (Qiagen). All qPCR analyses were performed on QuantStudio 3 (Thermo Fisher). To rule out contamination, we analyzed patient samples along with their corresponding reverse transcriptase non-template controls. Data analysis was performed according to the manufacturer's instructions. The assay that we used has two major differences from Antonarakis et al.'s method [11]. First, detection of RNA transcripts of PSA, PSMA, ARFL, and ARV7 in Adna ARV7 Test was performed exclusively by quantitative real-time PCR, while in Antonarakis et al.'s study PSA and PSMA transcripts were analyzed by an end-point PCR by Bioanalyzer. Second, while Antonarakis et al. determined absolute levels of ARFL and ARV7 by RT-qPCR, ARFL and ARV7 expression in Adna ARV7 Test was analyzed by RT-qPCR after an additional amplification by multiplex PCR.

4.5. Analysis of PSA and PSMA Expression in EpCAM+ CTCs by Adna Detect

The AdnaTest Prostate Cancer Detect kit (Adna Detect, Qiagen) was used to isolate mRNA from isolated CTCs and Sensiscript RT kit (Qiagen) for cDNA synthesis. cDNA was stored at −20 °C. cDNA was amplified by Multiplex PCR (Prostate Detect) for detection of PSA, PSMA, and actin as follows: Initial activation at 95 °C for 15 min followed by 42 cycles (94 °C for 30 s, 61 °C for 30 s, and 72 °C for 30 s) and a final step at 72 °C for 10 min. Actin served as an internal PCR control. Analysis of PCR products was performed with the Agilent 2100 Bioanalyzer (Agilent Technology) on a DNA 1000 Lab Chip according to manufacturer's instructions. The sample was considered positive if the PCR band had a concentration ≥10 ng/μL and the control gene actin had been detected.

4.6. Cell Spiking—Validation Experiments

To determine the specificity and sensitivity of the Adna ARV7 Test for CTC detection, we performed a series of spiking experiments with defined numbers of cultured human prostate cancer cell lines that are known to either express PSA, PSMA, ARFL, and ARV7 (VCaP, LNCaP) or not (DU145, PC3). Cells were spiked into healthy donor blood. LNCaP, DU145, and PC3 cell lines were routinely passaged in Roswell Park Memorial Institute culture medium (RPMI 1640), and VCaP cell line in Dulbecco's Modified Eagle's Medium (DMEM) medium. All media were supplemented with 10% (DU145, PC3, VCaP) or 20% (LNCaP) fetal bovine serum (FBS) (Invitrogen), and 1% penicillin/streptomycin solution (BioConcept). Cell lines were cultured in a humidified incubator with 5% CO_2 at 37 °C. For spiking experiments, all cell lines were cultured for at least two days after passaging and gently detached with detachin (Genlantis) and washed with phosphate buffered saline (PBS, Gibco). The spiked input cell numbers of 5, 10, and 68 cells were manually picked and transferred to 6 mL of peripheral blood of a presumably healthy female donor. The spiked input cell numbers of 200 and 50,000 were determined by dilution and transferred into 6 mL blood of a presumably healthy female donor. Blood samples with five LNCaP95 cells were kindly provided by Dr Siegfried Hauch, Qiagen, R&D Department, Hilden, Germany. Blood samples without spiked cells served as a negative control. Additionally, blood from a presumably healthy male donor ($n = 5$) was used to determine background levels of PSA, PSMA, ARFL, and ARV7 in peripheral whole blood. Spiked blood was stored under the same conditions as patient samples to minimize any bias and analyzed by a standard procedure that includes AdnaTest Prostate Cancer Select for CTC Isolation, AdnaTest Prostate Cancer Detect for mRNA isolation and cDNA synthesis, preamplification, and qPCR as described above (Section 4.4).

5. Conclusions

In summary, we comprehensively validated and helped to improve a commercially available EpCAM-based immunomagnetic enrichment method of CTCs in whole blood samples from mCRPC patients, aiming to assess mRNA expression of prostate-specific markers. To our knowledge, this is the first report on the Adna ARV7 Test in mCRPC patients. Apart from a few limitations, the Adna ARV7 Test provides an objective result of CTCs molecular profile without the need for CTC imaging and enumeration. The major advantages of this approach are the possibility to simultaneously assess the expression of several PC markers (PSMA, AR, and ARV7) on EpCAM+ CTCs and a straightforward workflow that is possible to establish in the majority of diagnostic laboratories.

Author Contributions: Conceptualization, L.B.; T.V.; S.G.; R.C.; and C.R. (Christian Rothermundt); methodology, L.B.; T.V.; C.R. (Christian Ruiz) and I.B.H.; validation, I.B.H.; formal analysis, I.B.H. and J.H.; investigation, I.B.H. and L.C.; resources, L.B., and T.V.; data curation, I.B.H., and J.H.; writing I.B.H., J.H.; L.B.; and T.V.; supervision, L.B., M.T., T.V.; project administration, I.B.H., H.P. and C.R. (Christian Ruiz); project management, E.K.; funding acquisition, L.B., T.V., C.R. (Christian Rothermundt), S.G., R.C. and C.R. (Christian Ruiz); patient recruitment, S.G., R.C., C.R. (Christian Rothermundt), N.F., T.H., W.M., R.P.M., P.V.B.; manuscript editing, S.G., R.C., C.R. (Christian Rothermundt).

Acknowledgments: We would like to acknowledge the help of Siegfried Hauch and Dr. Jens van der Flierdt (Qiagen, Hilden) in validation experiments, and for providing us with samples spiked with LNCaP95 cells and healthy donor controls. Qiagen provided 30% discount on the assay reagents for this study. We thank Stefanie Hayoz for critical review of the manuscript.

Conflicts of Interest: The following authors have affiliations with organizations with direct or indirect interest in the subject matter discussed in the manuscript: R.C.: Advisory role at Janssen, Astellas, Bayer, MSD, BMS, Roche, Novartis, Pfizer, Sanofi, Astra Zeneca and Speaker honorarium at Debiopharm, Astellas; C.R. (Christian Rothermundt): Consulting and/or advisory role at Novartis, Pfizer, Astellas Pharma, Eisai, PharmaMar, Bristol-Myers Squibb, MSD Oncology; SIGI: Advisory role and Speakers Bureau at AAA International, Active Biotech, Amgen, Astellas Pharma, Bayer, Bristol-Myers Squibb, CellSearch, Clovis, CureVac, Dendreon, ESSA Pharmaceuticals, Ferring, Innocrin Pharmaceuticals, Janssen Cilag, MaxiVAX SA, Millenium, Nectar, Novartis, Orion, Pfizer, ProteoMediX, Roche, Sanofi. Co-inventor on patent application (WO 2009138392 A1) for a method for biomarker discovery (granted in China, Europe, Japan and the US). Deputy of the ESMO guidelines committee for GU cancers, member of the EAU guideline panel for prostate cancer, past chair of the EORTC GU group. Member of the STAMPEDE trial management group; T.H.: advisory function for Bayer and MSD.

References

1. Negoita, S.; Feuer, E.J.; Mariotto, A.; Cronin, K.A.; Petkov, V.I.; Hussey, S.K.; Benard, V.; Henley, S.J.; Anderson, R.N.; Fedewa, S.; et al. Annual Report to the Nation on the Status of Cancer, part II: Recent changes in prostate cancer trends and disease characteristics: Recent Changes in Prostate Cancer Trends. *Cancer* **2018**, *124*, 2801–2814. [CrossRef] [PubMed]
2. Chen, J.; Li, L.; Yang, Z.; Luo, J.; Yeh, S.; Chang, C. Androgen-deprivation therapy with enzalutamide enhances prostate cancer metastasis via decreasing the EPHB6 suppressor expression. *Cancer Lett.* **2017**, *408*, 155–163. [CrossRef] [PubMed]
3. James, N.D.; Spears, M.R.; Clarke, N.W.; Dearnaley, D.P.; De Bono, J.S.; Gale, J.; Hetherington, J.; Hoskin, P.J.; Jones, R.J.; Laing, R.; et al. Survival with Newly Diagnosed Metastatic Prostate Cancer in the "Docetaxel Era": Data from 917 Patients in the Control Arm of the STAMPEDE Trial (MRC PR08, CRUK/06/019). *Eur. Urol.* **2015**, *67*, 1028–1038. [CrossRef] [PubMed]
4. Hussain, M.; Tangen, C.M.; Higano, C.; Schelhammer, P.F.; Faulkner, J.; Crawford, E.D.; Wilding, G.; Akdas, A.; Small, E.J.; Donnelly, B.; et al. Absolute Prostate-Specific Antigen Value After Androgen Deprivation Is a Strong Independent Predictor of Survival in New Metastatic Prostate Cancer: Data From Southwest Oncology Group Trial 9346 (INT-0162). *J. Clin. Oncol.* **2006**, *24*, 3984–3990. [CrossRef] [PubMed]
5. Mohler, J.L.; Armstrong, A.J.; Bahnson, R.R.; D'Amico, A.V.; Davis, B.J.; Eastham, J.A.; Enke, C.A.; Farrington, T.A.; Higano, C.S.; Horwitz, E.M.; et al. Prostate Cancer, Version 1.2016. *J. Natl. Compr. Cancer Netw. JNCCN* **2016**, *14*, 19–30. [CrossRef]
6. Gillessen, S.; Attard, G.; Beer, T.M.; Beltran, H.; Bossi, A.; Bristow, R.; Carver, B.; Castellano, D.; Chung, B.H.; Clarke, N.; et al. Management of Patients with Advanced Prostate Cancer: The Report of the Advanced Prostate Cancer Consensus Conference APCCC 2017. *Eur. Urol.* **2018**, *73*, 178–211. [CrossRef] [PubMed]
7. Galletti, G.; Leach, B.I.; Lam, L.; Tagawa, S.T. Mechanisms of resistance to systemic therapy in metastatic castration-resistant prostate cancer. *Cancer Treat. Rev.* **2017**, *57*, 16–27. [CrossRef]
8. Guo, Z.; Yang, X.; Sun, F.; Jiang, R.; Linn, D.E.; Chen, H.; Chen, H.; Kong, X.; Melamed, J.; Tepper, C.G.; et al. A Novel Androgen Receptor Splice Variant Is Up-regulated during Prostate Cancer Progression and Promotes Androgen Depletion-Resistant Growth. *Cancer Res.* **2009**, *69*, 2305–2313. [CrossRef] [PubMed]
9. Hu, R.; Dunn, T.A.; Wei, S.; Isharwal, S.; Veltri, R.W.; Humphreys, E.; Han, M.; Partin, A.W.; Vessella, R.L.; Isaacs, W.B.; et al. Ligand-Independent Androgen Receptor Variants Derived from Splicing of Cryptic Exons Signify Hormone-Refractory Prostate Cancer. *Cancer Res.* **2009**, *69*, 16–22. [CrossRef]

10. Antonarakis, E.S.; Lu, C.; Luber, B.; Wang, H.; Chen, Y.; Zhu, Y.; Silberstein, J.L.; Taylor, M.N.; Maughan, B.L.; Denmeade, S.R.; et al. Clinical Significance of Androgen Receptor Splice Variant-7 mRNA Detection in Circulating Tumor Cells of Men With Metastatic Castration-Resistant Prostate Cancer Treated With First- and Second-Line Abiraterone and Enzalutamide. *J. Clin. Oncol.* **2017**, *35*, 2149–2156. [CrossRef]
11. Antonarakis, E.S.; Lu, C.; Wang, H.; Luber, B.; Nakazawa, M.; Roeser, J.C.; Chen, Y.; Mohammad, T.A.; Chen, Y.; Fedor, H.L.; et al. AR-V7 and Resistance to Enzalutamide and Abiraterone in Prostate Cancer. *N. Engl. J. Med.* **2014**, *371*, 1028–1038. [CrossRef] [PubMed]
12. Sciarra, A.; Gentilucci, A.; Silvestri, I.; Salciccia, S.; Cattarino, S.; Scarpa, S.; Gatto, A.; Frantellizzi, V.; Von Heland, M.; Ricciuti, G.P.; et al. Androgen receptor variant 7 (AR-V7) in sequencing therapeutic agents for castratrion resistant prostate cancer: A critical review. *Medicine (Baltimore)* **2019**, *98*, e15608. [CrossRef] [PubMed]
13. Steinestel, J.; Luedeke, M.; Arndt, A.; Schnoeller, T.J.; Lennerz, J.K.; Wurm, C.; Maier, C.; Cronauer, M.V.; Schrader, A.J. Detecting predictive androgen receptor modifications in circulating prostate cancer cells. *Oncotarget* **2015**, *10*, 4213–4223. [PubMed]
14. Steinestel, J.; Bernemann, C.; Schrader, A.J.; Lennerz, J.K. Re: Emmanuel S. Antonarakis, Changxue Lu, Brandon Luber, et al. Clinical Significance of Androgen Receptor Splice Variant-7 mRNA Detection in Circulating Tumor Cells of Men with Metastatic Castration-resistant Prostate Cancer Treated with First- and Second-line Abiraterone and Enzalutamide. J Clin Oncol **2017**; 35, 2149–56: AR-V7 Testing: What's in it for the Patient? *Eur. Urol.* **2017**, *72*, e168–e169. [PubMed]
15. Sharp, A.; Welti, J.C.; Lambros, M.B.K.; Dolling, D.; Rodrigues, D.N.; Pope, L.; Aversa, C.; Figueiredo, I.; Fraser, J.; Ahmad, Z.; et al. Clinical Utility of Circulating Tumour Cell Androgen Receptor Splice Variant-7 Status in Metastatic Castration-resistant Prostate Cancer. *Eur. Urol.* **2019**. [CrossRef] [PubMed]
16. Liu, L.; Dong, X. Complex Impacts of PI3K/AKT Inhibitors to Androgen Receptor Gene Expression in Prostate Cancer Cells. *PLoS ONE* **2014**, *9*, e108780. [CrossRef] [PubMed]
17. Korenchuk, S.; Lehr, J.E.; MClean, L.; Lee, Y.G.; Whitney, S.; Vessella, R.; Lin, D.L.; Pienta, K.J. VCaP, a cell-based model system of human prostate cancer. *In Vivo Athens Greece* **2001**, *15*, 163–168.
18. Scher, H.I.; Lu, D.; Schreiber, N.A.; Louw, J.; Graf, R.P.; Vargas, H.A.; Johnson, A.; Jendrisak, A.; Bambury, R.; Danila, D.; et al. Association of AR-V7 on Circulating Tumor Cells as a Treatment-Specific Biomarker With Outcomes and Survival in Castration-Resistant Prostate Cancer. *JAMA Oncol.* **2016**, *2*, 1441–1449. [CrossRef]
19. Bernemann, C.; Schnoeller, T.J.; Luedeke, M.; Steinestel, K.; Boegemann, M.; Schrader, A.J.; Steinestel, J. Expression of AR-V7 in Circulating Tumour Cells Does Not Preclude Response to Next Generation Androgen Deprivation Therapy in Patients with Castration Resistant Prostate Cancer. *Eur. Urol.* **2017**, *71*, 1–3. [CrossRef]
20. Armstrong, A.J.; Halabi, S.; Luo, J.; Nanus, D.M.; Giannakakou, P.; Szmulewitz, R.Z.; Danila, D.C.; Healy, P.; Anand, M.; Rothwell, C.J.; et al. Prospective Multicenter Validation of Androgen Receptor Splice Variant 7 and Hormone Therapy Resistance in High-Risk Castration-Resistant Prostate Cancer: The PROPHECY Study. *J. Clin. Oncol.* **2019**, *37*, 1120–1129. [CrossRef]
21. Scher, H.I.; Graf, R.P.; Schreiber, N.A.; McLaughlin, B.; Lu, D.; Louw, J.; Danila, D.C.; Dugan, L.; Johnson, A.; Heller, G.; et al. Nuclear-specific AR-V7 Protein Localization is Necessary to Guide Treatment Selection in Metastatic Castration-resistant Prostate Cancer. *Eur. Urol.* **2017**, *71*, 874–882. [CrossRef] [PubMed]
22. Antonarakis, E.S. Predicting treatment response in castration-resistant prostate cancer: Could androgen receptor variant-7 hold the key? *Expert Rev. Anticancer Ther.* **2015**, *15*, 143–145. [CrossRef] [PubMed]
23. Onstenk, W.; Sieuwerts, A.M.; Kraan, J.; Van, M.; Nieuweboer, A.J.M.; Mathijssen, R.H.J.; Hamberg, P.; Meulenbeld, H.J.; De Laere, B.; Dirix, L.Y.; et al. Efficacy of Cabazitaxel in Castration-resistant Prostate Cancer Is Independent of the Presence of AR-V7 in Circulating Tumor Cells. *Eur. Urol.* **2015**, *68*, 939–945. [CrossRef] [PubMed]
24. Danila, D.C.; Samoila, A.; Patel, C.; Schreiber, N.; Herkal, A.; Anand, A.; Bastos, D.; Heller, G.; Fleisher, M.; Scher, H.I. Clinical Validity of Detecting Circulating Tumor Cells by AdnaTest Assay Compared With Direct Detection of Tumor mRNA in Stabilized Whole Blood, as a Biomarker Predicting Overall Survival for Metastatic Castration-Resistant Prostate Cancer Patients. *Cancer J.* **2016**, *22*, 315–320. [CrossRef] [PubMed]
25. Khetawat, G.; Faraday, N.; Nealen, M.L.; Vijayan, K.V.; Bolton, E.; Noga, S.J.; Bray, P.F. Human megakaryocytes and platelets contain the estrogen receptor beta and androgen receptor (AR): Testosterone regulates AR expression. *Blood* **2000**, *95*, 2289–2296. [PubMed]

26. Van der Toom, E.E.; Axelrod, H.D.; de la Rosette, J.J.; de Reijke, T.M.; Pienta, K.J.; Valkenburg, K.C. Prostate-specific markers to identify rare prostate cancer cells in liquid biopsies. *Nat. Rev. Urol.* **2019**, *16*, 7–22. [CrossRef] [PubMed]
27. Abu, E.O.; Horner, A.; Kusec, V.; Triffitt, J.T.; Compston, J.E. The Localization of Androgen Receptors in Human Bone. *J. Clin. Endocrinol. Metab.* **1997**, *82*, 3493–3497. [CrossRef]
28. Danila, D.C.; Heller, G.; Gignac, G.A.; Gonzalez-Espinoza, R.; Anand, A.; Tanaka, E.; Lilja, H.; Schwartz, L.; Larson, S.; Fleisher, M.; et al. Circulating Tumor Cell Number and Prognosis in Progressive Castration-Resistant Prostate Cancer. *Clin. Cancer Res.* **2007**, *13*, 7053–7058. [CrossRef]
29. De Bono, J.S.; Scher, H.I.; Montgomery, R.B.; Parker, C.; Miller, M.C.; Tissing, H.; Doyle, G.V.; Terstappen, L.W.W.M.; Pienta, K.J.; Raghavan, D. Circulating Tumor Cells Predict Survival Benefit from Treatment in Metastatic Castration-Resistant Prostate Cancer. *Clin. Cancer Res.* **2008**, *14*, 6302–6309. [CrossRef]
30. Heller, G.; McCormack, R.; Kheoh, T.; Molina, A.; Smith, M.R.; Dreicer, R.; Saad, F.; de Wit, R.; Aftab, D.T.; Hirmand, M.; et al. Circulating Tumor Cell Number as a Response Measure of Prolonged Survival for Metastatic Castration-Resistant Prostate Cancer: A Comparison With Prostate-Specific Antigen Across Five Randomized Phase III Clinical Trials. *J. Clin. Oncol.* **2018**, *36*, 572–580. [CrossRef]
31. Thalgott, M.; Heck, M.M.; Eiber, M.; Souvatzoglou, M.; Hatzichristodoulou, G.; Kehl, V.; Krause, B.J.; Rack, B.; Retz, M.; Gschwend, J.E.; et al. Circulating tumor cells versus objective response assessment predicting survival in metastatic castration-resistant prostate cancer patients treated with docetaxel chemotherapy. *J. Cancer Res. Clin. Oncol.* **2015**, *141*, 1457–1464. [CrossRef] [PubMed]
32. Okegawa, T.; Itaya, N.; Hara, H.; Tambo, M.; Nutahara, K. Circulating tumor cells as a biomarker predictive of sensitivity to docetaxel chemotherapy in patients with castration-resistant prostate cancer. *Anticancer Res.* **2014**, *34*, 6705–6710. [PubMed]
33. Perner, S.; Hofer, M.D.; Kim, R.; Shah, R.B.; Li, H.; Möller, P.; Hautmann, R.E.; Gschwend, J.E.; Kuefer, R.; Rubin, M.A. Prostate-specific membrane antigen expression as a predictor of prostate cancer progression. *Hum. Pathol.* **2007**, *38*, 696–701. [CrossRef] [PubMed]
34. Minner, S.; Wittmer, C.; Graefen, M.; Salomon, G.; Steuber, T.; Haese, A.; Huland, H.; Bokemeyer, C.; Yekebas, E.; Dierlamm, J.; et al. High level PSMA expression is associated with early psa recurrence in surgically treated prostate cancer. *Prostate* **2011**, *71*, 281–288. [CrossRef] [PubMed]
35. Joung, J.Y.; Cho, K.S.; Chung, H.S.; Cho, I.-C.; Kim, J.E.; Seo, H.K.; Chung, J.; Park, W.S.; Choi, M.K.; Lee, K.H. Prostate Specific Membrane Antigen mRNA in Blood as a Potential Predictor of Biochemical Recurrence after Radical Prostatectomy. *J. Korean Med. Sci.* **2010**, *25*, 1291–1295. [CrossRef] [PubMed]
36. Gorges, T.M.; Riethdorf, S.; von Ahsen, O.; Nastały, P.; Röck, K.; Boede, M.; Peine, S.; Kuske, A.; Schmid, E.; Kneip, C.; et al. Heterogeneous PSMA expression on circulating tumor cells—A potential basis for stratification and monitoring of PSMA-directed therapies in prostate cancer. *Oncotarget* **2016**, *7*, 34930–34941. [CrossRef]
37. Lowes, L.E.; Goodale, D.; Xia, Y.; Postenka, C.; Piaseczny, M.M.; Paczkowski, F.; Allan, A.L. Epithelial-to-mesenchymal transition leads to disease-stage differences in circulating tumor cell detection and metastasis in pre-clinical models of prostate cancer. *Oncotarget* **2016**, *7*, 76125–76139. [CrossRef]

5

Long-Term Dynamics of Three Dimensional Telomere Profiles in Circulating Tumor Cells in High-Risk Prostate Cancer Patients Undergoing Androgen-Deprivation and Radiation Therapy

Landon Wark [1], Harvey Quon [2], Aldrich Ong [2], Darrel Drachenberg [2], Aline Rangel-Pozzo [1,*] and Sabine Mai [1,*]

1 Cell Biology, Research Institute of Oncology and Hematology, University of Manitoba, CancerCare Manitoba, Winnipeg, MB R3E 0V9, Canada

2 Manitoba Prostate Center, Cancer Care Manitoba, Section of Urology, Department of Surgery, University of Manitoba, Winnipeg, MB R3E 0V9, Canada

* Correspondence: aline.rangelpozzo@umanitoba.ca (A.R.P.); sabine.mai@umanitoba.ca (S.M.);

Abstract: Patient-specific assessment, disease monitoring, and the development of an accurate early surrogate of the therapeutic efficacy of locally advanced prostate cancer still remain a clinical challenge. Contrary to prostate biopsies, circulating tumor cell (CTC) collection from blood is a less-invasive method and has potential as a real-time liquid biopsy and as a surrogate marker for treatment efficacy. In this study, we used size-based filtration to isolate CTCs from the blood of 100 prostate cancer patients with high-risk localized disease. CTCs from five time points: +0, +2, +6, +12 and +24 months were analyzed. Consenting treatment-naïve patients with cT3, Gleason 8-10, or prostate-specific antigen > 20 ng/mL and non-metastatic prostate cancer were included. For all time points, we performed 3D telomere-specific quantitative fluorescence in situ hybridization on a minimum of thirty isolated CTCs. The patients were divided into five groups based on the changes of number of telomeres vs. telomere lengths over time and into three clusters based on all telomere parameters found on diagnosis. Group 2 was classified as non-respondent to treatment and the Cluster 3 presented more aggressive phenotype. Additionally, we compared our telomere results with the PSA levels for each patient at 6 months of ADT, at 6 months of completed RT, and at 36 months post-initial therapy. CTCs of patients with PSA levels above or equal to 0.1 ng/mL presented significant increases of nuclear volume, number of telomeres, and telomere aggregates. The 3D telomere analysis of CTCs identified disease heterogeneity among a clinically homogeneous group of patients, which suggests differences in therapeutic responses. Our finding suggests a new opportunity for better treatment monitoring of patients with localized high-risk prostate cancer.

Keywords: androgen deprivation therapy; radiotherapy; localized high-risk prostate cancer; circulating tumor cells; three-dimensional (3D) telomere profiling

1. Introduction

Androgen deprivation therapy (ADT) combined with radiotherapy (RT) is a standard treatment for patients with localized high-risk prostate cancer (PCa) [1]. However, many patients will eventually regress after therapy and develop metastatic disease [2]. Therefore, there is an urgent need for a biomarker, which can reliably identify patients at high risk for recurrence and metastases [2]. For those patients, intensification of treatment beyond standard combination of ADT with radiation may be required [2].

All the options of intensification of therapy, including longer durations of ADT, combination with intense androgen blockade, chemotherapy or novel agents to target androgen activity through different pathways [3], are associated with additional toxicity. Therefore, it is critically important to ascertain the subgroup of patients in need of that next step. While clinical factors such as prostate-specific antigen (PSA) values, T-category, and Gleason scores have traditionally been used to risk-stratify prostate cancer patients, their accuracy is low (25%–40%) to predict important end points such as progression or metastases after radiation and ADT [4,5]. Due to the lack of a better surrogate, PSA measurement continues to be the main method to monitor treatment response and recurrence after treatment for prostate cancer [6]. However, this is fraught with difficulty as there is no reliable method to differentiate PSA produced by tumor vs. normal prostate tissue [5].

Many studies have evaluated the use of circulating tumor cells (CTCs) as a biomarker to predict disease progression and survival in patients with metastatic, advanced, or even early-stage PCa, as well as an endpoint marker in clinical trials [7–12]. As CTCs are responsible for distant metastasis, their analysis could potentially provide information about treatment response [7]. High CTC numbers are associated with aggressive disease, increased metastasis, and decreased time to relapse in men with castration-resistant and metastatic prostate cancer [10,12–14]. However, the value of CTCs detection in men with localized high-risk prostate cancer is unknown. Even though CTC collection from blood is a less-invasive method and can be used as a real-time liquid biopsy during regular follow-up [7], there are many challenges for the use of CTCs as a prognostic and/or predictive biomarker. For example, the number of CTCs found in patient samples depends on the isolation method used, because of CTCs immunophenotype heterogeneity, CTCs derived from the same tumor can present different expression of epithelial markers, such as EpCAM [15]. This difference might limit CTC detection by EpCAM-dependent technologies, like CellSearch [15]. In addition, many apoptotic CTC cells are also isolated and analyzed which may not necessarily be representative of potential metastatic cells; and, finally, CTCs can be absent in some non-metastatic PCa patients [15,16]. For a better clinical use of CTCs in PCa as a prognostic and/or predictive biomarker, a combination of enrichment (isolation), detection (identification), and characterization strategies (such as molecular profile), are necessary to improve our ability to identify high-risk lethal prostate cancer in patients with clinically localized high-risk prostate cancers [17,18].

In previous studies, we have demonstrated the potential of single-cell analysis of CTCs, combining a filtration-based CTC isolation technology with prostate cancer cell-specific antibodies, followed by the use of 3D telomere profiling to identify PCa patient subgroups [18,19]. Telomere shortening is one of the earliest molecular genomic events in prostate cancer tumorigenesis and can generate genomic instability [20]. The detection of shorter telomeres is associated with increased occurrence of lethal prostate cancer and decreased survival [21]. Additionally, androgen receptor (AR) inactivation by knockdown, androgen deprivation, or treatment with bicalutamide in LNCaP cells (prostate cancer cell line) can induce telomere breaks and telomere fusion [22]. However, telomere dysfunction was not observed following bicalutamide treatment in the AR-negative PC-3 prostate cell line [23]. Clinical studies assessing the effects of ADT on telomeres using prostate cancer patient samples are limited. In 2017, Cheung et al. reported no evidence that ADT deprivation accelerates telomerase shortening in men who have been diagnosed with prostate cancer [24]. However, leucocyte DNA was used for this analysis [24].

In a previous work, we began to follow the early dynamics of CTCs using 3D telomere analysis during ADT (a pilot study composed of 20 patients, where consenting treatment-naïve patients with cT3, Gleason 8-10, or prostate-specific antigen > 20 ng/mL and non-metastatic prostate cancer were included) [19]. We analyzed CTCs from high-risk prostate cancer patient´s samples at different time points: before ADT and RT (+0 month, untreated), after 2 months of ADT (+2 months) but prior RT, and 2 months after the final fraction of RT (+6 months). ADT begun 2 months before the start of RT and continued after RT was completed. At each time point, we enumerated CTCs from the blood, collected PSA values and investigated the nuclear 3D telomere architecture in CTCs derived from patients with non-metastatic high-risk prostate cancer before, post-ADT, and post-RT [19]. Contrary to CTC enumeration and PSA serum levels, we showed that nuclear 3D telomere architectural analysis is highly sensitive in detecting cellular events that affect the genome stability in CTCs, and we described three distinct telomere signatures in CTCs. Our previous data also indicated that only one-third (6/20, 30%) of patients with non-metastatic high-risk prostate cancer may be able to fully benefit from a synergistic ADT/radiotherapy treatment. However, a 2-month post-RT time point cutoff was too early to conclude disease outcome in our previous study and for a complete assessment of the effects of ADT and RT on 3D telomere architecture of CTCs, those patients need to be followed up for longer period.

Therefore, in the current study, we assessed if the 3D CTC telomere profiles can predict response to treatment when compared with PSA (the standard evaluation). We are comparing different PSA end points as early surrogates for tumor response, such as six-months PSA end levels after ADT, six-month PSA end levels after RT, and twelve-months PSA end levels after completed ADT (+36 months). The cutoff values were chosen on the basis of previous reports in which PSA end levels above or equal to 0.1 ng/mL after radiotherapy and long-course androgen deprivation therapy are associated with an increased risk of recurrence [6,25–27].

2. Results

2.1. *High-Risk Prostate Cancer CTCs were Selected Based on Their Androgen Receptor Staining*

One hundred PCa patients had their CTCs collected every six months for 2 years (Figure 1). The CTCs were collected using a size-based filtration technique (ScreenCell), which allows for the isolation of PCa CTCs in patients with low-, intermediate-, and high-risk disease [28]. Figure 1 shows a timeline summary of the treatment and PSA/CTC collection points over the course of the study. CTCs were collected and analyzed at different time points until 24 months and PSA levels at 6 months after continued ADT, 6 months after finished RT, and 12 months post-initial treatment (36 months) were used as early surrogates of treatment response.

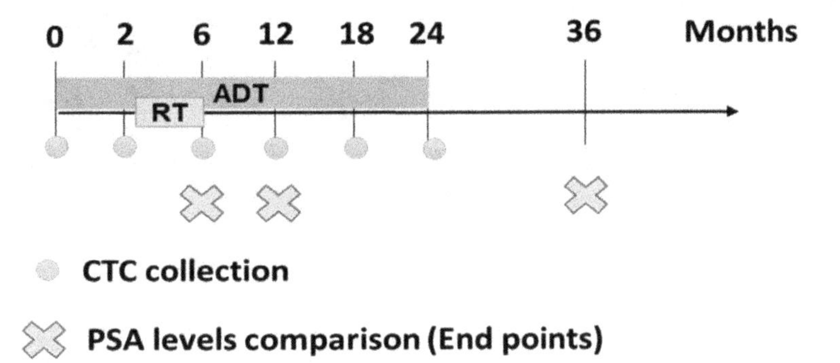

Figure 1. Summary timeline of treatment and PSA/CTC collection points over the course of the study. CTCs were collected and analyzed at 0 m (untreated), 2 m, 6 m, 12 m, 18 m and 24 months and PSA end levels at 6 months ADT, 6 months after finished RT and 36 months after initial treatment were used as early surrogates of treatment response.

We identified the prostate cancer CTCs based on their positive immunostaining for androgen receptor (AR). All analyzed samples contained androgen receptor-positive cells (Figure 2). Figure 2A shows an isolated CTC stained with AR antibody conjugated with Alexa Fluor 488, AR stains both the intracytoplasmic region and the cell membrane. The expression of AR on the isolated CTCs was found to be heterogeneous. We observed both intersample and intrasample variability in the level of fluorescent intensity. Our results were consistent with those described previously [19]. The AR expression can change during ADT treatment and a mixed population of CTCs with different expression levels of AR can also be found [29].

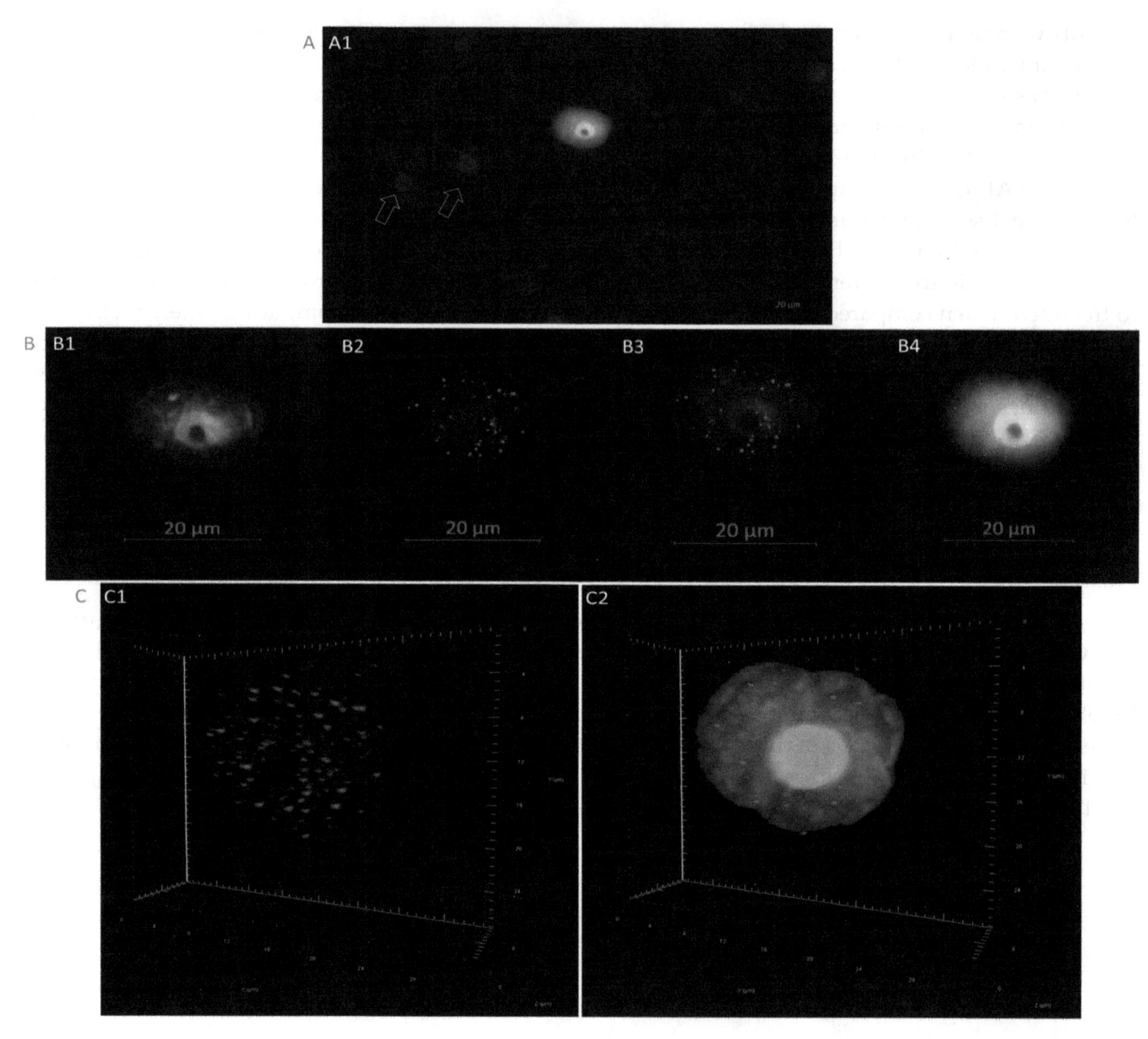

Figure 2. Example of a circulating tumor cell from a high-risk localized prostate cancer patient captured on top of a filter pore (**A**) (The arrows show empty filter pores). The prostate cancer CTCs are recognized based of their AR positive staining (**B**). (**B1**) Two-dimensional image showing a CTC AR+ in FITC (green); (**B2**) CTC with the telomeres labeled with telomere-specific Cy3-labeled probe (red); (**B3**) Merge between FITC and telomeres; and (**B4**) CTC counterstained with DAPI in blue. In C (**C1** and **C2**), the same cell is shown in three-dimensional representation. Red spots represent telomere signals; and the blue is DAPI.

2.2. CTCs Dynamics of the 3D Telomere Architecture Stratified Patients into 5 Distinct Subgroups

Cancer cells commonly exhibit genomic instability with the telomeres often being shorter than those in normal cells [27,30,31]. Based on the dynamics of the 3D telomere architecture of their CTCs (changes in number of short telomeres over time) at +0, +2, +6, +12, +18 and +24 months, patients were classified into five groups. In Figure 3, telomere length (signal intensity, x-axis) is plotted against the number of telomeres (y-axis) for all CTCs analyzed at each time point. Signals are grouped by their intensity level and this gives a picture of the CTCs telomere distribution in each sample or time point. In Figure 3A, at +0 month, approximately 220 telomeres have the same intensity, less than 10,000 a.u (arbitrary units of relative fluorescence intensity), which is of relatively low intensity and can be correlated with short telomeres. For normal lymphocytes, for example, this plot usually has small peaks between 0 and 20,000 a. u, in which the number of telomeres per nucleus on the y-axis ranges between 5 and 25 [16] and most of the telomere signals have high relative intensities, with signals detected up to 120,000 a.u [16].

Two Teloview^{TM} parameters were used to analyze the CTCs dynamics over time-number of telomere vs telomere length/intensity [32]. The first patient group (Group 1), at baseline (+0 month, untreated), showed a large number (peak count 220) of shorter telomeres (< 10,000 AU) which started to decrease post-ADT (+2 months) and remained at this level from post-RT (+6 months) until 24 months (Figure 3A). The second patient group (Group 2) showed CTC with a moderate number (peak count between 80 and 90) of shorter telomeres at baseline (+0 month) and the telomere length remained stable before and after treatment (Figure 3B). Patient from group 3 displayed CTC with small number (peak count between 40 and 50) of shorter telomeres at baseline (+0 month) and remained stable during all treatment procedures. However, in this group, the number of short telomeres started to increase at +18 months and reached a high peak at +24 months (Figure 3C). The fourth group (Group 4), at baseline (+0 month, untreated), showed a small to moderate number of short telomeres at baseline (+0 month). Nevertheless, when ADT started at +2 months, the effects were shown with a temporally increase of short telomeres that does not continue until the next time point (+6 months) and the number of short telomeres remained low until 24 months (Figure 3D). Patients from our last group (Group 5) displayed CTC with high number of short telomeres at baseline (+0 month) with a peak count between 180 and 200, which decreased during and after treatment. However, the number of short telomeres in Group 5 started to increase after 24 months (Figure 3E). Only patients in Group 2 showed no changes in 3D telomere architecture in response to ADT plus RT. Groups 1 and 4 presented the best dynamics of CTCs with an increase of telomere length in a long-term follow-up and, consequently, decrease of genomic instability. Even though Group 4 showed a peak of short telomeres at +2 months, this is probably due to the start of the ADT therapy, since ADT can induce telomere breaks and telomere fusion [20]. The number of short telomeres continued at low intensities until 24 months. Patients from Groups 3 and 5 exhibited the worst dynamics of CTCs after 24 months, i.e., both presented a high peak of short telomeres at the last time point (+24 months). Supplementary Table S1 shows all patients classified in the five subgroups. The most prevalent being Group 2 with 30%, followed by Group 1 with 25%, Group 5 with 18%, Group 3 with 15% and Group 4 with 12%. Supplementary Figure S1 shows the inter-sample variability of representative individual samples. Lymphocytes for each patient were used as an internal control (Supplementary Figure S2), and representative 3D images for each time point are shown in the Supplementary Figure S3.

2.3. The Telomere Parameters Can Predict PSA Increase at 6 Months of ADT, RT and at 36 Months after Initial Therapy

First, we compared the PSA end values of two time points: after 6 months of ADT and post 6 months of completed RT. Our aim was to identify which baseline telomere parameter (+0 month, untreated) could predict response to treatment. We used 6 months PSA end values (after ADT and after RT) and 36 months after initial therapy as an early surrogate for treatment response. We compare for example, patients with PSA end values below 0.1 ng/ml at 6 months (ADT or RT) vs the group of patients which PSA end value ≥ 0.1 ng/mL. Fifthy-nine patients (59/100) had a stable or decrease in PSA end values (< 0.1 ng/mL) and forty-one had an increased PSA end value above 0.1 ng/mL (≥ 0.1 ng/mL) at 6 months after ADT (Figure 4). Figure 4 shows a comparison between the two groups. The group with PSA ≥ 0.1 ng/mL had significantly higher nuclear volume ($p = 0.0007$), increased total number of signal ($p < 0.001$)—which can indicate an increase of aneuploidy- and increased number of telomere aggregates ($p = 0.01$)—which represent more telomere fusion- in comparison to the group with PSA < 0.1 ng/mL at 6 months after ADT. The total fluorescent telomere intensity, which is equivalent to telomere length, also decreased ($p = 0.0047$) in the group with PSA ≥ 0.1 ng/mL.

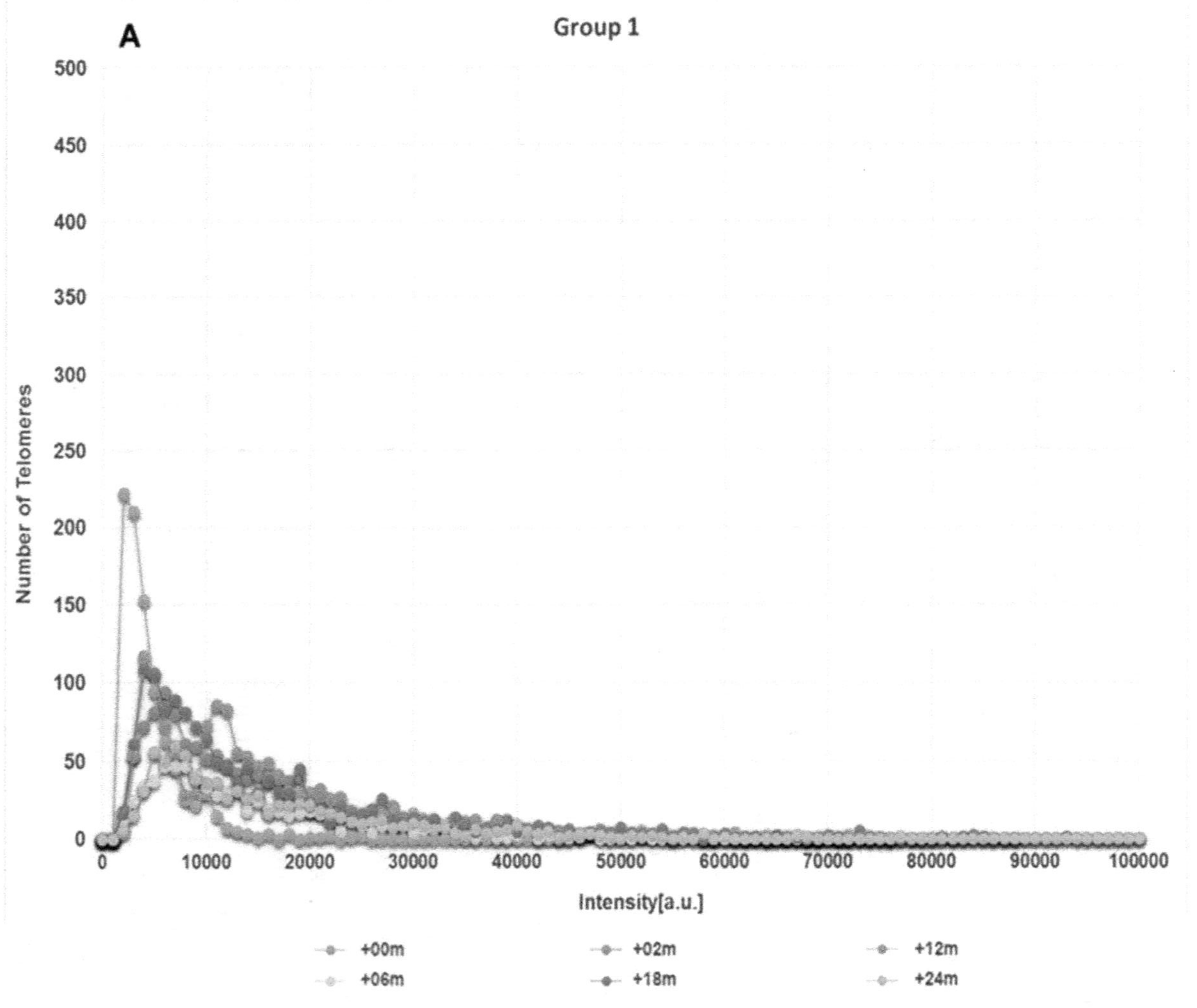

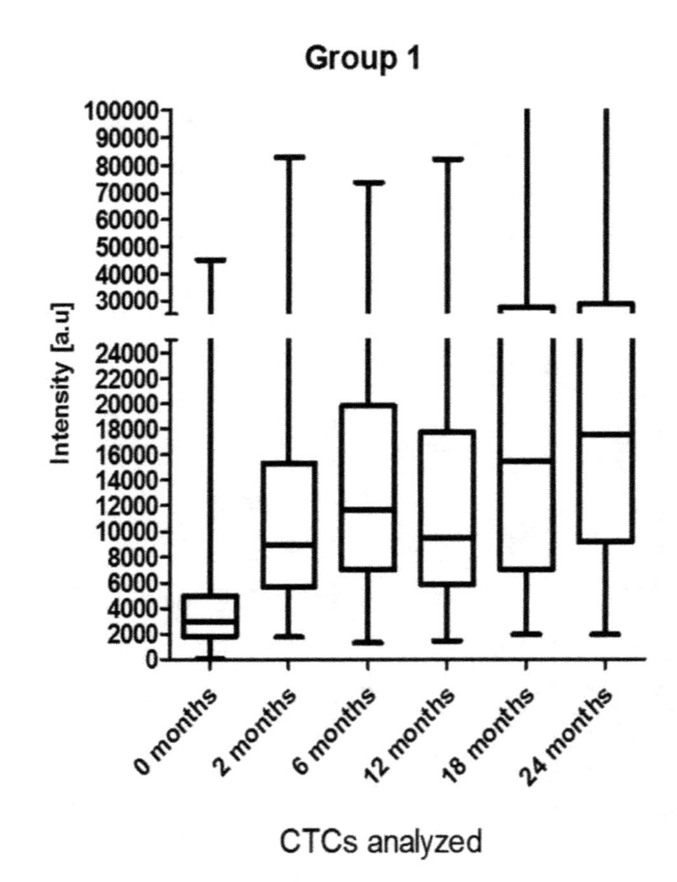

Figure 3. *Cont.*

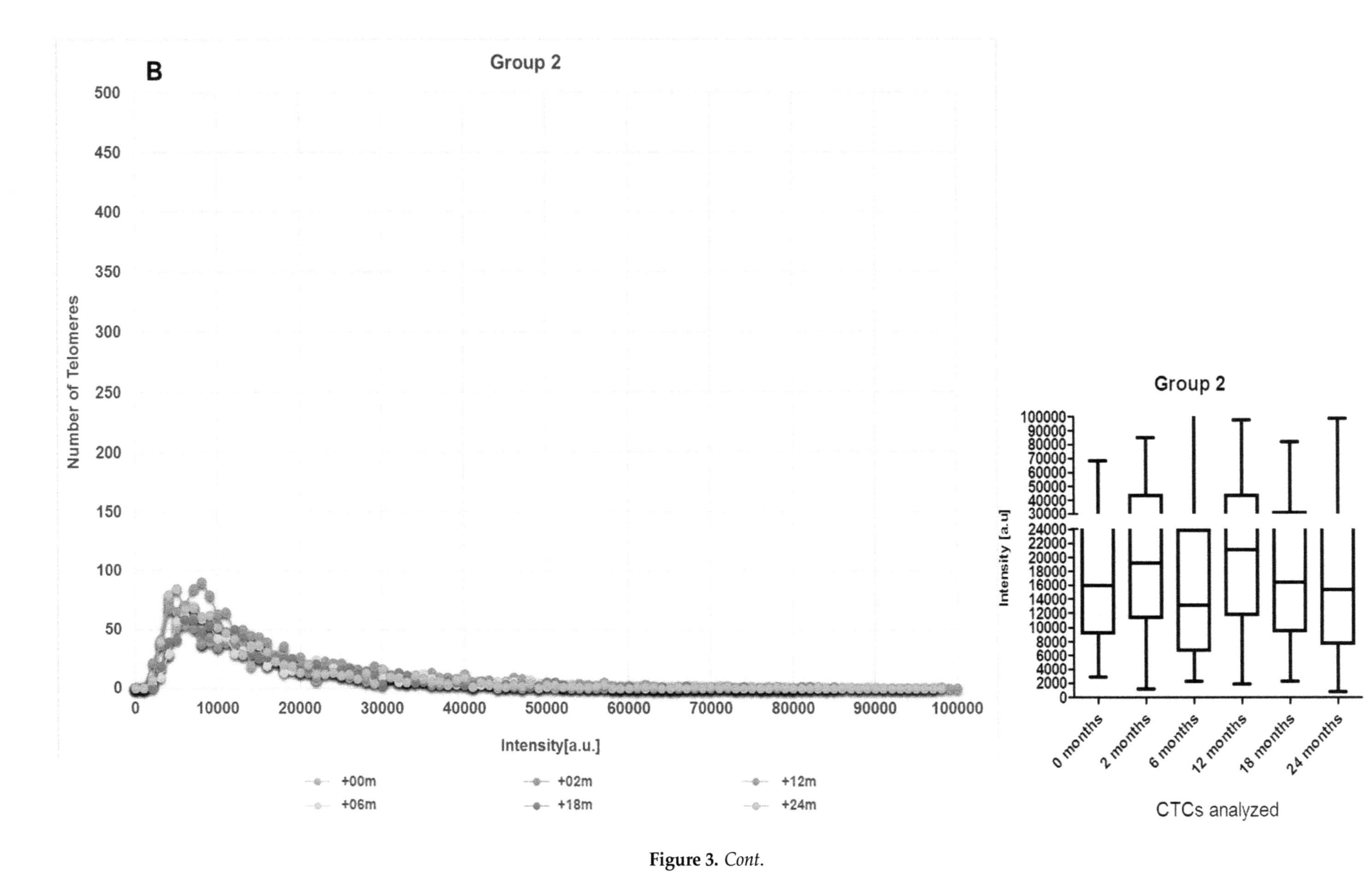

Figure 3. *Cont.*

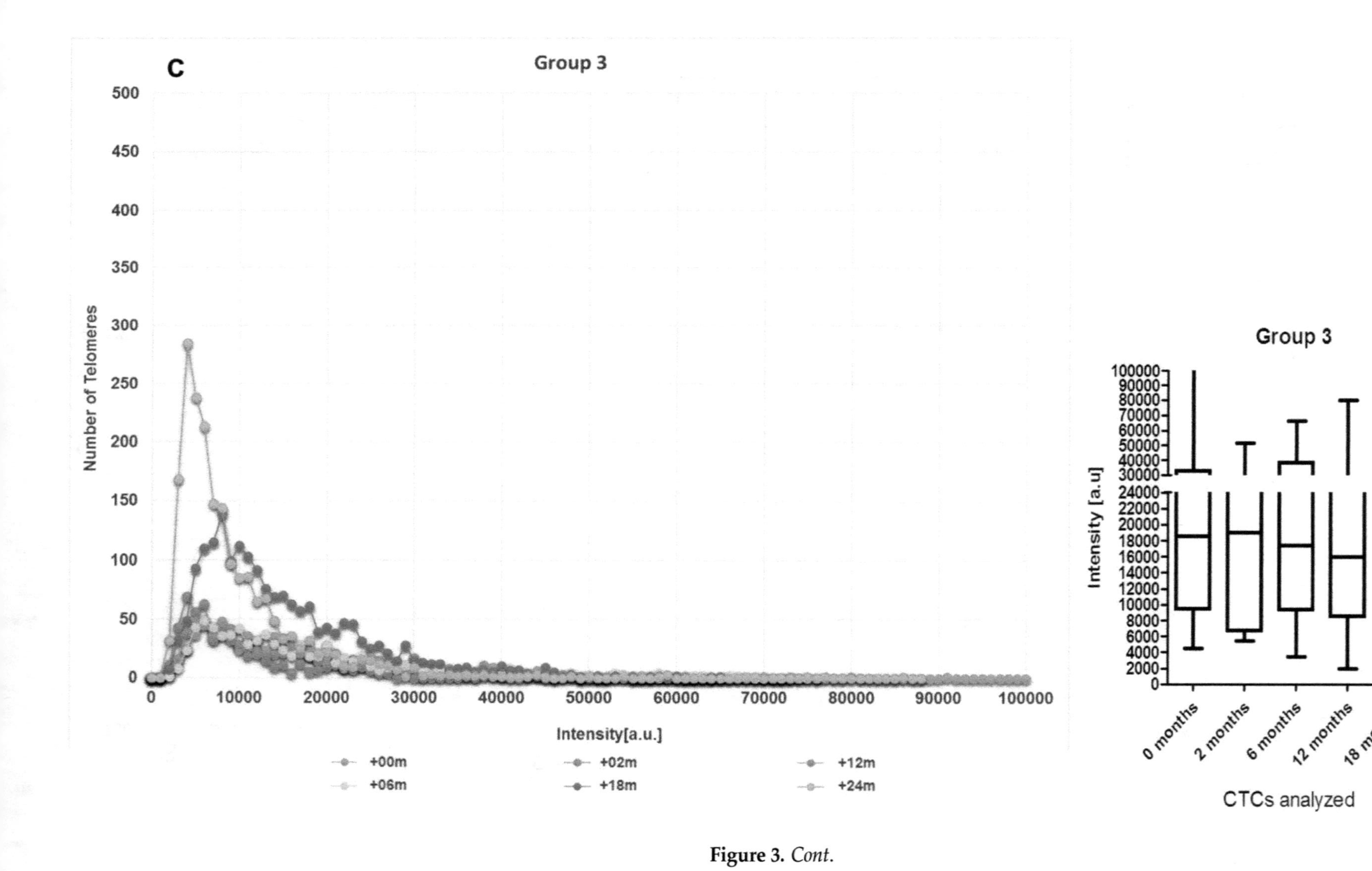

Figure 3. *Cont.*

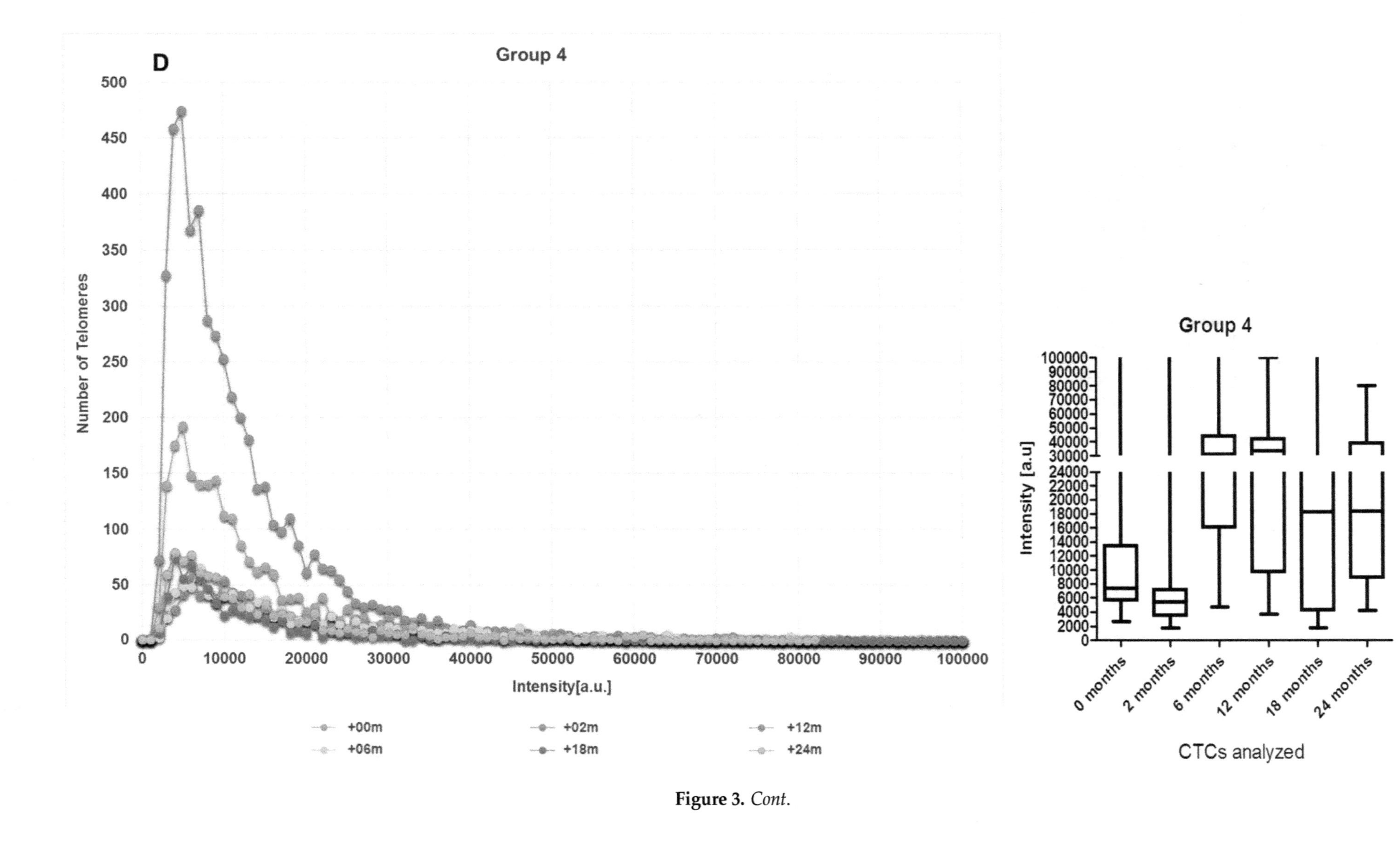

Figure 3. *Cont.*

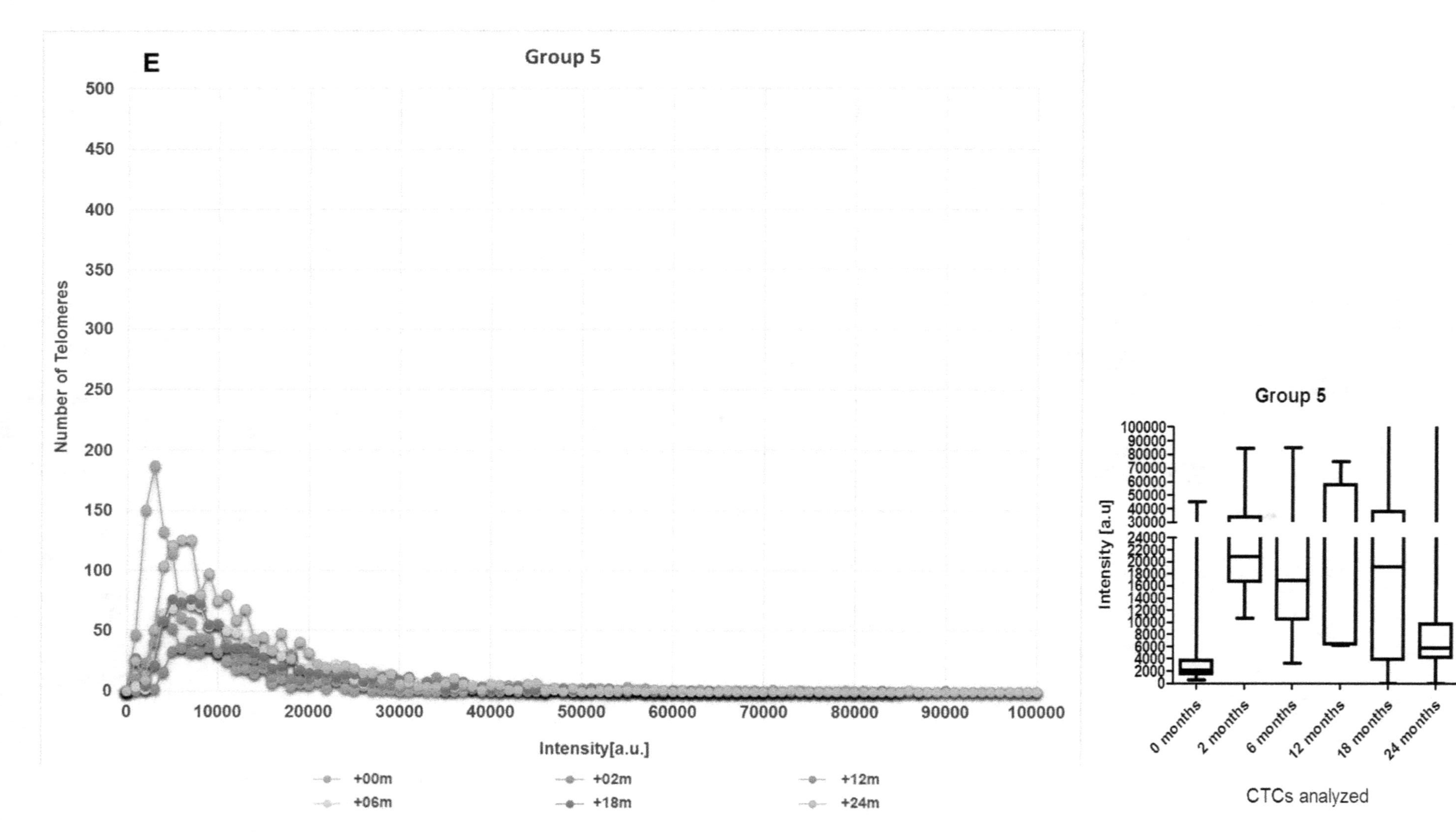

Figure 3. (**A**) Representative examples of the CTCs dynamics of telomere length profiles over time for patients assigned to Group 1 (**A**), Group 2 (**B**), Group 3 (**C**), Group 4 (**D**), and Group 5 (**E**). In each graph, the telomere length is shown in arbitrary units of fluorescence (AU). Baseline profile (+0 month, untreated) and other time point (2, 6, 12, 18, 24 months) are demarked with colors. Bars plot were used to illustrate inter-sample variability of representative individual samples in the groups.

A

Variable	PSA (N)	Mean	Std Dev	Std Err	Minimum	Maximum	Equality of Variances	
							F Value	Pr > F
Nuclear volume	< 0.1 (59)	138606	232066	299596	569.5	1021551	2.62	**0.0007**
	≥ 0.1 (41)	199430	375943	587124	433.7	1572381		
	Diff 1-2	-608243	298663	60516.5	-	-		
Total number of signals	< 0.1 (59)	284000	106041	13690	170000	820000	4.87	**<0.0001**
	≥ 0.1 (41)	325732	234034	36550	195000	171000		
	Diff 1-2	-41732	169798	34405	-	-		
Total intensity	< 0.1 (59)	461213	159236	205572	176763	1090104	2.28	**0.0038**
	≥ 0.1 (41)	431933	240626	375795	111972	1616141		
	Diff 1-2	-292799	196228	39875.8	-	-		
Total number of aggregates	< 0.1 (59)	28333	15061	0.1977	10000	90000	2.01	**0.0146**
	≥ 0.1 (41)	34390	21336	0.3303	20000	1550000		
	Diff 1-2	-0.6057	17864	0.3620	-	-		

B

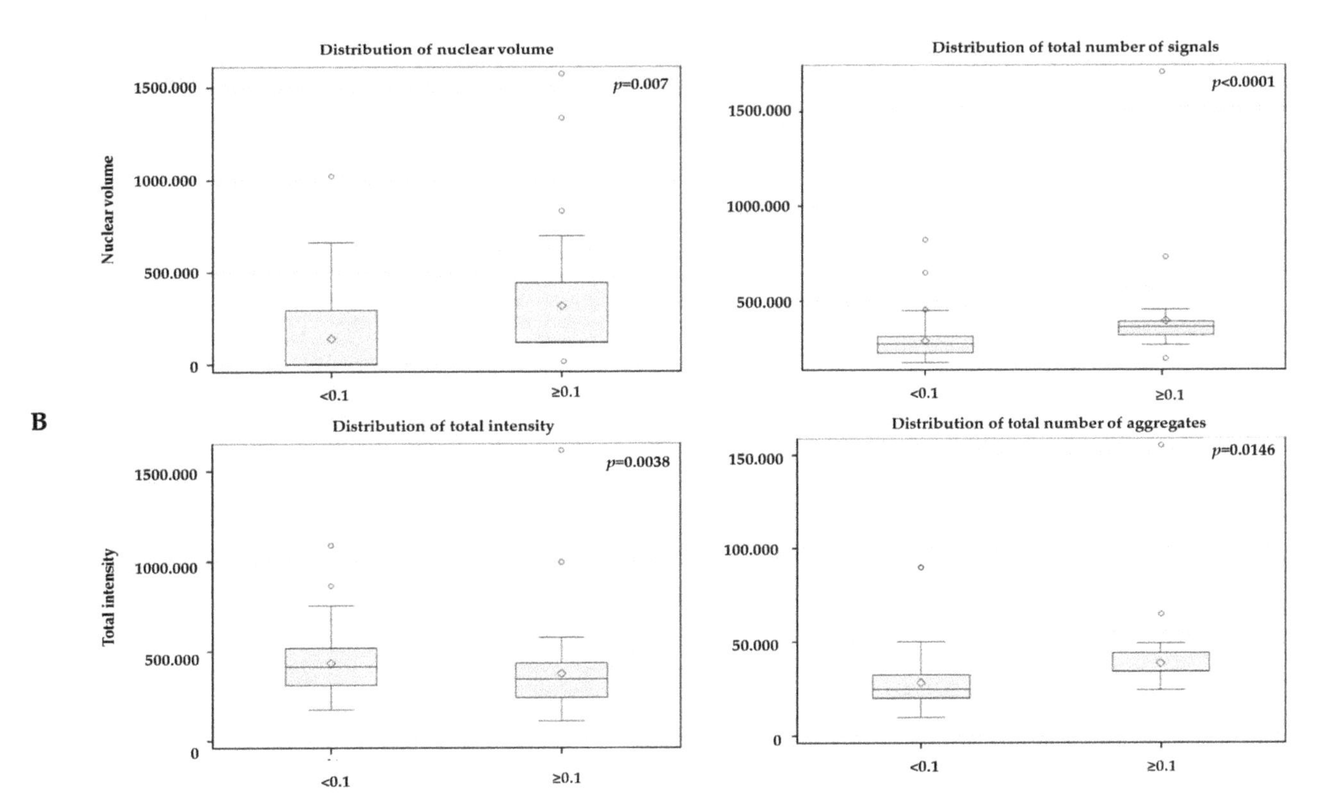

Figure 4. Statistical analysis comparing the telomeres parameters at 0 month with PSA end value at 6 months after androgen deprivation therapy using 0.1 ng/mL cutoff (**A**). (**B**) Box-plots generated by Statistical Analysis Software v 9.4 (SAS, Cary, NC USA). The box is divided in the following way: —the median is the middle line, the 50th percentile- the top of box is the 75th percentile- the bottom box is the 25th percentile. Concerning the whiskers, the upper top of whiskers represent the max observation 1.5× (interquartile range (IQR)—75th percentile minus 25th percentile), while the bottom of whiskers represent the minimum observation 1.5× (1QR from 25th). The observations plotted are outliers—beyond the 1.5× IQR or below. The sign in the box is the mean. The box indicates where 50 percent of the observations lies, extending to the whiskers indicates where most of the data lies and the points outside are extremes.

Using the 6 months PSA end value after RT as an early surrogate for treatment response, 78% of the patients (78/100) had PSA end value below the 0.1 ng/mL threshold at 6 months after RT (< 0.1 ng/mL) and 22% of the patients (22/100) had PSA end values above the 0.1 threshold (≥ 0.1 ng/mL).

In Figure 5, we compare the telomeres parameters at 0 month (untreated) with the PSA groups. The nuclear volume ($p = 0.02$), total number of telomere signals ($p = 0.0006$) and formation of telomeres aggregates ($p = 0.003$) also increased in the group with PSA end ≥ 0.1 ng/mL after RT. However, the total intensity did not decrease significantly. In addition, we found no association between the PSA end values and the 3D telomere groups (CTCs dynamics) over time ($p = 0.38$).

A

Variable	PSA (N)	Mean	Std Dev	Std Err	Minimum	Maximum	Equality of Variances	
							F Value	Pr > F
Nuclear volume	< 0.1 (78)	159694	271771	307721	5125	1332784	1.99	**0.0294**
	≥ 0.1 (23)	175516	383122	798864	4337	1572381		
	Diff 1-2	-15822.3	300108	712077	-	-		
Total number of signals	< 0.1 (78)	293696	187231	21200	170000	1710000	3.94	**0.0006**
	≥ 0.1 (23)	303077	94367	19677	195000	660000		
	Diff 1-2	-0.7715	171.563	40.707	-	-		
Total intensity	< 0.1 (78)	446043	197712	223865	176763	1616141	1.05	0.9444
	≥ 0.1 (23)	436276	193213	402877	111972	995235		
	Diff 1-2	97670	196721	466768	-	-		
Total number of aggregates	< 0.1 (78)	30577	19672	0.2242	10000	155000	3.19	**0.0033**
	≥ 0.1 (23)	31522	11020	0.2298	20000	65000		
	Diff 1-2	-0.1137	18.219	0.4323	-	-		

B

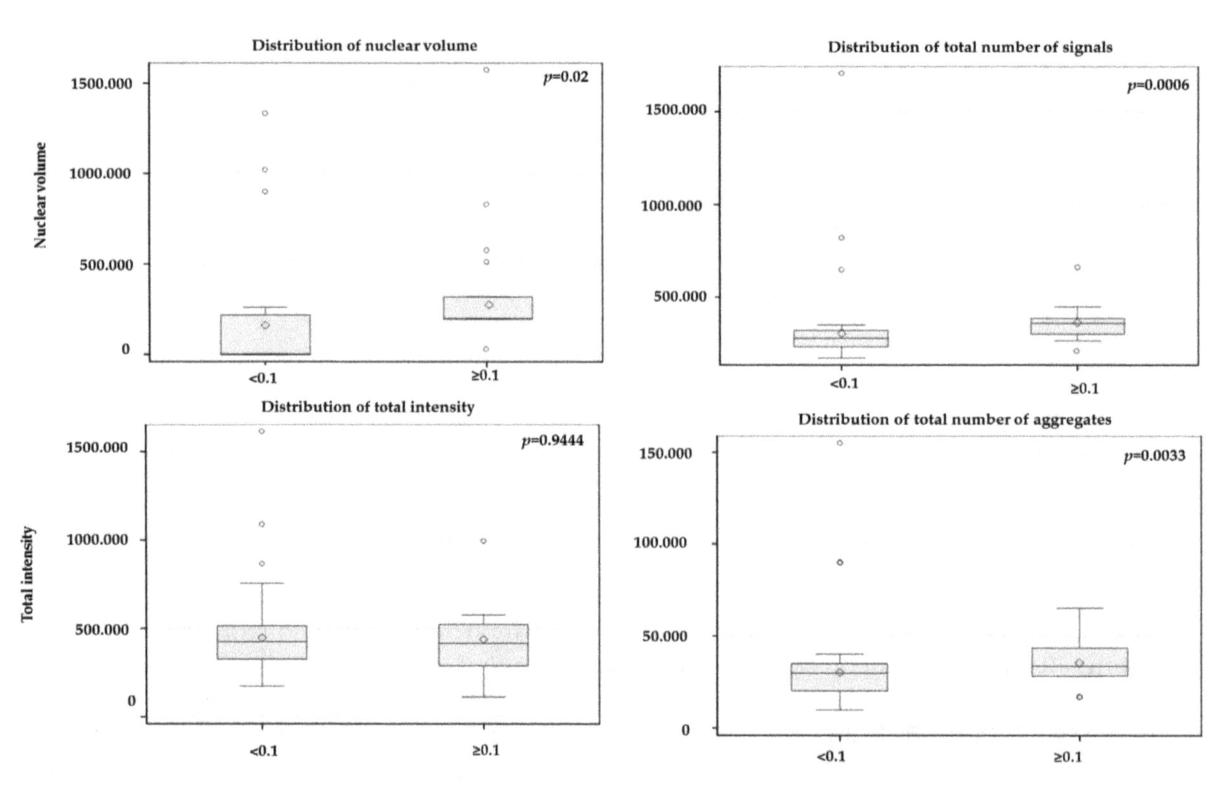

Figure 5. Statistical analysis comparing the telomeres parameters at 0m with PSA end value at 6 months after radiotherapy using 0.1 ng/mL cutoff (**A**). (**B**) Box-plots generated by Statistical Analysis Software v. 9.4. The box is divided in the following way: —the median is the middle line, the 50th percentile- the top of box is the 75th percentile- the bottom box is the 25th percentile. Concerning the whiskers, the upper top of whiskers represent the max observation 1.5× (interquartile range (IQR)—75th percentile minus 25th percentile), while the bottom of whiskers represent the minimum observation 1.5×(1QR from 25th). The observations plotted are outliers—beyond the 1.5× IQR or below. The sign in the box is the mean. The box indicates where 50 percent of the observations lies, extending to the whiskers indicates where most of the data lies and the points outside are extremes.

The 36-month PSA end after initial treatment, which means one year after the treatment was completed, was used as an early surrogate for tumor response. 78% of the patients (78/100) had PSA end value below the 0.1 ng/mL threshold at 36 months (< 0.1 ng/mL) and 22% of the patients (22/100) had PSA end values above the 0.1 threshold (≥ 0.1 ng/mL). In Figure 6, we compare the telomeres parameters at 0 m (untreated) with the PSA groups. The nuclear volume ($p = 0.0003$), total number of telomere signals ($p < 0.0001$), and formation of telomeres aggregates ($p = 0.0001$) increased in the group with PSA ≥ 0.1 ng/mL after 36 months of initial therapy. However, the total intensity significantly increased this time ($p = 0.0001$). We attribute the changes in total intensity to the increase in the number of telomere signals and formation of telomeres aggregates visualized since the first time point (6 months post-ADT) in the group with PSA end value above to ≥ 0.1 ng/mL.

A

Variable	PSA (N)	Mean	Std Dev	Std Err	Minimum	Maximum	Equality of Variances	
							F Value	Pr > F
Nuclear volume	< 0.1 (78)	131153	244411	276741	4337	1021551	3.13	**0.0003**
	≥ 0.1 (22)	263250	432364	921802	5255	1572381		
	Diff 1-2	-132098	294948	712011	-	-		
Total number of signals	< 0.1 (78)	287179	104293	11809	175000	820000	8.72	**<0.0001**
	≥ 0.1 (22)	350000	307946	65654	170000	1710000		
	Diff 1-2	-62821	169903	41015	-	-		
Total intensity	< 0.1 (78)	431623	160884	182165	176763	1090104	3.29	**0.0001**
	≥ 0.1 (22)	485978	291843	622212	111972	1616141		
	Diff 1-2	-543545	196439	47420.9	-	-		
Total number of aggregates	< 0.1 (78)	29487	14110	0.1598	10000	90000	9.11	**<0.0001**
	≥ 0.1 (22)	35682	28001	0.5970	15000	155000		
	Diff 1-2	-0.6195	18012	0.4348				

B

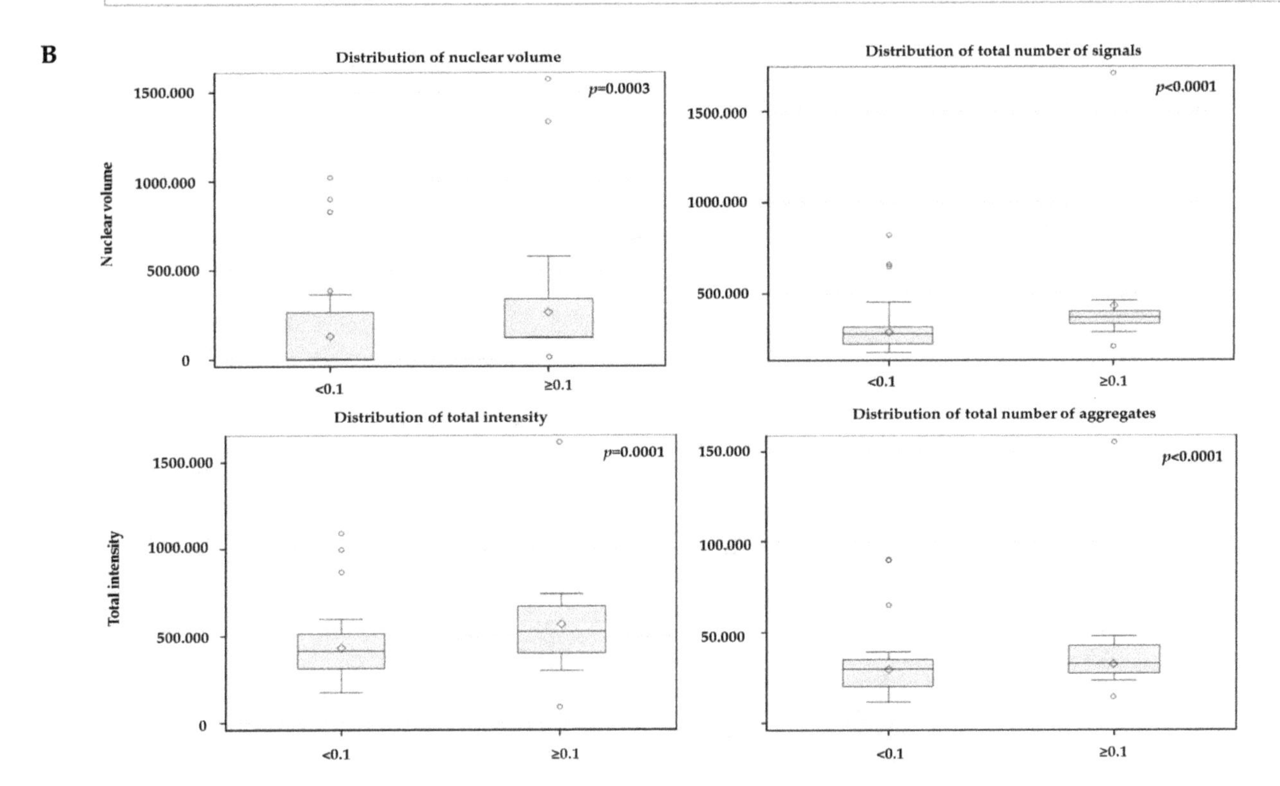

Figure 6. Statistical analysis comparing the telomeres parameters at 0 month with PSA value at 36 months after initial therapy using 0.1 ng/mL cutoff (**A**). (**B**) Box-plots generated by Statistical Analysis Software v. 9.4. The box is divided in the following way: —the median is the middle line, the 50th percentile- the top of box is the 75th percentile- the bottom box is the 25th percentile. Concerning the whiskers, the upper top of whiskers represent the max observation 1.5× (interquartile range (IQR)—75th

percentile minus 25th percentile), while the bottom of whiskers represent the minimum observation 1.5× (1QR from 25th). The observations plotted are outliers- beyond the 1.5× IQR or below. The sign in the box is the mean. The box indicates where 50 percent of the observations lies, extending to the whiskers indicates where most of the data lies and the points outside are extremes

Second, we performed a hierarchical centroid cluster analysis, which combines all TeloView™ data for each patient. The TeloView™ data includes all parameters (nuclear volume, total number of signals, total intensity and total number of aggregates) provided for each telomere. Three patients were excluded from the analysis. The remaining ninety-seven patients were grouped into three subgroups (clusters 1-3) (Figure 7). The three clusters identified from 3D telomere profiling data, after hierarchical centroid cluster analysis, distinguished patients with different levels of genomic instability and different risk of future prostate mortality, based on their PSA end values after 6 months of ADT, 6 months of RT, and 36 months after initial treatment (Figure 7). Supplementary Table S1 shows all patients classified in the three clusters. Cluster 3 contains predominantly high-risk patient for future prostate mortality. Cluster 3 had approximately 30% of patients with PSA end value above 0.1 ng/mL after 6 months of ADT which decreased to 16.67% after RT and return to 30%, after 36 months of initial treatment. In contrast, clusters 1 and 2 had lost the fluctuation in their PSA end values after ADT, RT, and 36 months of ADT. Cluster 3 comprises 18.55% of the patients in this study. Supplementary Table S1 shows all patients classified into the three clusters. Additionally, patients classified as clusters 1, 2, and 3, had different disease aggressiveness (cluster 3 > cluster 1 > cluster 2) based on their genomic instability pattern detected by 3D telomere analysis.

A

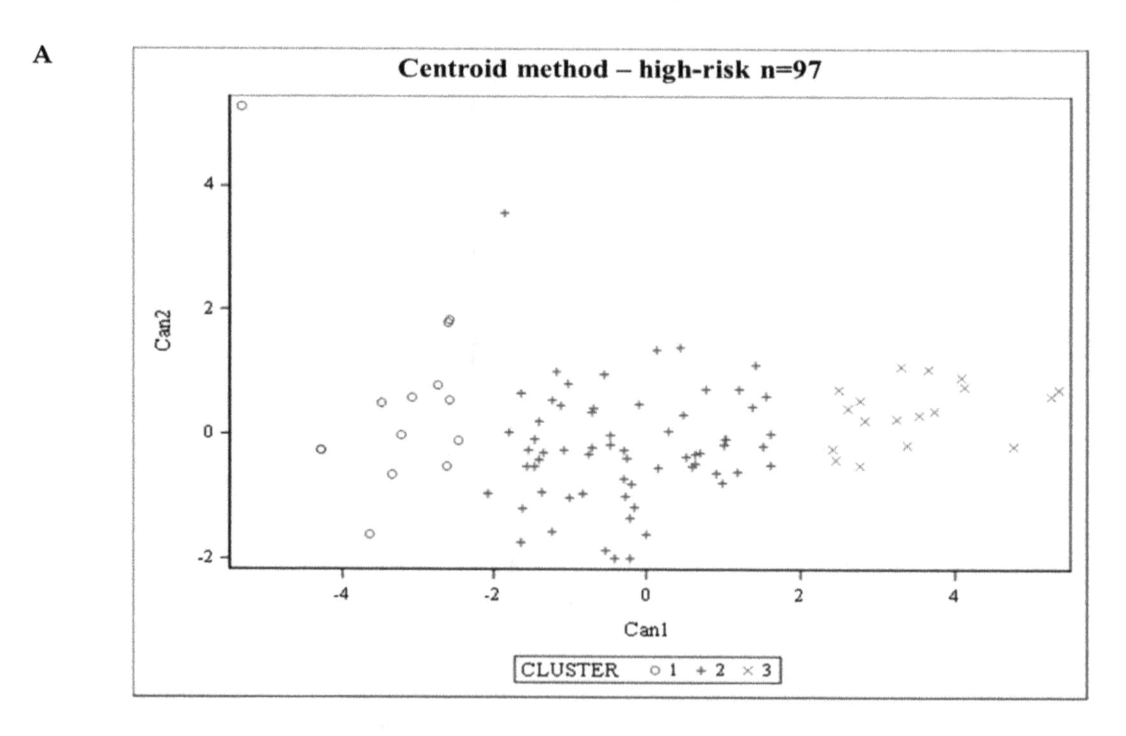

Figure 7. *Cont.*

Cluster groups and PSA values at 6 months ADT, RT and at 36 months after initial therapy using 0.1 ng/mL cutoff

B

CLUSTER	psa6adtg[1] (%, #pat)		psa6radg[2] (%, #pat)		psa36mgrp[3] (%, #pat)		Total (#pat)
	< 0.1ng/mL	≥ 0.1ng/mL	< 0.1ng/mL	≥ 0.1ng/mL	< 0.1ng/mL	≥ 0.1ng/mL	
1	28.57 (4)	71.43 (10)	57.14 (8)	42.86 (6)	71.43 (10)	28.57 (4)	14
	7.02 (4)	25.00 (10)	10.67 (8)	27.27 (6)	13.16 (10)	19.05 (4)	
2	63.08 (41)	36.92 (24)	80.00 (52)	20.00 (13)	83.08 (54)	16.92 (11)	65
	71.93 (41)	60.00 (24)	69.33 (52)	59.09 (13)	71.05 (54)	52.38 (11)	
3	66.67 (12)	33.33 (6)	83.33 (15)	16.67 (3)	66.67 (12)	33.33 (6)	18
	21.05 (12)	15.00 (6)	20.00 (15)	13.64 (3)	15.79 (12)	28.57 (6)	
Total	57	40	75	22	76	21	97

Figure 7. Centroid cluster analysis of 3D nuclear profiling of CTCs from 97 patients with high-risk prostate cancer (**A**). The combination of telomere parameters (Materials and Methods) allows the stratification of patients into clusters. Each cluster possesses a different level of genomic instability and different risk of future prostate mortality, based on their PSA end values after 6 months of ADT 6 months of RT, and 36 months after initial treatment (**B**). Patients in cluster 3 (green) had the highest percentage of patients with PSA ≥ 0.1ng/mL after treatment, while those in cluster 2 (red) and cluster 1 (blue) had an intermediate- to low percentage of patients with PSA ≥ 0.1ng/mL after treatment.

3. Discussion

In most cancers, early detection allows for improved outcomes, but for localized high-risk prostate cancer patients, early detection can also result in overdiagnosis and overtreatment [29]. In addition, localized high-risk prostate cancer patients face a more serious problem, a reliable prognostic tool capable of predicting whether the cancer will eventually develop into a lethal metastatic disease [18]. Although PSA levels are used for disease monitoring, the predictive value of PSA testing and screening is low (around 35%) and have been associated with a high rate of overdiagnosis/overtreatment in clinical trials [32–34]. In addition, other clinical parameters such as clinical stage and Gleason score tumor grade have limitations to detect and predict disease outcome [35]. This scenario demonstrates the importance for novel and less-invasive biomarkers that can reliably identify patients at high risk for recurrence and metastases.

Telomere shortening is one of the earliest events in prostate cancer tumorigenesis and continue during tumor progression [18]. Since detection of shorter telomeres is associated with increased occurrence of lethal prostate cancer and decreased survival times, the 3D telomere assessment could potentially improve prostate cancer screening by adding, to the current approaches, prognostic information to better stratify patients requiring active surveillance or more definitive treatment, such as surgical castration [18,31].

To our knowledge, the current study is the first to investigate the dynamics of the nuclear 3D telomere architecture in CTCs derived from patients with non-metastatic high-risk prostate cancer before, post-ADT, and post-RT (until 36 months after initial treatment). CTCs isolated before treatment was divided into five distinct telomere signatures. Remarkably, CTC analysis showed distinct dynamic changes in their 3D telomere signatures, which were unique to each group during ADT+RT treatment. Recent studies have provided insights to the clinical value of CTCs collected from blood in prostate cancer [17,36–38]. In the present study, we have used the ScreenCell filter device, which allows a size-based separation of CTCs from whole blood of patients with non-metastatic high-risk prostate cancer [28]. Captured CTCs underwent 3D telomere analysis to determine their nuclear 3D telomere profile before treatment and to investigate dynamic changes to their 3D telomere profiles during and after treatment with ADT and RT. The CTCs can be detected in blood before the occurrence of clinically relevant metastases providing insight into the genetics of the primary prostate tumor [39–45]. In this study, we show that nuclear 3D telomere architectural analysis is highly sensitive in detecting telomere changes that affect the genome stability in CTCs and may prove to be a valuable tool in monitoring treatment response in patients with non-metastatic high-risk prostate cancer.

As stated above, in a previous study, we showed the usefulness of PSA serum levels and CTC blood counts after treatment [17]. In studies using the CellSearch®system to enrich and isolate CTC in localized high-risk prostate cancer, no correlation between CTC count and other clinical-pathological parameters was found [46–48]. This highlights the importance of molecular analysis of CTC instead of only CTC count to provide additional information about the tumor. For example, AR-V7 expression in CTCs was found to be a predictor of response for treatment with abiraterone/enzalutamide and disease outcome [46].

Here, we assessed the effects of ADT and RT on 3D telomere architecture of captured CTCs until 24 months. We found that the 100 high-risk patients could be stratified into five distinct telomere signatures based on telomere numbers and telomere length (intensity). Furthermore, each of the five CTC groups responded to the combined treatment with different changes to telomere profiles, thus, providing unique insight about the complexity by which high-risk prostate cancer cells can adapt to those treatments. We also showed in previous studies the potential of 3D telomere architecture in predicting disease outcome and patient survival [16,17,47–49]. Based on our 3D telomere analysis, Group 2 may qualify as a non-responder, since the telomere profiles were stable throughout treatment. However, for Group 1, the treatment resulted in a dramatic decrease in the number of telomeres with an intensity less than 10,000 a.u. Intriguingly, this 3D telomere profile remained unaltered despite radiation-induced cell death. For Group 4, all patients presented a peak after ADT started, which decreased in later time points. Zhou et al. have already demonstrated that AR inactivation by androgen deprivation, in LNCaP cells (prostate cancer cell line) can induce telomere breaks and telomere fusion [20]. The peak that we observed in +2m can be a consequence of the ADT treatment, which induce genomic instability and consequently death. Similar to Groups 1 and 4, the treatment for Groups 3 and 5 also had the number of short telomeres decreased. However, at later time points, the population with short telomeres started to increase reflecting a positive selective pressure of ADT plus RT in favor of resistant prostate cancer clones.

The ability of cancer cells to survive specific treatments, such as ADT and RT, involves changes in 3D telomere architecture and reflects the effects of complex cellular processes in which genomic stability, instead of causing death, ensure tumor cell survival. The effect of ADT plus RT on specific 3D telomere profiles may reflect the evolution of heterogeneous prostate tumor sub clones, as showed at later time points for Groups 3 and 5. The cellular mechanisms responsible for the dynamic telomere alterations in these patients are currently unknown. It is important to recognize that the observed heterogeneity in telomere phenotype was limited to five unique 3D telomere signatures in 100 localized high-risk patient samples. The effect of these profiles on patient survival awaits future analysis. However, we used PSA after 6 months of ADT, after 6 months of finished RT, and 36-months after initial treatment as an early surrogate for tumor response. In all time points, increase of nuclear volume, total number of signals, and total number of aggregates were significantly different between the two patient population (< 0.1 ng/mL and ≥ 0.1 ng/mL PSA end). The cutoff value was chosen on the basis of previous reports in which increasing levels of PSA above 0.1 ng/mL after ADT and radiotherapy was associated with an increased risk of recurrence [6,22–25]. Additionally, the total intensity (associated with telomere length) decreased at 6 months of continued ADT and increased after 36 months after initial treatment. We attributed this to a process where the decrease of telomere length leads to a decrease of total intensity; however, as the formation of telomere aggregates (clusters of telomeres) continues, the resulting high intensity values of these clusters ensued an increase in the total intensity measurements in 36 months. It is important to highlight that we compared our data with the only approved biomarker guiding for treatment decisions in PCa [7]. However, PSA values often do not represent the current tumor status, potentially misleading therapeutic decisions [7].

Nevertheless, we found no association between the PSA end values and the 3D profile with CTCs dynamics over time ($p = 0.38$). Our data also predicted that only 32% (Groups 1 and 4) of the patients with non-metastatic high-risk prostate cancer could benefit from the combination of ADT/radiotherapy treatment. Additionally, 3D telomere analysis of circulating tumor cells offers a non-invasive method to follow up prostate patients during their treatment cycle.

The centroid cluster analysis identified three clusters, using all Teloview^{TM} parameters (number of telomeres, total intensity/length, telomere aggregates, and nuclear volume), which separated patients with different levels of genomic instability. Cluster 3 seems to correspond to somewhat more aggressive phenotypes than Cluster 2 and 1. We observed that in PSA 6 months after ADT, 33.33% of patients had PSA values above 0.1 ng/ml. This percentage decrease in the second time point (PSA 6 after RT), however, in the third time point (PSA after 36 months), cluster 3 is the only group that return to the same scenario found in the first time point (PSA 6 months after ADT), with 33.33% of the patients above 0.1 ng/ml (PSA). In the other clusters (1 and 2), the percentage of patients above 0.1 ng/ml after 36 months of treatment is lower than previous time points. In spite of a 3-years follow-up be too short to detect prostate cancer related mortality, our results demonstrate that CTCs retain important genetic information that could be used as a real time liquid biopsy to guide therapeutic decision and avoid overtreatment.

The current study has two important limitation. First, the 3-years follow-up was too short to detect prostate cancer related mortality, which affects the correlation of our data with a "real" clinical end-point. Second, our results were correlated with PSA end levels. Although PSA measurement after radiotherapy and androgen deprivation for localized prostate cancer has been proposed as an early prognostic biomarker [50,51]. PSA positive predictive value is only 25–40% [52]. We found that our telomere parameters are significantly different between PSA groups, which highlights the potential of our biomarker to be equal or superior to PSA. However, the effect of our findings on patient survival still awaits future analysis.

4. Materials and Methods

4.1. Patient Samples, Treatment and Study Design

The study cohort included one hundred (100) men who were treated between 2014 and 2016 with long-course ADT combined RT for clinically localized and non-metastatic high-risk prostate cancer. High-risk prostate cancer patients defined as having either cT3, Gleason score 8–10, or PSA > 20 ng/mL with no history of prior ADT, RT, or chemotherapy. Tumors were considered non-metastatic in patients with negative results in bone scan (Tc-99m-methylene diphosphonate (MDP)) and CT of the abdomen, and pelvis). This study was approved by the University ethics committee (University of Manitoba Ethics Protocol Reference No. H2011:336). All men were initially treated with 24 months of androgen deprivation therapy in the form of oral bicalutamide (given at a dose of 50 mg every day) (Figure 1) in combination with injections of goserelin or leuprolide.

Radiotherapy doses were 78 Gy after 2 months of initial therapy with ADT (for 3 months). This study was design for a clinical assessment and laboratory testing (CTCs collection plus PSA) in different time points: 0, 2, 6, 12, 18 and 24 months, being 0 m at diagnosis (Figure 1). All patients outside this time interval were excluded. We consider as an end-point PSA level < 0.1 ng/mL vs. ≥ 0.1 ng/mL after 6 months of ADT, after 6 months of RT and after 12 months of complete ADT (+36 months).

4.2. CTC Isolation and Androgen Receptor (AR) Staining

Patient blood was processed using the ScreenCellR filter method for the separation of prostate CTC [28]. Briefly, 3 mL of patient blood were incubated with 4 mL of ScreenCell buffer for 8 min. This buffer lyses the red blood cells, prefixes all nucleated cells present in the blood sample while preserving their architecture and enabling their fixation onto the filter membrane of the device.

Thereafter, this mix is filtrated passing through the microporous membrane filter (7.50 m pore size). This technique results in an average of 91.2% recovery CTC rate of 91.2% [28].

Filters containing captured cells were fixed for 10 min in 3.7% formaldehyde/1× PBS (Sigma, Oakville, ON, Canada) at room temperature. Filters were washed 3× 5 min each in 1× phosphate buffered saline (PBS)/50 mM $MgCl_2$ and blocked at 37 °C for 30 min in 4× SSC (0.6 M NaCl; 0.06 M sodium citrate)/4% bovine serum albumin (BSA, all Sigma-Aldrich, St. Louis, MO, USA). A FITC labelled mouse monoclonal antibody raised against amino acid residues 299-315 of the human androgen receptor (AR-441; Santa Cruz Biotechnology, Dallas, Texas, USA) was applied at 1:50 (20 ng/L) and allowed to incubate for 10 min at 37 °C in a humidified chamber. Excess antibody was removed in 3× washes with 1× PBS/50 mM $MgCl_2$ for 5 min each at RT. Filters were dehydrated in an ethanol series (70%, 90%, 100%), air dried and attached to microscopy slides with clear nail polish. Slides were then counterstained with DAPI (Sigma-Aldrich) and mounted with Vectashield (Vector Laboratories, Burlington, ON, Canada).

4.3. Telomere Three-dimensional Quantitative Fluorescent in situ Hybridization (3D-QFISH)

For 3D-QFISH [53], cells on the filters were incubated in 1× PBS for 5 min followed by a 10 min fixation in 3.7% formaldehyde/1× PBS and 3× washes in 1× PBS for 5 min each. Filters were treated with (50 g/mL pepsin (Sigma) in 0.01M HCl for 10 min at 37 °C, 1× washed in 1× PBS for 5 min followed by post-fixation for 10 min in 3.7% formaldehyde/1× PBS and 3× washes in 1× PBS for 5 min each. Filters were dehydrated in an ethanol series and air dried. Fluorochrome-coupled (Cy3) Telomere specific peptide nucleic acid (PNA) probe (DAKO, Agilent Technologies, Santa Clara, CA, USA) was applied (5 µL probe/slide) and, following denaturation at 80 °C for 3 min, hybridization was done for 2 h at 30 °C. Slides were washed in 70% deionized formamide (Sigma-Aldrich) in 10 mM Tris pH 7.4 for 15 min, rinsed in 1× PBS and once each in 2× SSC (5 min at 55 °C), 0.1× SSC and 2× SSC/0.05% Tween-20 at RT. Filters were again dehydrated and air dried. Filters were removed from the metal support ring using an 8 mm biopsy punch, placed on a new slide, DAPI stained, mounted with Vectashield (Vector Laboratories) with a coverslip.

4.4. Imaging & Analysis

Slides were imaged on a Zeiss AxioImager Z2 microscope with a Zeiss AxioCam MRmm Rev 3 digital camera using AxioVision Release 4.8.2 (Zeiss, Jena, Germany). A Cy3 filter was used to detect the Cy3 probe nuclear hybridization to telomeric repeats at an exposure time of 500 ms for all samples examined. A FITC filter was used to determine the presence of AR antibodies. Exposure times for the DAPI filter differed between slides. Eighty focal planes spaced 200 nm apart were imaged to create a three dimensional nuclear images of the circulating tumor cells and lymphocytes on the filter. Images were deconvolved using a constrained iterative algorithm [32]. For each patient sample, 30 CTC and 30 lymphocyte nuclei were analyzed using TeloView™ software [33] (used with permission of Telo Genomics Corp Inc. Toronto, ON, Canada). Each cell was analyzed for intensity of signal, presence of telomere aggregates (two or more signals that cannot be resolved due to proximity and defined as a signal with intensity above the standard deviation of signal intensity for that cell), number of signals per nucleus and nuclear volume. These measurements were determined for CTCs from each patient isolated at different time points during their treatment.

4.5. Statistical Analysis

For statistical analysis comparing PSA levels, we set a threshold of PSA end of 0.1 ng/L of PSA at 6 months after continues ADT, 6 months of after completed RT and 36 months after initial treatment. The CTCs were analyzed in five time points (+0, +2, +6, +12 and +24 m). For each time point, thirty cells were ranked and quartiles calculated for each telomere parameter measured by TeloView™ and compared between patients falling on both sides of the threshold. The telomeric parameters (number, length, telomere aggregates, nuclear volume, a/c ratio, etc.) were compared using a nested factorial analysis of variance followed by a least-square means multiple comparison. Graphical presentations indicated the p-value for the overall test of differences across all time points. Chi-square analysis compared the percentage of interphase telomeric signals intensities at defined quartile cut-offs. A significance level was set at 0.05. The Box-plots generated by Statistical Analysis Software v. 9.4 (SAS, Cary, NC USA.

5. Conclusions

Our study identified CTCs with three unique 3D telomere profiles among patients with localized high-risk prostate cancer. The distinct CTC telomere dynamics in each patient group in response to treatment provides a strong rationale for the use of 3D telomere analysis on CTCs as a way to monitor treatment response.

Supplementary Materials: Table S1: List of all patients with corresponding Gleason score and TMN staging at baseline, PSA levels at diagnosis, corresponding assigned telomere profile grouping and cluster for each patient. Supplementary Figure S1. Representative bar plots to illustrate inter-sample variability of representative individual samples in the Groups 1, Group 2, Group 3, Group 4, and Group 5. Supplementary Figure S2. Representative examples of the lymphocytes (internal control) dynamics of telomere length profiles over time for patients assigned to Group 1, Group 2, Group 3, Group 4 and Group 5. Supplementary Figure S3. Example of a circulating tumor cell from a high-risk localized prostate cancer patient captured on top of a filter pore for each time point. Three-dimensional representation of a CTC with the telomeres labeled with telomere-specific Cy3-labeled probe (red) and Merge between telomeres; and the counterstained DAPI (blue).

Author Contributions: Experimental part, L.W.; Analysis, L.W., A.O., D.D., A.R.-P.; Writing—Original Draft Preparation, A.R.-P.; Writing—Review & Editing, L.W., H.Q., D.D., S.M.; Supervision, S.M.; Project Administration, S.M.; Ethics approval, S.M.; Funding Acquisition, S.M.

Acknowledgments: The authors would like to thank the prostate cancer patients who contributed to this study in Manitoba/Canada and the research nurse, Paula Sitarik, for blood collection. We thank 3D Telo Genomics Corp. for the use of TeloView, Mary Cheang for statistical analysis and Elizabete Cruz for helping in the manuscript preparation. This study was supported by the Manitoba Tumor Bank, Winnipeg, Manitoba, funded in part by the CancerCare Manitoba Foundation and the Canadian Institutes of Health Research and is a member of the Canadian Tissue Repository Network. The authors also thank the Genomic Centre for Cancer Research and Diagnosis (GCCRD) for imaging. The GCCRD is funded by the Canada Foundation for Innovation and supported by CancerCare Manitoba Foundation, the University of Manitoba and the Canada Research Chair Tier 1 (S.M.). The GCCRD is a member of the Canadian National Scientific Platforms (CNSP) and of Canada BioImaging.

References

1. Mohler, J.L.; Armstrong, A.J.; Bahnson, R.R.; D'Amico, A.V.; Davis, B.J.; Eastham, J.A.; Enke, C.A.; Farrington, T.A.; Higano, C.S.; Horwitz, E.M.; et al. Prostate cancer, version 1. 2016. *J. Natl. Compr. Cancer Netw.* **2016**, *14*, 19–30. [CrossRef]
2. Sternberg, C.N.; Baskin-Bey, E.S.; Watson, M.; Worsfold, A.; Rider, A.; Tombal, B. Treatment patterns and characteristics of European patients with castration-resistant prostate cancer. *BMC Urol.* **2013**, *13*, 58. [CrossRef] [PubMed]
3. Merseburger, A.S.; Alcaraz, A.; von Klot, C.A. Androgen deprivation therapy as backbone therapy in the management of prostate cancer. *Onco Targets Ther.* **2016**, *9*, 7263–7274. [CrossRef] [PubMed]
4. Nguyen, P.L.; Martin, N.E.; Choeurng, V.; Palmer-Aronsten, B.; Kolisnik, T.; Beard, C.J.; Orio, P.F.; Nezolosky, M.D.; Chen, Y.W.; Shin, H.; et al. Utilization of biopsy-based genomic classifier to predict distant metastasis after definitive radiation and short-course ADT for intermediate and high-risk prostate cancer. *Prostate Cancer Prostatic Dis.* **2017**, *20*, 186–192. [CrossRef] [PubMed]
5. Roach, M.; Hanks, G.; Thames, H.; Schellhammer, P.; Shipley, W.U.; Sokol, G.H.; Sandler, H. Defining biochemical failure following radiotherapy with or without hormonal therapy in men with clinically localized prostate cancer: Recommendations of the RTOG-ASTRO Phoenix Consensus Conference. *Int. J. Radiat. Oncol.* **2006**, *65*, 965–974. [CrossRef] [PubMed]

6. D'Amico, A.V.; Chen, M.-H.; De Castro, M.; Loffredo, M.; Lamb, D.S.; Steigler, A.; Kantoff, P.W.; Denham, J.W. Surrogate endpoints for prostate cancer-specific mortality after radiotherapy and androgen suppression therapy in men with localised or locally advanced prostate cancer: An analysis of two randomised trials. *Lancet Oncol.* **2012**, *13*, 189–195. [CrossRef]
7. De Bono, J.S.; Scher, H.I.; Montgomery, R.B.; Parker, C.; Miller, M.C.; Tissing, H.; Doyle, G.V.; Terstappen, L.W.; Pienta, K.J.; Raghavan, D. Circulating Tumor Cells Predict Survival Benefit from Treatment in Metastatic Castration-Resistant Prostate Cancer. *Clin. Cancer Res.* **2008**, *14*, 6302–6309. [CrossRef]
8. Scher, H.I.; Jia, X.; de Bono, J.S.; Fleisher, M.; Pienta, K.J.; Raghavan, D.; Heller, G. Circulating tumour cells as prognostic markers in progressive, castration-resistant prostate cancer: A reanalysis of IMMC38 trial data. *Lancet Oncol.* **2009**, *10*, 233–239. [CrossRef]
9. Olmos, D.; Arkenau, H.T.; Ang, J.E.; Ledaki, I.; Attard, G.; Carden, C.P.; Reid, A.H.; A'Hern, R.; Fong, P.C.; Oomen, N.B.; et al. Circulating tumour cell (CTC) counts as intermediate end points in castration-resistant prostate cancer (CRPC): A single-centre experience. *Ann. Oncol.* **2009**, *20*, 27–33. [CrossRef]
10. Danila, D.C.; Heller, G.; Gignac, G.A.; Gonzalez-Espinoza, R.; Anand, A.; Tanaka, E.; Lilja, H.; Schwartz, L.; Larson, S.; Fleisher, M.; et al. Circulating Tumor Cell Number and Prognosis in Progressive Castration-Resistant Prostate Cancer. *Clin. Cancer Res.* **2007**, *13*, 7053–7058. [CrossRef]
11. Nagrath, S.; Sequist, L.V.; Maheswaran, S.; Bell, D.W.; Irimia, D.; Ulkus, L.; Smith, M.R.; Kwak, E.L.; Digumarthy, S.; Muzikansky, A.; et al. Isolation of rare circulating tumour cells in cancer patients by microchip technology. *Nature* **2007**, *450*, 1235–1239. [CrossRef] [PubMed]
12. Lee, R.J.; Saylor, P.J.; Michaelson, M.D.; Rothenberg, S.M.; Smas, M.E.; Miyamoto, D.T.; Gurski, C.A.; Xie, W.; Maheswaran, S.; Haber, D.A.; et al. A dose-ranging study of cabozantinib in men with castration-resistant prostate cancer and bone metastases. *Clin. Cancer Res.* **2013**, *19*, 3088–3094. [CrossRef] [PubMed]
13. Chen, J.-F.; Ho, H.; Lichterman, J.; Lu, Y.-T.; Zhang, Y.; Garcia, M.A.; Chen, S.-F.; Liang, A.-J.; Hodara, E.; Zhau, H.E.; et al. Sub-classification of prostate cancer circulating tumor cells (CTCs) by nuclear size reveals very-small nuclear CTCs in patients with visceral metastases. *Cancer* **2015**, *121*, 3240–3251. [CrossRef] [PubMed]
14. Dorff, T.B.; Groshen, S.; Tsao-Wei, D.D.; Xiong, S.; Gross, M.E.; Vogelzang, N.; Quinn, D.I.; Pinski, J.K. A Phase II Trial of a Combination Herbal Supplement for Men with Biochemically Recurrent Prostate Cancer. *Prostate Cancer Prostatic Dis.* **2014**, *17*, 359–365. [CrossRef] [PubMed]
15. Budna-Tukan, J.; Świerczewska, M.; Mazel, M.; Cieślikowski, W.A.; Ida, A.; Jankowiak, A.; Antczak, A.; Nowicki, M.; Pantel, K.; Azria, D.; et al. Analysis of Circulating Tumor Cells in Patients with Non-Metastatic High-Risk Prostate Cancer before and after Radiotherapy Using Three Different Enumeration Assays. *Cancers (Basel)* **2019**, *11*, 802. [CrossRef] [PubMed]
16. Hong, B.; Zu, Y. Detecting circulating tumor cells: Current challenges and new trends. *Theranostics* **2013**, *3*, 377–394. [CrossRef] [PubMed]
17. Liu, W.; Yin, B.; Wang, X.; Yu, P.; Duan, X.; Liu, C.; Wang, B.; Tao, Z. Circulating tumor cells in prostate cancer: Precision diagnosis and therapy. *Oncol. Lett.* **2017**, *14*, 1223–1232. [CrossRef] [PubMed]
18. Awe, J.A.; Saranchuk, J.; Drachenberg, D.; Mai, S. Filtration-based enrichment of circulating tumor cells from all prostate cancer risk groups. *Urol. Oncol.* **2017**, *35*, 300–309. [CrossRef]
19. Wark, L.; Klonisch, T.; Awe, J.; LeClerc, C.; Dyck, B.; Quon, H.; Mai, S. Dynamics of three-dimensional telomere profiles of circulating tumor cells in patients with high-risk prostate cancer who are undergoing androgen deprivation and radiation therapies. *Urol. Oncol.* **2017**, *35*, 112.e1–112.e11. [CrossRef]
20. Graham, M.K.; Meeker, A. Telomeres and telomerase in prostate cancer development and therapy. *Nat. Rev. Urol.* **2017**, *14*, 607–619. [CrossRef]
21. Heaphy, C.M.; Gaonkar, G.; Peskoe, S.B.; Joshu, C.E.; De Marzo, A.M.; Lucia, M.S.; Goodman, P.J.; Lippman, S.M.; Thompson, I.M.; Platz, E.A.; et al. Prostate stromal cell telomere shortening is associated with risk of prostate cancer in the placebo arm of the Prostate Cancer Prevention Trial. *Prostate* **2015**, *75*, 1160–1166. [CrossRef] [PubMed]
22. Zhou, J.; Richardson, M.; Reddy, V.; Menon, M.; Barrack, E.R.; Reddy, G.P.; Kim, S.H. Structural and functional association of androgen receptor with telomeres in prostate cancer cells. *Aging (Albany NY)* **2013**, *5*, 3–17. [CrossRef] [PubMed]

23. Kim, S.H.; Richardson, M.; Chinnakannu, K.; Bai, V.U.; Menon, M.; Barrack, E.R.; Reddy, G.P. Androgen receptor interacts with telomeric proteins in prostate cancer cells. *J. Biol. Chem.* **2010**, *285*, 10472–10476. [CrossRef] [PubMed]
24. Cheung, A.S.; Yeap, B.B.; Hoermann, R.; Hui, J.; Beilby, J.P.; Grossmann, M. Effects of androgen deprivation therapy on telomere length. *Clin. Endocrinol.* **2017**, *87*, 381–385. [CrossRef] [PubMed]
25. Zelefsky, M.J.; Lyass, O.; Fuks, Z.; Wolfe, T.; Burman, C.; Ling, C.C.; Leibel, S.A. Predictors of improved outcome for patients with localized prostate cancer treated with neoadjuvant androgen ablation therapy and three-dimensional conformal radiotherapy. *J. Clin. Oncol.* **1998**, *16*, 3380–3385. [CrossRef] [PubMed]
26. Lamb, D.S.; Denham, J.W.; Joseph, D.; Matthews, J.; Atkinson, C.; Spry, N.A.; Duchesne, G.; Ebert, M.; Steigler, A.; Delahunt, B.; et al. A Comparison of the Prognostic Value of Early PSA Test-Based Variables Following External Beam Radiotherapy, With or Without Preceding Androgen Deprivation: Analysis of Data From the TROG 96.01 Randomized Trial. *Int. J. Radiat. Oncol.* **2011**, *79*, 385–391. [CrossRef] [PubMed]
27. Benchikh, E.; Fegoun, A.; Villers, A.; Moreau, J.L.; Richaud, P.; Rebillard, X.; Beuzeboc, P. PSA and follow-up after treatment of prostate cancer. *Prog. Urol.* **2008**, *18*, 137–144. [CrossRef] [PubMed]
28. DeSitter, I.; Guerrouahen, B.S.; Benali-Furet, N.; Wechsler, J.; Jänne, P.A.; Kuang, Y.; Yanagita, M.; Wang, L.; Berkowitz, J.A.; Distel, R.J.; et al. A new device for rapid isolation by size and characterization of rare circulating tumor cells. *Anticancer Res.* **2011**, *31*, 427–441.
29. Miyamoto, D.T.; Lee, R.J.; Stott, S.L.; Ting, D.T.; Wittner, B.S.; Ulman, M.; Smas, M.E.; Lord, J.B.; Brannigan, B.W.; Trautwein, J.; et al. Androgen receptor signaling in circulating tumor cells as a marker of hormon- ally responsive prostate cancer. *Cancer Discov.* **2012**, *2*, 995–1003. [CrossRef]
30. Meeker, A.K.; Hicks, J.L.; Iacobuzio-Donahue, C.A.; Montgomery, E.A.; Westra, W.H.; Chan, T.Y.; Ronnett, B.M.; De Marzo, A.M. Telomere length abnormalities occur early in the initiation of epithelial carcinogenesis. *Clin. Cancer Res.* **2004**, *10*, 3317–3326. [CrossRef]
31. Schaefer, L.H.; Schuster, D.; Herz, H. Generalized approach for accelerated maximum likelihood based image restoration applied to three-dimensional fluorescence microscopy. *J. Microsc.* **2001**, *204*, 99–107. [CrossRef] [PubMed]
32. Hugosson, J.; Carlsson, S.; Aus, G.; Bergdahl, S.; Khatami, A.; Lodding, P.; Pihl, C.G.; Stranne, J.; Holmberg, E.; Liljia, H. Mortality results from the Göteborg randomised population-based prostate-cancer screening trial. *Lancet Oncol.* **2010**, *11*, 725–732. [CrossRef]
33. Schröder, F.H.; Hugosson, J.; Roobol, M.J.; Tammela, T.L.; Ciatto, S.; Nelen, V.; Kwiatkowski, M.; Lujan, M.; Lilja, H.; Zappa, M.; et al. Screening and Prostate-Cancer Mortality in a Randomized European Study. *N. Engl. J. Med.* **2009**, *360*, 1320–1328. [CrossRef] [PubMed]
34. Andriole, G.L.; Crawford, E.D.; Grubb, R.L.; Buys, S.S.; Chia, D.; Church, T.R.; Fouad, M.N.; Gelmann, E.P.; Kvale, P.A.; Reding, D.J.; et al. Mortality Results from a Randomized Prostate-Cancer Screening Trial. *N. Engl. J. Med.* **2009**, *360*, 1310–1319. [CrossRef] [PubMed]
35. Stamey, T.A.; McNeal, J.E.; Yemoto, C.M.; Sigal, B.M.; Johnstone, I.M. Biological Determinants of Cancer Progression in Men With Prostate Cancer. *JAMA* **1999**, *281*, 1395–1400. [CrossRef] [PubMed]
36. Vermolen, B.J.; Garini, Y.; Mai, S.; Mougey, V.; Fest, T.; Chuang, A.Y.-C.; Wark, L.; Young, I.T.; Chuang, T.C.-Y.; Chuang, T.C. Characterizing the three-dimensional organization of telomeres. *Cytom. Part A* **2005**, *67*, 144–150. [CrossRef] [PubMed]
37. Kolostova, K.; Zhang, Y.; Hoffman, R.M.; Bobek, V. In Vitro Culture and Characterization of Human Lung Cancer Circulating Tumor Cells Isolated by Size Exclusion from an Orthotopic Nude-Mouse Model Expressing Fluorescent Protein. *J. Fluoresc.* **2014**, *24*, 1531–1536. [CrossRef]
38. Chen, C.L.; Mahalingam, D.; Osmulski, P.; Jadhav, R.R.; Wang, C.M.; Leach, R.J.; Chang, T.C.; Weitman, S.D.; Kumar, A.P.; Sun, L.; et al. Single cell analysis of circulating tumor cells identifies cumulative expression patterns of EMT-related genes in metastatic prostate cancer. *Prostate* **2013**, *73*, 813–826. [CrossRef]
39. Raimondi, C.; Nicolazzo, C.; Gradilone, A. Circulating tumor cells isolation: The "post-EpCAM era". *Chin. J. Cancer Res.* **2015**, *27*, 461–470. [CrossRef]
40. Pantel, K.; Alix-Panabières, C. Functional studies on viable circulating tumor cells. *Clin. Chem.* **2016**, *62*, 328–334. [CrossRef]
41. Paris, P.L.; Kobayashi, Y.; Zhao, Q.; Zeng, W.; Sridharan, S.; Fan, T.; Adler, H.L.; Yera, E.R.; Zarrabi, M.; Zucker, S.; et al. Functional phenotyping and genotyping of circulating tumor cells from patients with castration resistant prostate cancer. *Cancer Lett.* **2009**, *277*, 164–173. [CrossRef] [PubMed]

42. Li, J.; Gregory, S.G.; Garcia-Blanco, M.A.; Armstrong, A.J. Using circulating tumor cells to inform on prostate cancer biology and clinical utility. *Crit. Rev. Clin. Lab. Sci.* **2015**, *52*, 191–210. [CrossRef] [PubMed]
43. Helzer, K.T.; Barnes, H.E.; Day, L.; Harvey, J.; Billings, P.R.; Forsyth, A. Circulating Tumor Cells Are Transcriptionally Similar to the Primary Tumor in a Murine Prostate Model. *Cancer Res.* **2009**, *69*, 7860–7866. [CrossRef] [PubMed]
44. Thalgott, M.; Rack, B.; Maurer, T.; Souvatzoglou, M.; Eiber, M.; Kreß, V.; Heck, M.M.; Andergassen, U.; Nawroth, R.; Gschwend, J.E.; et al. Detection of circulating tumor cells in different stages of prostate cancer. *J. Cancer Res. Clin. Oncol.* **2013**, *139*, 755–763. [CrossRef] [PubMed]
45. Davis, J.W.; Nakanishi, H.; Kumar, V.S.; Bhadkamkar, V.A.; McCormack, R.; Fritsche, H.A.; Handy, B.; Gornet, T.; Babaian, R.J. Circulating Tumor Cells in Peripheral Blood Samples From Patients With Increased Serum Prostate Specific Antigen: Initial Results in Early Prostate Cancer. *J. Urol.* **2008**, *179*, 2187–2191. [CrossRef] [PubMed]
46. Okegawa, T.; Ninomiya, N.; Masuda, K.; Nakamura, Y.; Tambo, M.; Nutahara, K. AR-V7 in circulating tumor cells cluster as a predictive biomarker of abiraterone acetate and enzalutamide treatment in castration-resistant prostate cancer patients. *Prostate* **2018**, *78*, 576–582. [CrossRef] [PubMed]
47. Gadji, M.; Adebayo, J.; Rodrigues, P.; Kumar, R.; Houston, D.S.; Klewes, L.; Dièye, T.N.; Rego, E.M.; Passetto, R.F.; de Oliveira, F.M.; et al. Profiling threedimensional nuclear telomeric architecture of myelodysplastic syndromes and acute myeloid leukemia defines patient subgroups. *Clin. Cancer Res.* **2012**, *18*, 3293–3304. [CrossRef] [PubMed]
48. Rangel-Pozzo, A.; De Souza, D.C.; Schmid-Braz, A.T.; De Azambuja, A.P.; Ferraz-Aguiar, T.; Borgonovo, T.; Mai, S. 3D Telomere Structure Analysis to Detect Genomic Instability and Cytogenetic Evolution in Myelodysplastic Syndromes. *Cells* **2019**, *8*, 304. [CrossRef]
49. Caria, P.; Dettori, T.; Frau, D.V.; Lichtenzstejn, D.; Pani, F.; Vanni, R.; Mai, S. Characterizing the three-dimensional organization of telomeres in papillary thyroid carcinoma cells. *J. Cell Physiol.* **2019**, *234*, 5175–5185. [CrossRef]
50. Williams, S. Surrogate endpoints in early prostate cancer research. *Transl. Androl. Urol.* **2018**, *7*, 472–482. [CrossRef]
51. Bryant, A.K.; D'Amico, A.V.; Nguyen, P.L.; Einck, J.P.; Kane, C.J.; McKay, R.R.; Simpson, D.R.; Mundt, A.J.; Murphy, J.D.; Rose, B.S. Three-month posttreatment prostate-specific antigen level as a biomarker of treatment response in patients with intermediate-risk or high-risk prostate cancer treated with androgen deprivation therapy and radiotherapy. *Cancer* **2018**, *124*, 2939–2947. [CrossRef] [PubMed]
52. Schröder, F.H.; Carter, H.B.; Wolters, T.; Bergh, R.C.V.D.; Gosselaar, C.; Bangma, C.H.; Roobol, M.J. Early Detection of Prostate Cancer in 2007. *Eur. Urol.* **2008**, *53*, 468–477. [CrossRef] [PubMed]
53. Hultdin, M.; Grönlund, E.; Norrback, K.; Eriksson-Lindström, E.; Just, T.; Roos, G. Telomere analysis by fluorescence in situ hybridization and flow cytometry. *Nucleic Acids Res.* **1998**, *26*, 3651–3656. [CrossRef] [PubMed]

6

Circulating Tumor Cells and Circulating Tumor DNA Detection in Potentially Resectable Metastatic Colorectal Cancer

François-Clément Bidard [1,2,3], Nicolas Kiavue [1,*], Marc Ychou [4,5], Luc Cabel [1,2,3], Marc-Henri Stern [6], Jordan Madic [2], Adrien Saliou [2], Aurore Rampanou [2], Charles Decraene [2,7], Olivier Bouché [8], Michel Rivoire [9], François Ghiringhelli [10], Eric Francois [11], Rosine Guimbaud [12], Laurent Mineur [13], Faiza Khemissa-Akouz [14], Thibault Mazard [4], Driffa Moussata [15], Charlotte Proudhon [2], Jean-Yves Pierga [1,2,16], Trevor Stanbury [17], Simon Thézenas [18] and Pascale Mariani [19]

1 Department of Medical Oncology, Institut Curie, PSL Research University, 75005 Paris, France; francois-clement.bidard@curie.fr (F.-C.B.); luc.cabel@curie.fr (L.C.); jean-yves.pierga@curie.fr (J.-Y.P.)
2 Circulating Tumor Biomarkers Laboratory, Institut Curie, PSL Research University, 75005 Paris, France; jordanmadic@yahoo.fr (J.M.); saliou.adrien@gmail.com (A.S.); aurore.rampanou@curie.fr (A.R.); charles.decraene@curie.fr (C.D.); charlotte.proudhon@curie.fr (C.P.)
3 UVSQ, Paris Saclay University, 92210 Saint Cloud, France
4 Department of Digestive Oncology, ICM Regional Cancer Institute of Montpellier, 34298 Montpellier, France; marc.ychou@icm.unicancer.fr (M.Y.); t-mazard@chu-montpellier.fr (T.M.)
5 Department of Oncology, Montpellier University, 34000 Montpellier, France
6 INSERM U830, Institut Curie, PSL Research University, 75005 Paris, France; marc-henri.stern@curie.fr
7 CNRS UMR144, Institut Curie, PSL Research University, 75005 Paris, France
8 Department of Medical Oncology, Hôpital Robert Debré, Reims University Hospital, 51100 Reims, France; obouche@chu-reims.fr
9 Department of Digestive Oncology, Centre Léon Bérard, 69008 Lyon, France; michel.rivoire@lyon.unicancer.fr
10 INSERM U866, Centre Georges-François Leclerc, 21000 Dijon, France; fghiringhelli@cgfl.fr
11 Department of Medical Oncology, Centre Antoine Lacassagne, 06189 Nice, France; eric.francois@nice.unicancer.fr
12 Department of Digestive Oncology, CHU de Toulouse, 31059 Toulouse, France; guimbaud.r@chu-toulouse.fr
13 Department of Digestive Oncology, Institut Sainte Catherine, 84000 Avignon, France; l.mineur@isc84.org
14 Department of Gastroenterology, Hôpital Saint Jean, 66000 Perpignan, France; faiza.khemissa@ch-perpignan.fr
15 Department of Gastroenterology, CHRU de Tours, 37044 Tours, France; d.moussata@chu-tours.fr
16 Université Paris Descartes, 75270 Paris, France
17 UCGI Group, R&D UNICANCER, 75654 Paris, France; t-stanbury@unicancer.fr
18 Biometrics Unit, ICM Regional Cancer Institute of Montpellier, 34298 Montpellier, France; simon.thezenas@icm.unicancer.fr
19 Department of Surgical Oncology, Institut Curie, PSL Research University, 75005 Paris, France; pascale.mariani@curie.fr
* Correspondence: nicolas.kiavue@curie.fr

Abstract: The management of patients with colorectal cancer (CRC) and potentially resectable liver metastases (LM) requires quick assessment of mutational status and of response to pre-operative systemic therapy. In a prospective phase II trial (NCT01442935), we investigated the clinical validity of circulating tumor cell (CTC) and circulating tumor DNA (ctDNA) detection. CRC patients with potentially resectable LM were treated with first-line triplet or doublet chemotherapy combined with targeted therapy. CTC (Cellsearch®) and Kirsten RAt Sarcoma (KRAS) ctDNA (droplet digital polymerase chain reaction (PCR)) levels were assessed at inclusion, after 4 weeks of therapy and before LM surgery. 153 patients were enrolled. The proportion of patients with high CTC counts (≥3 CTC/7.5mL) decreased during therapy: 19% (25/132) at baseline, 3% (3/108) at week 4 and 0/57 before surgery. ctDNA detection sensitivity at baseline was 91% (N=42/46) and also decreased during treatment. Interestingly, persistently detectable KRAS ctDNA ($p = 0.01$) at 4 weeks was associated with a lower R0/R1 LM resection rate. Among patients who had a R0/R1 LM resection, those with detectable ctDNA levels before liver surgery had a shorter overall survival ($p < 0.001$). In CRC patients with limited metastatic spread, ctDNA could be used as liquid biopsy tool. Therefore, ctDNA detection could help to select patients eligible for LM resection.

Keywords: circulating tumor cells; circulating tumor DNA; liquid biopsy; metastatic colorectal cancer; FOLFIRINOX

---

1. Introduction

While most patients diagnosed with metastatic colorectal cancer (CRC) have unresectable metastases [1], some can benefit from liver surgery after conversion of unresectable disease to resectable disease by chemotherapy and targeted therapy [1,2]. In this regard, triplet chemotherapy (FOLFOXIRI) may improve the metastasis resection rate and overall survival (OS) [3]. The PRODIGE-14 trial (NCT01442935) was a randomized phase II trial intended to compare prospectively the efficacy of first-line triplet (FOLFIRINOX) versus doublet chemotherapy (FOLFOX: fluorouracil, leucovorin and oxaliplatin or FOLFIRI: fluorouracil, leucovorin and irinotecan), combined with a targeted therapy (bevacizumab in RAt Sarcoma (RAS)-mutated tumors, cetuximab in RAS wild-type tumors), in CRC patients diagnosed with potentially resectable liver metastases (LM). Results of this study have been reported elsewhere [4].

In metastatic CRC, Circulating Tumor Cell (CTC) count by the CellSearch® system is known to be an independent prognostic factor in large studies, using a threshold of ≥3 CTC/7.5 mL of blood [5–7]; these findings were confirmed in a meta-analysis including heterogeneous detection techniques [8] and similar results were reported in other gastro-intestinal cancers [9,10]. Moreover, dynamic changes of CTC levels have been shown to be associated with progression-free survival (PFS) and OS: metastatic CRC patients with persistently elevated CTC levels after one month of chemotherapy had shorter PFS and OS than patients with decreasing CTC counts (PFS: 1.6 vs 6.2 months, $p = 0.02$; OS: 3.7 vs 11.0, $p = 0.0002$) [5].

Similarly, circulating tumor DNA (ctDNA) has proven to be useful for theranostic detection of tumor mutations [11]. While ctDNA analysis has been approved for epidermal growth factor receptor (EGFR) mutation detection in metastatic non-small cell lung cancer [12], it has been suggested as a tool for liquid biopsy in CRC. Preliminary studies addressed the overall concordance between archived tumor tissue and liquid biopsy at any stage of the metastatic disease [13]; more recent results strongly suggested that KRAS mutant subclones may be selected during anti-EGFR therapy, decreasing the overall concordance between nominal (archived tumor tissue-based) and the actual (liquid biopsy-based) KRAS status [14,15]. Furthermore, in RAS-mutated tumors, the RAS mutation may not be detectable in the plasma at first, but may later become detectable under anti-EGFR treatment [16]. In addition to these liquid biopsy applications, ctDNA levels could possibly monitor tumor dynamics [17] with early changes in ctDNA during chemotherapy in CRC associated with tumor response [18]. Detection of a residual disease by ctDNA after surgery in stage II CRC was also associated with early recurrences and poor outcome [19].

While the above-mentioned results were mostly obtained in metastatic CRC patients with heterogeneous clinical settings, we investigated the clinical validity of CTC and ctDNA detection specifically in CRC patients diagnosed with potentially resectable LM and included in the PRODIGE-14 trial. We observed a decrease of CTC and ctDNA detection rates during systemic therapy. We confirmed the prognostic value of CTC detection at baseline and during treatment, and showed that ctDNA detection was associated with a lower R0/R1 LM resection rate.

2. Materials and Methods

2.1. Patients and Treatment

The main trial, identified as NCT01442935, and its ancillary study on circulating tumor biomarkers were approved by a French ethics committee (Comité de Protection des Personnes). All subjects gave their informed consent for inclusion before they participated in the study. The study was conducted in accordance with the Declaration of Helsinki. The patients could accept to participate in the main trial but refuse the ancillary study.

The main inclusion criterion was histologically proven CRC with LM ineligible for curative resection at inclusion and without metastatic spread to other sites (except for up to 3 resectable pulmonary metastases). Other inclusion criteria were: having provided informed consent, good performance status (0–1), known exon 2 KRAS mutational status (as determined locally by standard routine technique on tumor tissue; the clinical trial was later amended to account for other KRAS and Neuroblastoma RAt Sarcoma (NRAS) mutations), adequate hematological, kidney, and liver functions, and no prior therapy for LM. Patients were randomized to either triplet (FOLFIRINOX) [20] or standard doublet (FOLFOX or FOLFIRI) chemotherapy regimens. Chemotherapy was administered in combination with cetuximab in patients with RAS wild-type cancers or with bevacizumab in patients with RAS-mutated cancers.

The main trial objective was to demonstrate the superiority of triplet chemotherapy over doublet chemotherapy in terms of complete (R0/R1) surgical resection of LM and has already been reported [4]. A R0 resection was defined as a microscopically margin-negative resection with a distance between the margins and the tumor ≥1 mm. A R1 resection indicates a macroscopically margin-negative resection but with a distance between the margins and the tumor < 1mm. The R0/R1 resection rate was defined as the number of patients who underwent R0/R1 resection divided by the total number of patients included (R0/R1 resection, R2 resection, or no LM surgery).

2.2. Circulating Tumor Biomarker Detection

Three blood draws were required for this ancillary study: before starting treatment, after 1 month of systemic therapy (all patients), and before any surgical resection of LM (in patients referred to surgery after the shrinkage of LM). Blood samples were sent, within 24h, to a central laboratory (Institut Curie, Paris, France).

CTC counts were performed by experienced readers in 7.5 mL of blood (collected in CellSave® tubes) using the CellSearch® system (Menarini Silicon Biosystems), which has previously been reported [21]. The use of different CTC positivity thresholds was planned in order to compare the classical threshold of ≥3 CTC [5] to other thresholds and find the optimal cutoff.

For ctDNA analysis, 4 mL of plasma was thawed and cell-free DNA (cfDNA) extracted using the QIAamp® Circulating Nucleic Acid Kit (Qiagen®), after two centrifugations as per routine procedures [22,23]. According to the manufacturer's protocol, digital droplet PCR (ddPCR) reactions were prepared using commercially available primers and TaqMan® probes (Bio-Rad®) with 10 ng of cfDNA. ddPCR mastermix solutions (20 µL) were transferred to a DG8 droplet generator cassette (Bio-Rad®) with 70 µL of oil. Emulsified PCR reactions were then transferred to a 96-well PCR plate and run on a C1000 thermal cycler (Bio-Rad®). Plates were analyzed on a QX-100 droplet reader (Bio-Rad®) with the QuantaSoft v1.7.4 software. Positivity threshold was defined as per manufacturer's instructions, ensuring 0.1% sensitivity. Samples with a variant allele frequency <0.1% were classified as ctDNA-negative. Negative controls were used to minimize the risk of false positive. The assay could detect the G12S, G12R, G12C, G12D, G12A, G12V, and G13D mutations.

In the PRODIGE-14 trial, patients were allocated targeted therapies (cetuximab or bevacizumab) based on the KRAS exon 2 mutational analysis by local assessment on tumor tissue. While extended KRAS exon 3 and 4 and NRAS screening became mandatory in the course of the PRODIGE 14 trial, we confined our ctDNA analysis to KRAS exon 2 mutations. After one month of systemic therapy and before any surgical resection of LM, the ctDNA detection assay was only performed in patients with a known exon 2 KRAS mutation in a tumor tissue sample.

2.3. Statistical Analyses

The main objective of this study was to evaluate CTC and ctDNA detection rates at each time point in mCRC patients. The proportion of patients with detectable ctDNA (using the KRAS exon 2 mutation assay in cfDNA) and with detectable CTC was assessed at baseline, after one month of therapy and before LM surgery, if any. Secondary objectives were to assess the associations of circulating tumor biomarkers and baseline patient characteristics with R0/R1 LM resection and OS. Prespecified analyses were planned accordingly. Analyses conducted with the ctDNA variable (binary: detected or not detected) were also conducted with ctDNA concentration (as a continuous variable: number of mutant KRAS (KRASmut) copies per milliliter), but only for patients with KRAS exon 2 mutated tumors, as determined by routine local assessment on tumor tissues. This hypothesis-generating study had no prespecified power because the detection of circulating tumor biomarkers was done whenever possible, in patients who agreed to participate in the ancillary study. Circulating tumor biomarker detections were blinded to patients and clinicians. Patient characteristics and outcomes were prospectively collected in case report forms for all patients included in the PRODIGE-14 study. OS was defined as time from inclusion to death from any cause. Differences between categorical variables were analyzed by a chi2 test or Fisher's exact test. Continuous variables were analyzed by a Kruskal–Wallis test. Survival curves were plotted according to the Kaplan–Meier method. Statistical significance between survival curves was assessed using the logrank test. Multivariate analysis was done by the Cox proportional hazards model with prognostic factors with a p-value of ≤ 0.10 in univariate analysis. Patients with one or more missing covariable were not included in the multivariate analysis. For all analyses, a p-value of ≤ 0.05 was considered to be statistically significant. This report was written in accordance with the REporting of tumor MARKer studies guidelines.

3. Results

3.1. Patient Characteristics

Between February 2011 and April 2015, 153 patients were enrolled. Patients characteristics are displayed in Table 1. At time of data analysis (01/2017), median follow-up was 37.2 months (IC95% (34–39); range 0–55.3 months); 96 patients were referred to surgery after undergoing a blood draw for circulating biomarker analysis, 91 patients (59%) had a R0/R1 LM resection after chemotherapy and targeted therapy, while 65 deaths (42%) had occurred.

3.2. CTC Detection: Correlation with R0/R1 Resection and Outcome

At baseline, blood samples from 132 patients were available for CTC detection (Figure 1, Supplementary Table S1). ≥1 CTC was detected in 7.5 mL of blood in 42% (N=56/132) of patients at baseline and associated with the percentage of liver infiltrated by metastases at baseline ($p = 0.003$). Using the validated ≥3 CTC/7.5 mL threshold, elevated CTC counts were observed in 19% (N = 25/132) of patients (Figure 2), and associated with the percentage of liver infiltrated by metastases at baseline ($p = 0.001$) and the synchronicity of LM (p = 0.04). CTC detection at baseline (≥1 or ≥3 CTC) was not associated with the trial's main objective, the R0/R1 resection of LM ($p = 0.37$ and $p = 0.18$). Associations of CTCs (≥3 CTC) or ctDNA detection with baseline clinicopathological characteristics of patients are displayed in Supplementary Tables S2 and S3, respectively.

Table 1. Patients characteristics. N = 153 patients included in the study.

Characteristics	Median Value or Number of Patients
Age, years	Median: 60 Range: 25–75
Performance Status	
0	95 (63%)
1	57 (37%)
Prior resection of the primary tumor	
No	103 (67%)
Yes	50 (33%)
Synchronous liver metastases	
No	19 (12%)
Yes	134 (88%)
% of liver infiltrated by metastases	
0–25%	41 (45%)
26–50%	28 (30%)
51–75%	15 (16%)
>75%	8 (9%)
CEA	
Normal	17 (11%)
>upper limit of normal	134 (89%)
CA19.9	
Normal	39 (37%)
>upper limit of normal	66 (63%)
KRAS exon 2 mutation in tumor sample	
No	94 (61%)
Yes	59 (39%)
Chemotherapy	
Doublet + targeted therapy	75 (49%)
Triplet + targeted therapy	78 (51%)
R0/R1 resection of liver metastases	
No	62 (41%)
Yes	91 (59%)

At 4 weeks, 108 patients were analyzed for CTC detection. ≥1 and ≥3 CTC were detected in 11% (N = 12/108) and 3% (N = 3/108) of patients, respectively. CTC counts decreased significantly during therapy ($p < 0.0001$), this decrease being similar in the treatment arms (doublet versus triplet, $p = 0.98$). CTC detection at 4 weeks (≥1 or ≥3 CTC) was not significantly associated with the eventual R0/R1 resection of LM, although none of the 3 patients with ≥3 CTC achieved a R0/R1 resection ($p = 0.06$).

Among patients referred to liver surgery, 57 patients were analyzed for CTC detection. In this selected population, ≥1 CTC was detected in 7% (N = 4/57) of patients and no patient had ≥3 CTC detected. CTC detection before surgery was not associated with R0/R1 resection ($p = 0.37$).

In regards to the prognostic impact of CTC, ≥3 CTC at baseline (HR = 2.2, CI95% [1.2;3.9], $p = 0.01$) and at 4 weeks (HR = 10.9, 95%CI [3.2;36.9]; $p < 0.001$) were correlated with shorter OS (Figure 3). ≥1 CTC was significantly associated with shorter OS at 4 weeks ($p = 0.04$), but not at baseline ($p = 0.38$) or before liver surgery ($p = 0.71$). In multivariate analysis, ≥3 CTC was found to be an independent prognostic factor for OS at both baseline and at 4 weeks (Supplementary Table S4).

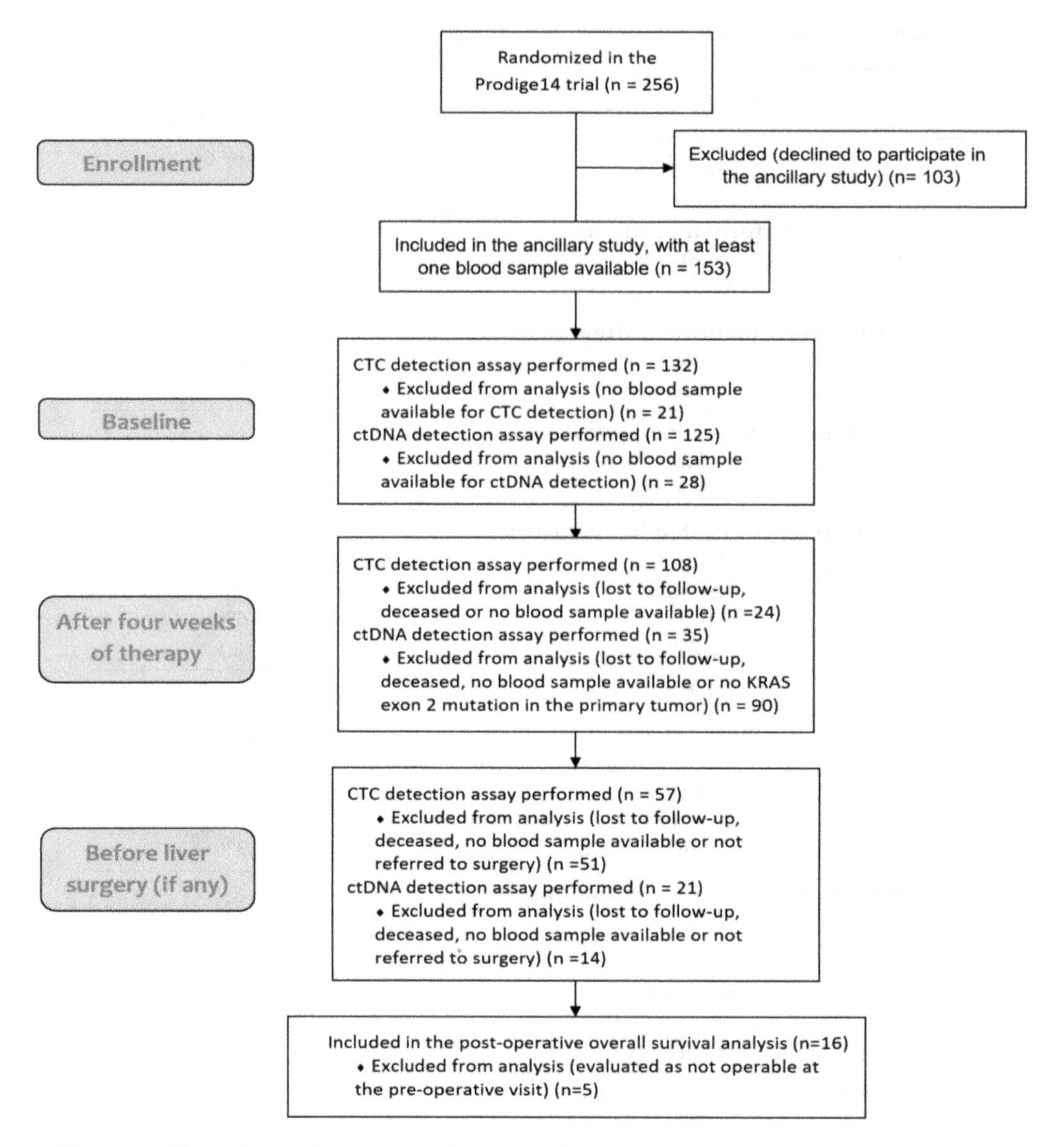

Figure 1. Flow chart of patients included in the analyses at the different time points.

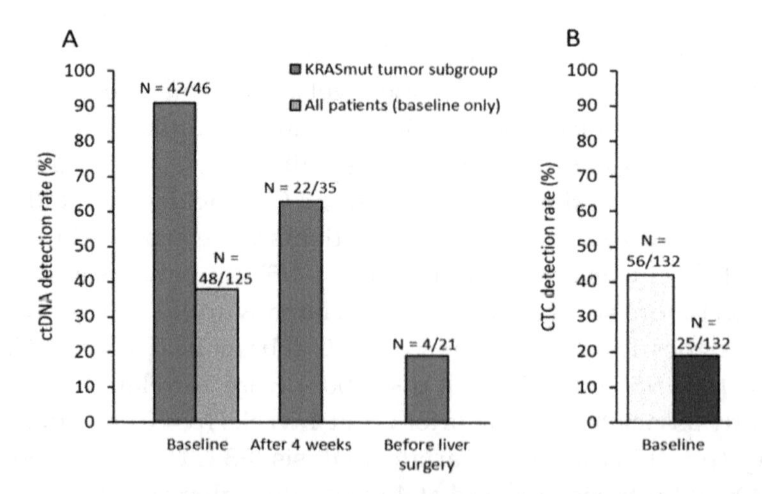

Figure 2. (**A**) ctDNA detection rate (KRAS exon 2 mutation with a variant allele frequency ≥ 0.1%) in all patients at baseline, and in the subgroup of patients with a KRAS exon 2 mutation as determined by routine local assessment on tumor tissues, at baseline, after 4 weeks and before liver surgery (if any). (**B**) CTC detection rate at each timepoint, with the ≥1CTC or the ≥3CTC/7.5mL of blood.

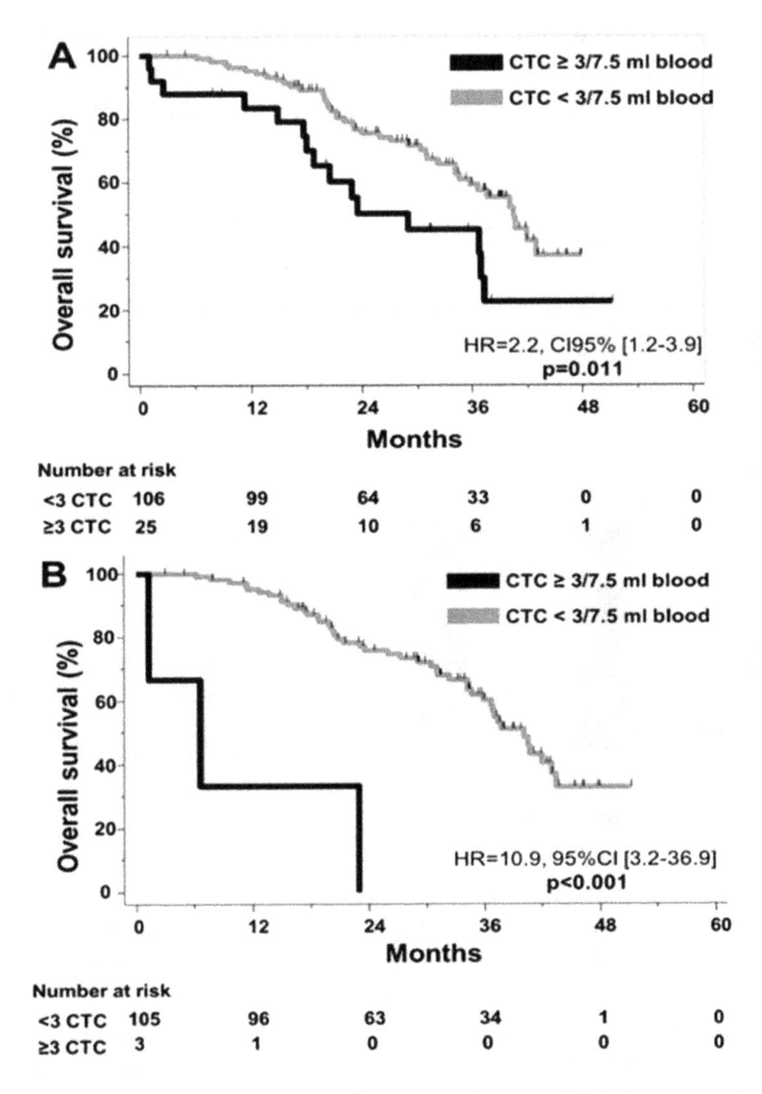

Figure 3. Kaplan–Meier curves for Overall Survival according to CTC detection (**A**) at baseline. (**B**) at 4 weeks.

3.3. KRAS Mutation: Correlation between Liquid and Solid Biopsy

At baseline, blood samples from 125 patients were available for KRAS exon 2 status assessment on plasma as part of our study; 46 of these 125 patients had a KRAS exon 2 mutated tumor according to their medical files (i.e., determined by routine local assessment; Table S1). Among these 46 patients, KRASmut ctDNA was detected at baseline in 42 patients (sensitivity of the liquid biopsy = 0.91, 95%CI [0.79;0.96]). The median number of KRASmut copies/mL plasma in all 46 patients was 378 (range [0;25380]). Among the 79 patients with KRAS wild-type tumors per local assessment, 6 patients (8%) had detectable KRASmut ctDNA (nominal specificity=92%, 95%CI [0.84;0.96]). However, all 6 patients displayed high levels of ctDNA (>150 KRASmut copies/mL plasma), suggesting the actual presence of a KRAS mutation rather than a lack of specificity of the liquid biopsy.

3.4. Dynamic Changes of ctDNA Levels, Correlation with R0/R1 Resection and Outcome

The following analyses were performed in the subgroup of patients with KRAS exon 2 mutated tumors, as determined by routine local assessment on tumor tissues (except at baseline for ctDNA detection as a dichotomized variable, because all patients underwent the ctDNA detection assay at this timepoint).

percentile minus 25th percentile), while the bottom of whiskers represent the minimum observation 1.5× (1QR from 25th). The observations plotted are outliers- beyond the 1.5× IQR or below. The sign (CA19.9: p = 0.24; CEA: p = 0.25). Baseline ctDNA concentration was however correlated with a lower R0/R1 resection rate (p = 0.05, Figure 4).

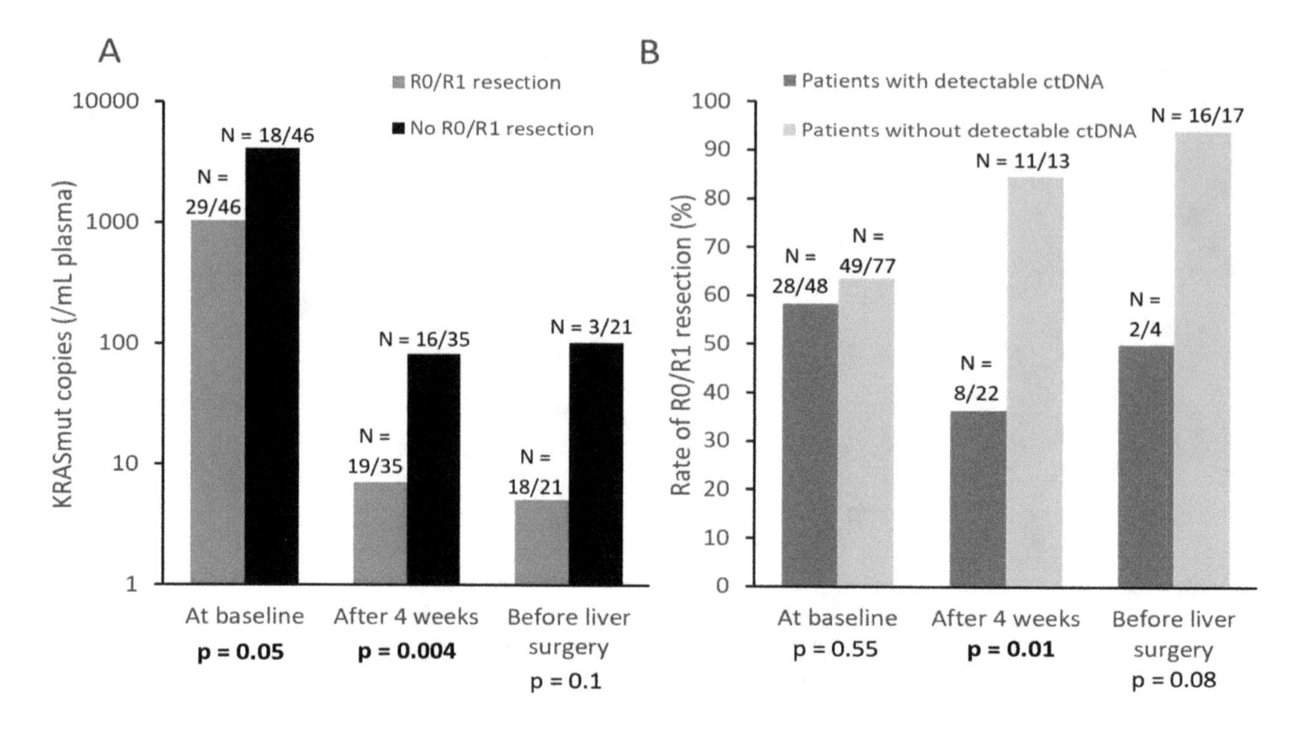

Figure 4. (**A**) Mean number of KRASmut copies per mL of plasma (continuous variable) at baseline, after four weeks, and before LM resection. N indicates the number of patients who achieved or did not achieve R0/R1 resection, among patients (with a KRAS mutated tumor) available for KRASmut assessment at each time point. (**B**) Rate of R0/R1 resection for patients with or without detectable ctDNA (dichotomized variable). N indicates the number of patients who achieved R0/R1 resection according to their ctDNA detection status, among patients who underwent the ctDNA detection assay at each timepoint.

KRAS ctDNA levels significantly decreased during therapy (p = 0.0001). 63% (N = 22/35) of patients with KRAS mutated tumors displayed detectable ctDNA at 4 weeks, while this ctDNA positivity rate dropped to 19% (N = 4/21) before surgery (Figure 2). At 4 weeks, lower ctDNA levels (as a continuous variable) were significantly correlated with eventual R0/R1 resection (p = 0.004, Figure 4). Similar results were observed with ctDNA detection as a dichotomized variable: patients with still detectable ctDNA after 4 weeks of systemic therapy had a lower R0/R1 resection rate than those with no ctDNA detected (36% vs 85%, p = 0.01, Figure 4).

In terms of OS, the 4 patients with no detectable ctDNA levels at baseline had an excellent prognosis (p = 0.05, HR not available; Figure 5A). ctDNA detection at 4 weeks had no prognostic impact (p = 0.31, Figure 5B, Supplementary Figure S1). However, among patients referred to LM resection, the detection of residual ctDNA levels before surgery was significantly associated with a short OS (HR = 31 CI95% [3.2;317], p < 0.001) (Figure 5C). A similar association was found with a short post-operative OS (Figure 5D).

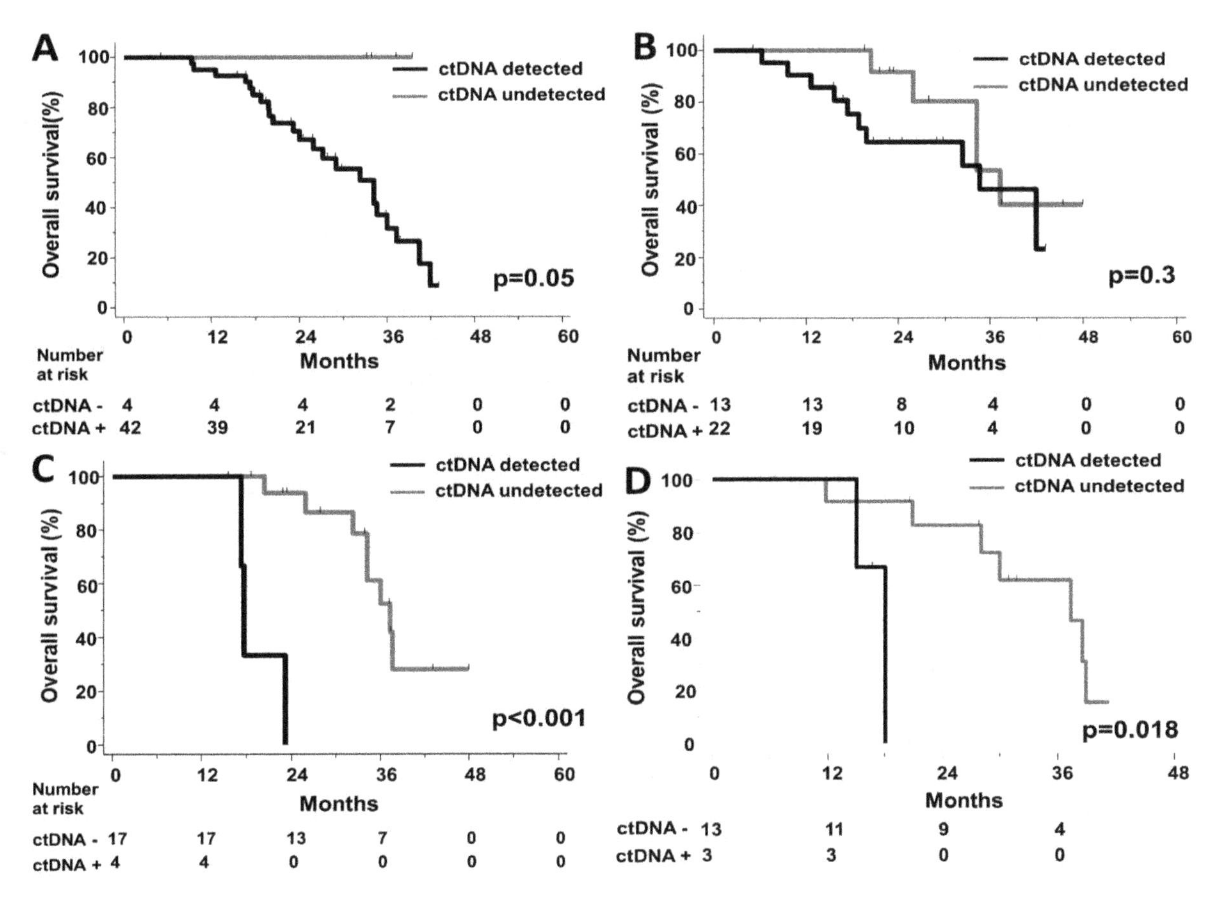

Figure 5. Kaplan–Meier curves for Overall Survival according to ctDNA detection (**A**) at baseline, (**B**) at 4 weeks, (**C**) before liver surgery (**D**) Kaplan–Meier curve for post-operative Overall Survival according to ctDNA detection before liver surgery.

4. Discussion

This is the first study to investigate the clinical validity of both CTC and ctDNA in patients with potentially resectable LM of CRC in a prospective clinical trial. These patients should be treated with intensive first-line systemic therapy combining poly-chemotherapy and the most appropriate targeted therapy (anti-EGFR antibodies in RAS wild-type tumors). Our study found that before the start of systemic therapy, the results of the assessment of tumor mutation status using ctDNA were closely correlated with those of local testing. Even more interestingly, we identified a few patients considered KRAS wild-type by tumor tissue sequencing that had significant KRAS mutant levels in their blood. A similar discrepancy was observed, but at much higher rates, in studies that focused on heavily pre-treated patients [15], suggesting a role of prior anti-EGFR therapies in the emergence of KRAS mutants subclones. Importantly, the probable benefit of anti-EGFR therapy is very limited in such cases [14]. In chemotherapy-naïve patients, the recent RAS Mutation Testing in the Circulating Blood of Patients With Metastatic Colorectal Cancer (RASANC) study [24] found that 8 of 412 patients had a RAS mutation in plasma but not in the primary tumor by local assessment. However, the authors performed central re-analysis on 6 of 8 tumor samples, by next-generation sequencing (NGS) or ddPCR, and found RAS mutations in all six samples. These results, and the shorter testing time, strongly suggest that ctDNA analysis might become a valuable theranostic tool in patients diagnosed with potentially resectable LM.

In addition to liquid biopsy applications at baseline, the clinical validity of ctDNA quantification was investigated at different time points during therapy. First, systemic therapy induced a significant decrease in ctDNA levels, highlighting that liquid biopsy has a very limited sensitivity once therapy has been initiated. We also found that ctDNA levels at different time points yielded significant prognostic information: undetectable ctDNA levels at baseline tended to be a prognostic factor, as demonstrated in other cancers such as metastatic lung cancer [25], ctDNA being correlated with tumor burden in various cancer types [26]. More interestingly, the absence of ctDNA at 4 weeks was correlated with a very high R0/R1 resection rate of LM (85%), suggesting that this biomarker could help decide whether liver surgery is appropriate for patients.

Finally, in patients referred to surgery for LM resection, persistently detectable ctDNA levels before surgery was associated with short post-surgical OS, suggesting that LM were not fully responding to therapy and/or that extra-hepatic micro-metastases were present. A study by Narayan and colleagues [27] in 59 metastatic CRC patients who underwent LM resection found an association between worse disease-specific survival and the detection of circulating mutant *TP53* copies during surgery (but not with ctDNA). However, blood samples were only obtained during and after surgery. Recent studies have found an association between dynamic changes in ctDNA detection and outcome in CRC, in the adjuvant setting, or in the metastatic setting. In the adjuvant setting [28], change of ctDNA status (as a dichotomized variable: detected or not detected) from positive to negative or from negative to positive was associated with respectively superior or lower recurrence-free survival. In the metastatic setting, a recent study [29], using a composite marker evaluating the decrease of ctDNA levels during chemotherapy, demonstrated that it could be used to predict response, progression-free survival, and OS.

If confirmed by further studies, we hypothesize that the absence of detectable ctDNA might become an important criterion prior to any LM resection in this patient population.

Regarding ctDNA analyses, limitations of our study include the limited number of KRAS mutated tumors enrolled and the focus on KRAS exon 2 mutations, as predefined in the study protocol at time of initiation, with no assessment of other KRAS, NRAS, and Rapidly Accelerated Fibrosarcoma homolog B (BRAF) mutations. Of note, while assessing several mutation hotspots in a single assay is usually achieved by NGS; multiplex ddPCR [30] and, more recently, drop-off ddPCR [31] may allow the screening of several hotspots in a single reaction. Larger mutation panel or methylation patterns can be used to detect and quantify ctDNA in a larger proportion of patients [18,24,32]. In the RASANC study [24], plasma samples from chemotherapy-naïve metastatic CRC patients were analyzed by NGS combined with methylation ddPCR, which allowed for a high detection rate of ctDNA (329/425, 77%).

Regarding CTC detection, our study showed its correlation with ctDNA levels, as already reported in patients with uveal melanoma LM [33]. However, the CTC detection rate was lower in our patient population than in prior studies in non-resectable metastatic CRC patients [5,7], probably because of the limited tumor burden in patients included in this study. While our study confirmed the prognostic impact of CTC count at baseline, the number of patients with persistently elevated CTC counts during therapy appeared very limited and prevents any clinical utility in this clinical setting, despite a proven clinical validity. We propose that more sensitive CTC detection techniques [34] be investigated in metastatic CRC to assess the clinical utility of CTC level, such as those relying on microfluidics [35], on EpCAM-independent CTC detection [36] and/or on the screening of larger blood volume [37].

Lastly, newly developed circulating biomarkers such as free serum amino acids [38] could be compared to CTC or ctDNA detection for their prognostic value. Similarly, circulating extracellular matrix components have been evaluated as biomarkers for cancer diagnosis and prognosis in various tumor types [39].

This prospective study showed that CTC and ctDNA had different detection profiles in mCRC patients with potentially resectable LM, the latter demonstrating interesting validity with regards to liquid biopsy and pre-operative prognostic applications.

Supplementary Materials:
Table S1: blood samples available. Table S2: associations of CTCs (as a dichotomous variable: ≥3 or <3 CTC/7.5 mL) and clinicopathological characteristics of patients at baseline (when available), treatment arm, or primary endpoint. Table S3: associations between ctDNA detection (as a dichotomous variable) and clinicopathological characteristics of patients at baseline (when available), treatment arm, or primary endpoint. Table S4: multivariate Cox regression with CTC detection at baseline and at 4 weeks (Overall Survival). Figure S1: post-operative overall survival according to ctDNA detection at 4 weeks.

Author Contributions: Conceptualization, F.-C.B., J.-Y.P., T.S. and P.M.; methodology F.-C.B., J.-Y.P. and P.M., software, S.T.; validation, F.-C.B., J.M., A.S., A.R., C.D., C.P. and J.-Y.P.; formal analysis, S.T.; investigation, F.-C.B., M.Y., M.H.S., J.M., A.S., A.R., C.D., O.B., M.R., F.G., E.F., R.G, L.M., F.K.A., T.M., D.M., W.C., C.P., J.-Y.P. and P.M.; resources, F.-C.B., C.D., C.P. and J.-Y.P.; data curation, S.T.; writing—original draft preparation, F.-C.B., L.C. and N.K.; writing—review and editing, F.-C.B. and N.K.; visualization, F.-C.B., L.C. and N.K.; supervision, F.-C.B.; project administration, F.-C.B.; funding acquisition, F.-C.B., M.Y., T.S. and P.M.

References

1. Adam, R.; Delvart, V.; Pascal, G.; Valeanu, A.; Castaing, D.; Azoulay, D.; Giacchetti, S.; Paule, B.; Kunstlinger, F.; Ghémard, O.; et al. Rescue surgery for unresectable colorectal liver metastases downstaged by chemotherapy: A model to predict long-term survival. *Ann. Surg.* **2004**, *240*, 644–657. [CrossRef] [PubMed]
2. Delaunoit, T.; Alberts, S.R.; Sargent, D.J.; Green, E.; Goldberg, R.M.; Krook, J.; Fuchs, C.; Ramanathan, R.K.; Williamson, S.K.; Morton, R.F.; et al. Chemotherapy permits resection of metastatic colorectal cancer: Experience from Intergroup N9741. *Ann. Oncol.* **2005**, *16*, 425–429. [CrossRef] [PubMed]
3. Falcone, A.; Ricci, S.; Brunetti, I.; Pfanner, E.; Allegrini, G.; Barbara, C.; Crinò, L.; Benedetti, G.; Evangelista, W.; Fanchini, L.; et al. Phase III trial of infusional fluorouracil, leucovorin, oxaliplatin, and irinotecan (FOLFOXIRI) compared with infusional fluorouracil, leucovorin, and irinotecan (FOLFIRI) as first-line treatment for metastatic colorectal cancer: The gruppo oncologico nord ovest. *J. Clin. Oncol.* **2007**, *25*, 1670–1676.
4. Ychou, M.; Rivoire, M.; Thezenas, S.; Guimbaud, R.; Ghiringhelli, F.; Mercier-Blas, A.; Mineur, L.; Francois, E.; Khemissa, F.; Moussata, D.; et al. FOLFIRINOX combined to targeted therapy according RAS status for colorectal cancer patients with liver metastases initially non-resectable: A phase II randomized Study—Prodige 14–ACCORD 21 (METHEP-2), a unicancer GI trial. *J. Clin. Oncol.* **2016**, *34*, 3512. [CrossRef]
5. Cohen, S.J.; Punt, C.J.A.; Iannotti, N.; Saidman, B.H.; Sabbath, K.D.; Gabrail, N.Y.; Picus, J.; Morse, M.; Mitchell, E.; Miller, M.C.; et al. Relationship of circulating tumor cells to tumor response, progression-free survival, and overall survival in patients with metastatic colorectal cancer. *J. Clin. Oncol.* **2008**, *26*, 3213–3221. [CrossRef]
6. Sastre, J.; Vidaurreta, M.; Gómez, A.; Rivera, F.; Massutí, B.; López, M.R.; Abad, A.; Gallen, M.; Benavides, M.; Aranda, E.; et al. Prognostic value of the combination of circulating tumor cells plus kras in patients with metastatic colorectal cancer treated with chemotherapy plus bevacizumab. *Clin. Colorectal Cancer* **2013**, *12*, 280–286. [CrossRef]
7. Tol, J.; Koopman, M.; Miller, M.C.; Tibbe, A.; Cats, A.; Creemers, G.J.M.; Vos, A.H.; Nagtegaal, I.D.; Terstappen, L.W.M.M.; Punt, C.J.A. Circulating tumour cells early predict progression-free and overall survival in advanced colorectal cancer patients treated with chemotherapy and targeted agents. *Ann. Oncol.* **2010**, *21*, 1006–1012. [CrossRef] [PubMed]
8. Huang, X.; Gao, P.; Song, Y.; Sun, J.; Chen, X.; Zhao, J.; Xu, H.; Wang, Z. Meta-analysis of the prognostic value of circulating tumor cells detected with the CellSearch System in colorectal cancer. *BMC Cancer* **2015**, *15*, 202. [CrossRef]
9. Bidard, F.C.; Huguet, F.; Louvet, C.; Mineur, L.; Bouche, O.; Chibaudel, B.; Artru, P.; Desseigne, F.; Bachet, J.B.; Mathiot, C.; et al. Circulating tumor cells in locally advanced pancreatic adenocarcinoma: The ancillary CirCe 07 study to the LAP 07 trial. *Ann. Oncol.* **2013**, *24*, 2057–2061. [CrossRef]
10. Bidard, F.C.; Ferrand, F.R.; Huguet, F.; Hammel, P.; Louvet, C.; Malka, D.; Boige, V.; Ducreux, M.; Andre, T.; de Gramont, A.; et al. Disseminated and circulating tumor cells in gastrointestinal oncology. *Crit. Rev. Oncol. Hematol.* **2012**, *82*, 103–115. [CrossRef]
11. Bidard, F.-C.; Weigelt, B.; Reis-Filho, J.S. Going with the flow: From circulating tumor cells to DNA. *Sci. Transl. Med.* **2013**, *5*, 207ps14. [CrossRef]
12. Douillard, J.-Y.; Ostoros, G.; Cobo, M.; Ciuleanu, T.; Cole, R.; McWalter, G.; Walker, J.; Dearden, S.; Webster, A.; Milenkova, T.; et al. Gefitinib treatment in EGFR mutated caucasian NSCLC: Circulating-free tumor dna as a surrogate for determination of egfr status. *J. Thorac. Oncol.* **2014**, *9*, 1345–1353. [CrossRef]
13. Thierry, A.R.; Mouliere, F.; El Messaoudi, S.; Mollevi, C.; Lopez-Crapez, E.; Rolet, F.; Gillet, B.; Gongora, C.; Dechelotte, P.; Robert, B.; et al. Clinical validation of the detection of KRAS and BRAF mutations from circulating tumor DNA. *Nat. Med.* **2014**, *20*, 430–435. [CrossRef]

14. Siravegna, G.; Mussolin, B.; Buscarino, M.; Corti, G.; Cassingena, A.; Crisafulli, G.; Ponzetti, A.; Cremolini, C.; Amatu, A.; Lauricella, C.; et al. Clonal evolution and resistance to EGFR blockade in the blood of colorectal cancer patients. *Nat. Med.* **2015**, *21*, 795. [CrossRef]
15. Tabernero, J.; Lenz, H.-J.; Siena, S.; Sobrero, A.; Falcone, A.; Ychou, M.; Humblet, Y.; Bouché, O.; Mineur, L.; Barone, C.; et al. Analysis of circulating DNA and protein biomarkers to predict the clinical activity of regorafenib and assess prognosis in patients with metastatic colorectal cancer: A retrospective, exploratory analysis of the CORRECT trial. *Lancet Oncol.* **2015**, *16*, 937–948. [CrossRef]
16. Raimondi, C.; Nicolazzo, C.; Belardinilli, F.; Loreni, F.; Gradilone, A.; Mahdavian, Y.; Gelibter, A.; Giannini, G.; Cortesi, E.; Gazzaniga, P. Transient disappearance of RAS mutant clones in plasma: A counterintuitive clinical use of EGFR inhibitors in RAS mutant metastatic colorectal cancer. *Cancers* **2019**, *11*, 42. [CrossRef]
17. Diehl, F.; Schmidt, K.; Choti, M.A.; Romans, K.; Goodman, S.; Li, M.; Thornton, K.; Agrawal, N.; Sokoll, L.; Szabo, S.A.; et al. Circulating mutant DNA to assess tumor dynamics. *Nat. Med.* **2008**, *14*, 985–990. [CrossRef] [PubMed]
18. Tie, J.; Kinde, I.; Wang, Y.; Wong, H.L.; Roebert, J.; Christie, M.; Tacey, M.; Wong, R.; Singh, M.; Karapetis, C.S.; et al. Circulating tumor DNA as an early marker of therapeutic response in patients with metastatic colorectal cancer. *Ann. Oncol.* **2015**, *26*, 1715–1722. [CrossRef] [PubMed]
19. Tie, J.; Wang, Y.; Tomasetti, C.; Li, L.; Springer, S.; Kinde, I.; Silliman, N.; Tacey, M.; Wong, H.-L.; Christie, M.; et al. Circulating tumor DNA analysis detects minimal residual disease and predicts recurrence in patients with stage II colon cancer. *Sci. Transl. Med.* **2016**, *8*, 346ra92. [CrossRef] [PubMed]
20. Ychou, M.; Rivoire, M.; Thezenas, S.; Quenet, F.; Delpero, J.-R.; Rebischung, C.; Letoublon, C.; Guimbaud, R.; Francois, E.; Ducreux, M.; et al. A randomized phase II trial of three intensified chemotherapy regimens in first-line treatment of colorectal cancer patients with initially unresectable or not optimally resectable liver metastases. The METHEP trial. *Ann. Surg. Oncol.* **2013**, *20*, 4289–4297. [CrossRef]
21. Allard, W.J. Tumor cells circulate in the peripheral blood of all major carcinomas but not in healthy subjects or patients with nonmalignant diseases. *Clin. Cancer Res.* **2004**, *10*, 6897–6904. [CrossRef]
22. Lebofsky, R.; Decraene, C.; Bernard, V.; Kamal, M.; Blin, A.; Leroy, Q.; Rio Frio, T.; Pierron, G.; Callens, C.; Bieche, I.; et al. Circulating tumor DNA as a non-invasive substitute to metastasis biopsy for tumor genotyping and personalized medicine in a prospective trial across all tumor types. *Mol. Oncol.* **2015**, *9*, 783–790. [CrossRef]
23. Madic, J.; Kiialainen, A.; Bidard, F.-C.; Birzele, F.; Ramey, G.; Leroy, Q.; Frio, T.R.; Vaucher, I.; Raynal, V.; Bernard, V.; et al. Circulating tumor DNA and circulating tumor cells in metastatic triple negative breast cancer patients: ctDNA and CTC in metastatic triple negative breast cancer. *Int. J. Cancer* **2015**, *136*, 2158–2165. [CrossRef]
24. Bachet, J.B.; Bouché, O.; Taieb, J.; Dubreuil, O.; Garcia, M.L.; Meurisse, A.; Normand, C.; Gornet, J.M.; Artru, P.; Louafi, S.; et al. RAS mutation analysis in circulating tumor DNA from patients with metastatic colorectal cancer: The AGEO RASANC prospective multicenter study. *Ann. Oncol.* **2018**, *29*, 1211–1219. [CrossRef]
25. Pécuchet, N.; Zonta, E.; Didelot, A.; Combe, P.; Thibault, C.; Gibault, L.; Lours, C.; Rozenholc, Y.; Taly, V.; Laurent-Puig, P.; et al. Base-position error rate analysis of next-generation sequencing applied to circulating tumor dna in non-small cell lung cancer: A prospective study. *PLoS Med.* **2016**, *13*, e1002199. [CrossRef] [PubMed]
26. Bettegowda, C.; Sausen, M.; Leary, R.J.; Kinde, I.; Wang, Y.; Agrawal, N.; Bartlett, B.R.; Wang, H.; Luber, B.; Alani, R.M.; et al. Detection of circulating tumor dna in early- and late-stage human malignancies. *Sci. Transl. Med.* **2014**, *6*, 224ra24. [CrossRef] [PubMed]
27. Narayan, R.R.; Goldman, D.A.; Gonen, M.; Reichel, J.; Huberman, K.H.; Raj, S.; Viale, A.; Kemeny, N.E.; Allen, P.J.; Balachandran, V.P.; et al. Peripheral circulating tumor dna detection predicts poor outcomes after liver resection for metastatic colorectal cancer. *Ann. Surg. Oncol.* **2019**, *26*, 1824–1832. [CrossRef] [PubMed]
28. Tie, J.; Cohen, J.; Wang, Y.; Lee, M.; Wong, R.; Kosmider, S.; Ananda, S.; Cho, J.H.; Faragher, I.; McKendrick, J.J.; et al. Serial circulating tumor DNA (ctDNA) analysis as a prognostic marker and a real-time indicator of adjuvant chemotherapy (CT) efficacy in stage III colon cancer (CC). *J. Clin. Oncol.* **2018**, *36*, 3516. [CrossRef]
29. Garlan, F.; Laurent-Puig, P.; Sefrioui, D.; Siauve, N.; Didelot, A.; Sarafan-Vasseur, N.; Michel, P.; Perkins, G.; Mulot, C.; Blons, H.; et al. Early evaluation of circulating tumor dna as marker of therapeutic efficacy in metastatic colorectal cancer patients (PLACOL Study). *Clin. Cancer Res.* **2017** *23*, 5416–5425. [CrossRef]

30. Taly, V.; Pekin, D.; Benhaim, L.; Kotsopoulos, S.K.; Le Corre, D.; Li, X.; Atochin, I.; Link, D.R.; Griffiths, A.D.; Pallier, K.; et al. Multiplex picodroplet digital PCR to detect KRAS mutations in circulating DNA from the plasma of colorectal cancer patients. *Clin. Chem.* **2013**, *59*, 1722–1731. [CrossRef]
31. Decraene, C.; Silveira, A.B.; Bidard, F.-C.; Vallée, A.; Michel, M.; Melaabi, S.; Vincent-Salomon, A.; Saliou, A.; Houy, A.; Milder, M.; et al. Multiple hotspot mutations scanning by single droplet digital PCR. *Clin. Chem.* **2018**, *64*, 317–328. [CrossRef]
32. Garrigou, S.; Perkins, G.; Garlan, F.; Normand, C.; Didelot, A.; Le Corre, D.; Peyvandi, S.; Mulot, C.; Niarra, R.; Aucouturier, P.; et al. A Study of hypermethylated circulating tumor DNA as a universal colorectal cancer biomarker. *Clin. Chem.* **2016**, *62*, 1129–1139. [CrossRef]
33. Bidard, F.-C.; Madic, J.; Mariani, P.; Piperno-Neumann, S.; Rampanou, A.; Servois, V.; Cassoux, N.; Desjardins, L.; Milder, M.; Vaucher, I.; et al. Detection rate and prognostic value of circulating tumor cells and circulating tumor DNA in metastatic uveal melanoma: Circulating tumor DNA in uveal melanoma. *Int. J. Cancer* **2014**, *134*, 1207–1213. [CrossRef]
34. Ferreira, M.M.; Ramani, V.C.; Jeffrey, S.S. Circulating tumor cell technologies. *Mol. Oncol.* **2016**, *10*, 374–394. [CrossRef]
35. Saliba, A.-E.; Saias, L.; Psychari, E.; Minc, N.; Simon, D.; Bidard, F.-C.; Mathiot, C.; Pierga, J.-Y.; Fraisier, V.; Salamero, J.; et al. Microfluidic sorting and multimodal typing of cancer cells in self-assembled magnetic arrays. *Proc. Natl. Acad. Sci. USA* **2010**, *107*, 14524–14529. [CrossRef]
36. Kuhn, P.; Keating, S.; Baxter, G.; Thomas, K.; Kolatkar, A.; Sigman, C. Lessons learned: Transfer of the high-definition circulating tumor cell assay platform to development as a commercialized clinical assay platform. *Clin. Pharmacol. Ther.* **2017**, *102*, 777–785. [CrossRef]
37. Andree, K.C.; Mentink, A.; Zeune, L.L.; Terstappen, L.W.M.M.; Stoecklein, N.H.; Neves, R.P.; Driemel, C.; Lampignano, R.; Yang, L.; Neubauer, H.; et al. Toward a real liquid biopsy in metastatic breast and prostate cancer: Diagnostic LeukApheresis increases CTC yields in a European prospective multicenter study (CTCTrap): Toward a real liquid biopsy in metastatic breast and prostate cancer. *Int. J. Cancer* **2018**, *143*, 2584–2591. [CrossRef]
38. Vsiansky, V.; Svobodova, M.; Gumulec, J.; Cernei, N.; Sterbova, D.; Zitka, O.; Kostrica, R.; Smilek, P.; Plzak, J.; Betka, J.; et al. Prognostic significance of serum free amino acids in head and neck cancers. *Cells* **2019**, *8*, 428. [CrossRef]
39. Giussani, M.; Triulzi, T.; Sozzi, G.; Tagliabue, E. Tumor extracellular matrix remodeling: new perspectives as a circulating tool in the diagnosis and prognosis of solid tumors. *Cells* **2019**, *8*, 81. [CrossRef]

7

Never Travel Alone: The Crosstalk of Circulating Tumor Cells and the Blood Microenvironment

Simon Heeke [1,2], Baharia Mograbi [1,2], Catherine Alix-Panabières [3] and Paul Hofman [1,2,4,*]

1 Université Côte d'Azur, CHU Nice, FHU OncoAge, 06000 Nice, France
2 Université Côte d'Azur, CNRS UMR7284, Inserm U1081, Institute for Research on Cancer and Aging, Nice (IRCAN), FHU OncoAge, 06000 Nice, France
3 Laboratory of Rare Human Circulating Cells (LCCRH), University Medical Centre, EA2415, Montpellier University, 34093 Montpellier, France
4 Laboratory of Clinical and Experimental Pathology and Biobank BB-0033-00025, Pasteur Hospital, FHU OncoAge, 06000 Nice, France
* Correspondence: hofman.p@chu-nice.fr.

Abstract: Commonly, circulating tumor cells (CTCs) are described as source of metastasis in cancer patients. However, in this process cancer cells of the primary tumor site need to survive the physical and biological challenges in the blood stream before leaving the circulation to become the seed of a new metastatic site in distant parenchyma. Most of the CTCs released in the blood stream will not resist those challenges and will consequently fail to induce metastasis. A few of them, however, interact closely with other blood cells, such as neutrophils, platelets, and/or macrophages to survive in the blood stream. Recent studies demonstrated that the interaction and modulation of the blood microenvironment by CTCs is pivotal for the development of new metastasis, making it an interesting target for potential novel treatment strategies. This review will discuss the recent research on the processes in the blood microenvironment with CTCs and will outline currently investigated treatment strategies.

Keywords: circulating tumor cells; hematological cells; neutrophils; platelets; liquid biopsy

---

1. Introduction

Circulating tumor cells (CTCs) have been extensively studied over the last decades, in particular as they play a crucial role in the diagnosis and the prognosis in many solid tumors as well as due to their predictive value associated with cancer targeted therapies as well as with immunotherapies [1–3]. CTCs are present in the blood stream as isolated CTCs (iCTCs) or in clusters of variable sizes that are often referred to as circulating tumor microemboli (CTMs) [4]. Following their migration from the primary site of the tumor into the blood, the tumor cells are constrained to high pressure and turbulences due to the blood stream and have to develop mechanisms of resistance for survival to consequently be able to adhere to the endothelium for tissue invasion and development of metastases [5]. Moreover, some CTCs are also able to come back to the primary tumor site and, consequently, to participate to the tumor growth [6]. However, the physical characteristics allowing the CTCs to survive are only partially known. Nevertheless, the biological characteristics of these cells and the phenotypic, genetic, and epigenetic modifications occurring during their migration from the primary tumor site until the development of distant metastases are beginning to be unraveled.

CTCs need to undergo significant changes to survive in the bloodstream—a new different environment. Thus, CTCs are challenged by physical forces in the circulation, they have to avoid being detected and killed by the immune system and finally, they need to extravasate from the blood stream to become the seed of new metastatic site(s) [7]. Recent works demonstrated that most of CTCs are not single cells travelling the blood alone but are accompanied by a plethora of blood cells and other CTCs and that a close interaction in the blood microenvironment is certainly needed to establish novel metastasis [8]. Interfering with this new microenvironment might help to develop strategies reducing the metastatic potential of tumors [8].

The aim of this review is therefore to summarize current knowledge concerning the role of the blood microenvironment and the different biological mechanisms occurring during its cross talk with CTCs. Additionally, potential therapeutic strategies and clinical approaches are discussed.

2. Brief Background on the Pathophysiology of CTCs Into the Blood Stream

2.1. The CTCs and the Constraints Due to the Blood Circulation

CTCs derive from primary tumor and/or metastatic sites and are consequently not adapted to the manifold challenges in the blood stream. Importantly, the flow of the blood stream, especially when passing the heart chambers, exposes cells to high mechanical sheer forces that can either directly destroy non-adapted cells or induce apoptosis in them [5,9,10]. Interestingly, CTCs seem to be stiffer than blood cells demonstrating their low adaptation to the blood stream [11] and tumor cells seem to be sensitive to those sheer forces indicating that the majority of CTCs will undergo apoptosis rather than forming metastasis in patients [12]. However, the different hemodynamic forces are important to allow the extravasation of tumor cells as they also remodel the endothelium [13], and consequently more knowledge on the biophysical properties allowing the formation of metastasis are needed [11]. Additionally, CTCs are directly exposed to the immune system and consequently they need to evade the detection from immune cells. Interestingly, programmed death ligand 1 (PD-L1), a costimulatory molecule inhibiting immune response can be expressed on CTCs and is associated with worse prognosis in lung [14,15] and head and neck cancer patients [16]. This indicates the active modulation of the immune response of CTCs to survive in the blood stream.

Lastly, as CTCs from cancers are of epithelial origin, they are adapted to grow in a network with other cells and are tightly interconnected by transmembrane proteins called integrins [17,18]. CTCs that leave the primary tumor site and enter the bloodstream lose the tight interaction with the surrounding cells, which can induce apoptosis in those cells through a phenomenon called anoikis [19]. Consequently, suppression of anoikis is required for survival of CTCs in the bloodstream [20], either by interaction of CTCs with other blood cells or by internally suppressing anoikis by activation of integrin signaling independent of cell–cell contacts [21].

2.2. Isolated CTC and Circulating Tumor Microemboli

While single CTCs are travelling in the blood stream, it has been demonstrated that CTC clusters or circulating tumor microemboli (CTMs) have a dramatically increased metastatic potential, as demonstrated in lung [4] and breast cancers [22]. Interestingly, a recent study conducted in 43 breast cancer patients demonstrated that CTC clustering alters DNA methylation patterns and increases stemness and consequently metastasis [23]. Single-cell bisulfite sequencing of single CTCs and CTCs derived from clusters revealed that transcription factors that are associated with a stem cell-like phenotype, like OCT4, NANOG, or SOX2, where hypomethylated in CTC clusters compared to single CTCs [23]. Interestingly, the authors also performed a drug-screening using 2486 FDA approved drugs to analyze their ability to interfere with CTC clustering. Thirty-one drugs have been detected that could serve as novel treatment to reduce the metastatic potential in breast cancer patients [23]. Consequently, the metastatic potential of CTCs might be limited in isolated CTCs and more research focusing on CTC clusters (CTMs) rather than on single cells might allow the design of novel treatment strategies to interact with the formation of CTC cluster to avoid metastasis in cancer patients.

3. Interaction of CTCs with Neutrophils

The role of neutrophils in cancer progression has been extensively studied recently [24,25]. Previously, it has been demonstrated that increased levels of circulating neutrophils are associated with bad prognosis in advanced cancer patients [26,27]. Moreover, the neutrophil-to-lymphocyte ratio has been demonstrated to be a prognostic factor in solid tumors [28]. In a recent study, Szczerba et al. demonstrated that within white blood cells (WBCs)–CTCs clusters of breast carcinoma, CTCs are associated with neutrophils in the majority of the cases [29]. Interestingly, using single-cell RNA sequencing, the authors showed that the transcriptome profiles of CTCs associated with neutrophils are different from those of CTCs alone, with differentially expressed genes that outline cell cycle progression leading to a more efficient metastasis formation. The authors noted that WBC–CTC clusters are relatively rare (less than 3.5%), whereas iCTCs alone are present in 88% of cases and CTCs clusters in less than 9% of the cases [29]. Despite, neutrophils directly interact with CTCs via ICAM-1 and neutrophils bound to CTCs facilitate the interaction of CTCs with endothelial cells in the liver, thereby promoting extravasation and liver metastasis [30]. Consequently, neutrophils play a major role for the CTC extravasation across the endothelial barrier and the onset of metastases (Figure 1) [31].

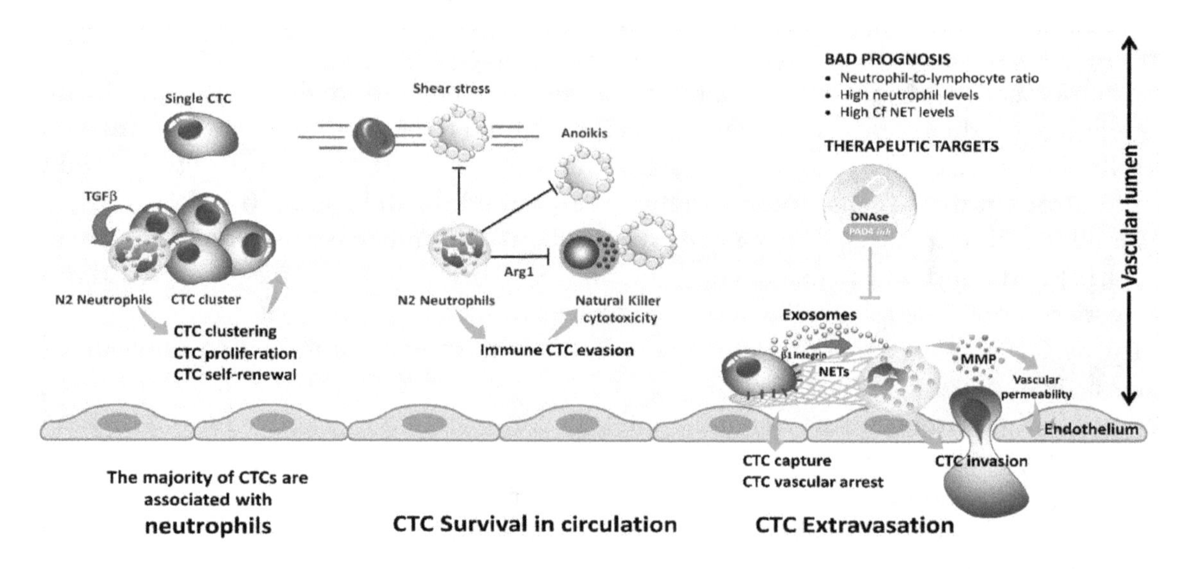

Figure 1. Compelling evidence indicates that blood neutrophils can offer proliferative and survival advantages to circulating tumor cells (CTCs) during their journey in the blood stream, rendering them more competent for metastasis development. Tumor-derived inflammatory factors strongly stimulate neutrophils to extrude chromatin webs called "neutrophil extracellular traps" (NETs). NETs, in turn, provide a niche to CTCs, arrest CTCs rolling, and promote metastasis. As such, understanding interaction of inflammatory N2 neutrophils with CTCs provides new potential therapeutic targets for disrupting these deadly metastatic seeds. As an example, blockade of NET formation using peptidylarginine deiminase 4 (PAD4) pharmacologic inhibitor or DNAse may decrease CTC colonization.

Neutrophils are able to generate neutrophil extracellular traps (NETs) by secreting their chromatin content during a process known as NETosis [32]. Initially, this process was described to be a mechanism to kill bacteria [33]. However, recent studies demonstrated that NETs are also promoting metastasis across various cancers [34–38]. Tohme et al. demonstrated that the NET formation induces a TLR-9 mediated response in cancer cells, which increased the migration and proliferation of CTCs [37]. Interestingly, it was shown that NETs promote extravasation of CTCs but in an IL-8 dependent manner. Consequently, blocking of IL-8 reduced the extravasation of tumor cells and neutrophils (Figure 1) [31]. Additionally, tumor-derived exosomes were also able to induce NET formation in neutrophils isolated from mice treated with granulocyte colony-stimulating factor (G-CSF) [39]. This phenomenon is even more important for tumors producing a large quantity of G-CSF associated with a high number of blood neutrophils [40,41]. Mechanistically, it has been demonstrated that the interaction of CTCs with NETs is mediated by β1-integrins expressed on tumor cells [42]. It is noteworthy that this integrin is physiologically overexpressed during infections and sepsis [42]. Generally, resolving NETs, for example, using DNAse I administration, reduced the number of metastasis making the NETs an interesting target for novel treatments reducing metastasis in patients (Figure 1) [34–38].

Different populations of neutrophils have different tumor promoting effects, and neutrophils are heavily modulated during cancer progression [43,44]. Consequently, extensive research is necessary to better decipher the role of the interaction of neutrophils and CTCs.

4. Interaction of CTCs with Myeloid-Derived Suppressor Cells

Myeloid-derived suppressor cells (MDSCs) are a heterogenous group of cells that are derived from the bone marrow and that are able to suppress the immune response via the suppression of T-cell response [45]. Commonly, MDSCs are classified in polymorphonuclear MDSCs (PMN-MDSC) and monocytic MDSCs (M-MDSC) [45]. As MDSCs are enriched in tumor tissue and able to suppress the immune system, it has been proposed that the interaction of CTCs with MDSCs might also promote metastasis [46]. Indeed, heterogenic clusters of CTCs and MDSCs have been reported in melanoma, pancreatic, and breast cancer patients [47,48], and co-culture of MDSCs with CTCs and T-cells demonstrated the T-cell suppressive effect of MDSCs [48]. Moreover, CTCs and MDSCs interact directly with each other, and increased reactive oxygen species (ROS) production in MDSCs induced NOTCH1 in CTCs, hereby promoting CTC proliferation [47]. Consequently, blocking the MDSC–CTC interaction might inhibit CTC proliferation and CTC immune evasion and might be an interesting target in anti-cancer therapy.

5. Interaction of CTCs with Platelets

In 1973, the role of platelets in cancer metastasis was already described, and the following work highlighted the role of platelets in cancer progression, especially during cancer metastases [49]. Indeed, different mechanisms occur during platelet–cancer cell interactions and crosstalk: (i) cancer cells can induce platelet activation; (ii) platelets support cancer metastasis and enhance cancer cell adhesion and arrest in vasculature; (iii) platelets assist immune evasion of cancer cells, and finally, (iv) platelets can enhance cancer evasion and tumor angiogenesis (Figure 2).

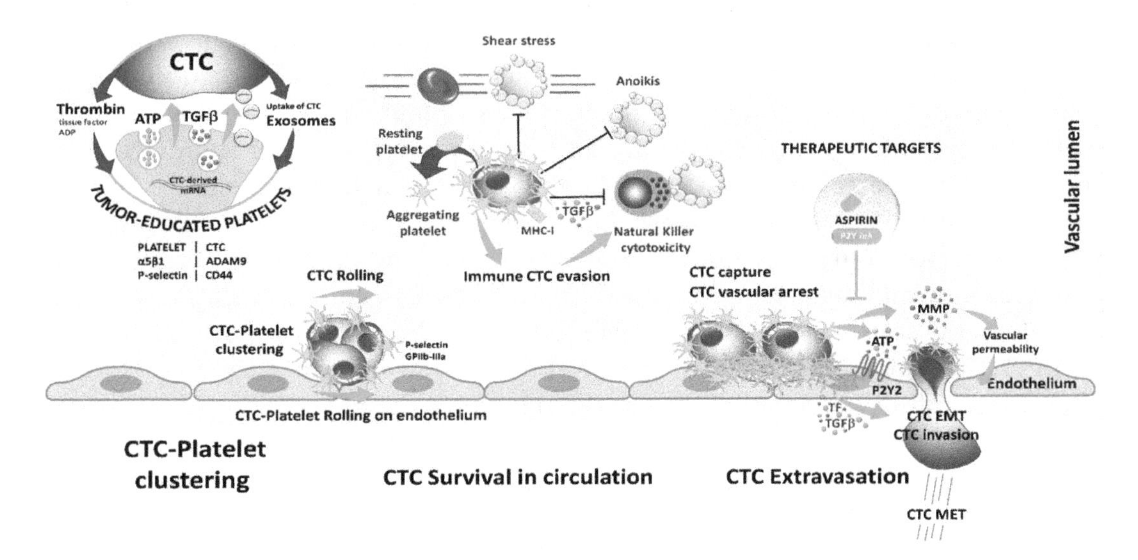

Figure 2. The dialogue between platelets and CTCs is reciprocal: CTCs activate and educate platelets while platelets contribute to CTCs' survival, escape from immune surveillance, tumor–endothelium interactions, and dissemination. Secretion of α-granules by activated platelets release high levels of TGF-β and ATP, a powerful activator of epithelial-to-mesenchymal transitioned (EMT) state and an endothelium relaxation factor (via P2Y), respectively. Inhibition of platelet α-granules secretion (Cox1) by aspirin, or of P2Y may abolish the metastatic potential of CTCs.

Interestingly, platelets can take up circulating mRNA from the CTCs, suggesting a possible modification in the platelet transcriptome that resembles the tumor profile [50]. In this context, platelets that circulate through and contact tumor sites can undergo modification due to the sequestration of RNA and biomolecules, which led to the concept of tumor-educated platelets (TEP) and may serve as an informative tool in cancer diagnosis [51,52]. This adherence could also help to decrease the impact of the pressure and the turbulence, especially in the heart chamber on the CTCs and can protect the CTCs against the physical stress in the blood stream (Figure 2) [53]. Indeed, platelets can form aggregates with CTCs, and CTCs induce platelet aggregation in a process known as tumor-cell-induced platelet aggregation (Figure 2) [54].

Platelets interact with tumor cells during blood dissemination leading to platelet activation and release of soluble mediators that alter the phenotype of the tumor cells and surrounding host cells [55]. However, the proximal events that initiate platelet activation are only partially characterized. It has been recently demonstrated that CD97 expressed on tumor cells may be involved in platelets activation [56]. CD97 is a g-protein coupled receptor that is undetectable in normal tissues except for smooth muscle cells but is abnormally expressed in different types of solid tumors. Ward et al. demonstrated that CD97 is able to activate platelets, which in turn secrete several mediators of the endothelial barrier, including ATP, which promotes evasion of CTCs off the blood stream and consequently promotes metastasis [56]. Additionally, the platelet P-selectin interacts with tumor CD44, and the fibrinogen receptor GPIIb-IIIa are involved in platelet rolling on CTCs and in platelet–CTC emboli (Figure 2) [57,58]. This leads to several alterations of platelets including protein synthesis, exosome release, blebbing of the membrane, and splicing of mRNAs [59]. Therefore, it seems that platelets form homotypic aggregates at the center of clusters that are surrounded by tumor cells at the periphery [54].

Despite the alterations induced by a direct contact between platelets and CTCs, the production of cytokines by platelets modifies the phenotype of CTCs. Molecular mechanisms by which CTCs maintain an epithelial-to-mesenchymal transitioned (EMT) state remain unclear. CTC clusters isolated from patients with advanced breast cancers highly exhibit mesenchymal markers and show an abundance of attached CD61-positive/platelets [60]. TGFβ1 secretion by alpha granules induces or increases the EMT observed in CTCs [55]. Likewise, platelets increase the tissue factor (TF) and the P2Y12 receptor activity, and both participate in EMT [61,62]. Moreover, platelets are involved in the adherence of CTCs to the endothelial barrier and to the transmigration of CTCs into the tissue for the development of metastasis [63]. One of the receptor ligand pair identified with such function is the ADAMA9 on CTCs that binds to the integrin α5β1 on the surface of platelets. This interaction is believed to promote platelet activation, granule secretion, and the transmigration of tumor cells through the endothelium [64]. Other mechanisms arising during the interaction between platelets and CTCs promote the migration of CTCs across the vasculature barrier: CTC-induced platelet aggregation leads to the release of ATP stored in dense granules; released ATP binds to the P2Y2 receptor stimulating cancer cell intravasation and metastatic dissemination [65]. Additionally, platelets and megacaryocytes play a major role in the survival of CTCs in the blood stream by different mechanisms. Platelets can protect certain CTCs against anoikis (a form of apoptosis that is induced when adherent cell lose contact to the surrounding cells) [66]. Even more, the adherence of platelets at the surface of CTCs may protect the CTCs to be recognized by some circulating immune cells, thereby promoting cell survival (Figure 2) [59,67]. Interestingly, platelets exert paracrine suppression of NK-mediated cytolytic activity. TGFβ released from activated platelets counteracts NK granule mobilization, cytotoxicity, and interferon-γ secretion [68]. Besides, platelet–CTC interaction can lead to the transfer of platelet major histocompatibility complex class I (MHC-I) to tumor cells preventing NK cell recognition via direct cell contacts [69]. This phenomenon is complex and not completely understood, and the platelets need to be activated at contact of CTCs entering in the circulation. Therefore, CTCs can release thrombin that attracts, activates, and aggregates platelets on their surface [54,59,70]. Several factors such as TF, thrombin, and ATP secreted by either platelets or CTCs induce platelets activation and formation of platelet–cancer cell aggregates [54,70].

Due to their multifaceted role in cancer metastasis, blocking of platelet–CTC interaction has also been studied as pharmacological target to reduce metastasis. Recently, Gareau et al. demonstrated that blocking this interaction using the P2Y12 inhibitor ticagrelor, reduced the number of metastasis and prolonged survival in a murine breast cancer model [71]. Additionally, in a clinical phase II study investigating the effect of aspirin on CTCs, less CTCs were detected in colorectal cancer patients upon aspirin treatment and the detected CTCs showed a more epithelial phenotype. Unfortunately, the results were not confirmed in a breast cancer model [72]. However, both aspirin and P2Y12 inhibitors inhibit platelet activation and demonstrate that modulation of platelets can reduce CTCs and metastasis, and the recent trials paved the way to actively investigate how the modulation of platelets can prevent metastasis-related cancer death [71,72].

6. Interaction of CTCs with Macrophages

Tumor-associated macrophages (TAMs) play a key role in activating dissemination and providing protection against the immune system [73]. However, the interplay of macrophages and CTCs is poorly understood. Previous works have been made to investigate the interaction between the macrophages and the CTCs in small cell lung carcinoma (SCLC) [74,75]. In these latter studies, different interactions have been observed by establishing SCLC cell lines and co-culture experiments with peripheral blood mononuclear cells (PBMCs). The authors showed that interaction of PBMCs with SCLC cells promote the differentiation of monocytes into macrophages, which express CD14, CD163, and CD68. These macrophages can secrete different cytokines such as osteopontin, monocyte chemoattractant protein-1, IL-8, chitinase 3-like 1, platelet factor, IL-1ra, and the matrix metalloproteinase-9 [74]. Likewise, PCa prostate cancer cells cultured with monocyte conditioned cell culture media, showed an increased invasion in vitro mediated by the IL-13Rα2 receptor expressed on cancer cells [76]. Additionally, Wei et al. demonstrated that the crosstalk of macrophages with tumor cells is necessary for the induction of EMT and release of CTCs into the blood stream. In their study, expression of IL6 of TAMs increased the secretion of CCL2 in tumor cells, which in turn recruited new macrophages [77].

The discussed studies demonstrate that the interaction of macrophages and tumor cells is not only important for progression at the primary tumor site but also for the promotion and differentiation of CTCs [77]. However, the interaction might be much closer than previously expected. Zhang et al. demonstrated that some circulating macrophages might be able to phagocyte apoptotic CTCs and incorporate the tumor DNA into their nuclei, consequently obtaining some malignant features like expression of epithelial markers (such as cytokeratins) and stem cell markers (e.g., OCT4) [78]. Consequently, circulating monocytes from solid cancer patients can express both CD163 and EpCAM. This led to the concept of "tumacrophages," which have the potential of invasive tumor cells but are protected against the immune system [78]. Even more, Gast et al. showed in a seminal work that viable tumor cells can fuse with macrophages to create hybrid cells, sharing markers of both tumor cells and macrophages [79]. Hybrid cells sharing epithelial (EpCAM expression) and hematological cell markers (CD45) were protected from detection by the immune system and correlated with disease stage and overall survival across several cancers [79]. Similar results were recently confirmed in a glioblastoma model where tumor cell–macrophage fusion cells demonstrated an increased invasive potential [80].

Consequently, the interplay of CTCs with macrophages is certainly important for metastasis and the discovery of tumor cell–macrophage fusion cells will help to develop novel biomarkers for cancer progression as well as novel potential therapeutic targets to block metastasis in patients.

7. Interaction of CTCs with Lymphocytes

Tumor cells constantly need to avoid being detected by immune cells to avoid being killed by them [81]. Likewise, CTCs in the blood stream are constantly required to avoid activation of the immune cells. The recent success of anti-cancer immunotherapy, most notably of checkpoint blocking antibodies, in several cancers have demonstrated that the immune system can be reactivated to target cancer cells [82]. Unfortunately, only limited studies have been carried out on the interaction of CTCs and lymphocytes. However, an inverse correlation between $CD3^+$, $CD4^+$, and $CD8^+$ peripheral T-lymphocytes and CTCs in NSCLC [83] and between $CD8^+$ peripheral lymphocytes in breast cancer [84] have been shown. Moreover, several studies demonstrated that regulatory T-cells infiltrating the tumor or detected in the peripheral blood are significantly more prevalent in breast cancer patients with CTCs than in patients without detectable CTCs [83–85]. While unfortunately mechanistic studies evaluating the interplay between T-lymphocytes and CTCs are lacking, the present studies, however, indicate that immune suppression by regulatory T-cells help in tumor cell dissemination in the blood stream. However, the responsible targets and mechanisms have to be unraveled to better understand the interplay of CTCs with lymphocytes.

Additionally, CTCs seem to be able to block interaction with lymphocytes by upregulating the programmed death ligand 1 (PD-L1) that inhibits the activation of T-lymphocytes [2]. This allows the CTCs to avoid being detected by the immune system and was indeed correlated with worse prognosis in NSCLC patients undergoing radio (chemo)-therapy [86]. Nevertheless, this might be only one of many mechanisms CTCs have to adapt to avoid detection by immune cells in the blood stream, and further research and clarification is needed.

8. Challenges and Perspectives

While CTCs have long been considered to be isolated cells floating in the blood stream, recent research demonstrated the close interaction of CTCs with the blood microenvironment. CTCs need to establish close interaction not only with platelets and neutrophils, but also with macrophages and endothelial cells to resist the physical stress in the blood stream and to evade detection by the immune system to finally leave the blood stream to establish new metastatic sites (Table 1).

Table 1. Summary of interactions of circulating tumor cells with other cell types in the blood microenvironment.

Interaction of CTCs with Other Cell Type	Interacting Targets/Processes	Effect	References
Neutrophils	ICAM-1	Facilitating interaction with endothelial cells and consequently extravasion off the blood stream.	[30]
	β1-integrin, tumor-derived exosomes	Formation of neutrophil extracellular traps (NETs) promoting proliferation and extravasion.	[31,37,39,42]
Myeloid-derived suppressor cells (MDSCs)	Reactive oxygen species (ROS) production by MDSCs	Increased proliferation of CTCs and inhibition of T-cells.	[47,48]
Blood platelets	Exosomes	Formation of tumor-educated platelets (TEPs).	[50–52]
	CD97, CD44, ADAMA9-α5β1 integrin, ATP	Modulation of endothelial cells by platelets leading to extravasion of CTCs.	[56,63–65]
	Cytokines produced by platelets	Induction of epithelial-to-mesenchymal transition in CTCs.	[55,60–62]
	TGFβ secreted by platelets	Suppression of cytolytic NK cells.	[68]
Macrophages	Cytokines produced by Macrophages	Increased invasion EMT of CTCs and immune suppression.	[74,76,77]
	Fusion with CTCs	Formation of "tumacrophages" that are protected from immune detection with invasive potential.	[78–80]
Lymphocytes	PD-L1	Suppression of cytotoxic T-cells.	[2]

While seminal research demonstrated well that the blood microenvironment is crucial for cell seeding, the mechanisms and interaction networks are not fully understood, and more research is needed. However, blocking the interaction of CTCs with platelets [71,72] as well as the resolving of NETs [34–38] demonstrated that targeting the interaction of CTCs with other cells is a promising therapeutic target and future research will certainly establish novel treatments to improve survival in cancer patients.

Author Contributions: Conceptualization, P.H.; writing—original draft preparation, S.H., P.H.; writing—review and editing, S.H., B.M., C.A.-P., P.H.; visualization, B.M.

References

1. Alix-Panabieres, C.; Pantel, K. Clinical Applications of Circulating Tumor Cells and Circulating Tumor DNA as Liquid Biopsy. *Cancer Discov.* **2016**, *6*, 479–491. [CrossRef] [PubMed]
2. Hofman, P.; Heeke, S.; Alix-Panabières, C.; Pantel, K. Liquid biopsy in the era of immune-oncology. Is it ready for prime-time use for cancer patients? *Ann. Oncol. Off. J. Eur. Soc. Med. Oncol.* **2019**. [CrossRef] [PubMed]
3. Pantel, K.; Alix-Panabières, C. Liquid biopsy and minimal residual disease—latest advances and implications for cure. *Nat. Rev. Clin. Oncol.* **2019**, *16*, 409–424. [CrossRef] [PubMed]
4. Carlsson, A.; Nair, V.S.; Luttgen, M.S.; Keu, K.V.; Horng, G.; Vasanawala, M.; Kolatkar, A.; Jamali, M.; Iagaru, A.H.; Kuschner, W.; et al. Circulating Tumor Microemboli Diagnostics for Patients with Non–Small-Cell Lung Cancer. *J. Thorac. Oncol.* **2014**, *9*, 1111–1119. [CrossRef] [PubMed]
5. Wirtz, D.; Konstantopoulos, K.; Searson, P.C. The physics of cancer: The role of physical interactions and mechanical forces in metastasis. *Nat. Rev. Cancer* **2011**, *11*, 512–522. [CrossRef] [PubMed]
6. Kim, M.-Y.; Oskarsson, T.; Acharyya, S.; Nguyen, D.X.; Zhang, X.H.-F.; Norton, L.; Massagué, J. Tumor Self-Seeding by Circulating Cancer Cells. *Cell* **2009**, *139*, 1315–1326. [CrossRef] [PubMed]
7. Strilic, B.; Offermanns, S. Intravascular Survival and Extravasation of Tumor Cells. *Cancer Cell* **2017**, *32*, 282–293. [CrossRef] [PubMed]
8. Guo, B.; Oliver, T.G. Partners in Crime: Neutrophil–CTC Collusion in Metastasis. *Trends Immunol.* **2019**. [CrossRef] [PubMed]
9. Phillips, K.G.; Kuhn, P.; McCarty, O.J.T. Physical Biology in Cancer. 2. The physical biology of circulating tumor cells. *Am. J. Physiol. Physiol.* **2014**, *306*, C80–C88. [CrossRef] [PubMed]
10. Barnes, J.M.; Nauseef, J.T.; Henry, M.D. Resistance to Fluid Shear Stress Is a Conserved Biophysical Property of Malignant Cells. *PLoS ONE* **2012**, *7*, e50973. [CrossRef] [PubMed]
11. Shaw Bagnall, J.; Byun, S.; Begum, S.; Miyamoto, D.T.; Hecht, V.C.; Maheswaran, S.; Stott, S.L.; Toner, M.; Hynes, R.O.; Manalis, S.R. Deformability of Tumor Cells versus Blood Cells. *Sci. Rep.* **2015**, *5*, 18542. [CrossRef] [PubMed]
12. Regmi, S.; Fu, A.; Luo, K.Q. High Shear Stresses under Exercise Condition Destroy Circulating Tumor Cells in a Microfluidic System. *Sci. Rep.* **2017**, *7*, 39975. [CrossRef] [PubMed]
13. Follain, G.; Osmani, N.; Azevedo, A.S.; Allio, G.; Mercier, L.; Karreman, M.A.; Solecki, G.; Garcia Leòn, M.J.; Lefebvre, O.; Fekonja, N.; et al. Hemodynamic Forces Tune the Arrest, Adhesion, and Extravasation of Circulating Tumor Cells. *Dev. Cell* **2018**, *45*, 33–52. [CrossRef] [PubMed]
14. Guibert, N.; Delaunay, M.; Lusque, A.; Boubekeur, N.; Rouquette, I.; Clermont, E.; Mourlanette, J.; Gouin, S.; Dormoy, I.; Favre, G.; et al. PD-L1 expression in circulating tumor cells of advanced non-small cell lung cancer patients treated with nivolumab. *Lung Cancer* **2018**, *120*, 108–112. [CrossRef] [PubMed]
15. Ilié, M.; Szafer-Glusman, E.; Hofman, V.; Chamorey, E.; Lalvée, S.; Selva, E.; Leroy, S.; Marquette, C.-H.; Kowanetz, M.; Hedge, P.; et al. Detection of PD-L1 in circulating tumor cells and white blood cells from patients with advanced non-small-cell lung cancer. *Ann. Oncol. Off. J. Eur. Soc. Med. Oncol.* **2018**, *29*, 193–199. [CrossRef] [PubMed]
16. Strati, A.; Koutsodontis, G.; Papaxoinis, G.; Angelidis, I.; Zavridou, M.; Economopoulou, P.; Kotsantis, I.; Avgeris, M.; Mazel, M.; Perisanidis, C.; et al. Prognostic significance of PD-L1 expression on circulating tumor cells in patients with head and neck squamous cell carcinoma. *Ann. Oncol. Off. J. Eur. Soc. Med. Oncol.* **2017**, *28*, 1923–1933. [CrossRef]
17. Harburger, D.S.; Calderwood, D.A. Integrin signalling at a glance. *J. Cell Sci.* **2009**, *122*, 159–163. [CrossRef]
18. Winograd-Katz, S.E.; Fässler, R.; Geiger, B.; Legate, K.R. The integrin adhesome: From genes and proteins to human disease. *Nat. Rev. Mol. Cell Biol.* **2014**, *15*, 273–288. [CrossRef]
19. Gilmore, A.P. Anoikis. *Cell Death Differ.* **2005**, *12*, 1473–1477. [CrossRef]
20. Kim, Y.-N.; Koo, K.H.; Sung, J.Y.; Yun, U.-J.; Kim, H. Anoikis Resistance: An Essential Prerequisite for Tumor Metastasis. *Int. J. Cell Biol.* **2012**, *2012*, 1–11. [CrossRef]

21. Alanko, J.; Mai, A.; Jacquemet, G.; Schauer, K.; Kaukonen, R.; Saari, M.; Goud, B.; Ivaska, J. Integrin endosomal signalling suppresses anoikis. *Nat. Cell Biol.* **2015**, *17*, 1412–1421. [CrossRef] [PubMed]
22. Aceto, N.; Bardia, A.; Miyamoto, D.T.; Donaldson, M.C.; Wittner, B.S.; Spencer, J.A.; Yu, M.; Pely, A.; Engstrom, A.; Zhu, H.; et al. Circulating Tumor Cell Clusters Are Oligoclonal Precursors of Breast Cancer Metastasis. *Cell* **2014**, *158*, 1110–1122. [CrossRef] [PubMed]
23. Gkountela, S.; Castro-Giner, F.; Szczerba, B.M.; Vetter, M.; Landin, J.; Scherrer, R.; Krol, I.; Scheidmann, M.C.; Beisel, C.; Stirnimann, C.U.; et al. Circulating Tumor Cell Clustering Shapes DNA Methylation to Enable Metastasis Seeding. *Cell* **2019**, *176*, 98–112. [CrossRef] [PubMed]
24. Shaul, M.E.; Fridlender, Z.G. Cancer-related circulating and tumor-associated neutrophils–subtypes, sources and function. *FEBS J.* **2018**, *285*, 4316–4342. [CrossRef] [PubMed]
25. Wu, L.; Saxena, S.; Awaji, M.; Singh, R.K. Tumor-Associated Neutrophils in Cancer: Going Pro. *Cancers* **2019**, *11*, 564. [CrossRef] [PubMed]
26. Dumitru, C.A.; Moses, K.; Trellakis, S.; Lang, S.; Brandau, S. Neutrophils and granulocytic myeloid-derived suppressor cells: Immunophenotyping, cell biology and clinical relevance in human oncology. *Cancer Immunol. Immunother.* **2012**, *61*, 1155–1167. [CrossRef] [PubMed]
27. Zhao, W.; Wang, P.; Jia, H.; Chen, M.; Gu, X.; Liu, M.; Zhang, Z.; Cheng, W.; Wu, Z. Neutrophil count and percentage: Potential independent prognostic indicators for advanced cancer patients in a palliative care setting. *Oncotarget* **2017**, *8*, 64499–64508. [CrossRef] [PubMed]
28. Templeton, A.J.; Knox, J.J.; Lin, X.; Simantov, R.; Xie, W.; Lawrence, N.; Broom, R.; Fay, A.P.; Rini, B.; Donskov, F.; et al. Change in Neutrophil-to-lymphocyte Ratio in Response to Targeted Therapy for Metastatic Renal Cell Carcinoma as a Prognosticator and Biomarker of Efficacy. *Eur. Urol.* **2016**, *70*, 358–364. [CrossRef]
29. Szczerba, B.M.; Castro-Giner, F.; Vetter, M.; Krol, I.; Gkountela, S.; Landin, J.; Scheidmann, M.C.; Donato, C.; Scherrer, R.; Singer, J.; et al. Neutrophils escort circulating tumour cells to enable cell cycle progression. *Nature* **2019**. [CrossRef]
30. Chow, S.C.; Spicer, J.D.; Kubes, P.; Giannias, B.; Cools-Lartigue, J.J.; Ferri, L.E.; McDonald, B. Neutrophils Promote Liver Metastasis via Mac-1–Mediated Interactions with Circulating Tumor Cells. *Cancer Res.* **2012**, *72*, 3919–3927. [CrossRef]
31. Chen, M.B.; Hajal, C.; Benjamin, D.C.; Yu, C.; Azizgolshani, H.; Hynes, R.O.; Kamm, R.D. Inflamed neutrophils sequestered at entrapped tumor cells via chemotactic confinement promote tumor cell extravasation. *Proc. Natl. Acad. Sci. USA* **2018**, *115*, 7022–7027. [CrossRef] [PubMed]
32. Papayannopoulos, V. Neutrophil extracellular traps in immunity and disease. *Nat. Rev. Immunol.* **2018**, *18*, 134–147. [CrossRef] [PubMed]
33. Brinkmann, V. Neutrophil Extracellular Traps Kill Bacteria. *Science* **2004**, *303*, 1532–1535. [CrossRef] [PubMed]
34. Al-Haidari, A.A.; Algethami, N.; Lepsenyi, M.; Rahman, M.; Syk, I.; Thorlacius, H. Neutrophil extracellular traps promote peritoneal metastasis of colon cancer cells. *Oncotarget* **2019**, *10*, 1238–1249. [CrossRef] [PubMed]
35. Cools-Lartigue, J.; Spicer, J.; McDonald, B.; Gowing, S.; Chow, S.; Giannias, B.; Bourdeau, F.; Kubes, P.; Ferri, L. Neutrophil extracellular traps sequester circulating tumor cells and promote metastasis. *J. Clin. Investig.* **2013**, *123*, 3446–3458. [CrossRef] [PubMed]
36. Park, J.; Wysocki, R.W.; Amoozgar, Z.; Maiorino, L.; Fein, M.R.; Jorns, J.; Schott, A.F.; Kinugasa-Katayama, Y.; Lee, Y.; Won, N.H.; et al. Cancer cells induce metastasis-supporting neutrophil extracellular DNA traps. *Sci. Transl. Med.* **2016**, *8*, 361ra138. [CrossRef] [PubMed]
37. Tohme, S.; Yazdani, H.O.; Al-Khafaji, A.B.; Chidi, A.P.; Loughran, P.; Mowen, K.; Wang, Y.; Simmons, R.L.; Huang, H.; Tsung, A. Neutrophil Extracellular Traps Promote the Development and Progression of Liver Metastases after Surgical Stress. *Cancer Res.* **2016**, *76*, 1367–1380. [CrossRef] [PubMed]
38. Huh, S.J.; Liang, S.; Sharma, A.; Dong, C.; Robertson, G.P. Transiently Entrapped Circulating Tumor Cells Interact with Neutrophils to Facilitate Lung Metastasis Development. *Cancer Res.* **2010**, *70*, 6071–6082. [CrossRef]
39. Leal, A.C.; Mizurini, D.M.; Gomes, T.; Rochael, N.C.; Saraiva, E.M.; Dias, M.S.; Werneck, C.C.; Sielski, M.S.; Vicente, C.P.; Monteiro, R.Q. Tumor-Derived Exosomes Induce the Formation of Neutrophil Extracellular Traps: Implications For The Establishment of Cancer-Associated Thrombosis. *Sci. Rep.* **2017**, *7*, 1–12. [CrossRef]
40. Cedervall, J.; Zhang, Y.; Huang, H.; Zhang, L.; Femel, J.; Dimberg, A.; Olsson, A.K. Neutrophil extracellular

traps accumulate in peripheral blood vessels and compromise organ function in tumor-bearing animals. *Cancer Res.* **2015**, *75*, 2653–2662. [CrossRef]

41. Thålin, C.; Demers, M.; Blomgren, B.; Wong, S.L.; von Arbin, M.; von Heijne, A.; Laska, A.C.; Wallén, H.; Wagner, D.D.; Aspberg, S. NETosis promotes cancer-associated arterial microthrombosis presenting as ischemic stroke with troponin elevation. *Thromb. Res.* **2016**, *139*, 56–64. [CrossRef] [PubMed]
42. Najmeh, S.; Cools-Lartigue, J.; Rayes, R.F.; Gowing, S.; Vourtzoumis, P.; Bourdeau, F.; Giannias, B.; Berube, J.; Rousseau, S.; Ferri, L.E.; et al. Neutrophil extracellular traps sequester circulating tumor cells via β1-integrin mediated interactions. *Int. J. Cancer* **2017**, *140*, 2321–2330. [CrossRef] [PubMed]
43. Patel, S.; Fu, S.; Mastio, J.; Dominguez, G.A.; Purohit, A.; Kossenkov, A.; Lin, C.; Alicea-Torres, K.; Sehgal, M.; Nefedova, Y.; et al. Unique pattern of neutrophil migration and function during tumor progression. *Nat. Immunol.* **2018**, *19*, 1236–1247. [CrossRef] [PubMed]
44. Engblom, C.; Pfirschke, C.; Zilionis, R.; Da Silva Martins, J.; Bos, S.A.; Courties, G.; Rickelt, S.; Severe, N.; Baryawno, N.; Faget, J.; et al. Osteoblasts remotely supply lung tumors with cancer-promoting SiglecF high neutrophils. *Science* **2017**, *358*, eaal5081. [CrossRef] [PubMed]
45. Veglia, F.; Perego, M.; Gabrilovich, D. Myeloid-derived suppressor cells coming of age. *Nat. Immunol.* **2018**, *19*, 108–119. [CrossRef] [PubMed]
46. Liu, Q.; Liao, Q.; Zhao, Y. Myeloid-derived suppressor cells (MDSC) facilitate distant metastasis of malignancies by shielding circulating tumor cells (CTC) from immune surveillance. *Med. Hypotheses* **2016**, *87*, 34–39. [CrossRef] [PubMed]
47. Sprouse, M.L.; Welte, T.; Boral, D.; Liu, H.N.; Yin, W.; Vishnoi, M.; Goswami-Sewell, D.; Li, L.; Pei, G.; Jia, P.; et al. PMN-MDSCs Enhance CTC Metastatic Properties through Reciprocal Interactions via ROS/Notch/Nodal Signaling. *Int. J. Mol. Sci.* **2019**, *20*, 1916. [CrossRef]
48. Arnoletti, J.P.; Fanaian, N.; Reza, J.; Sause, R.; Almodovar, A.J.O.; Srivastava, M.; Patel, S.; Veldhuis, P.P.; Griffith, E.; Shao, Y.P.; et al. Pancreatic and bile duct cancer circulating tumor cells (CTC) form immune-resistant multi-cell type clusters in the portal venous circulation. *Cancer Biol. Ther.* **2018**, *19*, 887–897. [CrossRef]
49. Gasic, G.J.; Gasic, T.B.; Galanti, N.; Johnson, T.; Murphy, S. Platelet—tumor-cell interactions in mice. The role of platelets in the spread of malignant disease. *Int. J. Cancer* **1973**, *11*, 704–718. [CrossRef]
50. Nilsson, R.J.A.; Balaj, L.; Hulleman, E.; van Rijn, S.; Pegtel, D.M.; Walraven, M.; Widmark, A.; Gerritsen, W.R.; Verheul, H.M.; Vandertop, W.P.; et al. Blood platelets contain tumor-derived RNA biomarkers. *Blood* **2011**, *118*, 3680–3683. [CrossRef]
51. Joosse, S.A.; Pantel, K. Tumor-Educated Platelets as Liquid Biopsy in Cancer Patients. *Cancer Cell* **2015**, *28*, 552–554. [CrossRef] [PubMed]
52. Best, M.G.; Sol, N.; Kooi, I.; Tannous, J.; Westerman, B.A.; Rustenburg, F.; Schellen, P.; Verschueren, H.; Post, E.; Koster, J.; et al. RNA-Seq of Tumor-Educated Platelets Enables Blood-Based Pan-Cancer, Multiclass, and Molecular Pathway Cancer Diagnostics. *Cancer Cell* **2015**, *28*, 666–676. [CrossRef] [PubMed]
53. Franco, A.T.; Corken, A.; Ware, J. Platelets at the interface of thrombosis, inflammation, and cancer. *Blood* **2015**, *126*, 582–588. [CrossRef] [PubMed]
54. Menter, D.G.; Tucker, S.C.; Kopetz, S.; Sood, A.K.; Crissman, J.D.; Honn, K.V. Platelets and cancer: A casual or causal relationship: Revisited. *Cancer Metastasis Rev.* **2014**, *33*, 231–269. [CrossRef] [PubMed]
55. Labelle, M.; Begum, S.; Hynes, R.O.O. Direct Signaling between Platelets and Cancer Cells Induces an Epithelial-Mesenchymal-Like Transition and Promotes Metastasis. *Cancer Cell* **2011**, *20*, 576–590. [CrossRef] [PubMed]
56. Ward, Y.; Lake, R.; Faraji, F.; Sperger, J.; Martin, P.; Gilliard, C.; Ku, K.P.; Rodems, T.; Niles, D.; Tillman, H.; et al. Platelets Promote Metastasis via Binding Tumor CD97 Leading to Bidirectional Signaling that Coordinates Transendothelial Migration. *Cell Rep.* **2018**, *23*, 808–822. [CrossRef]
57. Hanley, W.; McCarty, O.; Jadhav, S.; Tseng, Y.; Wirtz, D.; Konstantopoulos, K. Single molecule characterization of P-selectin/ligand binding. *J. Biol. Chem.* **2003**, *278*, 10556–10561. [CrossRef]
58. Camerer, E.; Qazi, A.A.; Duong, D.N.; Cornelissen, I.; Advincula, R.; Coughlin, S.R. Platelets, protease-activated receptors, and fibrinogen in hematogenous metastasis. *Blood* **2004**, *104*, 397–401. [CrossRef]
59. Gay, L.J.; Felding-Habermann, B. Contribution of platelets to tumour metastasis. *Nat. Rev. Cancer* **2011**, *11*, 123–134. [CrossRef]
60. Lu, Y.; Lian, S.; Ye, Y.; Yu, T.; Liang, H.; Cheng, Y.; Xie, J.; Zhu, Y.; Xie, X.; Yu, S.; et al. S-Nitrosocaptopril

prevents cancer metastasis in vivo by creating the hostile bloodstream microenvironment against circulating tumor cells. *Pharmacol. Res.* **2019**, *139*, 535–549. [CrossRef]

61. Wang, Y.; Sun, Y.; Li, D.; Zhang, L.; Wang, K.; Zuo, Y.; Gartner, T.K.; Liu, J. Platelet P2Y12 is involved in murine pulmonary metastasis. *PLoS ONE* **2013**, *8*, 1–12. [CrossRef] [PubMed]
62. Orellana, R.; Kato, S.; Erices, R.; Bravo, M.L.; Gonzalez, P.; Oliva, B.; Cubillos, S.; Valdivia, A.; Ibañez, C.; Brañes, J.; et al. Platelets enhance tissue factor protein and metastasis initiating cell markers, and act as chemoattractants increasing the migration of ovarian cancer cells. *BMC Cancer* **2015**, *15*, 290. [CrossRef] [PubMed]
63. Läubli, H.; Borsig, L. Selectins promote tumor metastasis. *Semin. Cancer Biol.* **2010**, *20*, 169–177. [CrossRef] [PubMed]
64. Mammadova-Bach, E.; Zigrino, P.; Brucker, C.; Bourdon, C.; Freund, M.; De Arcangelis, A.; Abrams, S.I.; Orend, G.; Gachet, C.; Mangin, P.H. Platelet integrin $\alpha 6\beta 1$ controls lung metastasis through direct binding to cancer cell–derived ADAM9. *JCI Insight* **2016**, *1*, 1–17. [CrossRef] [PubMed]
65. Schumacher, D.; Strilic, B.; Sivaraj, K.K.; Wettschureck, N.; Offermanns, S. Platelet-Derived Nucleotides Promote Tumor-Cell Transendothelial Migration and Metastasis via P2Y2 Receptor. *Cancer Cell* **2013**, *24*, 130–137. [CrossRef] [PubMed]
66. Velez, J.; Enciso, L.J.; Suarez, M.; Fiegl, M.; Grismaldo, A.; López, C.; Barreto, A.; Cardozo, C.; Palacios, P.; Morales, L.; et al. Platelets Promote Mitochondrial Uncoupling and Resistance to Apoptosis in Leukemia Cells: A Novel Paradigm for the Bone Marrow Microenvironment. *Cancer Microenviron.* **2014**, *7*, 79–90. [CrossRef]
67. Nieswandt, B.; Hafner, M.; Echtenacher, B.; Männel, D.N. Lysis of tumor cells by natural killer cells in mice is impeded by platelets. *Cancer Res.* **1999**, *59*, 1295–1300.
68. Palumbo, J.S.; Talmage, K.E.; Massari, J.V.; La Jeunesse, C.M.; Flick, M.J.; Kombrinck, K.W.; Hu, Z.; Barney, K.A.; Degen, J.L. Tumor cell-associated tissue factor and circulating hemostatic factors cooperate to increase metastatic potential through natural killer cell-dependent and -independent mechanisms. *Blood* **2007**, *110*, 133–141. [CrossRef]
69. Placke, T.; Örgel, M.; Schaller, M.; Jung, G.; Rammensee, H.G.; Kopp, H.G.; Salih, H.R. Platelet-derived MHC class I confers a pseudonormal phenotype to cancer cells that subverts the antitumor reactivity of natural killer immune cells. *Cancer Res.* **2012**, *72*, 440–448. [CrossRef]
70. Hu, L. Role of endogenous thrombin in tumor implantation, seeding, and spontaneous metastasis. *Blood* **2004**, *104*, 2746–2751. [CrossRef]
71. Gareau, A.J.; Brien, C.; Gebremeskel, S.; Liwski, R.S.; Johnston, B.; Bezuhly, M. Ticagrelor inhibits platelet–tumor cell interactions and metastasis in human and murine breast cancer. *Clin. Exp. Metastasis* **2018**, *35*, 25–35. [CrossRef] [PubMed]
72. Yang, L.; Lv, Z.; Xia, W.; Zhang, W.; Xin, Y.; Yuan, H.; Chen, Y.; Hu, X.; Lv, Y.; Xu, Q.; et al. The effect of aspirin on circulating tumor cells in metastatic colorectal and breast cancer patients: A phase II trial study. *Clin. Transl. Oncol.* **2018**, *20*, 912–921. [CrossRef] [PubMed]
73. Pathria, P.; Louis, T.L.; Varner, J.A. Targeting Tumor-Associated Macrophages in Cancer. *Trends Immunol.* **2019**, *40*, 310–327. [CrossRef] [PubMed]
74. Hamilton, G.; Rath, B.; Klameth, L.; Hochmair, M.J. Small cell lung cancer: Recruitment of macrophages by circulating tumor cells. *Oncoimmunology* **2016**, *5*, 1–9. [CrossRef] [PubMed]
75. Hamilton, G.; Rath, B. Circulating tumor cell interactions with macrophages: Implications for biology and treatment. *Transl. Lung Cancer Res.* **2017**, *6*, 418–430. [CrossRef] [PubMed]
76. Cavassani, K.A.; Meza, R.J.; Habiel, D.M.; Chen, J.F.; Montes, A.; Tripathi, M.; Martins, G.A.; Crother, T.R.; You, S.; Hogaboam, C.M.; et al. Circulating monocytes from prostate cancer patients promote invasion and motility of epithelial cells. *Cancer Med.* **2018**, *7*, 4639–4649. [CrossRef]
77. Wei, C.; Yang, C.; Wang, S.; Shi, D.; Zhang, C.; Lin, X.; Liu, Q.; Dou, R.; Xiong, B. Crosstalk between cancer cells and tumor associated macrophages is required for mesenchymal circulating tumor cell-mediated colorectal cancer metastasis. *Mol. Cancer* **2019**, *18*, 1–23. [CrossRef]
78. Zhang, Y.; Zhou, N.; Yu, X.; Zhang, X.; Li, S.; Lei, Z.; Hu, R.; Li, H.; Mao, Y.; Wang, X.; et al. Tumacrophage: Macrophages transformed into tumor stem-like cells by virulent genetic material from tumor cells. *Oncotarget* **2017**, *8*, 82326–82343. [CrossRef]

79. Gast, C.E.; Silk, A.D.; Zarour, L.; Riegler, L.; Burkhart, J.G.; Gustafson, K.T.; Parappilly, M.S.; Roh-Johnson, M.; Goodman, J.R.; Olson, B.; et al. Cell fusion potentiates tumor heterogeneity and reveals circulating hybrid cells that correlate with stage and survival. *Sci. Adv.* **2018**, *4*, eaat7828. [CrossRef]
80. Cao, M.-F.; Chen, L.; Dang, W.-Q.; Zhang, X.-C.; Zhang, X.; Shi, Y.; Yao, X.-H.; Li, Q.; Zhu, J.; Lin, Y.; et al. Hybrids by tumor-associated macrophages × glioblastoma cells entail nuclear reprogramming and glioblastoma invasion. *Cancer Lett.* **2019**, *442*, 445–452. [CrossRef]
81. Hanahan, D.; Weinberg, R.A. Hallmarks of Cancer: The Next Generation. *Cell* **2011**, *144*, 646–674. [CrossRef] [PubMed]
82. Ribas, A.; Wolchok, J.D. Cancer immunotherapy using checkpoint blockade. *Science* **2018**, *1355*, 1350–1355. [CrossRef] [PubMed]
83. Ye, L.; Zhang, F.; Li, H.; Yang, L.; Lv, T.; Gu, W.; Song, Y. Circulating Tumor Cells Were Associated with the Number of T Lymphocyte Subsets and NK Cells in Peripheral Blood in Advanced Non-Small-Cell Lung Cancer. *Dis. Markers* **2017**, *2017*, 1–6. [CrossRef] [PubMed]
84. Mego, M.; Gao, H.; Cohen, E.; Anfossi, S.; Giordano, A.; Sanda, T.; Fouad, T.; De Giorgi, U.; Giuliano, M.; Woodward, W.; et al. Circulating Tumor Cells (CTC) Are Associated with Defects in Adaptive Immunity in Patients with Inflammatory Breast Cancer. *J. Cancer* **2016**, *7*, 1095–1104. [CrossRef] [PubMed]
85. Xue, D.; Xia, T.; Wang, J.; Chong, M.; Wang, S.; Zhang, C. Role of regulatory T cells and CD8+ T lymphocytes in the dissemination of circulating tumor cells in primary invasive breast cancer. *Oncol. Lett.* **2018**, *16*, 3045–3053. [CrossRef] [PubMed]
86. Wang, Y.; Kim, T.H.; Fouladdel, S.; Zhang, Z.; Soni, P.; Qin, A.; Zhao, L.; Azizi, E.; Lawrence, T.S.; Ramnath, N.; et al. PD-L1 Expression in Circulating Tumor Cells Increases during Radio(chemo)therapy and Indicates Poor Prognosis in Non-small Cell Lung Cancer. *Sci. Rep.* **2019**, *9*, 566. [CrossRef] [PubMed]

8

Capture and Detection of Circulating Glioma Cells Using the Recombinant VAR2CSA Malaria Protein

Sara R. Bang-Christensen [1,2], Rasmus S. Pedersen [1], Marina A. Pereira [1], Thomas M. Clausen [1], Caroline Løppke [1], Nicolai T. Sand [1], Theresa D. Ahrens [1], Amalie M. Jørgensen [1], Yi Chieh Lim [3], Louise Goksøyr [1], Swati Choudhary [1], Tobias Gustavsson [1], Robert Dagil [1], Mads Daugaard [4], Adam F. Sander [1], Mathias H. Torp [5], Max Søgaard [6], Thor G. Theander [1], Olga Østrup [5], Ulrik Lassen [7], Petra Hamerlik [3], Ali Salanti [1,*] and Mette Ø. Agerbæk [1,*]

1 Centre for Medical Parasitology at Department for Immunology and Microbiology, Faculty of Health and Medical Sciences, University of Copenhagen and Department of Infectious Disease, Copenhagen University Hospital, 2200 Copenhagen, Denmark
2 VarCT Diagnostics, 2200 Copenhagen, Denmark
3 Danish Cancer Society Research Center, 2100 Copenhagen, Denmark
4 Department of Urologic Sciences, University of British Columbia, and Vancouver Prostate Centre, Vancouver, BC V6H 3Z6, Canada
5 Centre for Genomic Medicine, Copenhagen University Hospital, 2100 Copenhagen, Denmark
6 ExpreS²ion Biotechnologies, SCION-DTU Science Park, 2970 Hørsholm, Denmark
7 Department of Oncology, Copenhagen University Hospital, 2100 Copenhagen, Denmark
* Correspondence: salanti@sund.ku.dk (A.S.); mettea@sund.ku.dk (M.Ø.A.)

Abstract: Diffuse gliomas are the most common primary malignant brain tumor. Although extracranial metastases are rarely observed, recent studies have shown the presence of circulating tumor cells (CTCs) in the blood of glioma patients, confirming that a subset of tumor cells are capable of entering the circulation. The isolation and characterization of CTCs could provide a non-invasive method for repeated analysis of the mutational and phenotypic state of the tumor during the course of disease. However, the efficient detection of glioma CTCs has proven to be challenging due to the lack of consistently expressed tumor markers and high inter- and intra-tumor heterogeneity. Thus, for this field to progress, an omnipresent but specific marker of glioma CTCs is required. In this article, we demonstrate how the recombinant malaria VAR2CSA protein (rVAR2) can be used for the capture and detection of glioma cell lines that are spiked into blood through binding to a cancer-specific oncofetal chondroitin sulfate (ofCS). When using rVAR2 pull-down from glioma cells, we identified a panel of proteoglycans, known to be essential for glioma progression. Finally, the clinical feasibility of this work is supported by the rVAR2-based isolation and detection of CTCs from glioma patient blood samples, which highlights ofCS as a potential clinical target for CTC isolation.

Keywords: circulating tumor cells (CTCs); glioma; biomarker; rVAR2; malaria; enrichment and detection technologies

---

1. Introduction

Diffuse gliomas are the most common primary malignant brain tumors [1]. As the name implies, a general trait of these tumors is their diffuse invasion into the brain parenchyma, which impedes complete surgical resection and most likely explains the poor prognosis and frequent local recurrence [2]. A precise classification of diffuse gliomas is needed for the optimal diagnosis, stratification, and treatment of patients [3,4]. During the past decades, technologies for biopsy-based classification of diffuse gliomas have increased in their complexity [5–7]. However, repeated access to information regarding tumor progression remains challenging, due to the risk and inconvenience that are associated with performing patient brain biopsies. Several studies indicate that the cells constituting the infiltrative and invasive front of gliomas harbor tumor-initiating capacity and may be responsible for drug resistance and tumor recurrence [8–10]. Most likely, these migrating cells would also be the ones accessing the blood stream. Therefore, the isolation of circulating tumor cells (CTCs) from a liquid biopsy, such as blood or cerebrospinal fluid, could provide non-invasive, repeatable access to primary glioma cells for molecular analysis.

Tumors of the central nervous system were until recently not considered to be metastatic. However, organ recipients receiving organs from patients who succumbed to glioblastoma multiforme (GBM) have developed extracranial metastases, which strongly suggests that these organs harbored disseminated GBM cells [11]. In line with this, a few studies using different isolation and detection methods have detected CTCs in blood from glioma patients [12–15]. Taken together, these studies provide evidence that invasive glioma cells successfully intravasate to the blood circulation and may therefore potentially become an important and easily available source of information on the mutational and phenotypic state of the primary tumor. Molecular analysis of the circulating glioma cells could provide basis for the design and monitoring of personalized treatment strategies, as it has been the case with breast, prostate, and lung cancer [16–18]. However, the high degree of heterogeneity within gliomas constitutes a hindrance for the effective isolation and detection of such CTCs. The use of antibodies towards one or few protein surface markers will render the detection fragile to changes in the expression level of the selected marker. On the other hand, targeting several proteins by using an antibody cocktail increases the risk of false positives and high background levels due to healthy cells expressing one or more of the included markers. Hence, a single marker to distinguish a broad repertoire of glioma CTCs from healthy white blood cells (WBCs) is needed.

Notably, little attention has been given to cancer specific glycosylation patterns on CTCs and strategies for targeting these. Glycosaminoglycans (GAGs) are carbohydrate structures, which are added to proteins, called proteoglycans, as secondary modifications. Chondroitin sulfate (CS) is one type of GAG that is built up by repeated disaccharide units made up of *N*-acetyl-D-galactosamine and D-glucuronic acid units [19]. While the CS backbone structure is simple, an immense heterogeneity is achieved through additional modifications, such as alternate sulfation of component hydroxyl groups [20]. The long structures of repeated disaccharide units are implicated in the regulation of many oncogenic processes and CS up-regulation or modifications have been associated with cancer progression [21]. In the case of glioma, several chondroitin sulfate proteoglycans (CSPGs), including versican and NG2/CSPG4, have been shown to be up-regulated and involved in tumor cell growth, migration, and invasion, as well as in promoting angiogenesis [22–24].

We have previously shown that the recombinantly expressed VAR2CSA malarial protein (rVAR2) specifically binds a distinct CS structure, termed oncofetal chondroitin sulfate (ofCS), which is present in the placenta and on almost all cancer cells with limited expression in other normal tissues [25]. Although CS is present elsewhere in the vasculature, parasite infected erythrocytes that express VAR2CSA only bind in the placenta [26]. Thus, the protein has been evolutionary refined to specifically bind to ofCS and not to CS present in other organs.

We recently published a CTC isolation method demonstrating the use of rVAR2 protein on magnetic beads for the capture of CTCs from prostate, pancreatic, and hepatic cancer patient blood samples [27]. However, in terms of CTC detection after enrichment, this assay was still dependent on antibody staining using the epithelial marker cytokeratin (CK), thus limiting the applicability to cancers of epithelial origin. In this study, we investigate whether the rVAR2 protein can be applied in both the capture and detection step and thereby broaden the use of our CTC-isolation platform to include circulating glioma cells. We show that rVAR2 binds glioma cells of both adult and pediatric origin. We find that rVAR2 interacts with ofCS on several CSPGs that have shown to be up-regulated in GBM, including CD44, APLP2, CSPG4, PTPRZ1, versican, and syndecan 1. Furthermore, we confirm that rVAR2 binding is retained on a low-grade pediatric glioma cell line (Res259) after incubation with

Transforming Growth Factor beta (TGF-β). We validate that the rVAR2-based CTC capture enables capture of rare glioma cells spiked into blood, and show proof-of-concept of using rVAR2 for both the capture and downstream detection of such glioma cells. Importantly, we capture and detect glioma CTCs from glioma patient blood samples. Finally, CTCs from three patient samples are analyzed by whole exome sequencing (WES), which confirms the presence of glioma-associated mutations.

2. Materials and Methods

2.1. Production of Proteins

The recombinant DBL1-ID2a subunit or the shorter version, ID1-ID2a, of VAR2CSA (rVAR2) was expressed in SHuffle T7 Express Competent *E. coli* (NEB) and purified using affinity chromatography (HisTrap HP, GE Healthcare, Uppsala, Sweden), followed by cation exchange chromatography (HiTrap IMAC SP HP, GE Healthcare). Both constructs included a C-terminal 6x His-tag and V5-tag, as well as an N-terminal SpyTag. For the staining of CTCs, we produced the recombinant ID1-ID2a subunit of VAR2CSA in S2 insect, which encoded an N-terminal twin-strep affinity tag. Protein that was expressed in S2 cells was captured from the supernatant by Streptactin XT chromatography (Iba, GmbH, Germany) and polished by size exclusion (Superdex 200pg, GE).

Subsequently, purified monomeric proteins were identified by SDS-PAGE. All of the proteins were quality tested by decorin binding in ELISA and by ofCS binding on cancer cells using flow cytometry to ensure specificity.

The SpyCatcher domain was produced in *E. coli* BL21 as a soluble poly-HIS tagged protein, and purified using affinity chromatography (HisTrap, GE Healthcare), followed by anion exchange (HiTrap IMAC Q HP column, GE Healthcare). Purity was determined by SDS page and quality of protein was ensured by testing the capacity to form an isopeptide bond to the Spy-tagged rVAR2 protein. The SpyCatcher was biotinylated using NHS-biotin (Sigma-Aldrich, Steinheim, Germany). NHS-Biotin was dissolved in DMSO and added in 10 molar excess to the SpyCatcher. After a 1-h incubation at room temperature, the biotinylated SpyCatcher was purified using a zeba spin column with a 7 kDa cut off.

2.2. Cell Cultures

Janine Erler and Lara Perryman (Biotech Research & Innovation Centre, University of Copenhagen, Denmark) kindly provided the KNS-42, Res259, U87mg, and U118mg cell lines [28]. The U87mg cells were grown in EMEM, Res259 and KNS-42 were grown in DMEM/F12, and U118mg were grown in DMEM GlutaMAX. All culture media were supplemented with 10% fetal bovine serum, penicillin, streptomycin, and L-glutamine (except DMEM). The primary GBM cell, GBM02, was maintained as an in vivo model in NOG mice with ethical approval (2012-15-2934-00636). Tumor xenograft was dissociated using a papain dissociation kit (Worthington). Isolated ex-vivo GBM02 cells were authenticated by STR profiling and grown as neurospheres in Neurobasal media containing B-27 supplement (Gibco), GlutaMax (Gibco), 10 ng/mL EGF, and 10 ng/mL FGF, as described previously [29]. All of the cell lines were passaged at a regular basis and maintained at 5% CO_2 at 37 °C.

2.3. Flow Cytometry

The cells were grown to 70–80% confluency in appropriate growth media and then harvested in an EDTA detachment solution (Cellstripper®, Corning™). 100,000 cancer cells, WBCs from 100 µL RBC lysed blood, or a mixture of both (according to the description in the Results section) were added to each well in a 96 well plate. Cells were incubated with rVAR2 (400 nM–25 nM) for 30 min. at 4 °C. Subsequently, cells were washed twice and then incubated with FITC-labelled anti-V5 antibody (Invitrogen, 1:500) for 30 min. at 4 °C. Finally, the cells were washed twice and analyzed in a LSR-II (BD Biosciences) for staining intensity. Geometric mean fluorescent intensity (MFI) values were normalized to signals that were obtained when only adding the FITC-labelled anti-V5 antibody.

2.4. *TGF-β Treatment of Res259 Cells*

Res259 were seeded in a density of 2400–5200 cells/cm^2 in DMEM/F12 that was supplemented with 10% FBS in a T25 culture flask. Cells were allowed to attach for 24 h. After this, cells were treated with TGF-β (Cat. no. T7039, Sigma-Aldrich) at a concentration of 20 ng/mL or equal volumes of TGF-β suspension buffer as control (0.2 μm filtered distilled water) for 72 h to induce the mesenchymal transition. Transition was confirmed by changes in the expression of mesenchymal protein markers using western blot as well as changes in morphology.

For western blot analysis, the cells were lysed with EBC lysis buffer containing PhosSTOP (Sigma-Aldrich) and cOmplete EDTA-free Protease Inhibitor Cocktail (Roche) for 30 min. Protein extract was balanced using Bradford assay. An equal amount of protein lysates were loaded onto a NuPAGE 4–12% Bis-Tris gel (ThermoFisher Scientific), after which the samples were transferred to a nitrocellulose membrane (Biorad). Membranes were blocked in 5% skimmed milk powder in TBS-T. Anti-GAPDH (14C10) antibody (Cell Signaling, 1:1000), and anti-β-catenin (1:500), anti-*N*-cadherin (1:500), and anti-vimentin (1:1000) primary antibodies from the EMT Antibody Sampler Kit (Cell Signaling) were added to the membranes in TBST-T supplemented with 2% skimmed milk powder and incubated overnight at 4 °C. Following three washes in TBS-T, the membranes were incubated with HRP-linked goat anti-rabbit IgG (Cat. no. P0448, Dako, 1:2000) for 1 h at room temperature and the reactivity was detected using LumiGlo Reserve Chemiluminescent Substrate (KPL). Uncropped images of the membranes can be seen in Figure S6.

For fluorescent visualization of changes in morphology, the cells were grown on glass slides and fixed in 4% paraformaldehyde (PFA), washed three times in PBS, blocked with 1% BSA in PBS, and stained with Alexa Fluor® 594 Phalloidin (ThermoFisher, 1:40) for 20 min. at room temperature. Cells were subsequently stained with DAPI (Life Technologies) and mounted using FluorSave Reagent (Merck Millipore, Darmstadt, Germany). Staining was analyzed using Nikon TE2000-E C1 confocal microscope with 60× oil immersion objective lens (DIC).

For flow cytometry-based analysis of cells before and after induction with TGF-β, applying the same procedure as in "2.3. Flow Cytometry".

2.5. *Immunoprecipitation and Proteomics*

Membrane proteins were extracted by lysing the cells with EBC lysis buffer supplemented with a protease inhibitor cocktail (Roche). Biotinylated rVAR2 was added to the lysate and the mix was incubated overnight at 4 °C. The rVAR2 and bound protein was pulled down on streptavidin-coated dynabeads (MyOne C1, Invitrogen).

The pulled down lysate was dissolved in non-reducing LDS loading buffer (Invitrogen). The protein was reduced in 1 mM DTT and alkylated with 5.5 mM iodoacetamide. The samples were then run 1 cm into Bis-Tris gels and stained with coomasie blue. The protein was cut out, washed, and in-gel digested with trypsin. The resulting peptides were captured and washed using a C18 resin stage-tipping [30]. The peptides were sequenced using a Phusion Orbitrap Mass Spectrometer. Sample analysis and hit verification was performed using the MaxQuant software. All of the samples were verified against the control samples of cell lysates without rVAR2.

2.6. *Proximity Ligation Assay (PLA)*

The PLA protocol was run according to the manufacturer's instructions (Sigma-Aldrich). U87mg, U118mg, and KNS-42 cells were seeded on laminin-coated coverslips and fixed in 4% PFA. Unspecific binding of antibodies was minimized by incubating with a blocking solution with 1% BSA and 5% FBS in PBS for 1 h at room temperature. The samples were incubated with primary antibodies together with rVAR2 or rDBL4 over night at 4 °C in the following concentrations: rVAR2 (50 nM), rDBL4 (50 nM), anti-NRP1 (Cat. no. ab81321, 1:250), anti-NRP2 (Cat. no. sc-13117, 1:50), anti-PTPRZ1 (Cat. no. HPA015103, 1:61), anti-VCAN (Cat. no. HPA004726, 1:50), anti-CSPG4 (Cat. no. ab20156,

1:200), anti-DCN (Cat. no. PA5-27370, 1:100), anti-CD44 (Cat. no. BBA10, 1:200), anti-SDC1 (Cat. no. ab34164, 1:50), and anti-SDC4 antibody (Cat. no. HPA005716, 1:80). Between incubations, the cells were washed in Wash Buffer A (DUO82049, Sigma-Aldrich). An anti-V5 antibody (Invitrogen, 1:500) was used to detect rVAR2. The cells were then stained with Duolink® In Situ PLA® Probe Anti-Mouse MINUS (DUO92004) and Duolink® In Situ PLA® Probe Anti-Rabbit PLUS (DUO92002) diluted in Antibody Diluent (DUO82008). The cells were then treated with the ligation solution, followed by incubation with the amplification solution, which were provided with the kit Duolink® In Situ Detection Reagents Orange (DUO92007). The cells were washed with Wash Buffer B (DUO82048). The slides were mounted using Duolink® In Situ Mounting Medium with DAPI (DUO82040), and the results were then analyzed under a Nikon TE2000-E C1 confocal microscope with a 60× oil objective. A total of 75–100 cells were imaged per sample. The images were analyzed using the BlobFinder software (version 3.2.). Negative controls using the recombinant DBL4 domain of VAR2CSA are found in Figure S1.

2.7. *Immunocytochemistry of Cancer Cells Mixed with White Blood Cells (WBCs)*

Two mL blood from a healthy donor was drawn in a LBgard® vacutainer (BioMatrica). The blood sample was diluted 10 times in Red Blood Cell (RBC) lysis buffer resulting in a final concentration of 0.155M ammonium chloride, 0.01M potassium hydrogen carbonate and 0.1 mM EDTA, and incubated for 13 min. After centrifugation at 400× *g* for 8 min. the pelleted cells were resuspended in 2 mL Dulbecco's PBS (without Ca^{2+} and Mg^{2+}) supplemented with 2% Fetal Bovine Serum (FBS) (Gibco) and transferred to eppendorf tubes in aliquots of 0.5 mL. Res529, KNS-42 and U87mg were detached using 1 mL CellStripper (Corning™) and resuspended in their respective media. Approximately 2000 cells were added to each aliquot of WBCs. The samples were washed once, prior to incubation with a cocktail of 200 nM rVAR2, CF488-labelled anti-V5 (Cat.no. 20440, Biotium, 1:150), PE-labelled anti-CD45 [5B1] (Cat.no. 170-078-081, MACS Miltenyi Biotec, 1:40) and PE-labelled anti-CD66b [REA306] antibodies (Cat.no. 130-104-396, MACS Miltenyi BioTec, 1:40) in Dulbecco's PBS with 2% FBS for 30 min. at 4 °C. Finally, the cells were fixed with 4% PFA, stained with DAPI (Cat. no. D1306, Life Technologies) and mounted using Faramount Aquous Mounting Media (Dako). The slides were imaged using the 10× objective of Cytation 3 Cell Imaging Multi-Mode Reader (BioTek, Europe).

2.8. *Preparation of rVAR2-Coated Beads*

The Spy-tagged DBL1-ID2a or ID1-ID2a (rVAR2) was mixed with the biotinylated SpyCatcher in a 1.2:1 ratio and then incubated at room temperature for 1 h. After this step, the biotinylated rVAR2 protein was incubated with CELLection™ Biotin Binder Dynabeads® (4.5 µm) at room temperature for at least 30 min. resulting in rVAR2-coated beads (0.43 µg biotinylated protein per µL bead suspension). The remaining protein or antibody was removed by carefully washing the beads in Pierce™ Protein-Free (PBS) blocking buffer (Cat. no. 37572, ThermoFisher) three times, each time using a neodymium magnet (10 × 12 mm) for dragging beads into a pellet.

2.9. *Spike-In Experiments*

Prior to the spike-in experiments, the cancer cells were harvested with CellStripper (Corning™) or TrypLe (Cat. no. 12604013, Gibco) (only used for GBM02 cells) and resuspended in culture medium. For spike-in experiments measuring the efficiency of recovery, cancer cells were prestained using CellTracker™ Green CMFDA Dye (Cat. no. C7025, ThermoFisher), according to manufacturer's protocol for cells in suspension. Following staining, the cells were resuspended in complete growth media and incubated for 30 min. under normal growth conditions in order to recover.

Cell concentration was measured by manually counting the number of viable cells in a 1:1 mixture with Trypan Blue solution (Sigma-Aldrich). Subsequently, the cells were diluted to 10,000 cells/mL in Dulbecco's PBS and the desired number of cells were spiked into 3 mL blood. Triplicates of the spike-in volume (e.g., 10 µL for 100 cells) were placed on a glass slide and cells were manually counted under a light microscope (10× objective) in order to confirm the exact number of cells spiked into the blood. The average of the cell counts was used when calculating the percentage of recovery. Each of the spike-in experiments were repeated at least twice with 2–4 replicates per test.

When spiking in low cell numbers (5–10 cells), serial dilutions were made using cell culture media. The cell suspension was transferred to a 96-well plate and counted under a light microscope (10× objective) to ensure precise cell count before spike-in. Finally, cells were directly added from the well to the 3 mL blood sample.

After adding the cancer cells to the blood, the samples were immediately processed, as described in Section 2.11.

2.10. Patient Samples

Up to 9 mL blood samples from glioma patients were collected under ethical approval (journal no. H-3-2009-136). Informed written consent was obtained for all of the enrolled subjects. Blood was received in K2 EDTA-tubes and processed within 2 h of collection.

2.11. CTC Isolation from Blood

Three mL blood samples were lysed in 27 mL RBC lysis buffer reaching a final concentration of 0.155 M ammonium chloride, 0.01 M potassium hydrogen carbonate, and 0.1 mM EDTA for 13 min. After centrifugation at 400× *g* for 8 min., the cell pellet was gently washed in Dulbecco's PBS. The centrifugation step was repeated, and finally the cells were resuspended in 0.6 mL Pierce™ Protein-Free (PF) PBS blocking buffer (Cat. no. 37572, ThermoFisher) and then transferred to a low retention microcentrifuge tube (Fisherbrand). Under these conditions, the cells were incubated with ~1.8 million rVAR2-coated magnetic beads at 4 °C for 20 min. A neodymium cylinder magnet was used to drag cells bound to beads towards the side of the tube, enabling removal of supernatant. Cells were then fixed in 4% PFA for 5 min. and resuspended in Pierce™ Protein-Free (PBS) blocking buffer diluted 1:10 in Dulbecco's PBS.

2.12. CTC Staining and Enumeration

For spike-in experiments where the cells were prestained with CellTracker Green, cells were stained with DAPI (Cat. no. D1306, Life Technologies) diluted in Dulbecco's PBS with 0.5% BSA and 2 mM EDTA for 5 min. at room temperature. Following one wash, the bead-bound cells were added to a SensoPlate™ (24-well, glass bottom) (Cat. no. 662892, Greiner Bio-One). Excess liquid was removed by holding the bead-bound cells in place with a magnet underneath the plate, and the samples were mounted using Faramount Aqueous Mounting Media (DAKO). The entire well was scanned for DAPI and CellTracker signal using the Cytation™ 3 Cell Imaging Multi-Mode Reader and manually enumerated using the Gen5 software (BioTek).

For patient samples and spike-in experiments with non-prestained cells, the isolated cells were briefly blocked in Dulbecco's PBS containing 2% FBS, followed by incubation in a non-protein based blocking solution. The cells were then incubated with a mixture of 200 nM fluorophore-conjugated (Oregon Green® 488) rVAR2 and PE-labelled anti-CD45 [5B1] (Cat. no. 170-078-081, MACS Miltenyi Biotec, 1:10) and PE-labelled anti-CD66b [REA306] (Cat. no. 130-104-396, MACS Miltenyi BioTec, 1:20) antibodies for 30 min. at room temperature and then washed once in PBS with 2% FBS to remove excess staining reagents. Finally, the cells were DAPI-stained and mounted on a Sensoplate. Duplicates of 3 mL patient blood were imaged using the Cytation™ 3 Cell Imaging Multi-Mode Reader. Additional 1–2 mL blood was processed according to the above description, except that the exclusion marker that was used in this setup was APC-labelled anti-CD45 [HI30] antibody (Cat. no. 17-0459-42, ThermoFisher), and analyzed using the CellCelector™ (Automated Lab Solutions).

For spike-in experiments testing anti-CSPG4 antibody as a staining reagent after rVAR2 bead pull-down, cells were blocked in 10% Normal Donkey Serum (NDS), 0.5% BSA, and 2 mM EDTA in Dulbecco's PBS for 10 min. After this, the cells were incubated with anti-CSPG4 (Cat. no. ab20156,

Abcam, 1:100) antibody diluted in Dulbecco's PBS with 1% NDS, 0.5% BSA, and 2 mM EDTA for 30 min. at room temperature. The cells were washed once in Dulbecco's PBS containing 0.5% BSA and 2 mM EDTA and then incubated with anti-mouse IgG-FITC (Vector, Cat. no. FI-2000, 1:400) for 30 min. at room temperature. Finally, the cells were washed in Dulbecco's PBS containing 0.5% BSA and 2 mM EDTA, DAPI stained, and mounted on a Sensoplate.

2.13. Classification and Enumeration of rVAR2-Stained Cancer Cells or CTCs

The samples were scanned on a 10× objective using the Cytation™ 3 Cell Imaging Multi-Mode Reader and manually enumerated using the Gen5 software (BioTek). Putative CTCs were defined as DAPI+, CD45/CD66b−, and rVAR2+. The signal to noise ratios were adjusted according to the fluoresence of the CELLection Biotin Binder beads, such that a staining was only regarded as positive if the intensity was above the fluorescence from the beads. Furthermore, all the cells with a DAPI area below 4 μm were excluded from enumeration.

2.14. Single Cell Picking and Whole Genome Amplification

Cell samples were resuspended in 200 μL PBS and then loaded onto a CellCelector™ magnetic slide (Automated Lab Solutions) to align and preserve the localization of the magnetic beads and cells during scanning. Employing the CellCelector™, the samples were then screened for coinciding Origon Green® 488 and DAPI signals as well as absent APC fluorescent signals, thereby detecting potential CTCs. Single cells were picked by the CellCelector™ and then pooled into PCR tubes containing 5 μL lysis buffer and enzyme from the MALBAC® Single Cell WGA Kit (Yikon Genomics, Cat. no. EK100101210). Each tube was prepared with 10–20 cells. Whole genome amplification (WGA) was performed on picked CTCs or WBC controls, according to manufacturer's instructions. The quality of the WGA products was verified by agarose gel electrophoresis and concentrations were measured by Qubit™ dsDNA BR Assay Kit (Cat. no. Q32850, Thermofisher Scientific).

2.15. Whole Exome Sequencing

Whole exome sequencing on whole genome amplified DNA from isolated CTCs and patient-matched WBCs was performed as previously described [31]. The Exome sequencing data was aligned against the human reference genome (hg19/GRCh37) using bwa mem 0.7.15 and somatic SNVs and small indels were called using Mutect2 according to the GATK best practices for somatic short variant discovery using GATK 4.1.0.0. Variants outside a selected glioblastoma-related target region containing 95 candidate genes were excluded from the call set. Mutect2 was provided with data from picked, patient-specific WBCs as a matched reference sample to reduce the amount of germline variants in the call set, hence obtaining a list of somatic mutations only for the CTC samples. The most relevant mutations (described in somatic mutation databases or being frameshift/stop-gain) were further manually inspected by looking at aligned reads sequences. The Integrative Genomics Viewer (Broad Institute, UK) was used for the visualization of variants (Figure S4) [32].

3. Results

3.1. rVAR2 Binds to ofCS on Glioma Cells

We have previously shown that the rVAR2 protein interacts with ofCS present on cancer cell lines representing almost all known cancers [25,33,34]. We tested rVAR2 binding to a panel of cell lines, including low-grade (WHO grade II) diffuse glioma (Res259) as well as high-grade (WHO grade IV) GBM (U87mg, KNS-42, and U118mg), to test for the presence of ofCS in glioma. All of the cell lines were positive for rVAR2 binding by flow cytometry, indicating that the glioma lines expressed ofCS (Figure 1A and Figure S2).

Cell cultures poorly represent the phenotypic plasticity of cancer cells in vivo, where the tumor cells continuously respond to signals from the microenvironment. TGF-β, for instance, is known to enhance the migratory and invasive capability of glioma cells, most likely by pushing these cells towards a more mesenchymal phenotype [35,36]. Therefore, we evaluated whether TGF-β exposure of the low-grade Res259 glioma cell line affected the expression of ofCS, as measured by rVAR2 binding. It should be noted that, although glial cells originate from ectodermal tissue, these cells exhibit a more mesenchymal appearance, such as the expression of vimentin [28]. Thus, the transition is measured as an increased expression of mesenchymal markers, rather than a down-regulation of epithelial markers [36]. After incubation with TGF-β, the Res259 cells showed increased expression of the mesenchymal markers β-catenin and *N*-cadherin in accordance with a transition towards a more mesenchymal state (Figure 1B). This was accompanied by a clear change in morphology as cells tended to become more elongated, which confirms the occurrence of a transition [36,37] (Figure 1C). Importantly, when testing for rVAR2 binding to Res259 cells in flow cytometry before and after incubation with TGF-β, rVAR2 binding was not reduced (Figure 1D).

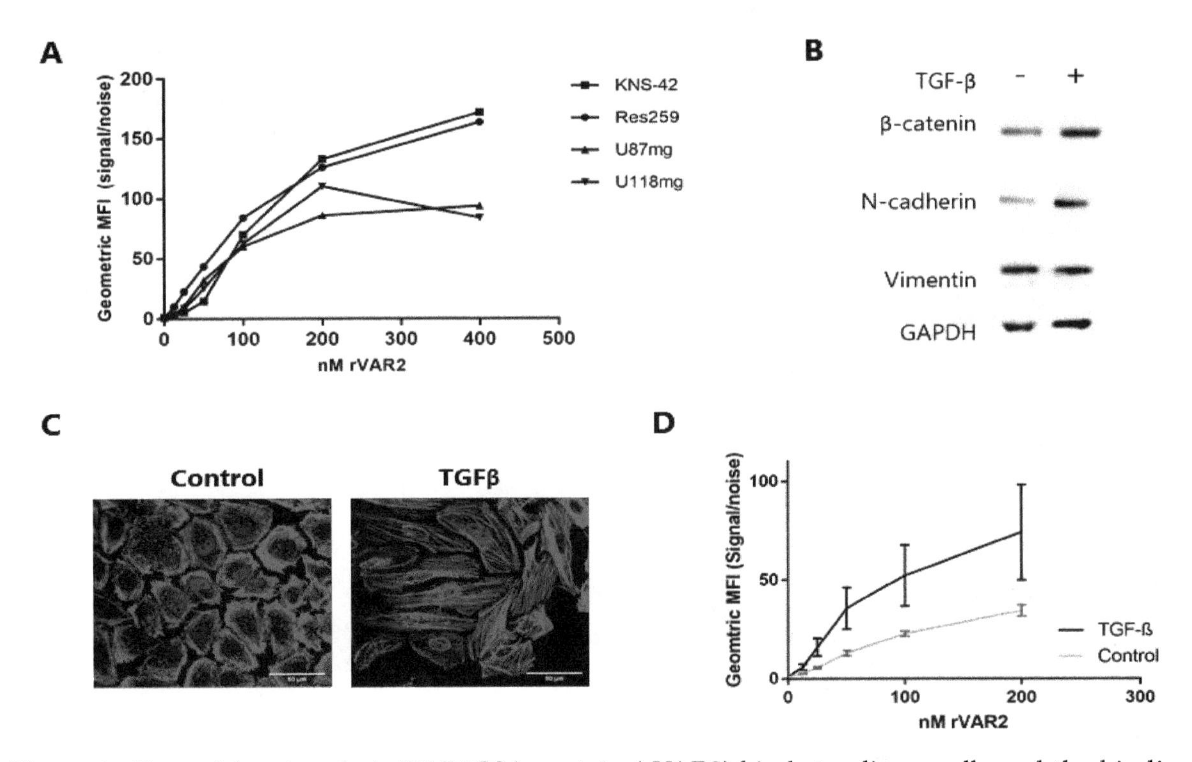

Figure 1. Recombinant malaria VAR2CSA protein (rVAR2) binds to glioma cells and the binding is unaffected by phenotypic changes. (**A**) rVAR2 binding to the glioma cell lines KNS-42, Res259, U87mg and U118mg was measured by flow cytometry using a FITC-conjugated anti-V5 antibody. Geometric Mean Fluorescence Intensity (MFI) was measured after incubation of cells with various rVAR2 concentrations. Results are displayed as signal/noise ratio. Figure represents data from one experiment, replicates are found in Figure S2. (**B**) Western blot of Res259 cell lysates after 72 h incubation with Transforming Growth Factor-beta (TGF-β) or buffer control. Membranes were incubated with anti-β-catenin, anti-*N*-cadherin, anti-Vimentin or anti-GAPDH antibodies and detected by anti-rabbit HRP antibody. (**C**) Representative images of fixed Res259 cells after 72 h incubation with TGF-β or buffer control. Cells were stained with phalloidin to stain F-actin (red) and DAPI to stain nuclei (blue). Scale bars, 50 μm. (**D**) rVAR2 binding to Res259 incubated with TGF-β or buffer control for 72 h measured by flow cytometry ($p < 0.001$, generalized least squares regression model). Geometric MFI was measured after incubation of cells with various rVAR2 concentrations and a FITC-conjugated anti-V5 antibody. Results are displayed as signal/noise ratio. Bars show standard deviation (n = 3).

3.2. rVAR2 Captures Glioma Cancer Cells Spiked Into Blood

To examine whether rVAR2 can be used for targeting circulating glioma cells, it is pivotal to ensure the specificity of rVAR2 binding to glioma cells in a background of normal white blood cells

(WBCs). Therefore, WBCs from 0.5 mL blood were mixed with 2000 U87mg, Res259, or KNS-42 cells and incubated with rVAR2 and a CF488-labeled anti-V5 antibody. rVAR2 binding showed a clear and specific membrane staining of all three cancer cell lines with minimal staining of the surrounding WBCs (Figure 2A). This was further tested by flow cytometry analysis showing specific binding to U87mg cells when mixed with WBCs (Figure 2B,C).

To examine whether rVAR2 binding to glioma cell lines could support magnetic capture and isolation of these cells from whole blood, 100 glioma cells were prestained using a CellTracker™ Green CMFDA Dye and spiked into a 3 mL blood sample from a healthy individual. This strategy allowed us to directly assess the recovery of spiked cells independent of downstream staining and detection biases. After lysis of the erythrocytes, cancer cells were isolated using rVAR2-coated magnetic beads (see Method section for details) [27]. By this procedure, we achieved an average recovery of 76%, 41%, 11%, and 64% for U87mg, Res259, KNS-42, and U118mg cells, respectively (Figure 2D). Furthermore, we spiked 3 mL blood samples with 100, 50, 10, or 5 U87mg cells to assess the sensitivity of the assay (Figure 2E). The average recovery ranged between 54–75% with no obvious association to the number of cells spiked into the blood ($p = 0.31$, one-way ANOVA). Although the recovery varied between the different cell lines, this data confirms that rVAR2 can be used as a capture molecule for the isolation of various glioma cell types.

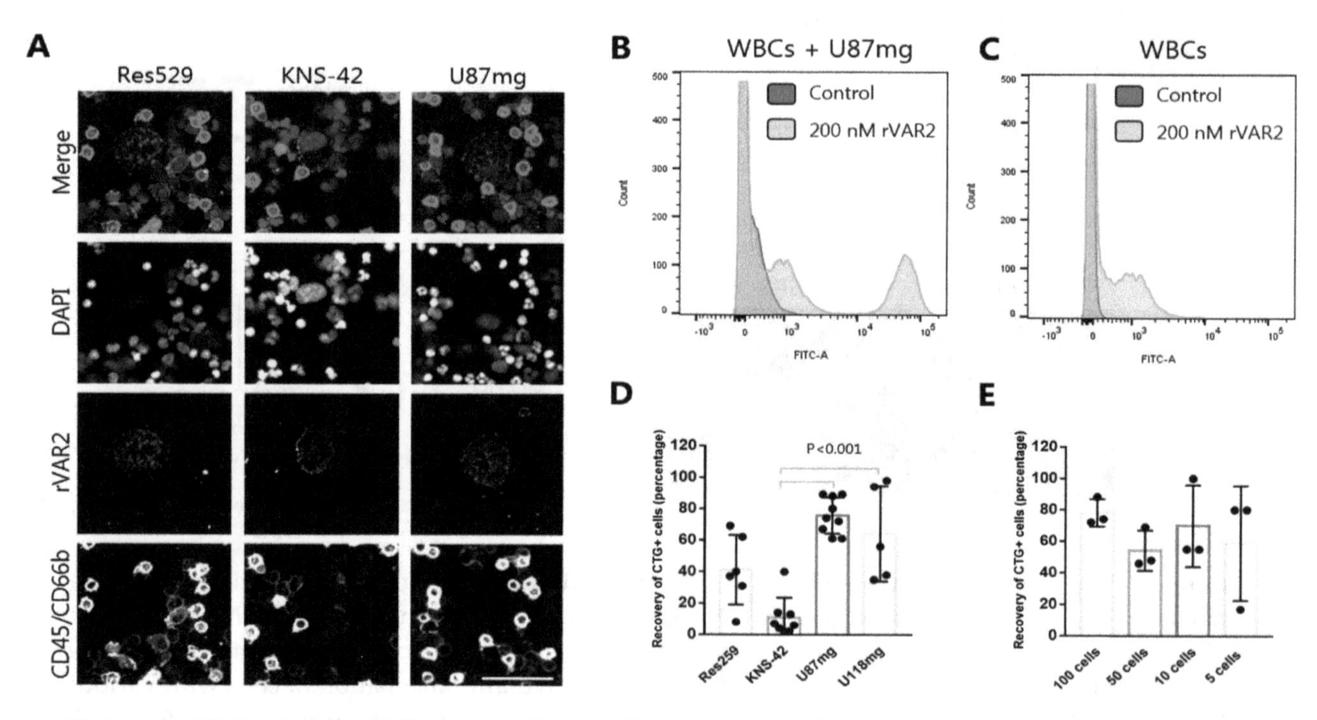

Figure 2. rVAR2 specifically binds to glioma cells and enables their retrieval from blood. (**A**) rVAR2 stains glioma cell lines in a background of white blood cells (WBCs). Res529 (left), KNS-42 (middle) and U87mg (right) cells were mixed with WBCs and stained with V5-tagged rVAR2 in combination with a CF488-conjugated anti-V5 antibody (green), PE-conjugated anti-CD45 and anti-CD66b antibodies (red), and DAPI (blue). Scale bar, 50 μm. (**B**) Flow cytometry analysis showing WBCs (100 μL RBC lysed blood) mixed with U87mg (50,000) cells and detected with either 200 nM rVAR2 and a FITC-conjugated anti-V5 antibody or with a FITC-conjugated anti-V5 antibody alone (control). (**C**) Same as in (**B**) but with no U87mg cells added. (**D**) Recovery of CellTracker Green-stained glioma cells from blood. 100 cells were spiked into 3 mL blood and recovered using rVAR2-coupled beads. Cells were stained with DAPI and scanned on the Cytation 3 Imager. Each dot represents the percentage of recovered cells from one sample. Bars represent mean recoveries and error bars show +/− standard deviation ($n \geq 2$) (one-way ANOVA with Bonferroni correction). (**E**) Recovery of CellTracker Green-stained U87mg cells from blood. The indicated number of U87mg cells was spiked into 3 mL blood and captured using rVAR2-coupled beads. Enumeration of cells and data presentation were done as in (**D**).

3.3. *rVAR2 Interacts with Several GBM-Associated Proteoglycans*

We have previously described that single cancer cells simultaneously display the ofCS modification on several proteoglycans [33]. We analyzed rVAR2-based pull-down of lysates from KNS-42, U118mg, and U87mg cell lines to investigate the proteoglycan display on glioma cells. The mass spectrometry results showed the pull-down of multiple key cancer-related proteoglycans (Table 1). Among the hits were several chondroitin sulfate proteoglycans (CSPGs) that have been described for GBM, such as CSPG4, CD44, APLP2, and SDC1 [38–41]. To validate these findings, we studied the co-localization of ofCS and selected protein cores from the pull-down proteomic list by proximity ligation assay (PLA) (Figure 3A). Indeed, compared to other proteoglycans rVAR2 binding and CD44 showed a strong co-localization on each evaluated glioma cell line ($p < 0.001$, one-way ANOVA) (Figure 3B). Despite the high PLA signal, CD44 was not further examined as a potential CTC marker, as anti-CD44 antibodies are also found to target a subset of healthy WBCs [42]. Similarly, ofCS and CSPG4 were clearly co-localizing on U87mg and U188mg cells ($p < 0.001$, one-way ANOVA) (Figure 3B). Interestingly, anti-CSPG4 antibodies are already being used for the capture and detection of circulating melanoma cells [43–46]. Since CSPG4 is also an emerging target for GBM CAR-T immunotherapy, we examined whether CSPG4 is still accessible for antibody staining after rVAR2-based capture [47,48]. Indeed, the captured U87mg cells showed clear and specific CSPG4 staining in a background of WBCs (Figure 3C). Hence, the capture of glioma CTC might be useful for predicting response to anti-CSPG4 CAR-T therapy.

Table 1. rVAR2-based protein pull-down hits from cell lysates.

Protein Name	Gene	Peptides Count	Seq. Coverage (%)	Ratio to Neg
	KNS-42			
Amyloid-like protein 2	APLP2	17	27	NA
CD44	CD44	9	37.4	41.17
Glypican 1	GPC1	10	23.7	NA
Glypican 4	GPC4	10	23	7.15
Integrin beta 1	ITGB1	11	15.2	7.50
Neuropilin 1	NRP1	12	21.1	NA
Neuropilin 2	NRP2	4	6	NA
Receptor-type tyrosine-protein phosphatase zeta	PTPRZ1	17	8.6	NA
Syndecan 1	SDC1	6	14.5	NA
Syndecan 2	SDC2	5	25.5	NA
Testican 1	SPOCK1	8	22.6	NA
Versican	VCAN	39	12.8	540.37
	U118mg			
Amyloid-like protein 2	APLP2	4	5.2	NA
CD44	CD44	8	37.4	17.26
Decorin	DCN	8	29.8	41.89
Neuropilin 1	NRP1	9	15.9	16.74
Versican	VCAN	13	4	NA
	U87mg			
Amyloid-like protein 2	APLP2	4	5.1	NA
Amyloid precursor protein	APP	3	4.3	NA
Carbonic anhydrase 9	CA9	1	3.3	NA
CD44	CD44	4	6.1	12.92
Chondroitin sulfate proteoglycan 4	CSPG4	7	3.1	NA
HLA class II histocompatibility antigen gamma chain	CD74	3	12.2	NA
Sushi repeat-containing protein SRPX	SRPX	7	15.7	NA
Syndecan-1	SDC1	3	15.2	9.71
Syndecan-4	SDC4	2	12.6	NA

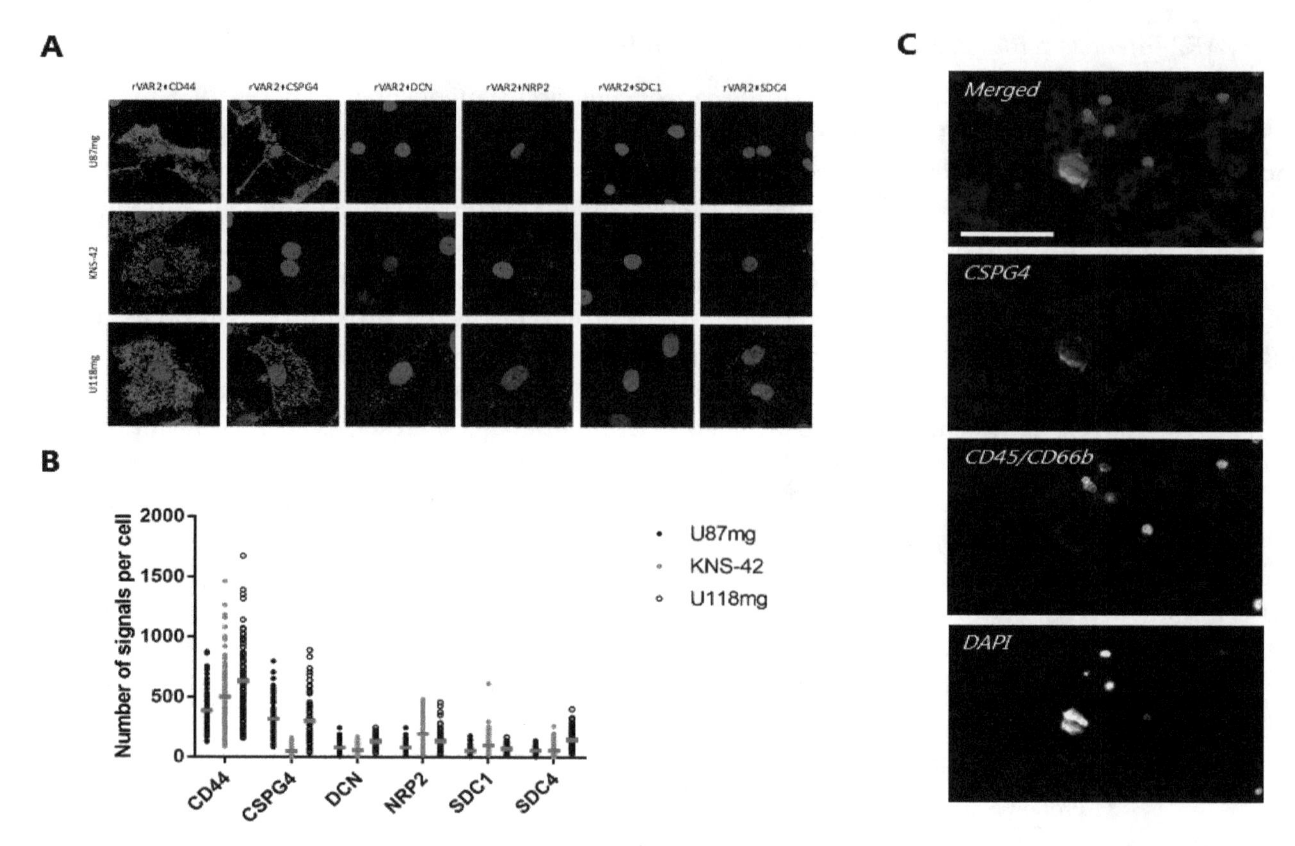

Figure 3. Evaluation of protein pull-down hits from mass spectrometry by proximity ligation assay (PLA). (**A**) Representative images of PLA assays on U87mg, KNS-42, and U118mg cells showing co-localization between rVAR2 and a panel of CSPGs as red dots. Cells were counterstained with DAPI (blue) and analyzed by confocal microscopy. All of the images are shown in same magnification using a 60× objective. (**B**) Quantification of the PLA co-localization signals between rVAR2 and each of the CSPGs analyzed. Data is shown as the number of signals per cell. Red bars represent the mean number of signals per cell. (**C**) Representative image showing specific CSPG4 staining (green) of an rVAR2-captured U87mg cell in a background of WBCs stained for CD45 and CD66b (both red). Cells were stained with DAPI (blue) and visualized on the Cytation 3 Imager with a 20× objective. Scale bar, 50 μm.

3.4. rVAR2 Detects Cancer Cells Spiked Into Blood Samples

All of the CTC capture protocols will, even in the best of circumstances, capture some normal WBCs along with the CTCs. A major challenge associated with the analysis of glioma CTCs is how to validate which captured cells are indeed cancer cells and not WBCs. With regard to carcinoma-derived CTCs the most widely used markers are EpCAM or CK. While gliomas are EpCAM negative, the results regarding CK-positivity are less consistent [12,49–51]. Thus, these markers are not optimal for the detection of glioma CTCs. Therefore, we established a platform where rVAR2-coupled beads were used for capturing, while a fluorophore-conjugated rVAR2 was used for microscopic detection of the captured cells. U87mg, Res259, and KNS-42 cells all showed rVAR2 staining after capture with rVAR2-coupled beads (Figure 4A and Figure S3). However, it was noticed that rVAR2-staining of U87mg after magnetic capture was somewhat reduced compared to the Res259 and KNS-42 cells. Next, we applied the same workflow to U87mg cells that were spiked into 3 mL healthy donor blood in order to mimic patient blood samples. The rVAR2 staining enabled detection of U87mg cells and their separation from CD45- and/or CD66b-positive WBCs (Figure 4B). The strategy was also effective with cells from a primary glioblastoma cell culture, GBM02 (Figure 4B).

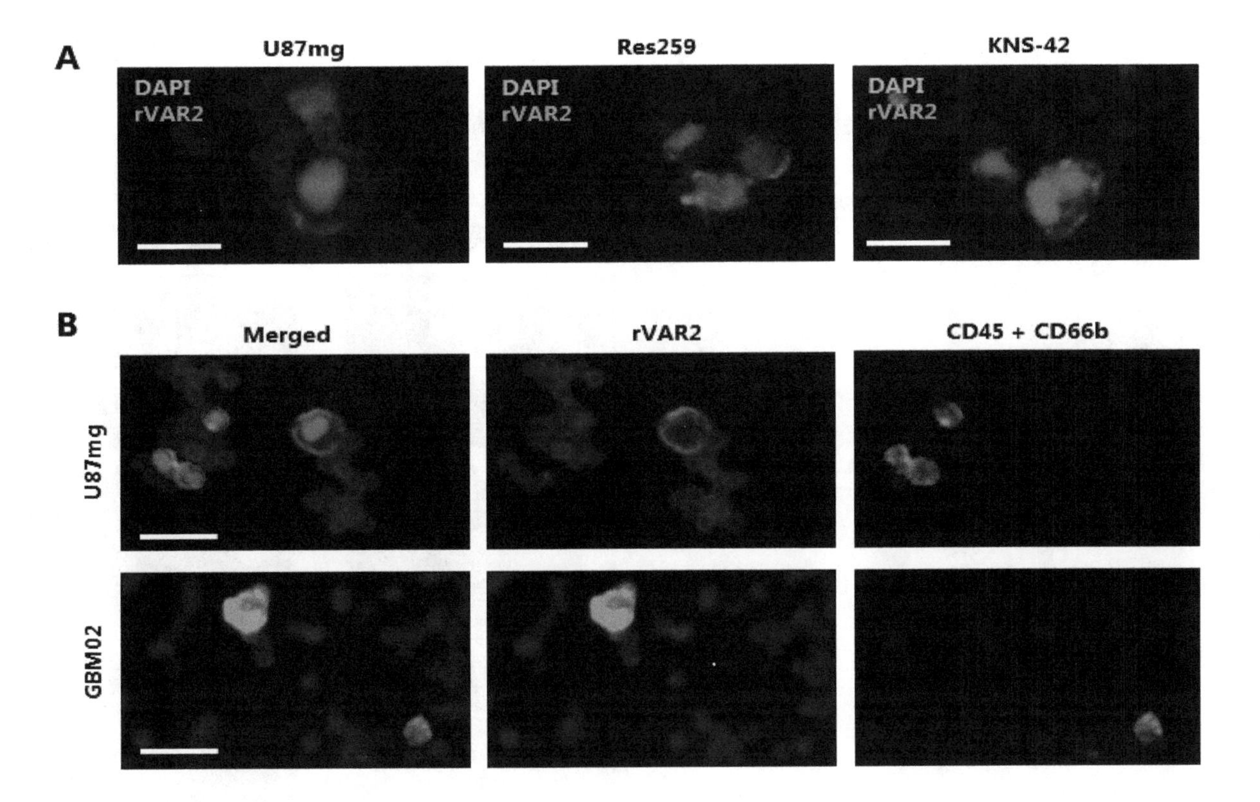

Figure 4. Using rVAR2 to stain glioma cells after capture with rVAR2-coupled beads. (**A**) Glioma cell lines (U87mg, Res259, and KNS-42) were incubated with rVAR2-coupled beads and stained with a fluorophore-conjugated rVAR2 (green) and DAPI (blue). Representative images were obtained using the Cytation 3 Imager with a 10× objective. Scale bars, 20 µm. (**B**) U87mg cells and GBM02 cells were spiked into 3 mL blood, retrieved using rVAR2-coupled beads, and stained using fluorescent rVAR2 (green), anti-CD45/CD66b antibodies (red) and DAPI (blue). Scale bars, 20 µm.

3.5. rVAR2 Captures and Detects CTCs in Glioma Patient Blood Samples

We tested for the presence of CTCs in blood samples from glioma patients using the combined rVAR2 capture and detection protocol. Duplicates of 3 mL blood samples from 10 glioma patients, suffering from oligodendroglioma (grade II), anaplastaic oligodendroglioma (grade III), or GBM (grade IV), were processed and visualized for enumeration. CTCs were manually enumerated as rVAR2+, CD45/CD66b−, and DAPI+ cells. The range of identified CTCs per 3 mL blood was 0.5–42 (Figure 5A). There was no obvious correlation between grade or type of diagnosis and CTC number. Interestingly, one patient who had progressed from an initial diagnosis of oligodendroglioma (grade II) to anaplastic oligodendroglioma (grade III) within a time span of 15 years had a relatively high number of CTCs (22 CTCs per 3 mL blood). In a patient with the reverse clinical history regressing from an initial diagnosis of anaplastic oligodendroglioma (grade III) to eight years later having oligodendroglioma (grade II) we detected an average of only 0.5 CTCs per 3 mL. Representative images of rVAR2+ cells from one of the GBM patients are shown (Figure 5B) and the full list of detected CTCs is found in Figure S5.

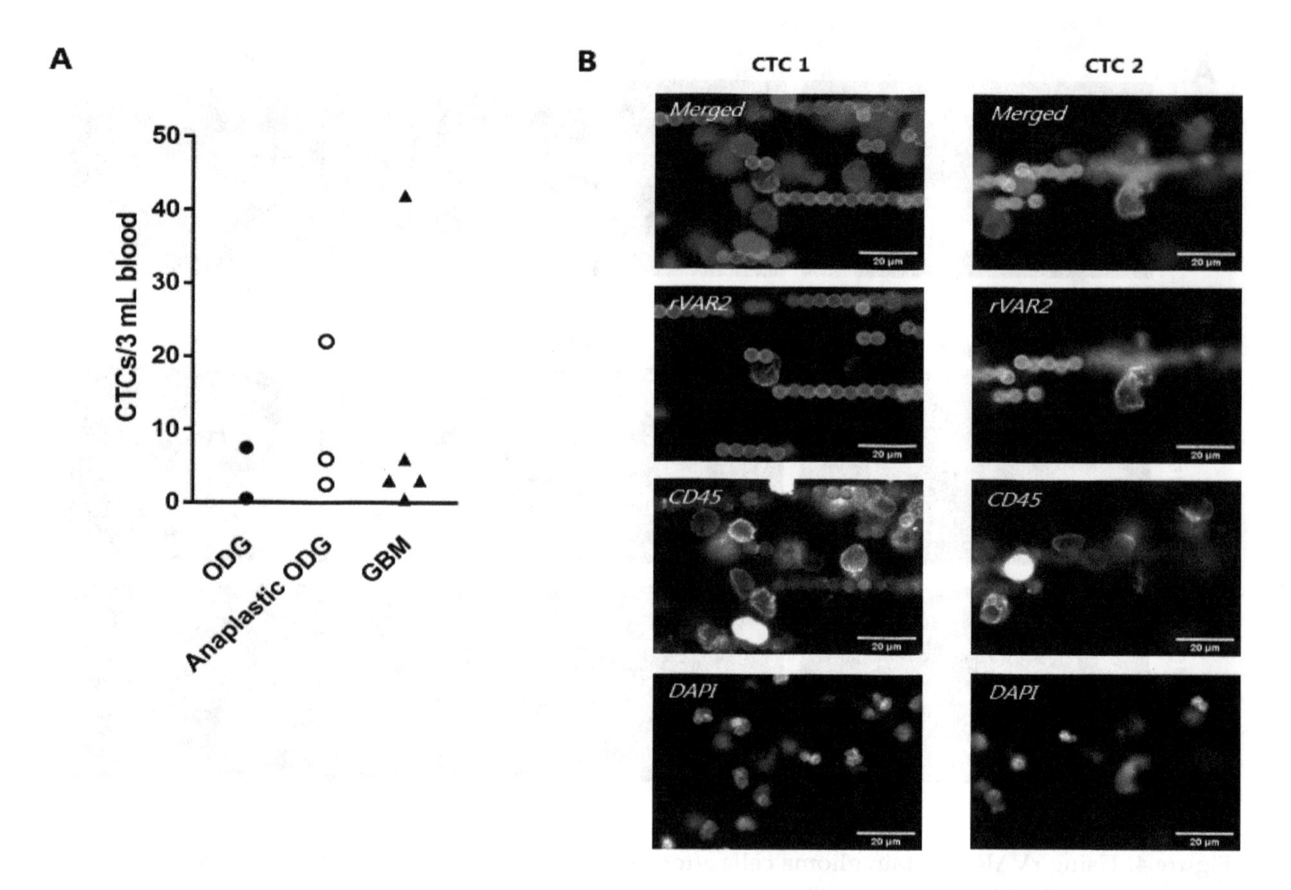

Figure 5. rVAR2 enables capture and detection of glioma circulating tumor cells (CTCs) from patient blood samples. (**A**) Average CTC count in blood samples from ten glioma patients. 3 mL blood samples were processed by using rVAR2-coupled beads followed by staining using a mixture of fluorophore-conjugated rVAR2 (green), anti-CD45/CD66b antibodies (red) and DAPI (blue). CTCs were defined as rVAR2+, CD45/CD66b−, DAPI+ cells. Each dot represents the average number of detected CTCs per 3 mL patient blood sample. The x-axis shows whether the patient was diagnosed with GBM, Anaplastic Oligodendroglioma (ODG), or Oligodendroglioma (ODG). (**B**) Representative images of identified CTCs from a patient diagnosed with GBM. The sample was stained with flurophore-conjugated rVAR2 (green), anti-CD45 antibody (magenta), and DAPI (blue). Images were obtained using the CellCelector™ (ALS) with a 40× objective. Scale bars, 20 µm.

3.6. Captured Glioma CTCs Show Cancer-Indicative Mutations

To confirm that the VAR2+, CD45− cells detected in the patient blood samples were indeed glioma-derived CTCs, we performed targeted whole exome sequencing (WES) searching for glioma relevant mutations. For three patient samples we single cell picked rVAR2+, CD45− cells, and patient-matched WBCs as germline controls using an ALS CellCelector™. For each patient, 2–4 CTCs were pooled into one sample and whole genome amplification (WGA) was performed (Figure 6). However, since WBCs were located close to some of the selected CTCs, the cell picking procedure resulted in samples containing CTCs together with some WBCs (Table 2). The WGA product was then used for WES. The WES results were filtered to only include glioma relevant mutations and each hit was visually confirmed by evaluating the IGV screen shots (Figure S4). Indeed, we identified genes with cancer-indicative mutations in all CTC samples: RB1, TP53/EPM2AIP1, and TP53/ALK for patient 1, 3, and 4, respectively (Table 2). Thus, the molecular profiling supports the tumor origin of the picked patient-derived CTCs.

Table 2. Mutations detected in patient CTCs by whole exome sequencing (WES).

Patient Information						Confirmed Mutation by WES					
ID	Diagnosis	Molecular Features in Tumor Biopsy	Sample	CTCs	WBCs	Gene	Transcript ID	Transcript Variant	Allele Fraction (%)	Protein Variant	Translation Impact
1	Anaplastic oligodendro-glioma	IDH1 mutation LOH 1p/19q MGMT methylation	1	2	19	RB1	NM_000321.2	c.1644delA	16.88	p.K548fs*3	Frameshift
			2	3	34	RB1	NM_000321.2	c.1644delA	1.59	p.K548fs*3	Frameshift
3	GBM	IDH1 wild type	1	2	23	TP53	NM_000546.5	c.892G>T	21.39	p.E298*	Stop-gain
						EPM2AIP1	NM_014805.3	c.128G>T	54.17	p.R43L	Missense
4	Anaplastic oligodendro-glioma	IDH1 mutation 1p/19q deletion MGMT methylation	1	4	14	TP53	NM_000546.5	c.493C>T	31.43	p.Q165*	Stop-gain
						ALK	NM_004304.4	c.3824G>T	33.33	p.R1275L	Missense

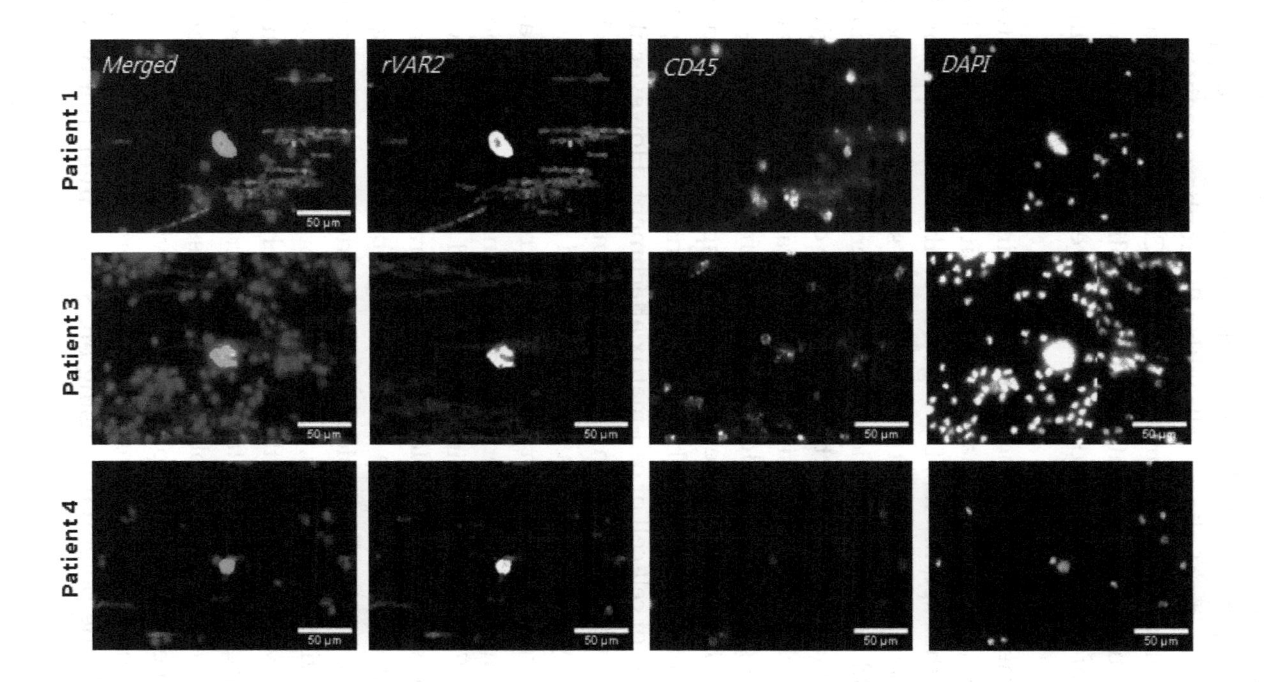

Figure 6. Identification of CTCs for whole exome sequencing (WES). Representative images of identified CTCs from three patients. Cells were stained with fluorophore-conjugated rVAR2 (green), anti-CD45 antibody (magenta) and DAPI (blue) and classified as CTCs if rVAR2+, CD45−, and DAPI+. Images were obtained using the CellCelector™ (ALS) with a 10× objective. Scale bars, 50 µm.

4. Discussion

The isolation and characterization of glioma CTCs have proven to be challenging, especially concerning the detection and validation of the tumor origin of the isolated cells. To date, only a few studies have shown the presence of circulating glioma cells utilizing either a single antibody marker or an antibody mixture for CTC detection [13,14]. Here, we present a novel strategy for glioma CTC capture and detection based on targeting the unique cancer-specific glycosaminoglycan structure ofCS. We show that rVAR2, which binds ofCS with high affinity, specifically targets a panel of glioma cell lines in a background of white blood cells (WBCs). Furthermore, we show that rVAR2 can be used for the capture of glioma cells that are spiked into blood by coupling the protein to magnetic beads. In addition, the staining of glioma cells with a fluorophore-conjugated rVAR2 after magnetic pull-down facilitates their detection and separation from WBCs. This workflow was applied to blood samples that were derived from ten glioma patients and established proof-of-concept for identification of glioma CTCs. In three of the patients, potential CTCs were picked and molecular analysis supported their tumor origin.

Flow cytometry analysis showed rVAR2 binding to all tested glioma cell lines. Interestingly, the cell lines showed varying maximum intensity at saturation indicating different levels of ofCS display. In all of the experiments, U87mg showed the lowest level of rVAR2-binding. This is interesting, since U87mg cells had the highest recovery when spiked into blood and isolated with rVAR2-coupled beads. This could indicate that efficiency of recovery, not only depends on the level of target expression but is also influenced by other factors, such as the capability of a given cell line to survive through the experimental workflow. When using spike-in of cancer cells in healthy donor blood, the experimental procedure, among others, includes detachment from the culture plate and exposure to the various components of a foreign immune system when spiked into blood. Patient-derived CTCs do indeed experience dramatic changes in physical conditions upon entering the circulation, such as shear stress forces and the loss of cell-cell or cell-matrix attachment. However, it could be debated how comparable this sequence of events is to the in vitro spike-in models, and thus how well cell line spike-in samples reflect the phenotypes of CTCs in patient-derived liquid biopsies.

Elevated levels of TGF-β in the tumor microenvironment and a mesenchymal phenotype of the glioma cells have independently been shown to be associated with a poor prognosis in glioma patients [36]. TGF-β is known to induce increased motility and invasive behavior, which underlines a potential link between cellular plasticity and intravasation of cancer cells into the bloodstream [52]. Notably, the expression of surface markers might be altered during such phenotypic changes, and this process should be taken into consideration when deciding on a capture and detection reagent for CTC capture. In line with other studies, we have previously shown that an EMT-like process can be induced in U87mg cells by incubating with TGF-β for 72 h [27]. Importantly, we also confirmed that rVAR2 binding to U87mg cells was maintained after the transition. However, several studies indicate that the EMT-like processes also play a role in the progression from low-grade to high-grade gliomas [53,54]. Here, we confirmed that ofCS display is retained when the Res259 low-grade glioma cell line is pushed towards a more mesenchymal morphology and protein expression pattern by incubation with TGF-β. This strengthens the potential of using ofCS as a target, not only for the capture, but also for the detection of glioma CTCs.

When considering previously published data showing that ofCS is presented by nearly all cancer cells of epithelial, mesenchymal, and hematopoietic origin, the use of rVAR2 staining reagent for CTC detection would be beneficial over traditional single-surface markers [25,27,33,34]. Here, we show that captured glioma CTCs can be identified by an rVAR2 stain. However, one should be cautious when using the same target for both capture and detection, as the general assumption is that the use of two independent markers would lead to a better exclusion of false positive hits. Furthermore, as CS is a common GAG that is displayed on all cell types, including WBCs, an extremely high degree of ofCS-specificity is needed to successfully capture and distinguish CTCs from WBCs. Interestingly, the naturally occurring VAR2CSA that is expressed by malaria-infected erythrocytes serves exactly this purpose, since binding to normal WBCs would result in parasite clearance [26]. However, the use of exclusion markers is highly important to exclude potential false positives. In the workflow presented here, we included CD45 and CD66b as exclusions markers to identify and reject a broad repertoire of WBCs. The optimized and combined rVAR2 capture and detection workflow enabled us to isolate and detect circulating glioma cells in glioma patients. In this very limited dataset, the number of CTCs detected did not correlate with type of diagnosis or WHO grade.

A potential clinical application of rVAR2-based CTC detection could be patient stratification based on the expression of therapeutically relevant CSPGs on CTC subsets. In this study, we sought to identify ofCS-modified proteoglycans in glioma by using rVAR2-based protein pull-down of lysates from KNS-42, U118mg, and U87mg cell lines. Indeed, the subsequent proteomics analysis showed the pull-down of multiple cancer-related proteoglycans with key roles in the pathogenesis of glioma. Unlike our previous study showing syndecan 1 to be the main VAR2CSA receptor in the placental syncytium [55], we found several interesting hits on the glioma cell lines, including syndicans, glypicans, neuropilins, decorin, versican, CSPG4, and PTPRZ1. The two last mentioned are currently being explored as potential anti-cancer targets in GBM [56–58]. In this study, we tested the use of anti-CSPG4 antibodies for staining and detection of cancer cells after rVAR2 capture, which could potentially be of future interest in the monitoring of anti-CSPG4 CAR-T therapies. Finally, CD44 was identified as a hit on all of the glioma cell lines. High CD44 expression is common in GBM and is used to identify GBM with particular poor survival chance [59,60]. Along this line, CD44 is expressed by GBM cancer stem cells, which promotes aggressive GBM growth [61]. Thus, adding a CD44-stain to rVAR2 captured CTCs could provide additional information regarding predicted outcome if sufficient exclusion markers are included. Another interesting application of the captured CTCs could be to culture and further characterize the CTCs in terms of responsiveness to relevant treatments. Liu et al. has shown proof-of-concept by culturing CTCs that are captured from a mouse GBM model [62]. To our knowledge, no one has to date been able to culture the sparse number of CTCs found in glioma patient blood samples.

We picked CTCs and performed WGA followed by WES against a panel of known glioma mutations to confirm that the detected rVAR2+, CD45−, and DAPI+ cells were actual CTCs derived from the brain tumors. Patient-matched WBCs were used as germline subtractions. Patient I, which was diagnosed with anaplastic oligodendroglioma, had CTCs with mutation in the *RB1* gene, which results in a frameshift with premature stop codon. Alterations in genes that are associated with the retinoblastoma pathway is a predictor of poor chance of survival in gliomas [63]. Interestingly, the somatic mutation pattern found in the tumor biopsy from this patient showed mutation of the *IDH1* gene, a common feature of lower grade gliomas [64], which was not detected in the CTC sample. However, CTCs could represent a minority of subclones in the primary tumor, which are not detectable by current standard NGS methods [65].

A *TP53* mutation was found in the CTCs from both patient 3 (GBM) and 4 (anaplastic oligodendroglioma). *TP53* encodes the p53 tumor suppressor protein, and this pathway is often deregulated in diffuse gliomas [66]. Another detected mutation in patient 3 was a missense mutation in the *EPM2AIP1* gene. The *EPM2AIP1* mutations have previously been described in different gastrointestinal cancers [67,68]. Interestingly, *EPM2AIP1* is part of a bidirectional promotor with *MLH1* and epimutations causing hypermethylation has been linked to hereditary colorectal cancers [69,70]. However, little is known regarding the functional role of *EPM2AIP1* silencing, as research has primarily focused on *MLH1*. In patient 4 the WES analysis also detected mutations in the *ALK* gene, which encodes a receptor tyrisone kinase. *ALK* is frequently mutated in neuroblastoma and indeed the detected NM_004304.4_p.R1275L variant is a described hot spot locus within the kinase domain. This hotspot mutation hinders the auto-inhibition of *ALK* and acts transformative. Consequently, neuroblastoma patients with *ALK* mutations show poorer overall survival [71]. Importantly, small molecules for targeted therapy of *ALK* have been developed and neuroblastoma cell lines harboring p.R1275 mutations show sensitivity towards *ALK* inhibitors, such as crizotinib [72,73]. Altogether, the specific detection of glioma-related mutation patterns in the CTC samples strongly indicates that the detected cells originate from a glioma site.

In summary, we present a method for enriching and staining CTCs from glioma patients. After a complete clinical validation the method could provide a powerful tool for non-invasive pheno- and genotyping of gliomas. Finally, the technology could potentially be used to monitor progression and recurrence in cancer patients.

Author Contributions: Conceptualization, Conceptualization, S.R.B.-C. and M.Ø.A.; Formal analysis, S.R.B.-C., T.D.A., M.H.T., T.G.T., O.Ø. and M.Ø.A.; Funding acquisition, A.S. and M.Ø.A.; Investigation, S.R.B.-C., R.S.P., M.A.P., T.M.C., C.L., N.T.S., T.D.A., A.M.J., M.H.T., O.Ø. and M.Ø.A.; Methodology, S.R.B.-C., M.A.P., T.M.C. and M.Ø.A.; Project administration, M.Ø.A.; Resources, Y.C.L., L.G., S.C., T.G., R.D., M.D., A.F.S., M.S., U.L. and P.H.; Supervision, S.R.B.-C., T.M.C., A.S. and M.Ø.A.; Validation, S.R.B.-C., R.S.P., C.L. and N.T.S.; Visualization, S.R.B.-C.; Writing—original draft, S.R.B.-C. and M.Ø.A.; Writing—review & editing, S.R.B.-C., T.M.C., T.G.T. and A.S.

Acknowledgments: The authors would like to thank Andreas Frederiksen and Sofie Amalie Schandorff for their great technical assistance. We would also like to thank Janine Erler and Lara Perryman (Biotech Research & Innovation Centre, University of Copenhagen, Denmark) for kindly providing cell lines for this study, and The Core Facility for Flow Cytometry (Faculty of Health and Medical Sciences, University of Copenhagen, Denmark) for the use of the LSR II flow cytometer.

References

1. Wesseling, P.; Kros, J.M.; Jeuken, J.W.J.D.H. The pathological diagnosis of diffuse gliomas: Towards a smart synthesis of microscopic and molecular information in a multidisciplinary context. *Diagn. Histopathol.* **2011**, *17*, 486–494. [CrossRef]
2. Claes, A.; Idema, A.J.; Wesseling, P. Diffuse glioma growth: A guerilla war. *Acta Neuropathol.* **2007**, *114*, 443–458. [CrossRef] [PubMed]
3. Olar, A.; Aldape, K.D. Using the molecular classification of glioblastoma to inform personalized treatment. *J. Pathol.* **2014**, *232*, 165–177. [CrossRef] [PubMed]
4. Tabatabai, G.; Stupp, R.; van den Bent, M.J.; Hegi, M.E.; Tonn, J.C.; Wick, W.; Weller, M. Molecular diagnostics of gliomas: The clinical perspective. *Acta Neuropathol.* **2010**, *120*, 585–592. [CrossRef] [PubMed]
5. Louis, D.N.; Perry, A.; Reifenberger, G.; von Deimling, A.; Figarella-Branger, D.; Cavenee, W.K.; Ohgaki, H.; Wiestler, O.D.; Kleihues, P.; Ellison, D.W. The 2016 World Health Organization Classification of Tumors of the Central Nervous System: A summary. *Acta Neuropathol.* **2016**, *131*, 803–820. [CrossRef]
6. Wang, Q.; Hu, B.; Hu, X.; Kim, H.; Squatrito, M.; Scarpace, L.; deCarvalho, A.C.; Lyu, S.; Li, P.; Li, Y.; et al. Tumor Evolution of Glioma-Intrinsic Gene Expression Subtypes Associates with Immunological Changes in the Microenvironment. *Cancer Cell* **2017**, *32*, 42–56. [CrossRef] [PubMed]
7. Ceccarelli, M.; Barthel, F.P.; Malta, T.M.; Sabedot, T.S.; Salama, S.R.; Murray, B.A.; Morozova, O.; Newton, Y.; Radenbaugh, A.; Pagnotta, S.M.; et al. Molecular Profiling Reveals Biologically Discrete Subsets and Pathways of Progression in Diffuse Glioma. *Cell* **2016**, *164*, 550–563. [CrossRef]
8. Piccirillo, S.G.; Dietz, S.; Madhu, B.; Griffiths, J.; Price, S.J.; Collins, V.P.; Watts, C. Fluorescence-guided surgical sampling of glioblastoma identifies phenotypically distinct tumour-initiating cell populations in the tumour mass and margin. *Br. J. Cancer* **2012**, *107*, 462–468. [CrossRef]
9. Mariani, L.; Beaudry, C.; McDonough, W.S.; Hoelzinger, D.B.; Kaczmarek, E.; Ponce, F.; Coons, S.W.; Giese, A.; Seiler, R.W.; Berens, M.E. Death-associated protein 3 (Dap-3) is overexpressed in invasive glioblastoma cells in vivo and in glioma cell lines with induced motility phenotype in vitro. *Clin. Cancer Res.* **2001**, *7*, 2480–2489.
10. Angelucci, C.; D'Alessio, A.; Lama, G.; Binda, E.; Mangiola, A.; Vescovi, A.L.; Proietti, G.; Masuelli, L.; Bei, R.; Fazi, B.; et al. Cancer stem cells from peritumoral tissue of glioblastoma multiforme: The possible missing link between tumor development and progression. *Oncotarget* **2018**, *9*, 28116–28130. [CrossRef]
11. Jimsheleishvili, S.; Alshareef, A.T.; Papadimitriou, K.; Bregy, A.; Shah, A.H.; Graham, R.M.; Ferraro, N.; Komotar, R.J. Extracranial glioblastoma in transplant recipients. *J. Cancer Res. Clin. Oncol.* **2014**, *140*, 801–807. [CrossRef] [PubMed]

12. Macarthur, K.M.; Kao, G.D.; Chandrasekaran, S.; Alonso-Basanta, M.; Chapman, C.; Lustig, R.A.; Wileyto, E.P.; Hahn, S.M.; Dorsey, J.F. Detection of brain tumor cells in the peripheral blood by a telomerase promoter-based assay. *Cancer Res.* **2014**, *74*, 2152–2159. [CrossRef] [PubMed]
13. Muller, C.; Holtschmidt, J.; Auer, M.; Heitzer, E.; Lamszus, K.; Schulte, A.; Matschke, J.; Langer-Freitag, S.; Gasch, C.; Stoupiec, M.; et al. Hematogenous dissemination of glioblastoma multiforme. *Sci. Transl. Med.* **2014**, *6*, 247ra101. [CrossRef] [PubMed]
14. Sullivan, J.P.; Nahed, B.V.; Madden, M.W.; Oliveira, S.M.; Springer, S.; Bhere, D.; Chi, A.S.; Wakimoto, H.; Rothenberg, S.M.; Sequist, L.V.; et al. Brain tumor cells in circulation are enriched for mesenchymal gene expression. *Cancer Discov.* **2014**, *4*, 1299–1309. [CrossRef] [PubMed]
15. Gao, F.; Cui, Y.; Jiang, H.; Sui, D.; Wang, Y.; Jiang, Z.; Zhao, J.; Lin, S. Circulating tumor cell is a common property of brain glioma and promotes the monitoring system. *Oncotarget* **2016**, *7*, 71330–71340. [CrossRef] [PubMed]
16. Appierto, V.; Di Cosimo, S.; Reduzzi, C.; Pala, V.; Cappelletti, V.; Daidone, M.G. How to study and overcome tumor heterogeneity with circulating biomarkers: The breast cancer case. *Semin. Cancer Biol.* **2017**, *44*, 106–116. [CrossRef] [PubMed]
17. Miyamoto, D.T.; Lee, R.J.; Stott, S.L.; Ting, D.T.; Wittner, B.S.; Ulman, M.; Smas, M.E.; Lord, J.B.; Brannigan, B.W.; Trautwein, J.; et al. Androgen receptor signaling in circulating tumor cells as a marker of hormonally responsive prostate cancer. *Cancer Discov.* **2012**, *2*, 995–1003. [CrossRef] [PubMed]
18. Pawlikowska, P.; Faugeroux, V.; Oulhen, M.; Aberlenc, A.; Tayoun, T.; Pailler, E.; Farace, F. Circulating tumor cells (CTCs) for the noninvasive monitoring and personalization of non-small cell lung cancer (NSCLC) therapies. *J. Thorac. Dis.* **2019**, *11* (Suppl. 1), S45–S56. [CrossRef]
19. Mikami, T.; Kitagawa, H. Biosynthesis and function of chondroitin sulfate. *Biochim. Biophys. Acta* **2013**, *1830*, 4719–4733. [CrossRef]
20. Gama, C.I.; Tully, S.E.; Sotogaku, N.; Clark, P.M.; Rawat, M.; Vaidehi, N.; Goddard, W.A., III; Nishi, A.; Hsieh-Wilson, L.C. Sulfation patterns of glycosaminoglycans encode molecular recognition and activity. *Nat. Chem. Biol.* **2006**, *2*, 467–473. [CrossRef]
21. Afratis, N.; Gialeli, C.; Nikitovic, D.; Tsegenidis, T.; Karousou, E.; Theocharis, A.D.; Pavao, M.S.; Tzanakakis, G.N.; Karamanos, N.K. Glycosaminoglycans: Key players in cancer cell biology and treatment. *FEBS J.* **2012**, *279*, 1177–1197. [CrossRef] [PubMed]
22. Onken, J.; Moeckel, S.; Leukel, P.; Leidgens, V.; Baumann, F.; Bogdahn, U.; Vollmann-Zwerenz, A.; Hau, P. Versican isoform V1 regulates proliferation and migration in high-grade gliomas. *J. Neurooncol.* **2014**, *120*, 73–83. [CrossRef] [PubMed]
23. Al-Mayhani, M.T.; Grenfell, R.; Narita, M.; Piccirillo, S.; Kenney-Herbert, E.; Fawcett, J.W.; Collins, V.P.; Ichimura, K.; Watts, C. NG2 expression in glioblastoma identifies an actively proliferating population with an aggressive molecular signature. *Neuro. Oncol.* **2011**, *13*, 830–845. [CrossRef] [PubMed]
24. Stallcup, W.B.; Huang, F.J. A role for the NG2 proteoglycan in glioma progression. *Cell Adh. Migr.* **2008**, *2*, 192–201. [CrossRef] [PubMed]
25. Salanti, A.; Clausen, T.M.; Agerbaek, M.O.; Al Nakouzi, N.; Dahlback, M.; Oo, H.Z.; Lee, S.; Gustavsson, T.; Rich, J.R.; Hedberg, B.J.; et al. Targeting Human Cancer by a Glycosaminoglycan Binding Malaria Protein. *Cancer Cell* **2015**, *28*, 500–514. [CrossRef] [PubMed]
26. Agerbaek, M.O.; Bang-Christensen, S.; Salanti, A. Fighting Cancer Using an Oncofetal Glycosaminoglycan-Binding Protein from Malaria Parasites. *Trends Parasitol.* **2019**, *35*, 178–181. [CrossRef] [PubMed]
27. Agerbaek, M.O.; Bang-Christensen, S.R.; Yang, M.H.; Clausen, T.M.; Pereira, M.A.; Sharma, S.; Ditlev, S.B.; Nielsen, M.A.; Choudhary, S.; Gustavsson, T.; et al. The VAR2CSA malaria protein efficiently retrieves circulating tumor cells in an EpCAM-independent manner. *Nat. Commun.* **2018**, *9*, 3279. [CrossRef] [PubMed]
28. Bax, D.A.; Little, S.E.; Gaspar, N.; Perryman, L.; Marshall, L.; Viana-Pereira, M.; Jones, T.A.; Williams, R.D.; Grigoriadis, A.; Vassal, G.; et al. Molecular and phenotypic characterisation of paediatric glioma cell lines as models for preclinical drug development. *PLoS ONE* **2009**, *4*, e5209. [CrossRef]
29. Rasmussen, R.D.; Gajjar, M.K.; Tuckova, L.; Jensen, K.E.; Maya-Mendoza, A.; Holst, C.B.; Mollgaard, K.; Rasmussen, J.S.; Brennum, J.; Bartek, J.; et al. BRCA1-regulated RRM2 expression protects glioblastoma cells from endogenous replication stress and promotes tumorigenicity. *Nat. Commun.* **2016**, *7*, 13398. [CrossRef]

30. Rappsilber, J.; Mann, M.; Ishihama, Y. Protocol for micro-purification, enrichment, pre-fractionation and storage of peptides for proteomics using StageTips. *Nat. Protoc.* **2007**, *2*, 1896–1906. [CrossRef]
31. Mogensen, M.B.; Rossing, M.; Ostrup, O.; Larsen, P.N.; Heiberg Engel, P.J.; Jorgensen, L.N.; Hogdall, E.V.; Eriksen, J.; Ibsen, P.; Jess, P.; et al. Genomic alterations accompanying tumour evolution in colorectal cancer: Tracking the differences between primary tumours and synchronous liver metastases by whole-exome sequencing. *BMC Cancer* **2018**, *18*, 752. [CrossRef] [PubMed]
32. Thorvaldsdottir, H.; Robinson, J.T.; Mesirov, J.P. Integrative Genomics Viewer (IGV): High-performance genomics data visualization and exploration. *Brief Bioinform.* **2013**, *14*, 178–192. [CrossRef] [PubMed]
33. Clausen, T.M.; Pereira, M.A.; Al Nakouzi, N.; Oo, H.Z.; Agerbaek, M.O.; Lee, S.; Orum-Madsen, M.S.; Kristensen, A.R.; El-Naggar, A.; Grandgenett, P.M.; et al. Oncofetal Chondroitin Sulfate Glycosaminoglycans Are Key Players in Integrin Signaling and Tumor Cell Motility. *Mol. Cancer Res.* **2016**, *14*, 1288–1299. [CrossRef] [PubMed]
34. Agerbaek, M.O.; Pereira, M.A.; Clausen, T.M.; Pehrson, C.; Oo, H.Z.; Spliid, C.; Rich, J.R.; Fung, V.; Nkrumah, F.; Neequaye, J.; et al. Burkitt lymphoma expresses oncofetal chondroitin sulfate without being a reservoir for placental malaria sequestration. *Int. J. Cancer* **2017**, *140*, 1597–1608. [CrossRef] [PubMed]
35. Platten, M.; Wick, W.; Weller, M. Malignant glioma biology: Role for TGF-beta in growth, motility, angiogenesis, and immune escape. *Microsc. Res. Tech.* **2001**, *52*, 401–410. [CrossRef]
36. Joseph, J.V.; Conroy, S.; Tomar, T.; Eggens-Meijer, E.; Bhat, K.; Copray, S.; Walenkamp, A.M.; Boddeke, E.; Balasubramanyian, V.; Wagemakers, M.; et al. TGF-beta is an inducer of ZEB1-dependent mesenchymal transdifferentiation in glioblastoma that is associated with tumor invasion. *Cell Death Dis.* **2014**, *5*, e1443. [CrossRef] [PubMed]
37. Shankar, J.; Messenberg, A.; Chan, J.; Underhill, T.M.; Foster, L.J.; Nabi, I.R. Pseudopodial actin dynamics control epithelial-mesenchymal transition in metastatic cancer cells. *Cancer Res.* **2010**, *70*, 3780–3790. [CrossRef] [PubMed]
38. Nevo, I.; Woolard, K.; Cam, M.; Li, A.; Webster, J.D.; Kotliarov, Y.; Kim, H.S.; Ahn, S.; Walling, J.; Kotliarova, S.; et al. Identification of molecular pathways facilitating glioma cell invasion in situ. *PLoS ONE* **2014**, *9*, e111783. [CrossRef] [PubMed]
39. Qiao, D.; Meyer, K.; Friedl, A. Glypican 1 stimulates S phase entry and DNA replication in human glioma cells and normal astrocytes. *Mol. Cell. Biol.* **2013**, *33*, 4408–4421. [CrossRef] [PubMed]
40. Xu, Y.; Yuan, J.; Zhang, Z.; Lin, L.; Xu, S. Syndecan-1 expression in human glioma is correlated with advanced tumor progression and poor prognosis. *Mol. Biol. Rep.* **2012**, *39*, 8979–8985. [CrossRef] [PubMed]
41. Schiffer, D.; Mellai, M.; Boldorini, R.; Bisogno, I.; Grifoni, S.; Corona, C.; Bertero, L.; Cassoni, P.; Casalone, C.; Annovazzi, L. The Significance of Chondroitin Sulfate Proteoglycan 4 (CSPG4) in Human Gliomas. *Int. J. Mol. Sci.* **2018**, *19*. [CrossRef] [PubMed]
42. Senbanjo, L.T.; Chellaiah, M.A. CD44: A Multifunctional Cell Surface Adhesion Receptor Is a Regulator of Progression and Metastasis of Cancer Cells. *Front. Cell Dev. Biol.* **2017**, *5*, 18. [CrossRef] [PubMed]
43. Faye, R.S.; Aamdal, S.; Hoifodt, H.K.; Jacobsen, E.; Holstad, L.; Skovlund, E.; Fodstad, O. Immunomagnetic detection and clinical significance of micrometastatic tumor cells in malignant melanoma patients. *Clin. Cancer Res.* **2004**, *10 Pt 1*, 4134–4139. [CrossRef] [PubMed]
44. Ulmer, A.; Schmidt-Kittler, O.; Fischer, J.; Ellwanger, U.; Rassner, G.; Riethmuller, G.; Fierlbeck, G.; Klein, C.A. Immunomagnetic enrichment, genomic characterization, and prognostic impact of circulating melanoma cells. *Clin. Cancer Res.* **2004**, *10*, 531–537. [CrossRef] [PubMed]
45. Rao, C.; Bui, T.; Connelly, M.; Doyle, G.; Karydis, I.; Middleton, M.R.; Clack, G.; Malone, M.; Coumans, F.A.; Terstappen, L.W. Circulating melanoma cells and survival in metastatic melanoma. *Int. J. Oncol.* **2011**, *38*, 755–760. [PubMed]
46. Gray, E.S.; Reid, A.L.; Bowyer, S.; Calapre, L.; Siew, K.; Pearce, R.; Cowell, L.; Frank, M.H.; Millward, M.; Ziman, M. Circulating Melanoma Cell Subpopulations: Their Heterogeneity and Differential Responses to Treatment. *J. Investig. Dermatol.* **2015**, *135*, 2040–2048. [CrossRef] [PubMed]
47. Pellegatta, S.; Savoldo, B.; Di Ianni, N.; Corbetta, C.; Chen, Y.; Patane, M.; Sun, C.; Pollo, B.; Ferrone, S.; DiMeco, F.; et al. Constitutive and TNFalpha-inducible expression of chondroitin sulfate proteoglycan 4 in glioblastoma and neurospheres: Implications for CAR-T cell therapy. *Sci. Transl. Med.* **2018**, *10*. [CrossRef]
48. Rodriguez, A.; Brown, C.; Badie, B. Chimeric antigen receptor T-cell therapy for glioblastoma. *Transl. Res.* **2017**, *187*, 93–102. [CrossRef]

49. Terada, T. Expression of cytokeratins in glioblastoma multiforme. *Pathol Oncol Res.* **2015**, *21*, 817–819. [CrossRef]
50. Goswami, C.; Chatterjee, U.; Sen, S.; Chatterjee, S.; Sarkar, S. Expression of cytokeratins in gliomas. *Indian J. Pathol. Microbiol.* **2007**, *50*, 478–481.
51. Fanburg-Smith, J.C.; Majidi, M.; Miettinen, M. Keratin expression in schwannoma; a study of 115 retroperitoneal and 22 peripheral schwannomas. *Mod. Pathol.* **2006**, *19*, 115–121. [CrossRef] [PubMed]
52. Pastushenko, I.; Brisebarre, A.; Sifrim, A.; Fioramonti, M.; Revenco, T.; Boumahdi, S.; Van Keymeulen, A.; Brown, D.; Moers, V.; Lemaire, S.; et al. Identification of the tumour transition states occurring during EMT. *Nature* **2018**, *556*, 463–468. [CrossRef] [PubMed]
53. Myung, J.; Cho, B.K.; Kim, Y.S.; Park, S.H. Snail and Cox-2 expressions are associated with WHO tumor grade and survival rate of patients with gliomas. *Neuropathology* **2010**, *30*, 224–231. [CrossRef] [PubMed]
54. Liu, Y.; Hu, H.; Wang, K.; Zhang, C.; Wang, Y.; Yao, K.; Yang, P.; Han, L.; Kang, C.; Zhang, W.; et al. Multidimensional analysis of gene expression reveals TGFB1I1-induced EMT contributes to malignant progression of astrocytomas. *Oncotarget* **2014**, *5*, 12593–12606. [CrossRef] [PubMed]
55. Ayres Pereira, M.; Mandel Clausen, T.; Pehrson, C.; Mao, Y.; Resende, M.; Daugaard, M.; Riis Kristensen, A.; Spliid, C.; Mathiesen, L.; Knudsen, L.E.; et al. Placental Sequestration of Plasmodium falciparum Malaria Parasites Is Mediated by the Interaction Between VAR2CSA and Chondroitin Sulfate A on Syndecan-1. *PLoS Pathog.* **2016**, *12*, e1005831. [CrossRef]
56. Foehr, E.D.; Lorente, G.; Kuo, J.; Ram, R.; Nikolich, K.; Urfer, R. Targeting of the receptor protein tyrosine phosphatase beta with a monoclonal antibody delays tumor growth in a glioblastoma model. *Cancer Res.* **2006**, *66*, 2271–2278. [CrossRef]
57. Higgins, S.C.; Bolteus, A.J.; Donovan, L.K.; Hasegawa, H.; Doey, L.; Al Sarraj, S.; King, A.; Ashkan, K.; Roncaroli, F.; Fillmore, H.L.; et al. Expression of the chondroitin sulphate proteoglycan, NG2, in paediatric brain tumors. *Anticancer Res.* **2014**, *34*, 6919–6924.
58. Wang, J.; Svendsen, A.; Kmiecik, J.; Immervoll, H.; Skaftnesmo, K.O.; Planaguma, J.; Reed, R.K.; Bjerkvig, R.; Miletic, H.; Enger, P.O.; et al. Targeting the NG2/CSPG4 proteoglycan retards tumour growth and angiogenesis in preclinical models of GBM and melanoma. *PLoS ONE* **2011**, *6*, e23062. [CrossRef]
59. Phillips, H.S.; Kharbanda, S.; Chen, R.; Forrest, W.F.; Soriano, R.H.; Wu, T.D.; Misra, A.; Nigro, J.M.; Colman, H.; Soroceanu, L.; et al. Molecular subclasses of high-grade glioma predict prognosis, delineate a pattern of disease progression, and resemble stages in neurogenesis. *Cancer Cell* **2006**, *9*, 157–173. [CrossRef]
60. Colman, H.; Zhang, L.; Sulman, E.P.; McDonald, J.M.; Shooshtari, N.L.; Rivera, A.; Popoff, S.; Nutt, C.L.; Louis, D.N.; Cairncross, J.G.; et al. A multigene predictor of outcome in glioblastoma. *Neuro Oncol.* **2010**, *12*, 49–57. [CrossRef]
61. Pietras, A.; Katz, A.M.; Ekstrom, E.J.; Wee, B.; Halliday, J.J.; Pitter, K.L.; Werbeck, J.L.; Amankulor, N.M.; Huse, J.T.; Holland, E.C. Osteopontin-CD44 signaling in the glioma perivascular niche enhances cancer stem cell phenotypes and promotes aggressive tumor growth. *Cell Stem Cell* **2014**, *14*, 357–369. [CrossRef] [PubMed]
62. Liu, T.; Xu, H.; Huang, M.; Ma, W.; Saxena, D.; Lustig, R.A.; Alonso-Basanta, M.; Zhang, Z.; O'Rourke, D.M.; Zhang, L.; et al. Circulating Glioma Cells Exhibit Stem Cell-like Properties. *Cancer Res.* **2018**, *78*, 6632–6642. [CrossRef] [PubMed]
63. Aoki, K.; Nakamura, H.; Suzuki, H.; Matsuo, K.; Kataoka, K.; Shimamura, T.; Motomura, K.; Ohka, F.; Shiina, S.; Yamamoto, T.; et al. Prognostic relevance of genetic alterations in diffuse lower-grade gliomas. *Neuro Oncol.* **2018**, *20*, 66–77. [CrossRef] [PubMed]
64. D'Alessio, A.; Proietti, G.; Sica, G.; Scicchitano, B.M. Pathological and Molecular Features of Glioblastoma and Its Peritumoral Tissue. *Cancers* **2019**, *11*. [CrossRef] [PubMed]
65. Heitzer, E.; Auer, M.; Gasch, C.; Pichler, M.; Ulz, P.; Hoffmann, E.M.; Lax, S.; Waldispuehl-Geigl, J.; Mauermann, O.; Lackner, C.; et al. Complex tumor genomes inferred from single circulating tumor cells by array-CGH and next-generation sequencing. *Cancer Res.* **2013**, *73*, 2965–2975. [CrossRef] [PubMed]
66. Zhang, Y.; Dube, C.; Gibert, M., Jr.; Cruickshanks, N.; Wang, B.; Coughlan, M.; Yang, Y.; Setiady, I.; Deveau, C.; Saoud, K.; et al. The p53 Pathway in Glioblastoma. *Cancers* **2018**, *10*. [CrossRef] [PubMed]
67. Muzny, D.M.; Bainbridge, M.N.; Chang, K.; Dinh, H.H.; Drummond, J.A.; Fowler, G.; Kovar, C.L.; Lewis, L.R.; Morgan, M.B.; Newsham, I.F.; et al. Comprehensive molecular characterization of human colon and rectal cancer. *Nature* **2012**, *487*, 330–337.
68. Dulak, A.M.; Stojanov, P.; Peng, S.Y.; Lawrence, M.S.; Fox, C.; Stewart, C.; Bandla, S.; Imamura, Y.;

Schumacher, S.E.; Shefler, E.; et al. Exome and whole-genome sequencing of esophageal adenocarcinoma identifies recurrent driver events and mutational complexity. *Nat. Genet.* **2013**, *45*, 478–486. [CrossRef]

69. Hitchins, M.; Williams, R.; Cheong, K.; Halani, N.; Lin, V.A.; Packham, D.; Ku, S.; Buckle, A.; Hawkins, N.; Burn, J.; et al. MLH1 germline epimutations as a factor in hereditary nonpolyposis colorectal cancer. *Gastroenterology* **2005**, *129*, 1392–1399. [CrossRef]
70. Hitchins, M.P.; Rapkins, R.W.; Kwok, C.T.; Srivastava, S.; Wong, J.J.L.; Khachigian, L.M.; Polly, P.; Goldblatt, J.; Ward, R.L. Dominantly Inherited Constitutional Epigenetic Silencing of MLH1 in a Cancer-Affected Family Is Linked to a Single Nucleotide Variant within the 5′ UTR. *Cancer Cell* **2011**, *20*, 200–213. [CrossRef]
71. Bresler, S.C.; Weiser, D.A.; Huwe, P.J.; Park, J.H.; Krytska, K.; Ryles, H.; Laudenslager, M.; Rappaport, E.F.; Wood, A.C.; McGrady, P.W.; et al. ALK mutations confer differential oncogenic activation and sensitivity to ALK inhibition therapy in neuroblastoma. *Cancer Cell* **2014**, *26*, 682–694. [CrossRef] [PubMed]
72. Chen, L.D.; Humphreys, A.; Turnbull, L.; Bellini, A.; Schleiermacher, G.; Salwen, H.; Cohn, S.L.; Bown, N.; Tweddle, D.A. Identification of different ALK mutations in a pair of neuroblastoma cell lines established at diagnosis and relapse. *Oncotarget* **2016**, *7*, 87301–87311. [CrossRef] [PubMed]
73. Bresler, S.C.; Wood, A.C.; Haglund, E.A.; Courtright, J.; Belcastro, L.T.; Plegaria, J.S.; Cole, K.; Toporovskaya, Y.; Zhao, H.Q.; Carpenter, E.L.; et al. Differential Inhibitor Sensitivity of Anaplastic Lymphoma Kinase Variants Found in Neuroblastoma. *Sci. Transl. Med.* **2011**, *3*, 108ra114. [CrossRef] [PubMed]

9

Bone Marrow Involvement in Melanoma: Potentials for Detection of Disseminated Tumor Cells and Characterization of their Subsets by Flow Cytometry

Olga Chernysheva *, Irina Markina, Lev Demidov, Natalia Kupryshina, Svetlana Chulkova, Alexandra Palladina, Alina Antipova and Nikolai Tupitsyn

FSBI "N.N. Blokhin Russian Cancer Research Center" of Ministry of Health of the Russian Federation, 115478 Moscow, Russia; irina160771@yandex.ru (I.M.); nntca@yahoo.com (L.D.); natalya-2511@yandex.ru (N.K.); chulkova@mail.ru (S.C.); alexandra.93@mail.ru (A.P.); a.s.antipova@gmail.com (A.A.); nntca@yahoo.com (N.T.)

* Correspondence: dr.chernysheva@mail.ru

Abstract: Disseminated tumor cells (DTCs) are studied as a prognostic factor in many non-hematopoietic tumors. Melanoma is one of the most aggressive tumors. Forty percent of melanoma patients develop distant metastases at five or more years after curative surgery, and frequent manifestations of melanoma without an identified primary lesion may reflect the tendency of melanoma cells to spread from indolent sites such as bone marrow (BM). The purpose of this work was to evaluate the possibility of detecting melanoma DTCs in BM based on the expression of a cytoplasmatic premelanocytic glycoprotein HMB-45 using flow cytometry, to estimate the influence of DTCs' persistence in BM on hematopoiesis, to identify the frequency of BM involvement in patients with melanoma, and to analyze DTC subset composition in melanoma. DTCs are found in 57.4% of skin melanoma cases and in as many as 28.6% of stage I cases, which confirms the aggressive course even of localized disease. Significant differences in the groups with the presence of disseminated tumor cells (DTCs^{+}) and the lack thereof (DTC^{-}) are noted for blast cells, the total content of granulocyte cells, and oxyphilic normoblasts of erythroid raw cells.

Keywords: bone marrow; melanoma; disseminated tumor cells; solid cancers; single-cell analysis; enrichment and detection technologies; flow cytometry; tumor stem cells; HMB-45; CD133

---

1. Introduction

Today oncology is a rapidly developing field of medicine. Every year novel target and immunological agents acting against cancer at the molecular level are added to clinical oncologists' practice, and many such agents are currently at various stages of clinical development. However, notwithstanding significant progress over the last decade and a broad variety of therapeutical options, several fundamental questions remain to be answered: What are causes of cancer development? What are mechanisms of metastasis and recurrence? At what stage of disease development can we influence these processes?

Over the last 150 years there were many theories to explain processes developing both in the tumor and in the patient's body. By the end of the first quarter of the 21st century the world medical community has passed a long way from the first publication by T.R. Ashworth in the *Medical Journal of Australia* in 1869 [1], where the author described for the first time circulating tumor cells in a cancer patient, and the 'seed and soil' theory proposed by Stephen Paget in 1889 [2], through the theory of late dissemination (linear progression) by William Stewart Halsted [3–5] to the theory of early metastasis (parallel progression) by Christophe Klein [6] and the concept of the premetastatic niche by Bethan Psaila and David Lyden in 2009 [7]. The key question in all of these theories was how tumor cells managed to overcome immune surveillance [8], to preserve their proliferative potential and to proliferate in alien environments even after several decades of latency [9,10].

It seems natural that bone marrow (BM) with its advanced capillary network and a cocktail of soluble protein factors, integrins, chemokines, cell adhesion molecules, and a variety of growth factors is the most attractive niche for tumor cells [11,12]. Being basically alien, BM makes its environment appropriate for disseminated tumor cell (DTC) persistence via sophisticated antigenic, immunogenic, and cellular mechanisms [13,14]. DTCs may have different fates in a new microenvironment. Most of them die within several weeks or months [15], while DTCs preserving their vitality without decrease or increase in their total number may enter latency and form so called dormant metastases.

Dormant tumor cells have three main differences from other tumor cells, i.e., the ability to survive in alien and even hostile environments for a long time, temporary but reversible growth arrest, and resistance to target cytostatic agents [16]. These DTC properties make them biologically closer to tumor stem cells, a minor primary tumor subset seeming to play a leading role in the self-maintenance and metastasis of malignancies [17].

BM involvement is described in multiple non-hemopoietic neoplasms and is shown to be an independent poor prognostic factor for overall and disease-free survival [18–23]. Interestingly, these publications mainly address cancers of the breast, stomach, lung, colon, or prostate, while studies of melanoma are few and require further analysis.

Observations of hematogenous metastases from melanoma after 10 [24] or even 40 [25] years after removal of the primary tumor and frequent melanoma manifestations without an identified primary may reflect melanoma cell tendency to spread from indolent sites [26,27] such as BM.

gp100—HMB-45, a cytoplasmatic premelanocytic glycoprotein is a reliable marker of melanoma cells. It was discovered as one of the first melanoma antigens to demonstrate high sensitivity (up to 93%) and specificity (up to 100%) [28] and is usable to identify DTCs.

The Hemopoiesis Immunology Laboratory (N.N. Blokhin Cancer Research Center, Russian Federation Health Ministry) has developed a procedure to identify DTCs by flow cytometry [29]. Flow cytometry has certain advantages as compared to cytology, immunohistochemistry, and molecular biology techniques. For instance, contemporary multicolor flow cytometry can analyze 12 or more parameters in a single cell and accumulate a large number of events with sensitivity close to that of PCR (10^{-4} to 10^{-6}) and allows most complete description of the DTC immunophenotype [30]. Besides pure quantification of DTCs, flow cytometry therefore helps to study DTC subsets such as tumor stem cells or to identify surface molecular targets for drugs (Her2/neu, PDL1).

The purpose of this work was to evaluate the possibility of detecting melanoma DTCs in BM based on the expression of HMB-45 using flow cytometry, to determine the frequency of BM involvement in patients with melanoma, to analyze DTC subset composition in melanoma as to the expression of CD56 and CD57 that were an additional criterion for melanoma immunological diagnosis, and to assess the proportion of tumor stem cells among DTCs based on the presence of CD133.

2. Materials and Methods

A total of 47 patients (23 males and 24 females) aged 20–72 (median 49.8) years managed at the N. N. Blokhin Russian Cancer Research Center for skin melanoma during 2018–2019 were enrolled in the study. The diagnosis was verified histologically in all patients. This study was approved by the institutional ethical committees (Local ethical committee N. N. Blokhin Russian Cancer Research Center of Ministry of Health of the Russian Federation; UDC 616-006, Reg. № AAAA-A16-116122210071-4, Inv. 479.) and was done with the informed consent of the patients. Most of the patients (42.6%) had stage IV disease based on complex examination. BM involvement was assessed by morphology and immunology at diagnosis. Table 1 demonstrates patient distribution by stage.

Table 1. The distribution of patients by disease stage.

Stage	Frequency	Percent (%)
I	7	14.9
IIa	1	2.1
IIb	5	10.6
IIc	3	6.4
III	11	23.4
IV	20	42.6
Total:	47	100

Morphological examination included myelogram count and identification of tumor cells on six Romanovsky-stained bone marrow smears by two morphologists in parallel. Immunological identification of DTCs in BM was done by flow cytometry. Samples were lysed using BD FACS lysing solution (Beckton Dickinson, Franklin Lakes, NJ, USA), then washed in phosphate-buffered saline (PBS), and re-suspended in 100 mL of PBS. Cells were incubated for 15–20 min in the dark at room temperature together with a cocktail of monoclonal antibodies directly conjugated with fluorescein isothiocyanate (FITC), phycoerythrin (PE), allophycocyanin (APC), and Horizon V500 and Horizon V450 fluorochromes (Table 2). All samples were processed within 24 h after collection. Antibody labeling was measured by multiparameter flow cytometry using FACS Canto II (Beckton Dickinson). Twenty million myelokaryocytes (or all cells in the sample) were collected to identify DTCs. Tumor cells were detected by the lack of expression of the common leukocyte antigen CD45 in combination with bright expression of HMB-45. To identify DTC subpopulations expression of CD133, CD56, and CD57 molecules was analyzed among the $CD45^{-}HMB\text{-}45^{+}$ cells.

Table 2. Monoclonal antibodies used in the study.

No.	MoAbs/Fluorochromes	Function	Manufacturer
1	Syto41	Nuclear dye	Thermo Fisher Scientific, Walthem, MA, USA
2	CD45	Leukocyte common antigen	Beckton Dickinson
3	HMB-45	Melanoma cell antigen gp100	Santa Cruz Biotechnology, Dallas, Tx, USA
4	CD56	Neuronal cell adhesion molecule (NCAM)	Beckton Dickinson
5	CD57	NK-cell molecule (HNK1)	Beckton Dickinson
6	CD133	Hematopoietic stem cell antigen	Beckton Dickinson

Results were analyzed using Kaluza Analysis v2.1 (Beckman Coulter, Brea, CA, USA) software. Statistical analysis of data used IBM-SPSS Statistics v.17 package (IBM, Armonk, NY, USA).

3. Results

Morphological analysis of BM biopsies included myelogram count and tumor cell identification.

In the analysis of hematopoiesis, we excluded cases with bone marrow dilution with peripheral blood. Comparison of the average bone marrow parameters according to the myelogram is shown in the Table 3.

Table 3. Comparison of the average bone marrow according to myelogram.

Myelogram Parameters	DTCs	*n*	Mean Value	Err^{std}mean	*p*
Cellularity	negative	19	67.0	6.51	NS*
	positive	20	67.3	7.87	
Blasts	negative	20	1.46	0.14	0.026
	positive	25	1.09	0.09	
Promyelocytes	negative	20	0.44	0.11	NS
	positive	25	0.37	0.08	
Neutrophilic myelocytes	negative	20	7.80	0.72	NS
	positive	25	8.95	0.54	
Neutrophilic metamyelocytes	negative	20	8.58	0.65	NS
	positive	25	7.83	0.53	
Band neutrophils	negative	20	16.50	0.91	NS
	positive	25	18.70	1.00	
Segmented neutrophils	negative	20	24.47	1.39	NS
	positive	25	27.266	1.71	
All granulocyte cells	negative	20	60.76	1.45	0.025
	positive	25	65.41	1.38	
Neutrophil maturation index	negative	20	0.43	0.034	NS
	positive	25	0.38	0.034	
Monocytes	negative	20	2.78	0.26	NS
	positive	25	3.30	0.24	
Lymphocytes	negative	20	12.85	0.79	NS
	positive	25	12.02	0.68	
Plasmocytes	negative	20	0.60	0.10	NS
	positive	25	0.77	0.15	
Basophilic normoblasts	negative	20	1.23	0.17	NS
	positive	25	0.97	0.13	
Polychromatophilic normoblasts	negative	20	11.16	0.91	NS
	positive	25	10.19	0.82	
Oxyphilic normoblasts	negative	20	9.08	0.88	0.006
	positive	25	6.25	0.52	
Sum of nucleated erythroid cells	negative	20	21.47	1.44	0.042
	positive	25	17.41	1.29	
Erythroid maturation index	negative	20	0.96	0.01	NS
	positive	25	0.96	0.01	
Leuco–erythroid ratio	negative	20	4.02	0.39	0.034
	positive	25	5.58	0.59	

* NS—not significant.

Significant differences in the groups with the presence of disseminated tumor cells (DTCs+) and the lack thereof (DTC−) were noted for blast cells, the total content of granulocyte cells, and erythroid germ indicators.

The level of blast cells was higher in patients with no DTCs: 1.46% ± 0.14% (n = 20) and 1.1% ± 0.09% (n = 25), p = 0.026.

On the contrary, the total content of granulocyte cells was higher in patients with DTCs in the BM: 65.4% ± 1.4% (n = 25) and 60.8% ± 1.5%: (n = 20), p = 0.025.

The most significant differences were obtained with respect to cells of the erythroid series. Thus, the percentage of oxyphilic normoblasts was significantly higher in patients with no DTCs in the BM: 9.1% ± 0.88% (n = 20) and 6.3% ± 0.52% (n = 25), p = 0.006. It should be noted that, in the group as a whole, the levels of oxyphilic normoblasts were increased compared to the norm in 67% of patients. Accordingly, the sum of nucleated erythroid cells was also higher in melanoma patients with no DTCs in the BM: 21.5% ± 1.4% (n = 20) and 17.4% ± 1.3% (n = 25), p = 0.042. This was reflected in a significantly higher leuco–erythroid ratio in patients with the presence of DTCs in the bone marrow: 5.6% ± 0.6% (n = 25) and 4.0% ± 0.4 (n = 20), p = 0.034.

It is interesting to note that when analyzing according to the tables of conjugacy of characters, only two indicators of the myelogram were reliably associated with the presence of DTCs in the BM—the total content of granulocyte cells and the level of oxyphilic normoblasts.

The relationship of DTCs with the total amount of granulocyte cells consisted in the fact that, in the presence of DTCs, a decrease in the total level of granulocyte cells was observed in only 8% of cases, while in the absence of DTCs a decrease in granulocyte cells was observed in 30% of cases, chi-square = 8.9; $p = 0.012$.

A different situation was noted with respect to oxyphilic normoblasts, whose normal content in the absence of DTCs was observed in 15% of cases, and in the presence of DTCs—three times more often—in 44% of cases. On the contrary, an increase in the level of oxyphilic normoblasts in the absence of DTCs occurred in 85% of cases, in the presence of DTCs, in 56%, chi-square = 4.4; $p = 0.037$. Melanoma cells were identified in BM by morphology in one of 47 cases only (Figure 1).

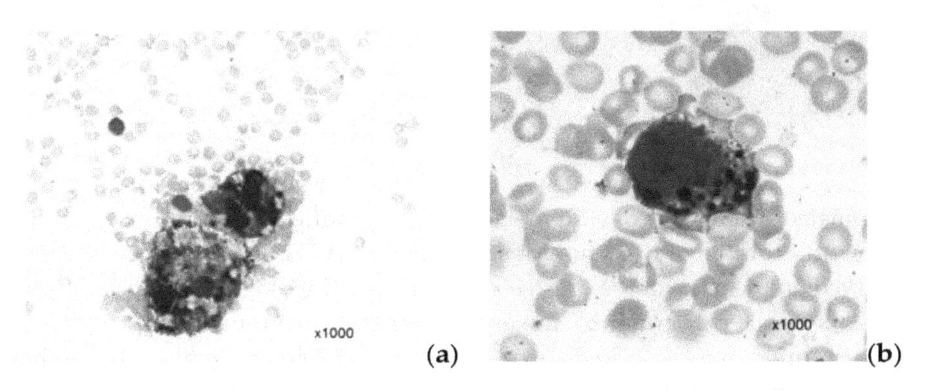

Figure 1. Melanoma disseminated tumor cells in bone marrow. This figure presents a case of detection of skin melanoma cells in bone marrow punctate ((**a**) and (**b**), ×1000 magnification). Punctate bone marrow is poor. Normal lines of myelopoiesis are depressed. Cell complexes of a non-hemopoietic nature are determined. Additionally, there are scattered, separately lying tumor cells. They are represented by cells of a large size, and basophilic colored pigment granules of various sizes are visible in the cytoplasm. The morphological picture of the bone marrow is characteristic of metastatic lesions in melanoma.

Immunological analysis of DTCs in BM was based on a threshold of one tumor cell ($Syto41^+CD45^-HMB\text{-}45^+$) per 10 million myelokaryocytes. A mean of 14,146,987 (±957,728) myelokaryocytes were analyzed in each sample. DTCs were found in 57.4% of BM samples ($n = 27$) based on the threshold level. Interestingly, flow cytometry of melanoma cells has specific features due to morphological characteristics of these cells such as a rather large size and the presence of pigmented inclusions of various diameters (from dust-like to large fused granules of different diameter). For instance, the melanoma DTCs have high direct and side light scatter characteristics and require adequate protocols for flow cytometer tuning.

There were no significant differences in DTC counts with respect to gender, age, or disease stage. What is important is that BM involvement was discovered at all disease stages (Table 4). This means that hematogenic tumor cell dissemination occurs already from clinically localized disease.

Table 4. Frequency of disseminated tumor cell (DTC) detection at various stages of melanoma.

Stage	Number of Patients	Frequency of DTCs + Cases
I	7	28.6%
II	9	55.6%
III	11	63.6%
IV	20	65.0%

DTCs were additionally characterized by CD56 and CD57 expression. In our study, CD56 and CD57 expression was assessed in 23 BM samples. Among them, DTCs were present in 54.2% (n = 13) though these cells did not express CD56. CD57 expression on DTCs was found in six cases (46.2%) (Figure 2). Importantly, not all 100% of DTCs in each BM sample demonstrated CD57 expression. On average 87.4% ± 5.8% of DTCs were CD57-positive. Of interest, 50% of CD57+ patients had stage IV, two of six had stage III, and one patient had stage IIc disease.

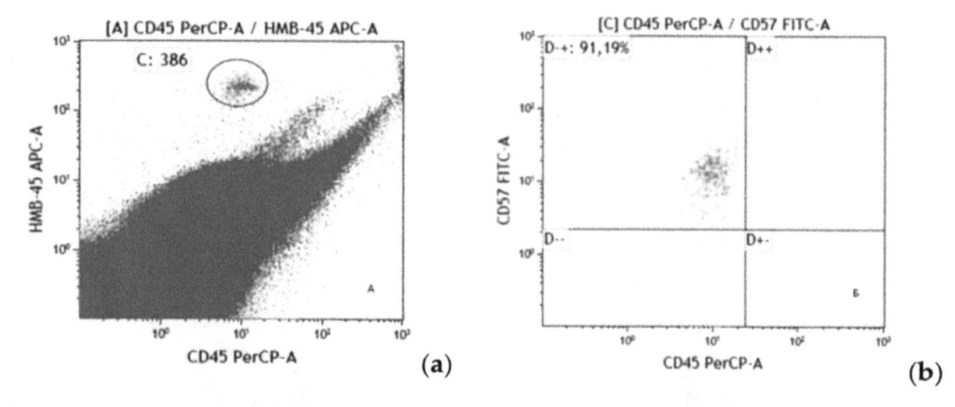

Figure 2. Disseminated tumor cells of skin melanoma as identified by immunological flow cytometry. This figure shows an example of detection of skin melanoma DTCs. On the cytogram (**a**) in gate C, DTCs were observed on the basis of the bright expression of HMB-45 (y-axis) and the absence of CD45 expression (x-axis). On the cytogram (**b**), the analysis of the subpopulation composition of DTCs in melanoma was performed in relation to the expression of the CD57 antigen. Cells are characterized by distinct CD57 expression (y-axis) and lack of CD45 expression (x-axis). Most DTCs (91.19%) are CD57+.

A minor tumor stem cell (TSC) subset with maximum resistance to conventional anticancer therapies plays a special role in metastasis. According to the literature, melanoma TSCs are characterized by expression of antigens such as CD44, CD271, and CD133. We identified TSCs among melanoma DTCs by CD133 expression.

CD133 expression was analyzed in 22 BM samples. Half of these BM samples were DTC-positive. There was a single DTC-positive sample containing a CD133+-DTC subset, which accounted for 1.38% of all DTCs in this case (Figure 3).

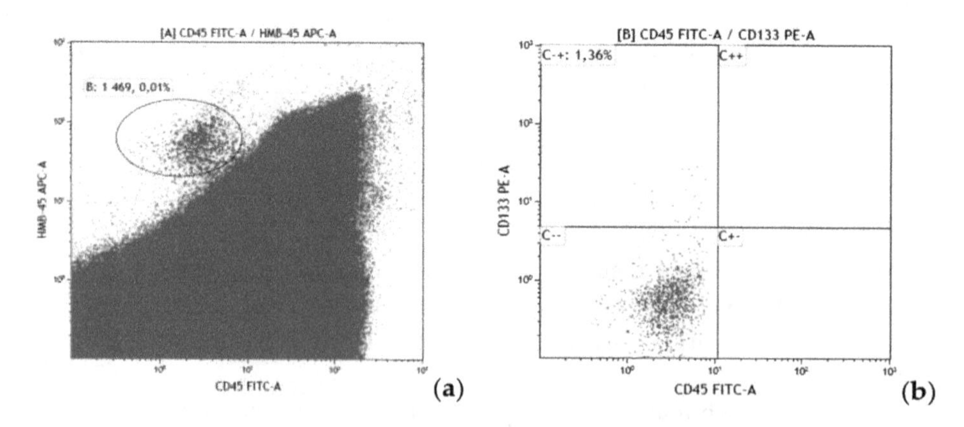

Figure 3. Identification of CD133-positive DTCs. This figure shows an example of the assessment of DTC subpopulations by the tumor stem cell marker CD133. On the cytogram (**a**), 0.01% DTCs was detected by the distinct expression of HMB-45 (y-axis) and the lack of expression of the pan-leukocyte antigen CD45 (x-axis). Within the DTCs, expression of CD133 evaluated. Cytogram (**b**) (x-axis is CD45, y-axis is CD133) shows that CD133+ cells make up 1.36% of all DTCs.

We have demonstrated that both the primary tumor and DTCs in BM may have a heterogeneous composition and express various antigens. The significance of this finding for the disease course and prognosis deserves assessment in further studies.

4. Discussion

BM as a niche for micrometastasis plays a key role in hematogeneous dissemination. By creating a unique microenvironment for tumor cells BM maintains their proliferative potential for many years. Disease recurrence decades after treatment of the primary is described for many entities, and skin melanoma is not an exception. Forty percent of skin melanoma patients develop distant metastases at five or more years after curative surgery [27], therefore finding novel factors for disease prognosis and markers of early tumor cell dissemination for personalization of early systemic treatment is of much importance.

As demonstrated in our study, flow cytometry with a specific antibody HMB-45 in combination with CD45 is a useful technique to assess BM involvement in melanoma. DTCs were found in 57.4% of skin melanoma cases. The DTCs were present in 28.6% of stage I disease, which confirms the aggressiveness of skin melanoma even in localized disease. The findings of CD57 and CD133 expression are evidence of DTCs heterogeneity and the complex hierarchical relations between the primary and the DTCs. The prognostic significance of our results will be assessed in further studies.

Thus, we can talk about the complex relationship of hematopoiesis in general and the development of skin melanoma. Both myelo- and erythropoiesis are involved in and reacting to the tumor process occurring in the body. Of particular interest are changes in the bone marrow hematopoiesis arising in the presence of DTCs. Perhaps they are a reflection of the reorganization of the microenvironment of the DTCs, contributing to their long-term persistence in the bone marrow. The role of these changes in the prognosis of the course of the disease remains to be assessed.

The 'seed and soil' theory therefore is still valuable after 150 years and requires further development using up-to-date diagnostic approaches.

Author Contributions: O.C.—conceptualization, methodology, formal analysis, investigation, writing—original draft preparation, visualization, and project administration; I.M.—conceptualization, methodology, data curation, resources, writing—original draft preparation, investigation, and project administration; L.D.—writing—review and editing, supervision, and project administration; N.K.—software, resources and investigation; S.C.—conceptualization and formal analysis; A.P.—resources and formal analysis; A.A.—resources; N.T.—conceptualization, methodology, formal analysis, review and editing, supervision, and project administration.

References

1. Ashworth, T.R. A case of cancer in which cells similar to those in the tumors were seen in the blood after death. *Med. J. Australia* **1869**, *14*, 146–147.
2. Paget, S. The distribution of secondary growths in cancer of the breast. *Lancet* **1889**, *1*, 571–573. [CrossRef]
3. Fidler, I.J.; Kripke, M.L. Metastasis results from preexisting variant cells within a malignant tumor. *Science* **1977**, *197*, 893–895. [CrossRef] [PubMed]
4. Fearon, E.R.; Vogelstein, B. A genetic model for colorectal tumorigenesis. *Cell* **1990**, *61*, 759–767. [CrossRef]
5. Fidler, I.J.; Hart, I.R. Biological diversity in metastatic neoplasms: origins and implications. *Science* **1982**, *217*, 998–1003. [CrossRef] [PubMed]
6. Klein, C.A. Parallel progression of primary tumours and metastases. *Nat. Rev. Cancer* **2009**, *9*, 302–312. [CrossRef] [PubMed]
7. Psaila, B.; Lyden, D. The metastatic niche: Adapting the foreign soil. *Nat. Rev. Cancer* **2009**, *9*, 285–293. [CrossRef] [PubMed]
8. Koebel, C.M.; Vermi, W.; Swann, J.B.; Zerafa, N.; Rodig, S.J.; Old, L.J.; Smyth, M.J.; Schreiber, R.D. Adaptive immunity maintains occult cancer in an equilibrium state. *Nature* **2007**, *450*, 903–907. [CrossRef]

9. Willis, R.A. *The Spread of Tumours in the Human Body*; J. & A. Churchill: London, UK, 1934.
10. Hadfield, G. The dormant cancer cell. *Br. Med. J.* **1954**, *2*, 607–610. [CrossRef]
11. Weilbaecher, K.N.; Guise, T.A.; McCauley, L.K. Cancer to bone: A fatal attraction. *Nat. Rev. Cancer* **2011**, *11*, 411–425. [CrossRef]
12. Jones, D.H.; Nakashima, T.; Sanchez, O.H.; Kozieradzki, I.; Komarova, S.V.; Sarosi, I.; Morony, S.; Rubin, E.; Sarao, R.; Hojilla, C.V.; et al. Regulation of cancer cell migration and bone metastasis by RANKL. *Nature* **2006**, *440*, 692–696. [CrossRef] [PubMed]
13. Aguirre-Ghiso, J.A. Models, mechanisms and clinical evidence for cancer dormancy. *Nat. Rev. Cancer* **2007**, *7*, 834–846. [CrossRef] [PubMed]
14. Sosa, M.S.; Bragado, P.; Aguirre-Ghiso, J.A. Mechanisms of disseminated cancer cell dormancy: An awakening field. *Nat. Rev. Cancer* **2014**, *14*, 611–622. [CrossRef] [PubMed]
15. Luzzi, K.J.; MacDonald, I.C.; Schmidt, E.E.; Kerkvliet, N.; Morris, V.L.; Chambers, A.F.; Groom, A.C. Multistep nature of metastatic inefficiency: Dormancy of solitary cells after successful extravasation and limited survival of early micrometastases. *Am. J. Pathol.* **1998**, *153*, 865–873. [CrossRef]
16. Ghajar, C.M. Metastasis prevention by targeting the dormant niche. *Nat. Rev. Cancer.* **2015**, *15*, 238–247. [CrossRef] [PubMed]
17. Malanchi, I.; Santamaria-Martínez, A.; Susanto, E.; Peng, H.; Lehr, H.A.; Delaloye, J.F.; Huelsken, J. Interactions between cancer stem cells and their niche govern metastatic colonization. *Nature* **2012**, *481*, 85–89. [CrossRef] [PubMed]
18. Janni, W.; Vogl, F.D.; Wiedswang, G.; Synnestvedt, M.; Fehm, T.; Jückstock, J.; Borgen, E.; Rack, B.; Braun, S.; Sommer, H. Persistence of disseminated tumor cells in the bone marrow of breast cancer patients predicts increased risk for relapse—A European pooled analysis. *Clin. Cancer Res.* **2011**, *17*, 2967–2976. [CrossRef]
19. Pantel, K.; Izbicki, J.; Passlick, B.; Angstwurm, M.; Häussinger, K.; Thetter, O.; Riethmüller, G. Frequency and prognostic significance of isolated tumor cells in bone marrow of patients with non-small-cell lung cancer without overt metastases. *Lancet* **1996**, *347*, 649–653. [CrossRef]
20. Lilleby, W.; Stensvold, A.; Mills, I.G.; Nesland, J.M. Disseminated tumor cells and their prognostic significance in nonmetastatic prostate cancer patients. *Int. J. Cancer* **2013**, *133*, 149–155. [CrossRef]
21. Flatmark, K.; Borgen, E.; Nesland, J.M.; Rasmussen, H.; Johannessen, H.O.; Bukholm, I.; Rosales, R.; Hårklau, L.; Jacobsen, H.J.; Sandstad, B.; et al. Disseminated tumour cells as a prognostic biomarker in colorectal cancer. *Br. J. Cancer* **2011**, *104*, 1434–1439. [CrossRef]
22. Besova, N.S.; Obarevich, E.S.; Davydov, M.M.; Beznos, O.A.; Tupitsyn, N.N. Prognostic values of the presence of disseminated tumor cells in the bone marrow in patients with disseminated stomach cancer before start of treatment with antitumor drugs. *Pharmateca* **2017**, *350*, 62–66.
23. Besova, N.S.; Obarevich, E.S.; Beznos, O.A.; Tupitsyn, N.N.; Davydov, M.M. Prognostic value of the dynamics of disseminated tumor cells in the bone marrow in patients with disseminated adenocarcinoma of the stomach or the esophagogastric junction. *Pharmateca* **2017**, *350*, 83–86.
24. Eskelin, S.; Pyrhonen, S.; Summanen, P.; Hahka-Kemppinen, M.; Kivelä, T. Tumor doubling times in metastatic malignant melanoma of the uvea: Tumor progression before and after treatment. *Ophthalmology* **2000**, *107*, 1443–1449. [CrossRef]
25. Coupland, S.E.; Sidiki, S.; Clark, B.J.; McClaren, K.; Kyle, P.; Lee, W.R. Metastatic choroidal melanoma to the contralateral orbit 40 years after enucleation. *Arch. Ophthalmol.* **1996**, *114*, 751–756. [CrossRef] [PubMed]
26. Damsky, W.; Micevic, G.; Meeth, K.; Muthusamy, V.; Curley, D.P.; Santhanakrishnan, M.; Erdelyi, I.; Platt, J.T.; Huang, L.; Theodosakis, N.; et al. mTORC1 activation blocks BrafV600E-induced growth arrest but is insufficient for melanoma formation. *Cancer Cell* **2015**, *27*, 41–56. [CrossRef] [PubMed]
27. Rocken, M. Early tumor dissemination, but late metastasis: Insights into tumor dormancy. *J. Clin. Investig.* **2010**, *120*, 1800–1803. [CrossRef] [PubMed]
28. Wick, M.R.; Swanson, P.E. Recognition of malignant melanoma by monoclonal antibody HMB-45. An immunohistochemical study of 200 paraffin-embedded cutaneous tumors. *J. Cutan. Pathol.* **1988**, *15*, 201–207. [CrossRef] [PubMed]
29. Davydov, M.I.; Tupitsin, N.N. Assessment of minimal bone marrow involvement by flow cytometry. *Hematopoiesis Immunol.* **2014**, *12*, 8–17.

10

Cooperative and Escaping Mechanisms between Circulating Tumor Cells and Blood Constituents

Carmen Garrido-Navas [1], Diego de Miguel-Pérez [1,2], Jose Exposito-Hernandez [3], Clara Bayarri [1,4], Victor Amezcua [3], Alba Ortigosa [1], Javier Valdivia [3], Rosa Guerrero [3], Jose Luis Garcia Puche [1], Jose Antonio Lorente [1,2] and Maria José Serrano [1,3,*]

1 GENYO, Centre for Genomics and Oncological Research (Pfizer/University of Granada/Andalusian Regional Government), PTS Granada Av. de la Ilustración, 114, 18016 Granada, Spain; carmen.garrido@genyo.es (C.G.-N.); diego.miguel@genyo.es (D.d.M.-P.); ci.bayarri@gmail.com (C.B.); albaortigosa@correo.ugr.es (A.O.); jlpuche@ugr.es (J.L.G.P.); jlorente@ugr.es (J.A.L.)
2 Laboratory of Genetic Identification, Department of Legal Medicine, University of Granada, Av. de la Investigación, 11, 18071 Granada, Spain
3 Integral Oncology Division, Virgen de las Nieves University Hospital, Av. Dr. Olóriz 16, 18012 Granada, Spain; jose.exposito.sspa@juntadeandalucia.es (J.E.-H.); victor.amezcua.md@gmail.com (V.A.); jvaldib@gmail.com (J.V.); mjs@ugr.es (R.G.)
4 Department of Thoracic Surgery, Virgen de las Nieves University Hospital, Av. de las Fuerzas Armadas, 2, 18014 Granada, Spain
* Correspondence: mjose.serrano@genyo.es

Abstract: Metastasis is the leading cause of cancer-related deaths and despite measurable progress in the field, underlying mechanisms are still not fully understood. Circulating tumor cells (CTCs) disseminate within the bloodstream, where most of them die due to the attack of the immune system. On the other hand, recent evidence shows active interactions between CTCs and platelets, myeloid cells, macrophages, neutrophils, and other hematopoietic cells that secrete immunosuppressive cytokines, which aid CTCs to evade the immune system and enable metastasis. Platelets, for instance, regulate inflammation, recruit neutrophils, and cause fibrin clots, which may protect CTCs from the attack of Natural Killer cells or macrophages and facilitate extravasation. Recently, a correlation between the commensal microbiota and the inflammatory/immune tone of the organism has been stablished. Thus, the microbiota may affect the development of cancer-promoting conditions. Furthermore, CTCs may suffer phenotypic changes, as those caused by the epithelial–mesenchymal transition, that also contribute to the immune escape and resistance to immunotherapy. In this review, we discuss the findings regarding the collaborative biological events among CTCs, immune cells, and microbiome associated to immune escape and metastatic progression.

Keywords: circulating tumor cells; tumor cell dissemination; immune system; microbiome

---

1. Introduction

The presence of circulating tumor cells (CTCs) in the peripheral blood has been largely associated with reduced disease-free and overall survival [1–3]. Even though metastasis is a highly inefficient process (tumor cell survival is less than 0.01%) [4], it is responsible for the majority of cancer-associated deaths [5]. In fact, it is accepted that CTCs are the initiator factor of metastatic relapse and their presence identifies patients with a higher risk of developing metastasis [6,7]. However, the complex biological processes enabling CTCs to survive and disseminate is not yet well understood and little is known about the cellular and genetic events involved both in the metastatic initiation and in its progression.

The success of the metastatic process is conditioned by the established relationship between tumor cells and the surrounding microenvironment. During the metastatic process, tumor cells interact with the immune system, which modulates this process [8]. The immune system has a dual role, both repressing but also promoting cancer progression. In fact, formation of CTC clusters or microemboli, not only composed of CTCs but also leukocytes, cancer-associated fibroblasts, endothelial cells, and platelets, was shown to facilitate the metastatic process and thus be related to poorer outcome in patients with breast [9] and gastric cancer [10], among others.

In this review, we will focus our interest in the "intimate friendship" between CTCs and the immune system. This private alliance benefits tumor progression through CTCs survival in this hostile microenvironment, the blood.

However, we cannot forget that the microenvironment is not only composed by immune system cells, stromal cells, and components of the extracellular matrix (ECM). CTCs and microbes co-evolve inside the ecosystem within our bodies [11,12] as will be further described in Section 3. This interaction influences the activity of the immune system on cell survival and expansion of CTCs [13].

In this review, we evaluate the current literature on interactions among CTCs, immune system cells, and microbiome in the tumor progression. We discuss how immune cells–CTC interactions contribute to the survival of these CTCs and how the microbiome can promote this positive association, finally supporting the metastatic process.

2. Promotion of Circulating Tumor Cells through the Immune System

The immune system is educated to eliminate the foreign and to respect the innate [14]. However, in the case of cancer, tumor cells are able to use the immune system to facilitate their own survival and migration. This phenomenon is known as concomitant immunity (CI) [15,16].

The plasticity of the immune system is well known and thus, according with the tumor type, the functional contribution of each immune cell can also be different [17,18]. However, some immune events are intimately associated with promotion of cancer, independently on the tumor type. Inflammation is one of these events and it is recognized as one of the "hallmarks" of cancer [19]. This process involves different types of immune cells, among which platelets, macrophages, and neutrophils can be highlighted [20,21].

Platelets are anucleated blood cells with a diameter of 2–4 μm originated during megakaryocytes maturation in the bone marrow and circulate in large numbers ($1.5–4.0 \times 10^9$/L) in the bloodstream [22,23]. Platelets are the main cells involved in thrombosis and hemostasis, thus, related with the physiological and pathophysiological processes occurring during inflammation [24]. Interestingly, several studies have reported their role in cancer progression, especially during cancer metastasis [25] as they actively promote the metastatic process. Metastasis-promoting mechanisms affected by platelets are related to both migration of tumor cells and cancer cell survival in circulation [26].

Regarding the migration process, platelets store large amounts of transforming growth factor β (TGFβ), which is associated with an increase of the invasion potential of tumor cells. Thus, tumor cells-conjugated platelets release mediators to modify blood vessels permeability, including dense granule-release, histamine, eicosanoid metabolites, or serotonin [23]. These mediators induce endothelial cell retraction, exposing the basement membrane, and thus facilitate cancer cell extravasation [27]. In addition, platelets activation by cancer cells lead to the generation of platelet-derived microparticles (PDMPs), which can also release mediators like TxA2 and 12-HETE. These metabolites may enhance cell migration and invasion, eventually increasing the metastatic potential of cancer cells [28].

However, self-migration ability of CTCs is not enough to complete a successful metastatic process, survival of these cells, once in the blood, depends on the formation of circulating microemboli [9] as well as the acquisition of resistant phenotypes to the surrounding microenvironment. The acquisition of these phenotypes involves a biological process known as epithelial to mesenchymal transition (EMT) [29]. The EMT process explains how tumor cells change their phenotype, allowing them to detach, invade, and metastasize through the blood or lymphatic systems. Among others, the EMT involves loss of E-cadherin, disrupting cell-to-cell adhesions and altering gene expression by increasing β-catenin nuclear localization [30]. In contrast, N-cadherin, which is highly expressed in mesenchymal cells, fibroblasts, neural tissue, and cancer cells, is elevated during EMT. This cadherin switch, from E-cadherin to N-cadherin, is closely associated with the increased invasiveness, motility and metastasis potential of tumor cells. Moreover, activated platelets induce EMT through secretion of growth factors and cytokines (e.g., TNFα and TGFβ) [31]. Interestingly, these cytokines are also associated with the inflammatory process as previously explained [32].

Furthermore, CTCs-conjugated platelets also coordinate the engagement of other immune cells during the dissemination process [33]. In fact, CTCs-conjugated platelets recruit neutrophils, macrophages, and other immune cells through release of chemokines, such as CXCL5 or CXCL7. Among white blood cells (WBC), neutrophils are recognized as the mediators of metastasis initiation [34,35]. Neutrophils promote tumor development by initiating an angiogenic switch and facilitating colonization of CTCs. In fact, some groups support the idea that WBC shape a protective cover around CTCs, avoiding their recognition and destruction by other immune cells [36]. It has long been known that circulating platelet–neutrophil complexes are present in a wide range of inflammatory conditions including cancer. In this interaction, neutrophils are responsible to activate platelets and it was shown that the neutrophils–platelets interaction initiates inflammatory responses [37]. Platelets interact with neutrophils by multiple intermediates including platelet P-selectin binding to neutrophil P-selectin glycoprotein ligand-1 (PSGL-1) [38]. In addition, it has been suggested that the neutrophils-platelets complexes interacting to CTCs bring the latter to the endothelium, which is an essential step in hematogenous dissemination metastasis [34]. Thus, platelets prime tumor cells to promote neutrophil extracellular traps (NETs) formation, which are also involved in endothelial activation [39].

However, the interaction between the platelets–tumor cells complex and immune cells is not only restricted to neutrophils. The release of CXCL12, which is highly present in platelets, allows recruitment of CXCR4-positive cells such as macrophages to prepare the metastatic niche for CTCs [40]. Neutrophils are the first leukocytes to be recruited in response to chemotactic signaling and are responsible for stimulating the repair process and initiating inflammation. This influx is followed by monocytes, which, upon entry into the tissue, differentiate into macrophages. These macrophages promote invasion and metastasis from the primary tumor site through their ability to engage cancer cells in an autocrine loop that promotes cancer cell [41]. This autocrine signaling triggers cancer cells to produce CSF-1, which promote epidermal growth factor (EGF) production by macrophages. Finally, cancer cells and macrophages co-migrate towards tumor blood vessels, where macrophage-derived VEGF-A promotes cancer cell intravasation [42]. In addition, tumor migration is upregulated by macrophage-derived cathepsins, SPARC, or CCL18, that enhance tumor cell adhesion to extracellular matrix proteins [41]. Finally, CTCs produce CCL2 that recruits inflammatory monocytes, which in turn increase vascular permeability and allow migration of these tumor cells [16].

Nevertheless, migration and survival of CTCs belong together, so the promotion of CTC migration alone is not enough to allow metastasis. Anoikis is a programmed cell death induced by cell detachment [43] and essential for CTC survival. Another effect of the collaboration between immune cells and CTCs includes the protection of CTCs from anoikis [44]. Platelets are involved in this protective mechanism as it was observed that they induce RhoA-(myosin phosphatase targeting subunit 1) and MYPT1-protein phosphatase (PP1)-mediated Yes-associated protein 1 (YAP1) dephosphorylation and nuclear translocation, resulting in apoptosis resistance [45]. Apoptosis signal-regulating kinase 1 (Ask1) is a stress-responsive Ser/Thr mitogen-activated protein kinase kinase kinase (MAP3K) in the Jun N-terminal kinases (JNK) and p38 pathways. Once Ask1 levels are reduced in platelets, active phosphorylation of protein kinase B (Akt), JNK and p38 is downregulated, and thus tumor metastasis is attenuated [46].

In conclusion, the fate of CTCs is not to survive alone but with help of their mates within the immune system and thus, survival of CTCs depends on their ability to interact with immune cells. However, this favorable interaction between CTC and immune cells depends also on the status of our gut microbiota that is intimately linked with the nature of the immune system.

3. Survival of Circulating Tumor Cells Through the Interaction of Microbiota with the Immune System

The evolution of any disease, including cancer, depends highly on the physiological status of the host. The gut microbiota has emerged as an important factor of health and disease [47]. Likewise, our microbiota conditions the status of our immune system [48]. In fact, gut microorganisms are involved in the immune system development and in the response of the host against different pathologies, like cancer. Taken together, tumor cells-microbiome-immune system interactions may improve the likelihood of cell survival and induce tumor cell migration (Figure 1) [49].

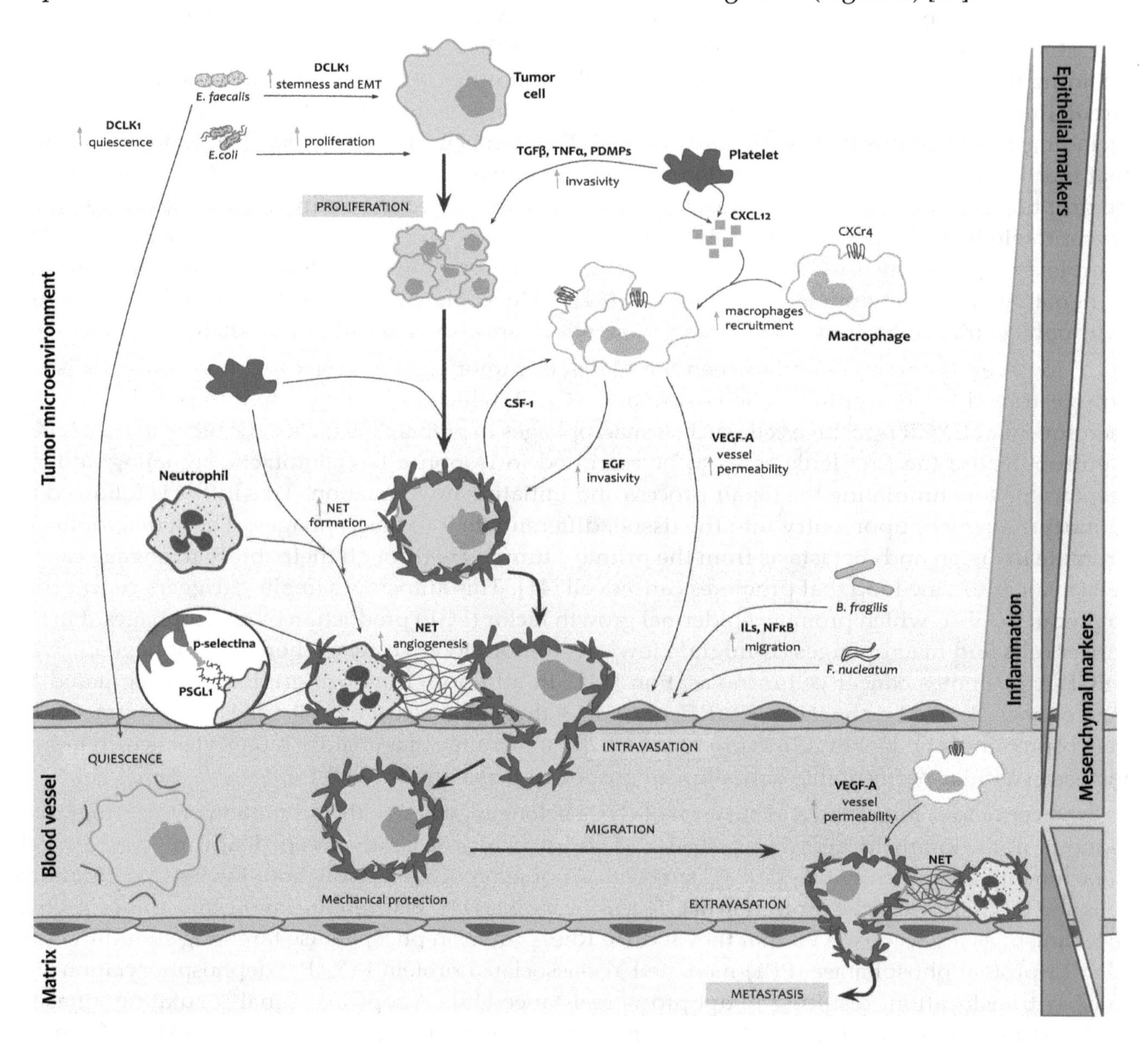

Figure 1. Interactions between circulating tumor cells (CTCs), immune system cells, and microbiome. Metabolites and cytokines produced by bacteria such as *Bacterioides fragilis*, *Enterococcus faecium*, *Escherichia coli*, and *Fusobacterium nucleatum* facilitate proliferation and migration of circulating tumor cells (CTCs), promote stemness and epithelial to mesenchymal transition (EMT), and help CTCs to enter quiescence. Furthermore, platelets interact with proliferating tumor cells directly, by formation of CTCs-platelet complexes allowing CTCs to escape the immune system but also indirectly, through three different ways: secretion of growth factors such as TFGβ, TNFα either alone or enclosed in platelets-derived microparticles (PDMPs) that increase invasivity of CTCs; secretion of chemokines such as CXCL12, increasing macrophages recruitment, what ultimately impact on invasivity and vessel permeability through epidermal growth factor (EGF) and VEFG-A, respectively; and formation of platelet-neutrophil

complexes (through P-selectin and PSGL1) that eventually generate neutrophil extracellular traps (NET) promoting angiogenesis and facilitating CTC intravasation to blood vessels. Finally, macrophages and NET also facilitate CTC extravasation from blood vessels to the extracellular matrix to produce metastasis.

As it has been mentioned before, the inflammatory process is an essential step in the development and progression of cancer. Microbes have a critical role in the initiation and maintenance of chronic inflammatory conditions [50,51]. However, how do microbes influence on the inflammation process and on the migration and survival of CTCs?

The gut microbiota contributes to cancer progression through different mechanisms. Recently, it was demonstrated that both, DNA-damaging superoxide radicals or genotoxins produced by the gut bacteria could initiate colon cancer. In addition, bacteria may induce cell proliferation through interactions with T-helper cells or Toll-like receptors, respectively [52]. In colon cancer patients, an increase of the *Escherichia coli* population was observed to induce colitis and colibactin synthesis and thereby, to promote inflammation.

Furthermore, it has been demonstrated that microbes as *Bacillus* sp., *Enterococcus faecium*, and *E. coli* produce peptides which alter host epithelial growth factor, activating intracellular pathways associated to migration. In a pioneer work, Wynendaele, E et al. [53] discovered that certain quorum sensing peptides produced by bacteria (molecules that microbes use to coordinate their gene expression and behavior) interact with cancer cells. This study demonstrated that Phr0662 (*Bacillus* sp.), EntF-metabolite (*E. faecium*), and EDF-derived (*E. coli*) peptides can initiate HCT-8/E11 colon cancer cell invasion. According to results of this group, the Phr0662 peptide targets epidermal growth factor receptors (EGFR and ErbB2). Upregulation of EGFR induces activation of the Ras/raf/MEK/MAPK, PI3K/Akt, and STAT intracellular signaling cascades [54] altering gene transcription and allowing migration of tumor cells. However, despite this work being extremely interesting, it is only a preliminary and exploratory in vitro assay and more exhaustive analyses including cancer patients should be carried out to validate these results.

Microbes can also alter cancer cell epigenetics through production of metabolites affecting gene expression [55]. This is the case of *Bifidobacterium* spp., which produces folate, one of the most powerful methyl donors involved in gene silencing. Thus, the gut microbiome is also involved in chromatin remodeling via acetylation and deacetylation of histones through butyrate production. Butyrate is a common metabolite of the microbiome, inducing cell differentiation via histone acetylation of the intestinal T reg cells [56].

Moreover, the microbiome plays an important role in the epithelial mesenchymal transition (EMT), an essential step for CTC migration and survival. In fact, microbes produce toxins that contribute to EMT [57]. Some of those microbes, as *Bacteroides fragilis*, *Fusobacterium nucleatum*, and *E. faecalis*, clear E-cadherin from epithelial cells, a transmembrane adhesion protein, leading to colonic epithelial proliferation [58]. Most of the studies on the interaction of the microbiome with cancer cells have been developed on colon cancer, murine models, or in vitro assays. Likewise, in a recent study including a murine model, colonic epithelial cells were transformed to express Ly6A/E, a stem cell marker implicating mesenchymal features, and Doublecortin-like kinase 1 (DCKL1), a marker of cancer, by the presence of *E. faecalis* [59]. DCLK1 is a member of the protein kinase super family and the doublecortin family, which is overexpressed in many cancers, including colon, pancreas, liver, esophageal, and kidney cancers. It is now suggested to be a master regulator of pluripotency factors, including Nanog, Oct4, Sox2, Klf4, and Myc, that are critical for stemness of cancer cells (CSC, cancer stem cells) and EMT transcriptional factors, including Snail, Slug, Twist, and Zeb 1 [60]. Interestingly, all these markers are involved in regulation of both EMT and CSCs and are controlled by DCLK1 expression in cancer models [61]. Furthermore, Westphalen, CB et al. [62], reported that DCLK1 induces quiescence of tumor cells. Quiescence is a common property of CSCs that is associated with the EMT process as a critical step for the migration and progression of tumor cells [63]. In consequence, EMT allows not only CTCs migration and tumor relapse, but also, induces the ability of CTCs to escape the immune system cancer treatments (Figure 1) [64].

Another biological mechanism used by the microbiome to enhance cancer progression includes the modulation of the immune system. Among all the microbes involved in this process, the enterotoxin *Bacteriodes fragilis* (ETBF) stands up due to the activation of STATA3 and T helper cells, both with an important role in the inflammatory process [65]. In fact, Chung, L et al. [65], demonstrated that *Bacteroides fragilis* toxin (BFT) can activate a pro-carcinogenic inflammatory cascade, related to IL-17R, NF-κB, and Stat3 signaling, in colonic epithelial cells (CECs). Likewise, the activation of NF-κB in these cells, induces other chemokines as CXCL1 that mediates recruitment of CXCR2-expressing polymorphonuclear immature myeloid cells, promoting ETBF-mediated distal colon tumorigenesis. Another bacterium associated with poor oncological outcomes is *Fuscobacterium nucleatum*. It has been suggested that *F. nucleatum* promotes tumorigenesis through both pro-inflammatory and immunosuppressive effects. Furthermore, *F. nucleatum* is associated with activation of cytokines IL-6, IL-12, IL-17, and TNF-α, which cooperatively upregulate NFκB, a critical regulator of cellular proliferation [66]. Some studies associated the presence of high levels of *F. nucleatum* with the EMT process [67]. In this context, Mima, K et al. [68], identified that *F. nucleatum* adheres to and invades epithelial cells mainly through the virulence factors, including Fusobacterium adhesin A, Fusobacterium autotransporter protein 2, and fusobacterial outer membrane protein. To the contrary, other studies raise their skepticism about the role of *F. nucleatum* on EMT, as it still remains unclear whether *F. nucleatum* triggers the colonic EMT process. Ma, CT, et al., showed that *F. nucleatum* infection did not affect expression levels of E-cadherin and β-catenin [69]. However, it was associated with proliferation and invasion of colon cancer cells as it significantly increased phosphorylation of p65 (a subunit of nuclear factor-κB), as well as expression of interleukin (IL)-6, IL-1β and matrix metalloproteinase (MMP)-13. Regardless of the fact that there are not any explicit studies evaluating the direct action of the microbiota on CTCs, we here reviewed some of the biological processes in which microbes alter tumorigenic pathways. As they are involved in inflammation or inducing EMT, both biological processes intimately associated with the ability of CTCs to migrate and to survive, we suggest the potential interaction between them.

4. Conclusions and Perspectives

In conclusion, the complex interactions between the microbiome, the immune system and CTCs may allow us to grasp the insights of the dissemination process occurring in cancer and the immune system´s mechanisms involved in this process. Therefore, the interactions among microbiome, immune system, and CTCs could aid the rational design of interventions that strengthen the antimetastatic potential of combined treatments to prevent appearance of metastasis. Moreover, emerging evidences may provide new mechanisms to control the dissemination process through the development of new therapeutic strategies with the microbiota as target. However, this topic is still an incipient area of research and further investigation is needed to clarify the association of the microbiome with the immune system and the dissemination process.

Author Contributions: Design and writing by M.J.S., C.G.-N., and J.E.-H.; Figures designed by A.O., C.B., J.A.L., and D.d.M.-P.; Discussion J.L.G.P., V.A., J.V., and R.G.

Acknowledgments: We would like to extend our gratitude to Hugh Ilyine for the English revision.

References

1. Bayarri-Lara, C.; Ortega, F.G.; De Guevara, A.C.L.; Puche, J.L.; Zafra, J.R.; De Miguel-Pérez, D.; Ramos, A.S.-P.; Giraldo-Ospina, C.F.; Gómez, J.A.N.; Delgado-Rodríguez, M.; et al. Circulating Tumor Cells Identify Early Recurrence in Patients with Non-Small Cell Lung Cancer Undergoing Radical Resection. *PLoS ONE* **2016**, *11*, e0148659. [CrossRef] [PubMed]

2. Delgado-Ureña, M.; Ortega, F.G.; De Miguel-Pérez, D.; Rodriguez-Martínez, A.; García-Puche, J.L.; Ilyine, H.; Lorente, J.A.; Exposito-Hernandez, J.; Garrido-Navas, M.C.; Delgado-Ramirez, M.; et al. Circulating tumor cells criteria (CyCAR) versus standard RECIST criteria for treatment response assessment in metastatic colorectal cancer patients. *J. Transl. Med.* **2018**, *16*, 251. [CrossRef] [PubMed]
3. Mamdouhi, T.; Twomey, J.D.; McSweeney, K.M.; Zhang, B. Fugitives on the run: Circulating tumor cells (CTCs) in metastatic diseases. *Cancer Metastasis Rev.* **2019**, *38*, 297–305. [CrossRef] [PubMed]
4. Rejniak, K.A. Circulating Tumor Cells: When a Solid Tumor Meets a Fluid Microenvironment. *Adv. Exp. Med. Biol.* **2016**, *936*, 93–106. [PubMed]
5. Chaffer, C.L.; Weinberg, R.A. A Perspective on Cancer Cell Metastasis. *Science* **2011**, *331*, 1559–1564. [CrossRef]
6. Tsai, W.-S.; Chen, J.-S.; Shao, H.-J.; Wu, J.-C.; Lai, J.-M.; Lu, S.-H.; Hung, T.-F.; Chiu, Y.-C.; You, J.-F.; Hsieh, P.-S.; et al. Circulating Tumor Cell Count Correlates with Colorectal Neoplasm Progression and Is a Prognostic Marker for Distant Metastasis in Non-Metastatic Patients. *Sci. Rep.* **2016**, *6*, 24517. [CrossRef]
7. Al-Mehdi, A.; Tozawa, K.; Fisher, A.; Shientag, L.; Lee, A.; Muschel, R. Intravascular origin of metastasis from the proliferation of endothelium-attached tumor cells: A new model for metastasis. *Nat. Med.* **2000**, *6*, 100–102. [CrossRef]
8. Balkwill, F.; Mantovani, A. Inflammation and cancer: Back to Virchow? *Lancet* **2001**, *357*, 539–545. [CrossRef]
9. Aceto, N.; Bardia, A.; Miyamoto, D.T.; Donaldson, M.C.; Wittner, B.S.; Spencer, J.A.; Yu, M.; Pely, A.; Engstrom, A.; Zhu, H.; et al. Circulating Tumor Cell Clusters Are Oligoclonal Precursors of Breast Cancer Metastasis. *Cell* **2014**, *158*, 1110–1122. [CrossRef]
10. Abdallah, E.A.; Braun, A.C.; Flores, B.C.; Senda, L.; Urvanegia, A.C.; Calsavara, V.; De Jesus, V.H.F.; Almeida, M.F.A.; Begnami, M.D.; Coimbra, F.J.; et al. The Potential Clinical Implications of Circulating Tumor Cells and Circulating Tumor Microemboli in Gastric Cancer. *Oncologist* **2019**, *24*, e854–e863. [CrossRef]
11. Whisner, C.M.; Athena Aktipis, C. The Role of the Microbiome in Cancer Initiation and Progression: How Microbes and Cancer Cells Utilize Excess Energy and Promote One Another's Growth. *Curr. Nutr. Rep.* **2019**, *8*, 42–51. [CrossRef] [PubMed]
12. Contreras, A.V.; Cocom-Chan, B.; Hernandez-Montes, G.; Portillo-Bobadilla, T.; Resendis-Antonio, O. Host-Microbiome Interaction and Cancer: Potential Application in Precision Medicine. *Front. Physiol.* **2016**, *7*, 606. [CrossRef] [PubMed]
13. Bose, M.; Mukherjee, P. Role of Microbiome in Modulating Immune Responses in Cancer. *Mediat. Inflamm.* **2019**, *2019*, 1–7. [CrossRef] [PubMed]
14. Mellman, I.; Coukos, G.; Dranoff, G. Cancer immunotherapy comes of age. *Nature* **2011**, *480*, 480–489. [CrossRef] [PubMed]
15. Janssen, L.M.; Ramsay, E.E.; Logsdon, C.D.; Overwijk, W.W. The immune system in cancer metastasis: Friend or foe? *J. Immunother. Cancer* **2017**, *5*, 79. [CrossRef] [PubMed]
16. Kitamura, T.; Qian, B.-Z.; Pollard, J.W. Immune cell promotion of metastasis. *Nat. Rev. Immunol* **2015**, *15*, 73–86. [CrossRef]
17. Blomberg, O.S.; Spagnuolo, L.; de Visser, K.E. Immune regulation of metastasis: Mechanistic insights and therapeutic opportunities. *Dis. Model. Mech.* **2018**, *11*, dmm036236. [CrossRef]
18. Leone, K.; Poggiana, C.; Zamarchi, R. The Interplay between Circulating Tumor Cells and the Immune System: From Immune Escape to Cancer Immunotherapy. *Diagnostics* **2018**, *8*, 59. [CrossRef]
19. Hanahan, D.; Weinberg, R.A. Hallmarks of Cancer: The Next Generation. *Cell* **2011**, *144*, 646–674. [CrossRef]
20. Morrell, C.N.; Aggrey, A.A.; Chapman, L.M.; Modjeski, K.L. Emerging roles for platelets as immune and inflammatory cells. *Blood* **2014**, *123*, 2759–2767. [CrossRef]
21. Hamilton, G.; Rath, B. Circulating tumor cell interactions with macrophages: Implications for biology and treatment. *Transl. Lung Cancer Res.* **2017**, *6*, 418–430. [CrossRef] [PubMed]
22. Machlus, K.R.; Italiano, J.E. The incredible journey: From megakaryocyte development to platelet formation. *J. Cell Biol.* **2013**, *201*, 785–796. [CrossRef] [PubMed]
23. Li, N. Platelets in cancer metastasis: To help the "villain" to do evil. *Int. J. Cancer* **2016**, *138*, 2078–2087. [CrossRef] [PubMed]
24. Margetic, S. Inflammation and haemostasis. *Biochem. Med.* **2012**, *22*, 49–62. [CrossRef]
25. Micalizzi, D.S.; Maheswaran, S.; Haber, D.A. A conduit to metastasis: Circulating tumor cell biology. *Genes Dev.* **2017**, *31*, 1827–1840. [CrossRef]

26. Yu, L.-X.; Yan, L.; Yang, W.; Wu, F.-Q.; Ling, Y.; Chen, S.-Z.; Tang, L.; Tan, Y.-X.; Cao, D.; Wu, M.-C.; et al. Platelets promote tumour metastasis via interaction between TLR4 and tumour cell-released high-mobility group box1 protein. *Nat. Commun.* **2014**, *5*, 5256. [CrossRef]
27. Van Zijl, F.; Krupitza, G.; Mikulits, W. Initial steps of metastasis: Cell invasion and endothelial transmigration. *Mutat. Res. Mutat. Res.* **2011**, *728*, 23–34. [CrossRef]
28. Plantureux, L.; Mège, D.; Crescence, L.; Dignat-George, F.; Dubois, C.; Panicot-Dubois, L. Impacts of Cancer on Platelet Production, Activation and Education and Mechanisms of Cancer-Associated Thrombosis. *Cancers (Basel)* **2018**, *10*, 441. [CrossRef]
29. Lim, J.; Thiery, J.P. Epithelial-mesenchymal transitions: Insights from development. *Development* **2012**, *139*, 3471–3486. [CrossRef]
30. Serrano, M.J.; Ortega, F.G.; Alvarez-Cubero, M.J.; Nadal, R.; Sánchez-Rovira, P.; Salido, M.; Rodriguez, M.; García-Puche, J.L.; Delgado-Rodríguez, M.; Sole, F.; et al. EMT and EGFR in CTCs cytokeratin negative non-metastatic breast cancer. *Oncotarget* **2014**, *5*, 7486–7497. [CrossRef]
31. Tsubakihara, Y.; Moustakas, A. Epithelial-Mesenchymal Transition and Metastasis under the Control of Transforming Growth Factor β. *Int. J. Mol. Sci.* **2018**, *19*, 3672. [CrossRef] [PubMed]
32. Wojdasiewicz, P.; Poniatowski, Ł.A.; Szukiewicz, D. The role of inflammatory and anti-inflammatory cytokines in the pathogenesis of osteoarthritis. *Mediat. Inflamm.* **2014**, *2014*, 561459. [CrossRef] [PubMed]
33. Gruber, I.; Landenberger, N.; Staebler, A.; Hahn, M.; Wallwiener, D.; Fehm, T. Relationship between circulating tumor cells and peripheral T-cells in patients with primary breast cancer. *Anticancer. Res.* **2013**, *33*, 2233–2238. [PubMed]
34. Tao, L.; Zhang, L.; Peng, Y.; Tao, M.; Li, L.; Xiu, D.; Yuan, C.; Ma, Z.; Jiang, B. Neutrophils assist the metastasis of circulating tumor cells in pancreatic ductal adenocarcinoma: A new hypothesis and a new predictor for distant metastasis. *Medicine (Baltimore)* **2016**, *95*, e4932. [CrossRef] [PubMed]
35. Zhang, J.; Qiao, X.; Shi, H.; Han, X.; Liu, W.; Tian, X.; Zeng, X. Circulating tumor-associated neutrophils (cTAN) contribute to circulating tumor cell survival by suppressing peripheral leukocyte activation. *Tumor Biol.* **2016**, *37*, 5397–5404. [CrossRef]
36. Uppal, A.; Wightman, S.C.; Ganai, S.; Weichselbaum, R.R.; An, G. Investigation of the essential role of platelet-tumor cell interactions in metastasis progression using an agent-based model. *Theor. Biol. Med. Model.* **2014**, *11*, 17. [CrossRef] [PubMed]
37. Sreeramkumar, V.; Adrover, J.M.; Ballesteros, I.; Cuartero, M.I.; Rossaint, J.; Bilbao, I.; Nácher, M.; Pitaval, C.; Radovanovic, I.; Fukui, Y.; et al. Neutrophils scan for activated platelets to initiate inflammation. *Science* **2014**, *346*, 1234–1238. [CrossRef]
38. Moore, K.L.; Patel, K.D.; Bruehl, R.E.; Li, F.; Johnson, D.A.; Lichenstein, H.S.; Cummings, R.D.; Bainton, D.F.; McEver, R.P. P-selectin glycoprotein ligand-1 mediates rolling of human neutrophils on P-selectin. *J. Cell Biol.* **1995**, *128*, 661–671. [CrossRef]
39. Abdol Razak, N.; Elaskalani, O.; Metharom, P. Pancreatic Cancer-Induced Neutrophil Extracellular Traps: A Potential Contributor to Cancer-Associated Thrombosis. *Int. J. Mol. Sci.* **2017**, *18*, 487. [CrossRef]
40. Sun, X.; Cheng, G.; Hao, M.; Zheng, J.; Zhou, X.; Zhang, J.; Taichman, R.S.; Pienta, K.J.; Wang, J.; Zhou, X. CXCL12/CXCR4/CXCR7 chemokine axis and cancer progression. *Cancer Metastasis Rev.* **2010**, *29*, 709–722. [CrossRef]
41. Sangaletti, S.; Di Carlo, E.; Gariboldi, S.; Miotti, S.; Cappetti, B.; Parenza, M.; Rumio, C.; Brekken, R.A.; Chiodoni, C.; Colombo, M.P. Macrophage-Derived SPARC Bridges Tumor Cell-Extracellular Matrix Interactions toward Metastasis. *Cancer Res.* **2008**, *68*, 9050–9059. [CrossRef] [PubMed]
42. Nielsen, S.R.; Schmid, M.C. Macrophages as Key Drivers of Cancer Progression and Metastasis. *Mediat. Inflamm.* **2017**, *2017*, 9624760. [CrossRef] [PubMed]
43. Kim, Y.-N.; Koo, K.H.; Sung, J.Y.; Yun, U.-J.; Kim, H. Anoikis Resistance: An Essential Prerequisite for Tumor Metastasis. *Int. J. Cell Biol.* **2012**, *2012*, 1–11. [CrossRef] [PubMed]
44. Heeke, S.; Mograbi, B.; Alix-Panabières, C.; Hofman, P. Never Travel Alone: The Crosstalk of Circulating Tumor Cells and the Blood Microenvironment. *Cells* **2019**, *8*, 714. [CrossRef]
45. Abylkassov, R.; Xie, Y. Role of Yes-associated protein in cancer: An update. *Oncol. Lett.* **2016**, *12*, 2277–2282. [CrossRef] [PubMed]
46. Kamiyama, M.; Shirai, T.; Tamura, S.; Suzuki-Inoue, K.; Ehata, S.; Takahashi, K.; Miyazono, K.; Hayakawa, Y.;

Sato, T.; Takeda, K.; et al. ASK1 facilitates tumor metastasis through phosphorylation of an ADP receptor P2Y12 in platelets. *Cell Death Differ.* **2017**, *24*, 2066–2076. [CrossRef] [PubMed]

47. Mohajeri, M.H.; Brummer, R.J.M.; Rastall, R.A.; Weersma, R.K.; Harmsen, H.J.M.; Faas, M.; Eggersdorfer, M. The role of the microbiome for human health: From basic science to clinical applications. *Eur. J. Nutr.* **2018**, *57*, 1–14. [CrossRef]
48. Cianci, R.; Pagliari, D.; Piccirillo, C.A.; Fritz, J.H.; Gambassi, G. The Microbiota and Immune System Crosstalk in Health and Disease. *Mediat. Inflamm.* **2018**, *2018*, 1–3. [CrossRef]
49. Li, W.; Deng, Y.; Chu, Q.; Zhang, P. Gut microbiome and cancer immunotherapy. *Cancer Lett.* **2019**, *447*, 41–47. [CrossRef]
50. Ong, H.S.; Yim, H.C.H. Microbial Factors in Inflammatory Diseases and Cancers. In *Advances in Experimental Medicine and Biology*; Springer: Singapore, 2017; pp. 153–174.
51. Sussman, D.A.; Santaolalla, R.; Strobel, S. Cancer in inflammatory bowel disease. *Curr. Opin. Gastroenterol.* **2012**, *28*, 327–333. [CrossRef]
52. Sieling, P.A.; Chung, W.; Duong, B.T.; Godowski, P.J.; Modlin, R.L. Toll-like receptor 2 ligands as adjuvants for human Th1 responses. *J. Immunol.* **2003**, *170*, 194–200. [CrossRef] [PubMed]
53. Wynendaele, E.; Verbeke, F.; D'Hondt, M.; Hendrix, A.; Van De Wiele, C.; Burvenich, C.; Peremans, K.; De Wever, O.; Bracke, M.; De Spiegeleer, B. Crosstalk between the microbiome and cancer cells by quorum sensing peptides. *Peptides* **2015**, *64*, 40–48. [CrossRef] [PubMed]
54. Wee, P.; Wang, Z. Epidermal Growth Factor Receptor Cell Proliferation Signaling Pathways. *Cancers (Basel)* **2017**, *9*, 52.
55. Gerhauser, C. Impact of dietary gut microbial metabolites on the epigenome. *Philos. Trans. R. Soc. Lond. B Biol. Sci.* **2018**, *373*, 20170359. [CrossRef]
56. Ji, J.; Shu, D.; Zheng, M.; Wang, J.; Luo, C.; Wang, Y.; Guo, F.; Zou, X.; Lv, X.; Li, Y.; et al. Microbial metabolite butyrate facilitates M2 macrophage polarization and function. *Sci. Rep.* **2016**, *6*, 24838. [CrossRef] [PubMed]
57. Gaines, S.; Williamson, A.J.; Kandel, J. How the microbiome is shaping our understanding of cancer biology and its treatment. *Semin. Colon Rectal Surg.* **2018**, *29*, 12–16. [CrossRef]
58. Sears, C.L.; Geis, A.L.; Housseau, F. Bacteroides fragilis subverts mucosal biology: From symbiont to colon carcinogenesis. *J. Clin. Investig.* **2014**, *124*, 4166–4172. [CrossRef] [PubMed]
59. Wang, X.; Yang, Y.; Huycke, M.M. Commensal bacteria drive endogenous transformation and tumour stem cell marker expression through a bystander effect. *Gut* **2015**, *64*, 459–468. [CrossRef]
60. Chandrakesan, P.; Weygant, N.; May, R.; Qu, D.; Chinthalapally, H.R.; Sureban, S.M.; Ali, N.; Lightfoot, S.A.; Umar, S.; Houchen, C.W. DCLK1 facilitates intestinal tumor growth via enhancing pluripotency and epithelial mesenchymal transition. *Oncotarget* **2014**, *5*, 9269–9280. [CrossRef]
61. Chandrakesan, P.; Panneerselvam, J.; Qu, D.; Weygant, N.; May, R.; Bronze, M.; Houchen, C. Regulatory Roles of Dclk1 in Epithelial Mesenchymal Transition and Cancer Stem Cells. *J. Carcinog. Mutagen.* **2016**, *7*, 1–8.
62. Westphalen, C.B.; Asfaha, S.; Hayakawa, Y.; Takemoto, Y.; Lukin, D.J.; Nuber, A.H.; Brandtner, A.; Setlik, W.; Remotti, H.; Muley, A.; et al. Long-lived intestinal tuft cells serve as colon cancer–initiating cells. *J. Clin. Investig.* **2014**, *124*, 1283–1295. [CrossRef]
63. Thiery, J.P.; Acloque, H.; Huang, R.Y.; Nieto, M.A. Epithelial-Mesenchymal Transitions in Development and Disease. *Cell* **2009**, *139*, 871–890. [CrossRef] [PubMed]
64. Alvarez-Cubero, M.J.; Vázquez-Alonso, F.; Puche-Sanz, I.; Ortega, F.G.; Martin-Prieto, M.; Garcia-Puche, J.L.; Pascual-Geler, M.; Lorente, J.A.; Cozar-Olmo, J.M.; Serrano, M.J. Dormant Circulating Tumor Cells in Prostate Cancer: Therapeutic, Clinical and Biological Implications. *Curr. Drug Targets* **2016**, *17*, 693–701. [CrossRef] [PubMed]
65. Chung, L.; Orberg, E.T.; Geis, A.L.; Chan, J.L.; Fu, K.; Shields, C.E.D.; Dejea, C.M.; Fathi, P.; Chen, J.; Finard, B.B.; et al. Bacteroides fragilis Toxin Coordinates a Pro-carcinogenic Inflammatory Cascade via Targeting of Colonic Epithelial Cells. *Cell Host Microbe* **2018**, *23*, 203–214.e5. [CrossRef] [PubMed]
66. Yang, Y.; Weng, W.; Peng, J.; Hong, L.; Yang, L.; Toiyama, Y.; Gao, R.; Liu, M.; Yin, M.; Pan, C.; et al. Fusobacterium nucleatum Increases Proliferation of Colorectal Cancer Cells and Tumor Development in Mice by Activating Toll-Like Receptor 4 Signaling to Nuclear Factor−κB, and Up-regulating Expression of MicroRNA-21. *Gastroenterology* **2017**, *152*, 851–866.e24. [CrossRef] [PubMed]

67. Rubinstein, M.R.; Wang, X.; Liu, W.; Hao, Y.; Cai, G.; Han, Y.W. Fusobacterium nucleatum Promotes Colorectal Carcinogenesis by Modulating E-Cadherin/β-Catenin Signaling via its FadA Adhesin. *Cell Host Microbe* **2013**, *14*, 195–206. [CrossRef] [PubMed]
68. Mima, K.; Nishihara, R.; Qian, Z.R.; Cao, Y.; Sukawa, Y.; Nowak, J.A.; Yang, J.; Dou, R.; Masugi, Y.; Song, M.; et al. *Fusobacterium nucleatum* in colorectal carcinoma tissue and patient prognosis. *Gut* **2016**, *65*, 1973–1980. [CrossRef]
69. Ma, C.; Luo, H.; Gao, F.; Tang, Q.; Chen, W. Fusobacterium nucleatum promotes the progression of colorectal cancer by interacting with E-cadherin. *Oncol. Lett.* **2018**, *16*, 2606–2612. [CrossRef]

11

Association between Microsatellite Instability Status and Peri-Operative Release of Circulating Tumour Cells in Colorectal Cancer

James W. T. Toh [1,2,3,*], Stephanie H. Lim [1], Scott MacKenzie [4], Paul de Souza [1,4], Les Bokey [4], Pierre Chapuis [3] and Kevin J. Spring [1,4,*]

[1] Medical Oncology, Ingham Institute of Applied Research, School of Medicine, Western Sydney University and SWS Clinical School, UNSW Sydney 2170, NSW, Australia; stephanie.lim@health.nsw.gov.au (S.H.L.); P.DeSouza@westernsydney.edu.au (P.d.S.)

[2] Division of Colorectal Surgery, Department of Surgery, Westmead Hospital, Sydney 2145, Australia

[3] Department of Colorectal Surgery, Concord Hospital and Discipline of Surgery, Sydney Medical School, University of Sydney, Sydney 2137, Australia; pierre.chapuis@sydney.edu.au

[4] Liverpool Clinical School, Western Sydney University, Sydney 2170, Australia; S.Mackenzie@westernsydney.edu.au (S.M.); L.Bokey@westernsydney.edu.au (L.B.)

* Correspondence: james.toh@usyd.edu.au (J.W.T.T.); k.spring@westernsydney.edu.au (K.J.S.).

Abstract: Microsatellite instability (MSI) in colorectal cancer (CRC) is a marker of immunogenicity and is associated with an increased abundance of tumour infiltrating lymphocytes (TILs). In this subgroup of colorectal cancer, it is unknown if these characteristics translate into a measurable difference in circulating tumour cell (CTC) release into peripheral circulation. This is the first study to compare MSI status with the prevalence of circulating CTCs in the peri-operative colorectal surgery setting. For this purpose, 20 patients who underwent CRC surgery with curative intent were enrolled in the study, and peripheral venous blood was collected at pre- (t1), intra- (t2), immediately post-operative (t3), and 14–16 h post-operative (t4) time points. Of these, one patient was excluded due to insufficient blood sample. CTCs were isolated from 19 patients using the Isoflux™ system, and the data were analysed using the STATA statistical package. CTC number was presented as the mean values, and comparisons were made using the Student *t*-test. There was a trend toward increased CTC presence in the MSI-high (H) CRC group, but this was not statistically significant. In addition, a Poisson regression was performed adjusting for stage (I-IV). This demonstrated no significant difference between the two MSI groups for pre-operative time point t1. However, time points t2, t3, and t4 were associated with increased CTC presence for MSI-H CRCs. In conclusion, there was a trend toward increased CTC release pre-, intra-, and post-operatively in MSI-H CRCs, but this was only statistically significant intra-operatively. When adjusting for stage, MSI-H was associated with an increase in CTC numbers intra-operatively and post-operatively, but not pre-operatively.

Keywords: circulating tumour cells; colorectal cancer; colorectal surgery; microsatellite instability

---

1. Introduction

Biomarkers in colorectal cancer (CRC) have had limited success in clinical application to date, but microsatellite instability (MSI) status is emerging as a biomarker of clinical relevance. It is known that CRC exhibiting high level MSI (MSI-H) is associated with increased tumour infiltrating lymphocytes (TILs) and is a marker of immunogenicity [1–3]. MSI-H CRCs are less likely to disseminate due to TILs as a protective factor, yet a double-edged sword exists in that MSI-H CRCs have more mutations and are associated with more adverse pathological features. However, what is not known is whether the abundance of TILs decreases the risk of tumour dissemination by reducing the release of circulating tumour cells (CTCs), which have metastatic potential.

Microsatellites are short tracts of repetitive sequence (1–6 base pairs or more that are generally repeated between 5 and 50 times) found disseminated throughout the genome. Due to the repetitive nature of microsatellites, these regions are prone to change (instability) during replication. In MSI-H CRC, the resultant microsatellite alterations result in frameshifts that truncate proteins and may lead to inactivation of affected coding regions. Usually, microsatellite alterations are sensed by mismatch repair (MMR) genes that act like spellcheckers or DNA damage sensors, which detect mutations and signal for repair or apoptosis. When there is a loss of DNA damage sensors, either through genetic or epigenetic inactivation of MMR genes, this leads to loss of appropriate signalling and an accumulation of genetic mutations. In clinical practice, MSI-H occurs in 10–15% of colorectal cancers and is defined by IHC staining demonstrating MMR deficiency (MMRD).

The serrated neoplastic pathway is one of the two sporadic pathways that result in MSI-H cancers, the other classical pathway being the adenoma-carcinoma pathway involving chromosomal instability (CIN) and that results in microsatellite stable (MSS) cancers. Interestingly, it appears that MSI-H colorectal cancers are less likely to progress to stage IV disease compared to their MSS counterpart [4,5]. However, it is unclear if the biology of this observation is associated with decreased release of CTC. The hypothesis we sought to test was that the abundance of TILs in MSI-H CRCs may reduce the release of CTCs and, by doing so, protect against the risk of dissemination.

CTCs were first identified by Dr. Thomas Ashworth in 1869 [6]. Under the microscope, it was observed that "cells identical with those of the cancer itself" were present in the blood of a man with metastatic cancer. Since then, CTCs have been shown to be both a predictive and prognostic biomarker, but have remained in the research domain rather than clinical application due to cost. Evaluating circulating tumour DNA (ctDNA) has also shown great utility as a diagnostic approach for cancer management. Instead of identifying cancer cells (CTCs) in the bloodstream, identifying ctDNA depends on DNA released into the bloodstream from the tumour cell nucleus as it dies and is replaced by new cancer cells. A recent study by Tie et al. investigated ctDNA in stage II colon cancer to detect patients at high risk of recurrence. In that study, they also assessed the association between post-operative ctDNA status and conventional high-risk clinicopathological factors, but were not able to show an association, albeit that the majority of patients in the study were ctDNA negative [7].

With the call for universal MSI testing in CRC, it is important to understand the immunobiology of MSI to understand its clinical implications and its role in guiding prognosis and adjuvant therapy. It is known that MSI is associated with TILs [8]. However, it is not yet known if MSI status affects the release of CTC. It is not clear if patients with abundant TILs have a reduction of CTC count and whether this is stage dependent. This pilot study is the first to investigate CTC count in elective colorectal surgery and to analyse possible differences in the pre-operative, intra-operative, and post-operative stage of treatment and correlate this with MSI status.

2. Materials and Methods

2.1. Patients and Blood Samples

Twenty patients undergoing elective laparoscopic or open colorectal surgery at either Liverpool or Westmead Hospitals were enrolled in the study approved by the South Western Sydney Local Health District Ethics (Ref: HREC/13/LPOOL/158). All patients gave informed written consent for blood collection and CTC analysis. Peripheral venous blood was collected at four time points: (t1) pre-operative blood collection in the anaesthetic bay of operating theatres; (t2) intra-operatively after mobilisation of bowel was completed; (t3) at time of completion of surgery; and (t4) fourteen-sixteen hours post-operatively. One patient had insufficient blood volume collected and was excluded from analysis.

2.2. CTC Enrichment and Enumeration

Quantification of CTCs was performed using the IsoFlux^{TM} instrument (Fluxion Biosciences Inc, Alameda, CA, USA). Peripheral venous blood was collected into 9 mL anti-coagulant K_2EDTA tubes (Vacuette 455036) and processed within 24 h of collection in accordance with the Isoflux protocol using the CTC enrichment (910-0091) and enumeration (910-0093) kits supplied by the manufacturer. Briefly, immuno-magnetic EpCAM linked beads was used to capture CTCs, and after processing through the Isoflux instrument, CTCs were identified by immune staining using anti-cytokeratin (CK-7, -8, -18, and -19), Hoechst 33342 dye, and anti-CD45. After transferring each sample to 24 well SensoPlates^{TM} (Cat. No. 662892, Geriner Bio-One, GmbH, Kremsminster, Austria) and applying coverslips, each sample was scanned and visualized using a 10× objective on a fluorescence Olympus IX71 inverted microscope. Putative CTCs were defined as CK^+, $DAPI^+$, and $CD45^-$, nucleated and morphologically intact cells.

2.3. Clinical and Histopathological Data

Carcinoembryonic antigen (CEA), tumour infiltrating lymphocytes (TILs), microsatellite instability (MSI), and BRAF status were recorded. TILs were reported by the pathologist as present when there were more than 5 intraepithelial lymphocytes/100 epithelial cells (assessed on minimum three high power (×400) fields) [8]. MSI and BRAF status was tested by immunohistochemistry.

Data on patient demographics, histopathological features of the tumour, and CTC count at four time periods were collected. Certified pathologists examined the tissue biopsy specimens post-operatively and provided the histopathology diagnosis. All patients had follow-up at one year with disease-free survival (DFS) being the main outcome.

2.4. Statistical Analysis

Analysis was performed on STATA (Stata MP, Version 15; StataCorp LP). The Student *t*-test was used to compare between groups, and a Poisson regression was used to adjust for stage. The Student *t*-test was used instead of the Wilcoxon–Mann–Whitney U-test and the Kruskal–Wallis tests as these non-parametric tests are better used to compare medians, whereas the Student *t*-test provides a better assessment of means.

3. Results

3.1. Clinical and Surgical Characteristics

In total, 80 samples from 20 patients with colorectal cancer who underwent elective open or laparoscopic colorectal surgery with curative intent were recruited for this study. However, four samples from one patient had insufficient blood volume collected, and this patient was excluded from analysis. CTC isolation and enrichment were performed using the IsoFlux^{TM} system. Of the nineteen patients who had CTCs enumerated, two patients had high-grade dysplasia without malignancy (these were 30 × 33 × 23 mm and 57 × 50 × 55 mm villous adenomas). Of the remaining 17 patients, 3 (17.6%) were stage I, 6 (35.3%) stage II, 7 (41.2%) stage III, and 1 (5.9%) stage IV. Nine (52.9%) were right-sided (caecal (n = 4), ascending colon (n = 4), and transverse colon (n = 1)), the rest (41.2%) were left-sided (rectum (n = 4), rectosigmoid (n = 1), sigmoid (n = 3)). The histopathology of all seventeen patients was adenocarcinoma. Further, two of the four patients with rectal cancer had neoadjuvant chemoradiotherapy. Patient demographic, clinical, and surgical characteristics are summarized in Table 1, and the CTC yield for each patient at the different time points are in Table S1 (Supplementary Material).

Table 1. Patient demographic, clinical, and surgical characteristics and circulating tumour cell (CTC) number. MSI-H, microsatellite instability-high; MSS, microsatellite stable.

Patient Characteristics	Microsatellite Status	
	MSI-H	MSS
Patient number	4	13
Age	85.5 (54–86)	66 (44–86)
Female:male	3:1	6:7
Right colon	4 (100%)	5 (38.5%)
Left colon	0	3 (23.1%)
Rectal/rectosigmoid	0	5 (38.5%)
Grade		
High	3 (75%)	1 (8.3%)
Moderate	1 (25%)	10 (83.3%)
Low	0	1 (8.3%)
BRAF mutant:wild-type	3:1	N/A
Stage		
I	0	3 (23.1%)
II	2 (50%)	4 (30.8%)
III	2 (50%)	5 (38.5%)
IV	0	1 (7.7%)
CTC number		
t1	10.5 (0–29)	1 (0–61)
t2	52 (44–189)	1 (0–74)
t3	23 (1–83)	1 (0–17)
t4	34 (6–65)	1 (0–12)

Continuous data shown as the mean with the range; count data presented as the frequencies and percentages. Three patients are not included in the Table as the histopathology was villous adenoma, with high-grade dysplasia for two patients, and one patient had insufficient samples.

3.2. CTC Yield in All Patients

First, we looked at the number of CTCs enumerated for all 19 patients (Table 1 and supplementary Table S1. CTC number was presented as the mean values and comparisons made using the Student *t*-test. CTCs were enumerated for 19 patients: microsatellite stable (MSS) ($n = 15$); microsatellite unstable (MSI-H) ($n = 4$), respectively. A Student *t*-test was used to test the difference in CTC number between MSS and MSI-H CRCs, respectively, at the four different time points: t1 (8.2 vs. 12.5, $p = 0.6191$); t2 (23.7 vs. 37.8, $p = 0.5893$); t3 (9.3 vs. 12.3, $p = 0.7798$); and t4 (8.1 vs. 18.8, $p = 0.3696$). It was apparent that at each of these time points, there was no significant difference between MSS and MSI-H patient groups.

3.3. CTC Yield in Cancer Patients Only

Excluding the two patients with villous adenoma and high-grade dysplasia, there was a trend towards higher CTC number for MSI-H CRC, but this was not statistically significant between the two groups: t1 (7.9 vs. 12.5, $p = 0.6191$); t2 (22.2 vs. 37.8, $p = 0.5893$); t3 (8.7 vs. 12.3, $p = 0.7798$); t4 (8.3 vs. 18.8, $p = 0.3696$). In addition, a Poisson regression was performed adjusting for stage (I-IV). This demonstrated no significant difference between the two MSI groups for t1. However, t2, t3, and t4 were all associated with an increase in CTC number for MSI-H CRCs.

3.4. MSI Status and CTC Number by Stage of CRC

For this analysis, there were no MSI-H patients in the stage I and IV groups; however, there were two MSI-H (caecum and ascending colon) patients in the stage II group; whereas there were four stage II MSS CRC (caecum, rectosigmoid and two rectum) patients. For the stage II patients, there was a significant spike at the t2 timepoint for the MSI-H group (Figure 1; Panel A and Panel B). However, the sample size was too small to perform reliable statistical analysis. For stage III patients, there appeared to be a trend towards higher CTC count for the MSI-H group, but this was not statistically significant: t1 (13.2 vs. 14.5, $p = 0.9540$); t2 (15.2 vs. 23.5, $p = 0.7700$); t3 (3.6 vs. 20, $p = 0.1981$); t4 (3.6 vs. 29, $p = 0.1589$); Figure 1; C and D. When combining data for all stage I-III patients, there was a statistically significant spike in t2 count in the MSI-H group. There was also a trend toward higher CTC number for t1, t3, and t4 in the MSI-H group, but this was not statistically significant (Figure 1; E and F): t1 (7.5 vs. 12.5, $p = 0.6027$); t2 (8.25 vs. 37.75, $p = 0.0328$); t3 (2.5 vs. 12.25, $p = 0.0878$); t4 (3.67 vs. 18.75, $p = 0.0604$).

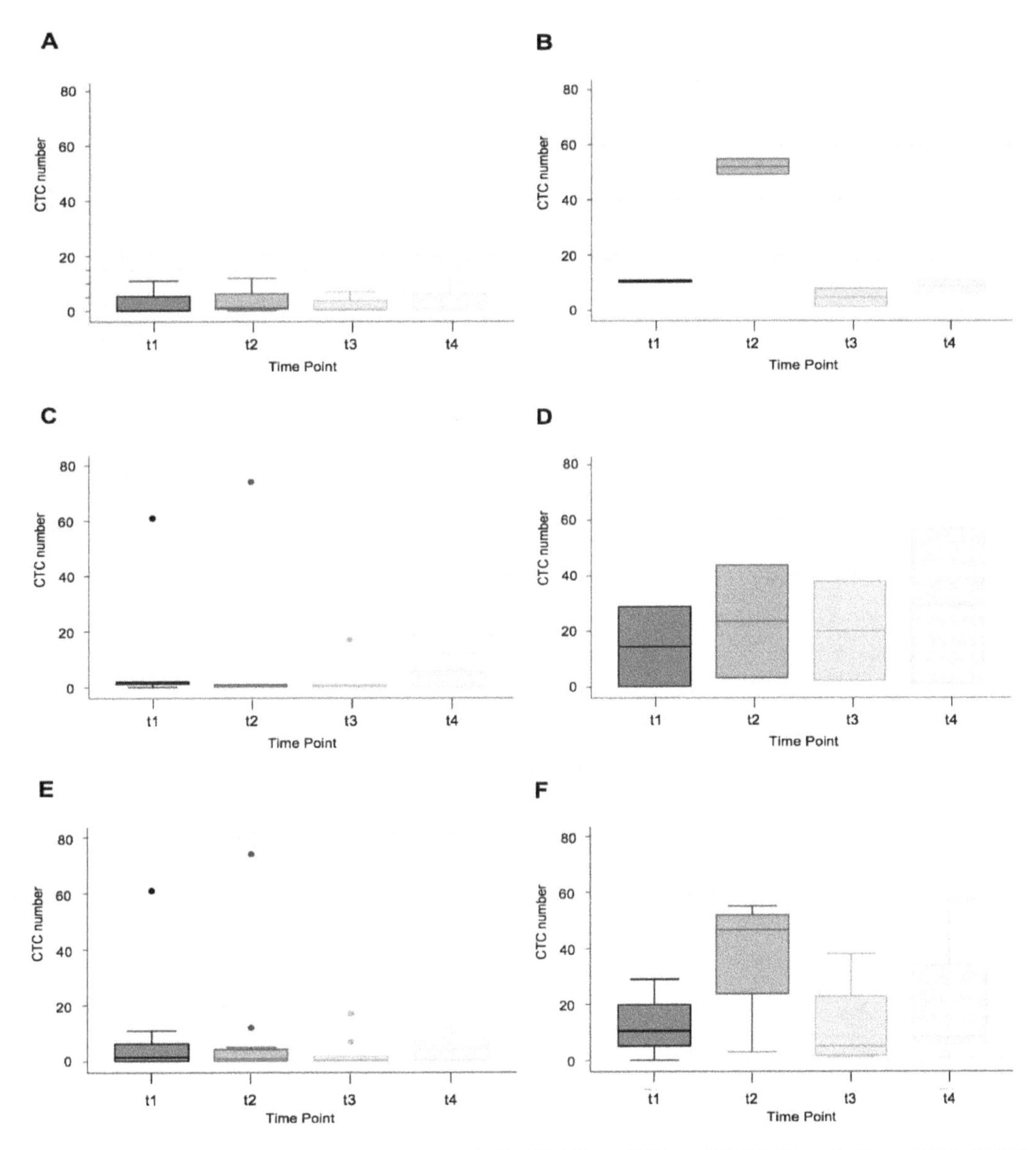

Figure 1. CTC number for stage II (Panels **A** and **B**), III (**C** and **D**) and I-III (Panels **E** and **F**) MSS (left) vs. MSI-H (right) colorectal cancer at different sample time points: t1 (pre-operative), t2 (intra-operative), t3 (immediate post-operative), t4 (14–16 h post-operative).

There was only one patient with stage IV colon cancer. This patient had a right-sided cancer that was MSS and had high CTCs that were persistently elevated with 13 CTCs detected at the pre-operative time point (t1), which increased to 189 during surgery and then remained high post-operatively with 83 and 65 CTCs detected, respectively (Figure 2).

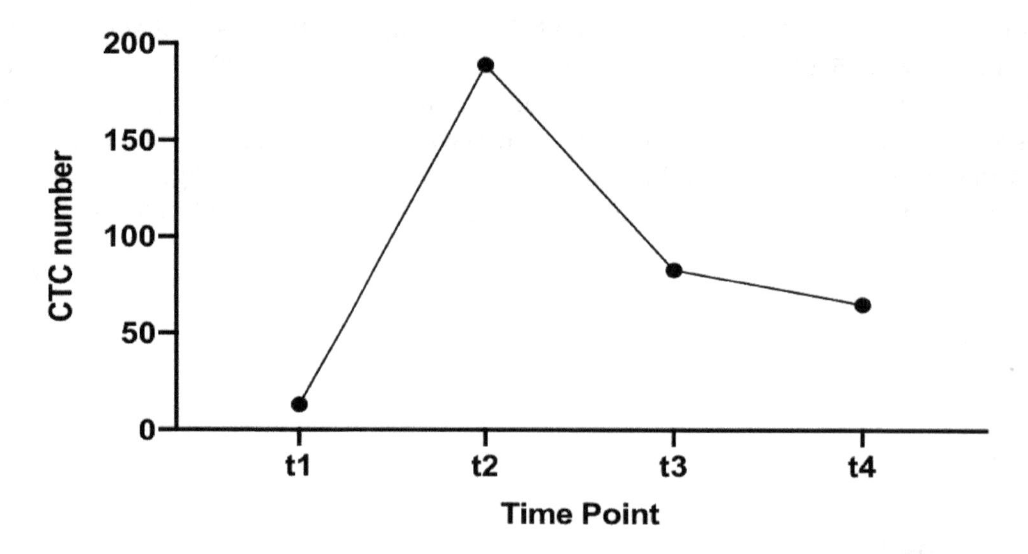

Figure 2. CTC number at different time points for patient S16 (stage IV, MSS CRC).

3.5. MSI Status and CTC by Side (Right vs. Left) Stage I-III Colon Cancer

There were no left-sided MSI-H CRCs in this study, so it was not possible to compare between MSI-H and MSS for left-sided colon and rectal cancer. However, there were four MSI-H and five MSS right-sided colon cancers. One MSS was excluded from analysis as it was a stage IV CRC. There was no statistically significant difference in CTC number for right-sided colon cancer by MSI status, but there was a trend for increased CTC number at t2, t3, and t4 time points (t1 (20 vs. 12.5, $p = 0.6375$); t2 (22.75 vs. 37.75, $p = 0.5001$); t3 (6.5 vs. 12.25, $p = 0.5677$); t4 (6 vs. 18.75, $p = 0.0942$); Figure 3, Panels A and B).

There were eight left-sided CRCs in this study, of which none exhibited MSI. Two patients received neoadjuvant therapy, whereas six patients did not receive neoadjuvant therapy. There were four rectal cancers, one rectosigmoid and three sigmoid cancers. The CTC number overall was low in this subgroup; however, it appeared there was a difference between those who received neoadjuvant therapy and those who did not (Figure 3, Panels C and D). Reliable statistical analysis was not performed due to the small sample size.

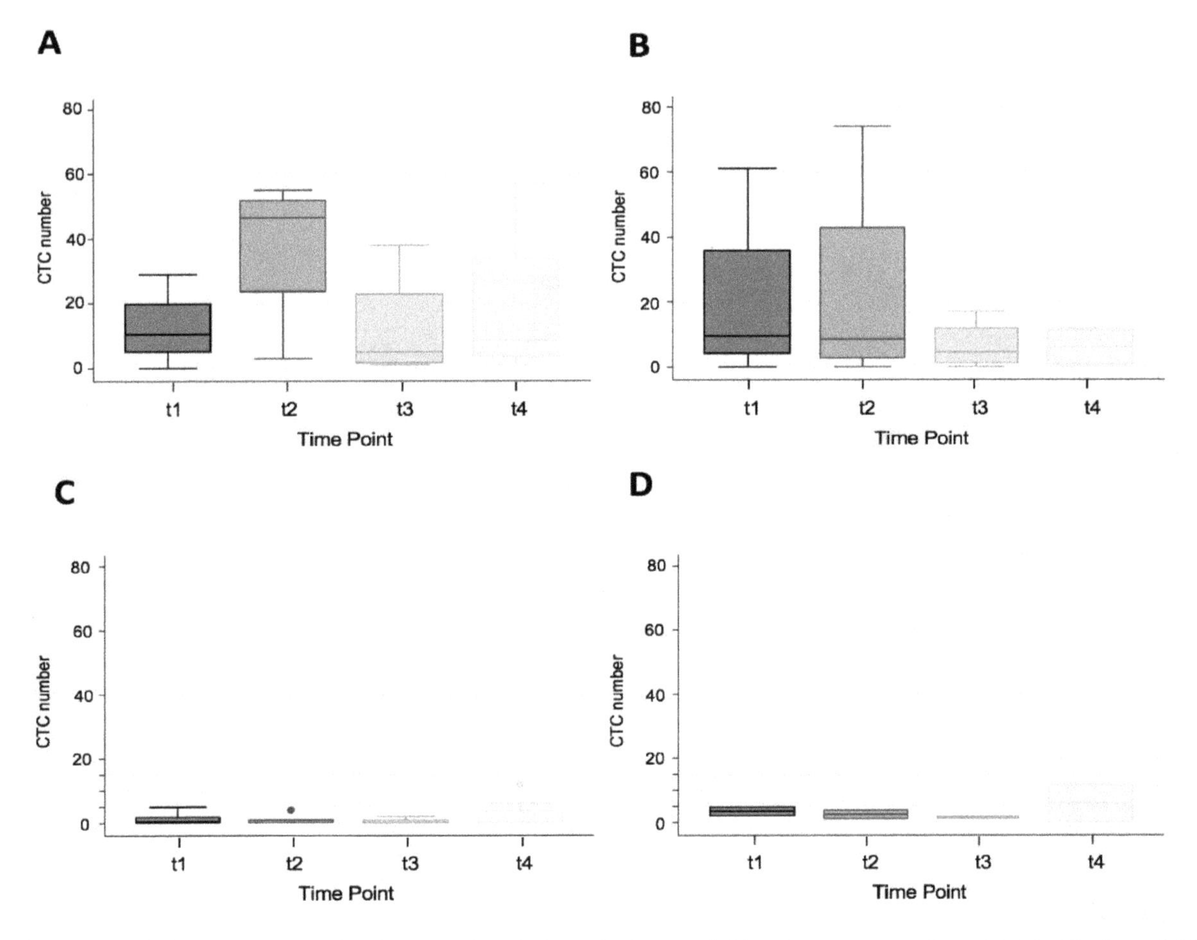

Figure 3. CTC number and MSI status for right-sided stage I-III colon cancer, (Panel **A**) MSS, (Panel **B**) MSI-H and CTC number for left-sided colorectal cancer, no neoadjuvant (Panel **C**), and neoadjuvant therapy (Panel **D**) at sample time points t1 (pre-operative), t2 (Intra-operative), t3 (immediate post-operative), t4 (14–16 h post-operative).

3.6. Poisson Regression Model with Post-Estimation Marginal Fundamental Analysis

A Poisson regression model was run with MSI and stage as independent variables. A post-estimation marginal means and marginal effects fundamental analysis was performed. This showed no difference in CTC number at the t1 time point, but a statistically significant difference in CTC number at the t2, t3, and t4 time points with stage as a covariate: t1: 7.96 (6.42–9.50) vs. 12.29 (8.88–15.71); t2: (20.99 (18.55–23.43) vs. 46.01 (38.18–53.83); t3: 7.72 (6.28–9.16) vs. 20.69 (13.88–27.49); t4: 8.03 (6.51–9.54) vs. 21.91 (16.62–27.18). In addition, a Poisson regression model was run for stage of CRC, and a post-estimation marginal fundamental analysis was performed. This showed a significant difference in CTC number at the t1, t2, t3, and t4 time points between stages independent of MSI status (Figure 4). Preoperatively, the CTC number at the t1 time point for stage I was 3.99 (2.48–5.49); for stage II 6.78 (5.33–8.22); for stage III 11.53 (9.58–13.48); for stage IV 19.61 (13.63–25.59).

3.7. Non-Parametric Wilcoxon–Mann–Whitney U-test

When comparing the median CTC number between MSS and MSI-H CRCs, there was again a trend toward increased CTC number with MSI-H CRC, but this was only statistically significant at the t2 time point when comparing stage I-III MSS and MSI-H CRC.

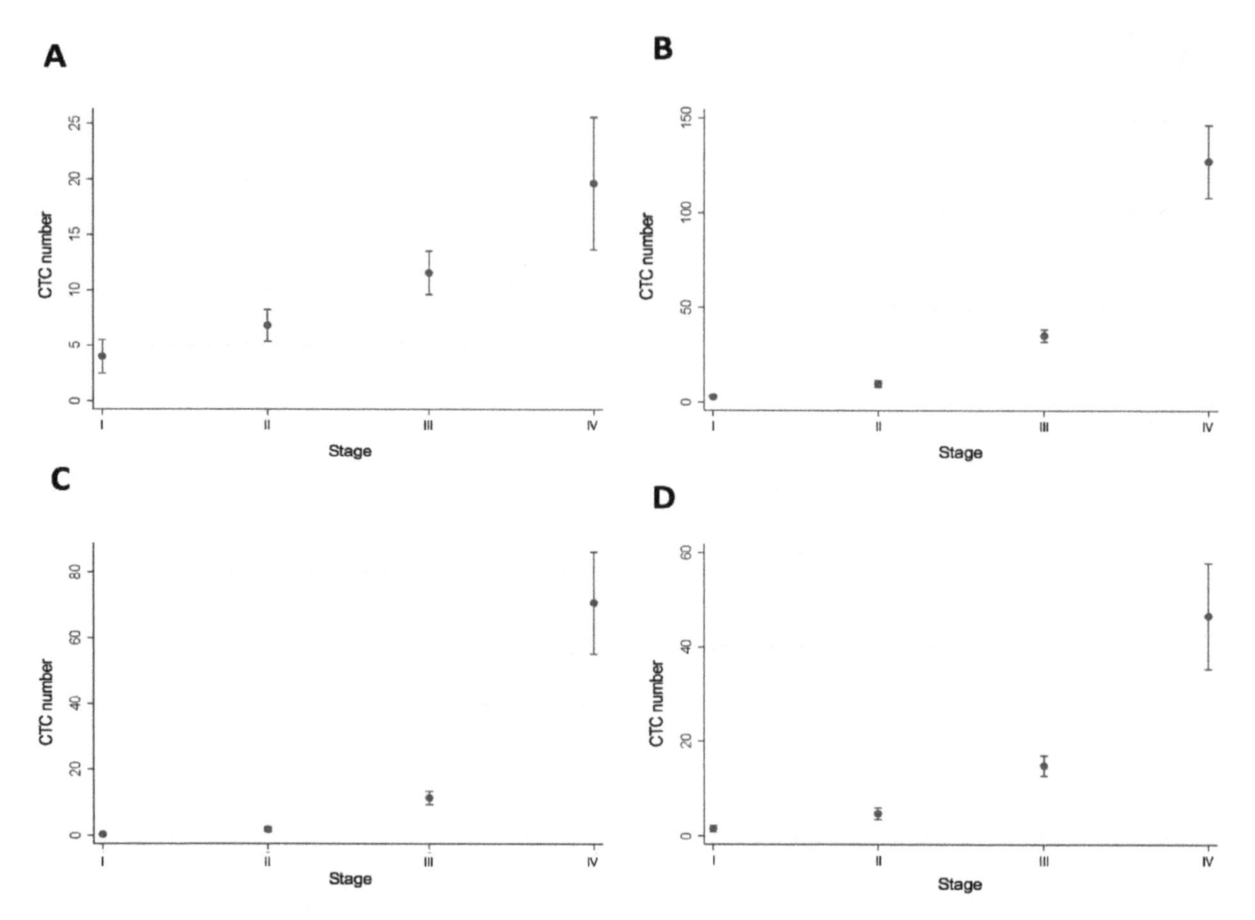

Figure 4. Poisson regression model with post-estimation marginal fundamental analysis of stage and CTC number at sample time points t1 and t2 (Panels **A** and **B**) and t3 and t4 (Panels **C** and **D**).

4. Discussion

There is an abundance of clinical data reporting on the association between MSI status and prognosis in CRC showing that MSI-H may be associated with better prognosis [3,4]. The evidence in the literature shows that CTCs may be important in prognostication of colon cancer [5] in predicting dissemination [6], overall and disease-free survival [7], and lymph node involvement [8,9]. Higher levels of CTCs have been associated with worse outcome and may predict for poor disease-free survival [10–12]. Most existing studies assessing the relationship between MSI and prognosis in CRC have been clinical studies or histopathological studies. We took a fundamentally different approach to instead examine the immune-biological characteristics of CRC based on MSI status.

Our study confirmed that CTC measurements correlated with dissemination and stage of disease. There was a statistically significant difference in CTC number with stage (I-IV) and across all time points (preoperatively, intraoperatively, and postoperatively, as shown in Figure 4). However, notably, our study showed that MSI-H CRCs (which have been reported to be immunogenic and associated with enhanced survival) [3,4] were associated with increased peri-operative release of CTCs (Figure 1). Further, both the mean and median CTC count at all time points were higher in the MSI-H group compared to the MSS group. Our hypothesis that increased TILs would decrease peri-operative release of CTCs in MSI-H CRCs was not supported by the data.

While most of the analyses performed were not statistically significant, overall, there was a trend towards increased CTC number at all time points (t1, t2, t3, and t4) in the MSI-H group. When analysing stage II CRC and stage I-III CRC, there was a statistically significant increase in the CTC number for the t2 time point: 3.5 vs. 52, $p = 0.0005$ and 8.25 vs. 37.75, $p = 0.0328$, respectively, in the MSI-H group. For all other comparisons, there was increased CTC number in the MSI-H group, but it was not statistically significant. A Poisson regression was performed to adjust for stage (I-IV). This demonstrated no significant difference between the two MSI groups for the t1 time point. However, the t2, t3, and t4 time points were all associated with increased CTC number for MSI-H CRCs.

The literature on CTCs has shown that a measurement of more than three CTCs per 7.5 mL peripheral blood may be associated with poor survival, although some studies suggest 1–2 CTCs per 7.5 mL may also be associated with a worse outcome [13]. Furthermore, higher post-operative CTC numbers may be associated with a higher risk of recurrence [14], whereas improvements in post-operative CTC number from pre-operative baselines has been associated with better survival [15]. In this study, the median CTC number for the t1, t2, t3, and t4 time points for MSS CRC was one, whereas the median for MSI-H CRC was significantly >3 in all the corresponding time points. However, with data in the literature showing enhanced survival in MSI-H CRC, the high CTC numbers for MSI-H CRCs and low CTC numbers for MSS CRCs found in this study did not correlate with the clinicopathological data reporting on survival existing in the literature. On the other hand, this study may have shown that the cut-off for prognostication of CRC by CTC measurement may actually be influenced by MSI status and that there should not be a single cut-off, but the benchmark may depend on characteristics such as MSI. One hypothesis was that CTCs released into the bloodstream of patients with MSI-H CRCs usually remain microsatellite unstable [16], which maintains an enhanced immunogenic response from circulating lymphocytes in the bloodstream [17]. Thus, its presence may not represent the same risk of metastases as CTCs associated with MSS CRCs [18,19].

What about the protective effect of MSI? The immunogenicity of MSI is believed to be associated with the presence of TILs. Collinearity between MSI-H status and TILs has been shown in clinicopathological data, including the data from our own cohort study [20], and has been traditionally associated with a better prognosis [3,4]. It is believed that the survival and clinical benefits of MSI may be due to its immunogenicity, with MSI associated with increased TILs [2,21–24] and TILs associated with a better prognosis [2,25–34]. MSI-H CRCs may also be associated not only with an abundance of intra-tumoral TILs, but also with a higher density of associated cytotoxic, helper, and regulatory T lymphocytes in peripheral blood [17], as well as increased activity in the bone marrow [35]. Studies have shown that both the immunogenic TILs response at the tumour site, as well as the circulatory system are believed to be associated with a decreased risk of lymph node metastases [19,36] and distant metastases [18] in MSI CRCs. The assumption, hence, would be that MSI-H CRCs should be associated with decreased CTC dissemination into the blood.

However, this study showed that the immuno-biology of MSI-H CRCs is more complex than this. As is well established in the literature, MSI-H CRCs are also associated with poor differentiation [19,37], are larger and more likely to be mucinous [38–40], as well as being higher grade tumours with a greater mutational burden and the mucinous phenotype more likely to be associated with signet cells. It is also believed that MSI-H CRCs are associated with a high number of frameshift mutation peptides (FSPs) [41,42] when compared to MSS CRCs [43,44]. The MSI hyper-mutational state is also thought to be the reason for poor differentiation and other adverse pathological features of the tumour. The higher grade of tumour associated with hyper-mutational state may be the reason for the increased peri-operative release of CTCs. On Poisson regression analysis, our study showed no difference in pre-operative (t1) CTC numbers, but a statistically significant difference in intra-operative (t2 and t3) and post-operative (t4) release of CTCs. From the literature, immunogenic TILs may reduce the risk of dissemination, but this study showed that they did not do so by preventing the release of CTCs. We currently do not have long term survival and recurrence data, nor CTC measurements outside of the 24 h window peri-operatively, but a study looking at recurrence, survival, and CTC number at 6 months, 1 year, 3 year, and 5 years would be interesting and would examine the role that CTCs may play in the post-operative management of CRCs including whether this may be used to predict recurrence or survival accurately.

Notably, this pilot study showed results that warrant further investigations. The increased CTC positivity seen in MSI-H CRCs is a significant point of difference from the currently understood immune-biological mechanisms associated with MSI-H CRCs. In practice, this means that the benchmark cut-off points for CTC enumeration may be influenced by tumour characteristic, and future clinical applications of CTC in CRC management may need to take this into consideration. Another consideration is that the improved survival associated with immunogenic MSI-H CRCs may not be as profound as once believed, and the increased peri-operative release of CTCs in MSI-H CRCs may be revisited with future studies with a larger patient cohort. This could prove a useful follow up of the suggestions made by the European Society of Medical Oncology (ESMO) guidelines that patients with stage II colon cancer with MSI are at very low risk of recurrence and unlikely to benefit from chemotherapy.

In this study, the overall patient number ($n = 20$) was low and a limitation for insightful statistical analysis. There was significant heterogeneity within the study with two patients receiving neoadjuvant chemoradiotherapy, no direct comparisons available for stage I and IV (no MSI-H CRC in these subgroups), and the inclusion of right colon, left colon, and rectal cancers. The incidence of MSI in CRC is 10–15%. In our cohort, there were only four MSI-H CRCs, of which two were stage II and two were stage III, and the majority of patients were MSS CRCs, being a limitation of our data. From a tumour biology perspective, the main implication of our findings is that immunogenic MSI-H CRCs did not protect against release of CTCs, and the protective effect against metastatic disease was not by reducing CTCs. Further, the main clinical implication of this study involves the utility of CTCs for monitoring and surveillance with potentially differential baseline levels of CTCs associated with the two subtypes of CRC.

5. Conclusions

There was a trend toward increased CTC release pre-, intra-, and post-operatively in MSI-H CRCs, but this was only statistically significant intra-operatively. When adjusting for stage, MSI-H was associated with an increase in CTC number intra-operatively and post-operatively, but not pre-operatively. This dataset was limited, and further studies are required. Finally, these data suggested that immunogenic MSI-H CRCs did not suppress CTC release and that different reference ranges may be required for CTC enumeration of MSI-H and MSS CRC.

Author Contributions: Conceptualization, J.W.T.T., K.J.S., and S.H.L.; formal analysis, J.W.T.T., K.J.S., and S.H.L.; investigation, J.W.T.T., S.M., S.H.L. and K.J.S.; resources, K.J.S. and P.d.S.; data curation, J.W.T.T., S.M. and S.H.L.; methodology, J.W.T.T., S.H.L. and K.J.S.; writing, original draft preparation, J.W.T.T. and K.J.S.; writing, review and editing, K.J.S., P.C., L.B. and P.d.S.; project administration, K.J.S.; funding acquisition, K.J.S. and P.d.S. All authors have read and agreed to the published version of the manuscript.

References

1. Tie, J.; Wang, Y.; Tomasetti, C.; Li, L.; Springer, S.; Kinde, I.; Silliman, N.; Tacey, M.; Wong, H.L.; Christie, M.; et al. Circulating tumor DNA analysis detects minimal residual disease and predicts recurrence in patients with stage II colon cancer. *Sci. Transl. Med.* **2016**, *8*, 346ra392. [CrossRef]
2. Michael-Robinson, J.M.; Biemer-Huttmann, A.; Purdie, D.M.; Walsh, M.D.; Simms, L.A.; Biden, K.G.; Young, J.P.; Leggett, B.A.; Jass, J.R.; Radford-Smith, G.L. Tumour infiltrating lymphocytes and apoptosis are independent features in colorectal cancer stratified according to microsatellite instability status. *Gut* **2001**, *48*, 360–366. [CrossRef] [PubMed]
3. Popat, S.; Hubner, R.; Houlston, R.S. Systematic review of microsatellite instability and colorectal cancer prognosis. *J. Clin. Oncol.* **2005**, *23*, 609–618. [CrossRef]
4. Guastadisegni, C.; Colafranceschi, M.; Ottini, L.; Dogliotti, E. Microsatellite instability as a marker of prognosis and response to therapy: A meta-analysis of colorectal cancer survival data. *Eur. J. Cancer* **2010**, *46*, 2788–2798. [CrossRef]

5. Steinert, G.; Scholch, S.; Koch, M.; Weitz, J. Biology and significance of circulating and disseminated tumour cells in colorectal cancer. *Langenbeck's Arch. Surg.* **2012**, *397*, 535–542. [CrossRef]
6. Hiraiwa, K.; Takeuchi, H.; Hasegawa, H.; Saikawa, Y.; Suda, K.; Ando, T.; Kumagai, K.; Irino, T.; Yoshikawa, T.; Matsuda, S.; et al. Clinical significance of circulating tumor cells in blood from patients with gastrointestinal cancers. *Ann. Surg. Oncol.* **2008**, *15*, 3092–3100. [CrossRef]
7. Liu, Y.; Qian, J.; Feng, J.G.; Ju, H.X.; Zhu, Y.P.; Feng, H.Y.; Li, D.C. Detection of circulating tumor cells in peripheral blood of colorectal cancer patients without distant organ metastases. *Cell. Oncol.* **2013**, *36*, 43–53. [CrossRef]
8. Gazzaniga, P.; Gianni, W.; Raimondi, C.; Gradilone, A.; Russo, G.L.; Longo, F.; Gandini, O.; Tomao, S.; Frati, L. Circulating tumor cells in high-risk nonmetastatic colorectal cancer. *Tumour Biol.* **2013**, *34*, 2507–2509. [CrossRef]
9. Katsuno, H.; Zacharakis, E.; Aziz, O.; Rao, C.; Deeba, S.; Paraskeva, P.; Ziprin, P.; Athanasiou, T.; Darzi, A. Does the presence of circulating tumor cells in the venous drainage of curative colorectal cancer resections determine prognosis? A meta-analysis. *Ann. Surg. Oncol.* **2008**, *15*, 3083–3091. [CrossRef]
10. Groot Koerkamp, B.; Rahbari, N.N.; Buchler, M.W.; Koch, M.; Weitz, J. Circulating tumor cells and prognosis of patients with resectable colorectal liver metastases or widespread metastatic colorectal cancer: A meta-analysis. *Ann. Surg. Oncol.* **2013**, *20*, 2156–2165. [CrossRef]
11. Rahbari, N.N.; Aigner, M.; Thorlund, K.; Mollberg, N.; Motschall, E.; Jensen, K.; Diener, M.K.; Büchler, M.W.; Koch, M.; Weitz, J. Meta-analysis shows that detection of circulating tumor cells indicates poor prognosis in patients with colorectal cancer. *Gastroenterology.* **2010**, *138*, 1714–1726. [CrossRef]
12. Thorsteinsson, M.; Jess, P. The clinical significance of circulating tumor cells in non-metastatic colorectal cancer—a review. *Eur J. Surg Oncol.* **2011**, *37*, 459–465. [CrossRef]
13. Gazzaniga, P.; Raimondi, C.; Gradilone, A.; Biondi Zoccai, G.; Nicolazzo, C.; Gandini, O.; Longo, F.; Tomao, S.; Lo Russo, G.; Seminara, P.; et al. Circulating tumor cells in metastatic colorectal cancer: Do we need an alternative cutoff? *J. Cancer Res. Clin. Oncol.* **2013**, *139*, 1411–1416. [CrossRef]
14. Peach, G.; Kim, C.; Zacharakis, E.; Purkayastha, S.; Ziprin, P. Prognostic significance of circulating tumour cells following surgical resection of colorectal cancers: A systematic review. *Br. J. Cancer* **2010**, *102*, 1327–1334. [CrossRef]
15. Yalcin, S.; Kilickap, S.; Portakal, O.; Arslan, C.; Hascelik, G.; Kutluk, T. Determination of circulating tumor cells for detection of colorectal cancer progression or recurrence. *Hepato-gastroenterology* **2010**, *57*, 1395–1398.
16. Steinert, G.; Scholch, S.; Niemietz, T.; Iwata, N.; García, S.A.; Behrens, B.; Voigt, A.; Kloor, M.; Benner, A.; Bork, U. Immune escape and survival mechanisms in circulating tumor cells of colorectal cancer. *Cancer Res.* **2014**, *74*, 1694–1704. [CrossRef]
17. Schwitalle, Y.; Kloor, M.; Eiermann, S.; Linnebacher, M.; Kienle, P.; Knaebel, H.P.; Tariverdian, M.; Benner, A.; Doeberitz, M.v.K. Immune response against frameshift-induced neopeptides in HNPCC patients and healthy HNPCC mutation carriers. *Gastroenterology* **2008**, *134*, 988–997. [CrossRef]
18. Buckowitz, A.; Knaebel, H.P.; Benner, A.; Bläker, H.; Gebert, J.; Kienle, P.; Doeberitz, M.v.K.; Kloor, M. Microsatellite instability in colorectal cancer is associated with local lymphocyte infiltration and low frequency of distant metastases. *Br. J. Cancer* **2005**, *92*, 1746–1753. [CrossRef]
19. Kazama, Y.; Watanabe, T.; Kanazawa, T.; Tanaka, J.; Tanaka, T.; Nagawa, H. Microsatellite instability in poorly differentiated adenocarcinomas of the colon and rectum: Relationship to clinicopathological features. *J. Clin. Pathol.* **2007**, *60*, 701–704. [CrossRef]
20. Toh, J.; Chapuis, P.H.; Bokey, L.; Chan, C.; Spring, K.J.; Dent, O.F. Competing risks analysis of microsatellite instability as a prognostic factor in colorectal cancer. *Br. J. Surg.* **2017**, *104*, 1250–1259. [CrossRef]
21. Kim, J.H.; Kang, G.H. Molecular and prognostic heterogeneity of microsatellite-unstable colorectal cancer. *World J. Gastroenterol. WJG* **2014**, *20*, 4230–4243. [CrossRef] [PubMed]
22. Greenson, J.K.; Bonner, J.D.; Ben-Yzhak, O.; Cohen, H.I.; Miselevich, I.; Resnick, M.B.; Trougouboff, P.; Tomsho, L.; Kim, E.; Low, M.; et al. Phenotype of microsatellite unstable colorectal carcinomas: Well-differentiated and focally mucinous tumors and the absence of dirty necrosis correlate with microsatellite instability. *Am. J. Surg. Pathol.* **2003**, *27*, 563–570. [CrossRef] [PubMed]

23. Smyrk, T.C.; Watson, P.; Kaul, K.; Lynch, H.T. Tumor-infiltrating lymphocytes are a marker for microsatellite instability in colorectal carcinoma. *Cancer* **2001**, *91*, 2417–2422. [CrossRef]
24. Tougeron, D.; Fauquembergue, E.; Rouquette, A.; Pessot, F.L.; Sesboüé, R.; Laurent, M.; Berthet, P.; Mauillon, J.; Fiore, F.D.; Sabourin, J.C. Tumor-infiltrating lymphocytes in colorectal cancers with microsatellite instability are correlated with the number and spectrum of frameshift mutations. *Mod. Pathol.* **2009**, *22*, 1186–1195. [CrossRef] [PubMed]
25. Pages, F.; Berger, A.; Camus, M.; Sanchez-Cabo, F.; Costes, A.; Molidor, R.; Mlecnik, B.; Kirilovsky, A.; Nilsson, M.; Damotte, D.; et al. Effector memory T cells, early metastasis, and survival in colorectal cancer. *New Engl. J. Med.* **2005**, *353*, 2654–2666. [CrossRef]
26. Mlecnik, B.; Tosolini, M.; Kirilovsky, A.; Berger, A.; Bindea, G.; Meatchi, T.; Bruneval, P.; Trajanoski, Z.; Fridman, W.H.; Page's, F.; et al. Histopathologic-based prognostic factors of colorectal cancers are associated with the state of the local immune reaction. *J. Clin. Oncol.* **2011**, *29*, 610–618. [CrossRef]
27. Ling, A.; Edin, S.; Wikberg, M.L.; Oberg, A.; Palmqvist, R. The intratumoural subsite and relation of CD8(+) and FOXP3(+) T lymphocytes in colorectal cancer provide important prognostic clues. *Br. J. Cancer* **2014**, *110*, 2551–2559. [CrossRef]
28. Svennevig, J.L.; Lunde, O.C.; Holter, J.; Bjorgsvik, D. Lymphoid infiltration and prognosis in colorectal carcinoma. *Br. J. Cancer* **1984**, *49*, 375–377. [CrossRef]
29. Jass, J.R. Lymphocytic infiltration and survival in rectal cancer. *J. Clin. Pathol.* **1986**, *39*, 585–589. [CrossRef]
30. Shunyakov, L.; Ryan, C.K.; Sahasrabudhe, D.M.; Khorana, A.A. The influence of host response on colorectal cancer prognosis. *Clin. Colorectal Cancer* **2004**, *4*, 38–45. [CrossRef]
31. Dahlin, A.M.; Henriksson, M.L.; Van Guelpen, B.; Stenling, R.; Öberg, A.; Rutegård, J.; Palmqvist, R. Colorectal cancer prognosis depends on T-cell infiltration and molecular characteristics of the tumor. *Mod. Pathol.* **2011**, *24*, 671–682. [CrossRef] [PubMed]
32. Chiba, T.; Ohtani, H.; Mizoi, T.; Naito, Y.; Nagura, H.; Ohuchi, A.; Ohuchi, K.; Shiiba, K.; Kurokawa, Y.; Satomi, S. Intraepithelial CD8+ T-cell-count becomes a prognostic factor after a longer follow-up period in human colorectal carcinoma: Possible association with suppression of micrometastasis. *Br. J. Cancer* **2004**, *91*, 1711–1717. [CrossRef] [PubMed]
33. Zlobec, I.; Lugli, A.; Baker, K.; Roth, S.; Minoo, P.; Hayashi, S.; Terracciano, L.; Jass, J.R. Role of APAF-1, E-cadherin and peritumoral lymphocytic infiltration in tumour budding in colorectal cancer. *J. Pathol.* **2007**, *212*, 260–268. [CrossRef] [PubMed]
34. Prall, F.; Duhrkop, T.; Weirich, V.; Ostwald, C.; Lenz, P.; Nizze, H.; Barten, M. Prognostic role of CD8+ tumor-infiltrating lymphocytes in stage III colorectal cancer with and without microsatellite instability. *Hum. Pathol.* **2004**, *35*, 808–816. [CrossRef]
35. Koch, M.; Beckhove, P.; Op den Winkel, J.; Autenrieth, D.; Wagner, P.; Nummer, D.; Specht, S.; Antolovic, D.; Galindo, L. Schmitz-Winnenthal, F.H. Tumor infiltrating T lymphocytes in colorectal cancer: Tumor-selective activation and cytotoxic activity in situ. *Ann. Surg.* **2006**, *244*, 986–992. [CrossRef]
36. Lamberti, C.; Lundin, S.; Bogdanow, M.; Pagenstecher, C.; Friedrichs, N.; Büttner, R.; Sauerbruch, T. Microsatellite instability did not predict individual survival of unselected patients with colorectal cancer. *Int. J. Colorectal Dis.* **2007**, *22*, 145–152. [CrossRef]
37. Xiao, H.; Yoon, Y.S.; Hong, S.M.; Ae Roh, S.; Cho, D.H.; Yu, C.S.; Kim, J.C. Poorly differentiated colorectal cancers: Correlation of microsatellite instability with clinicopathologic features and survival. *Am. J. Clin. Pathol.* **2013**, *140*, 341–347. [CrossRef]
38. Thibodeau, S.N.; Bren, G.; Schaid, D. Microsatellite instability in cancer of the proximal colon. *Science* **1993**, *260*, 816–819. [CrossRef]
39. Yoon, Y.S.; Kim, J.; Hong, S.M.; Lee, J.L.; Kim, C.W.; Park, I.J.; Lim, S.B.; Yu, C.S.; Kim, J.C. Clinical implications of mucinous components correlated with microsatellite instability in patients with colorectal cancer. *Colorectal Dis.* **2015**, *17*, O161–O167. [CrossRef]
40. Karahan, B.; Argon, A.; Yildirim, M.; Vardar, E. Relationship between MLH-1, MSH-2, PMS-2,MSH-6 expression and clinicopathological features in colorectal cancer. *Int. J. Clin. Exp. Pathol.* **2015**, *8*, 4044–4053.
41. Maby, P.; Tougeron, D.; Hamieh, M.; Mlecnik, B.; Kora, H.; Bindea, G.; Angell, H.K.; Fredriksen, T.; Elie, N.; Fauquembergue, E.; et al. Correlation between density of CD8+ T cell infiltrates in microsatellite unstable colorectal cancers and frameshift mutations: A rationale for personalized immunotherapy. *Cancer Res.* **2015**. [CrossRef] [PubMed]

42. Saeterdal, I.; Bjorheim, J.; Lislerud, K.; Gjertsen, M.K.; Bukholm, I.K.; Olsen, O.C.; Nesland, J.M.; Eriksen, J.A.; Møller, M.; Lindblom, A.; et al. Frameshift-mutation-derived peptides as tumor-specific antigens in inherited and spontaneous colorectal cancer. *Proc. Natl. Acad. Sci. USA* **2001**, *98*, 13255–13260. [CrossRef] [PubMed]
43. Sun, Z.; Yu, X.; Wang, H.; Zhang, S.; Zhao, Z.; Xu, R. Clinical significance of mismatch repair gene expression in sporadic colorectal cancer. *Exp. Ther. Med.* **2014**, *8*, 1416–1422. [CrossRef] [PubMed]
44. Vogelstein, B.; Papadopoulos, N.; Velculescu, V.E.; Zhou, S.; Diaz, L.A., Jr.; Kinzler, K.W. Cancer genome landscapes. *Science* **2013**, *339*, 1546–1558. [CrossRef]

12

Prognostic Significance of *TWIST1*, *CD24*, *CD44*, and *ALDH1* Transcript Quantification in EpCAM-Positive Circulating Tumor Cells from Early Stage Breast Cancer Patients

Areti Strati [1], Michail Nikolaou [2], Vassilis Georgoulias [3] and Evi S. Lianidou [1,*]

1. Analysis of Circulating Tumor Cells Lab, Department of Chemistry, University of Athens, 15771 Athens, Greece
2. Medical Oncology Unit, "Elena Venizelou" Hospital, 11521 Athens, Greece
3. Metropolitan General Hospital, 15562 Athens, Greece

* Correspondence: lianidou@chem.uoa.gr

Abstract: (1) Background: The aim of the study was to evaluate the prognostic significance of EMT-associated (*TWIST1)* and stem-cell (SC) transcript (*CD24, CD44, ALDH1*) quantification in EpCAM+ circulating tumor cells (CTCs) of early breast cancer patients. (2) Methods: 100 early stage breast cancer patients and 19 healthy donors were enrolled in the study. *CD24, CD44,* and *ALDH1* transcripts of EpCAM$^+$ cells were quantified using a novel highly sensitive and specific quadraplex RT-qPCR, while *TWIST1* transcripts were quantified by single RT-qPCR. All patients were followed up for more than 5 years. (3) Results: A significant positive correlation between overexpression of *TWIST1* and *CD24*$^{-/low}$/*CD44*^{high} profile was found. Kaplan–Meier analysis revealed that the ER/PR-negative (HR-) patients and those patients with more than 3 positive lymph nodes that overexpressed *TWIST1* in EpCAM$^+$ cells had a significant lower DFI (log rank test; $p < 0.001$, $p < 0.001$) and OS (log rank test; $p = 0.006$, $p < 0.001$). Univariate and multivariate analysis also revealed the prognostic value of *TWIST1* overexpression and *CD24*$^{-/low}$/*CD44*^{high} and *CD24*$^{-/low}$/*ALDH1*^{high} profile for both DFI and OS. (4) Conclusions: Detection of *TWIST1* overexpression and stem-cell (*CD24, CD44, ALDH1*) transcripts in EpCAM$^+$ CTCs provides prognostic information in early stage breast cancer patients.

Keywords: liquid biopsy; circulating tumor cells; epithelial–mesenchymal transition; stem cells; early breast cancer

---

1. Introduction

Circulating tumor cells (CTCs) are major players in liquid biopsy [1,2], and their molecular characterization is highly important for rational treatment decisions and for monitoring therapeutic response [3], whereas their analysis at the single cell level has the potential to reveal tumor heterogeneity in real time [4]. In breast cancer, a subpopulation of tumor cells that display stem cell-like properties [5] determines the aggressive characteristics and drug resistance of tumor clonal evolution [6]. Cancer stem cells (CSCs) that mediate tumor metastasis and therapeutic resistance have the capacity to transition between mesenchymal and epithelial-like states [7]. It has already been shown that breast cancer cells with the CD44+CD24−/low phenotype [8] that overexpress aldehyde dehydrogenase 1 (ALDH1+) [9] are able to form tumors in mice with high tumorigenic capacity. It has also been shown that disseminated tumor cells (DTCs) [10] and CTCs express the putative stem cell CD44+/CD24− and/or ALDH1+/CD24− phenotypical profile [11,12]. Moreover, in primary human luminal breast cancer, the metastasis-initiating cells containing CTC that express EPCAM, CD44, CD47, and the proto-oncogene MET are related with reduced overall survival (OS) [13]. In other types of cancer, various stem cell markers have also been identified and correlated with metastatic capacity [14] and poor prognosis [15].

It is now known that breast cancer stem cells exist in distinct mesenchymal-like (epithelial–mesenchymal transition [EMT]) as CD44+/CD24− and epithelial-like (mesenchymal-epithelial transition [MET]) states that express ALDH1. This transition between EMT- and MET-like states is highly important for their capacity to invade, disseminate, and grow at metastatic sites [16]. Many studies have already shown that a major proportion of CTC express both EMT and tumor stem cell characteristics [17–19]. Recently it was shown that an EpCAM-/ALDH1+/HER2+/EGFR+/HPSE+/Notch1+ profile in CTC drives these cells to metastasize to the brain [20]. At the single cell level, it has been shown that CTC that co-express the stem cell marker ALDH1 and the mesenchymal marker *TWIST1* may prevail during disease progression [21]. However, the prognostic significance of EMT and Stem cell (SC) markers in CTC has only been shown up to now in metastatic colorectal cancer [22] and metastatic breast cancer [23].

In early breast cancer, the molecular detection of cytokeratin 19 (CK-19) mRNA-positive cells in peripheral blood before [24], during [25], and after adjuvant therapy [26] is associated with worse prognosis, while their elimination seems to be an efficacy indicator of treatment [27]. The prognostic significance of CTC count using the CellSearch system in neoadjuvant [28] and adjuvant early breast cancer patients [29] has been also shown. Moreover, the administration of "secondary" adjuvant trastuzumab in patients with HER2(−) breast cancer can eliminate chemotherapy-resistant CK19 mRNA-positive CTCs [30], in contrast to the Treat CTC phase II trial that failed to prove the efficacy of trastuzumab in the detection rate of CTC [31]. However, in early breast cancer stages the early detection of recurrence remains a big challenge [32], and until now, there are not solid data proving the prognostic significance of EMT/SC(+) cells. The aim of the current study was to evaluate the prognostic significance of *TWIST1*, *CD24*, *CD44*, and *ALDH1* mRNA quantification in EpCAM-positive circulating tumor cells from early stage breast cancer patients with a long follow-up.

2. Materials and Methods

2.1. Cell Lines

The human mammary carcinoma cell line SKBR-3 was used as a positive control for the development of the quadraplex RT-qPCR assay for *CD24*, *CD44*, *ALDH1*, *HPRT*, while MDA-MB-231 cancer cell line was used as a positive control for the expression of *TWIST1* [33]. Cells were counted in a hemocytometer and their viability was assessed by trypan blue dye exclusion. cDNAs of all cancer cell lines were kept in aliquots at −20 °C and used for the analytical validation of the assay, prior to the analysis of patient's samples.

2.2. Patients

In total, 100 patients with non-metastatic breast cancer from the Medical Oncology Unit "Elena Venizelou" Hospital and IASO General hospital were enrolled in the study from September 2007 until January 2013. Peripheral blood (20 mL) was obtained from all these patients two weeks after the removal of the primary tumor and before the initiation of adjuvant chemotherapy. The chemotherapeutic adjuvant treatment for these patients has been previously reported [34]. The clinical characteristics for these patients at the time of diagnosis are shown in Supplementary Table S1. All patients signed an informed consent to participate in the study, which was approved by the Ethics and Scientific Committees of our Institutions. Peripheral blood (20 mL) was obtained from 19 healthy female blood donors (HD) and was analyzed in the same way as patients' samples (control group).

2.3. Isolation of EpCAM+ CTCs

To reduce blood contamination by epithelial cells from the skin, the first 5 mL of blood were discarded, and the blood collection tube was at the end disconnected before withdrawing the needle.

Peripheral blood (20 mL in EDTA) from (HD) and patients was collected and processed within 3 h in exactly the same manner. After collection, peripheral blood was diluted with 20 mL phosphate buffered saline (PBS, pH 7.3), and peripheral blood mononuclear cells (PBMCs) were isolated by gradient density centrifugation using Ficol-Paque TM PLUS (GE Healthcare, Bio-Sciences AB) at 670 g for 30 min at room temperature. The interface cells were removed and washed twice with 40 mL of sterile PBS (pH 7.3, 4 °C), at 530 g for 10 min. EpCAM+ cells were enriched using immunomagnetic Ber-EP4 coated capture beads (Dynabeads® Epithelial Enrich, Invitrogen, Carlsbad, CA, USA), according to the manufacturer's instructions [33].

2.4. RNA Extraction-cDNA Synthesis

Total RNA isolation was performed using TRIZOL-LS (ThermoFischer, Carlsbad, CA, USA). All RNA preparation and handling steps took place in a laminar flow hood under RNAse-free conditions. The isolated RNA from each fraction was dissolved in 20 μL of RNA storage buffer (Ambion, ThermoFischer, USA) and stored at −70 °C until use. RNA concentration was determined by absorbance readings at 260 nm using the Nanodrop-1000 spectrophotometer (NanoDrop, Technologies, Wilmington, DE, USA). mRNA was isolated from the total RNA using the Dynabeads mRNA Purification kit (ThermoFischer, USA), according to the manufacturer's instructions. cDNA synthesis was performed using the High capacity RNA-to-cDNA kit (ThermoFischer, USA) in a total volume of 20 μL, according to the manufacturer's instructions.

2.5. RT-qPCR

A novel quadraplex RT-qPCR assay was first developed for *CD24, CD44, ALDH1,* and *HPRT* (reference gene). Primers and dual hybridization probes were de novo in-silico designed, using Primer Premier 5.0 software (Premier Biosoft, Palo Alto, CA, USA). The specificity of all primer and hybridization probe sequences was first tested by homology searches in the nucleotide database (NCBI, nucleotide BLAST). Cross reaction between all oligonucleotide sequences was also examined. Each probe set included a 3′-fluorescein (F) donor probe and a 5′-LC acceptor probe that was different for each gene set: *CD24* (610 nm), *CD44* (640 nm), *ALDH1* (670 nm) and *HPRT* (705 nm). A color compensation test was performed by using pure dye spectra so that spectral overlap between dyes was corrected [35]. Quadraplex RT-qPCR reactions were performed in the LightCycler 2.0 (Roche, Mannheim, Germany). Component concentrations and the cycling conditions for the quadraplex RT-qPCR assay were optimized in detail. The amplification reaction mixture (10 μL) contained 1 μL of the PCR Synthesis Buffer (5X), 2.4 μL of $MgCl_2$ (25 mM), 0.2 μL dNTPs (10 mM), 0.8 μL BSA (10 μg/μL), 0.1 μL Hot Start DNA polymerase (HotStart, 5 U/μL, Promega, Madison, WI, USA), 0.5 μL of a mixture containing all eight primers (10 μM), 0.5 μL of a mixture containing all eight dual hybridization probes (3 μM), and H2O (added to the final volume). Cycling conditions of the *CD24, CD44, ALDH1, HPRT* quadraplex RT-qPCR assay were: 95 °C/2 min; 45 cycles of 95 °C/20 s, annealing at 59 °C/20 s, and extension at 72 °C/20 s. For the development and analytical validation of the novel quadraplex RT-qPCR assay, we generated individual PCR amplicons corresponding to the gene-targets studied that served as quantification calibrators, as we have previously described [33]. RT-qPCR for *TWIST1* was performed as previously described [33,36]. All data were evaluated in respect to *TWIST1, CD24, CD44,* and *ALDH1* expression by normalizing the EpCAM+ fraction of PBMCs to the expression of *HPRT* and the 2−ΔΔCt approach, as described in detail by Livak and Schmittgen [37]. A cut-off value was calculated as the mean of signals derived by samples of healthy individual analyzed in exactly the same way plus 2SD for *TWIST1, CD44,* and *ALDH1* transcripts and as the mean of signals derived by samples of healthy individual minus 2SD for *CD24*.

2.6. Statistical Analysis

Statistical analysis was performed using SPSS (SPSS Statistics 25.0, company, Armonk, NY, USA). The chi-square test of independence or Fisher exact test (SPSS, version 25.0) was used to make comparisons between groups. The DFI and OS rate were calculated by the Kaplan–Meier method and were evaluated by the log-rank test. Cox proportional hazards (PH) models were used to evaluate the relationship between EMT and Stem Cell status and event-time distributions, with tumor size, grade, number of involved lymph nodes, ER, PR, HER2, and age. Parametric and non-parametric tests were used to compare continuous variables between groups. All P-values are two-sided. A level of $p < 0.05$ is considered statistically significant.

3. Results

3.1. Analytical Validation of the Quadraplex RT-qPCR Assay for CD24, CD44, ALDH1, HPRT

The analytical specificity of the developed assay was checked by using all oligonucleotides in a common master mix in four different reactions in the presence of one individual gene target each time. Each primer pair and dual hybridization probe pair amplifies specifically only the corresponding target sequence and is detected only in the corresponding wavelength (Supplementary Figure S1A). The analytical sensitivity was determined for each individual gene target using a calibration curve. These calibration curves were generated using serial dilutions of individual gene-specific external standards in triplicate for each concentration, ranging from 10^5 copies/μL to 10 copies/μL. The analytical detection limit corresponded to 3 copies/μL while the quantification limit was equal to 9 copies/μL (Supplementary Figure S1B). The developed assay showed linearity over the entire quantification range and correlation coefficients greater than 0.99 in all cases, indicating a precise log-linear relationship. Intra and inter-assay variance: Repeatability or intra–assay variance of the quadraplex RT-qPCR was evaluated by repeatedly analyzing four cDNA samples corresponding to 1, 10, 100, and 1000 SKBR-3 cells in the same assay, in three parallel determinations. Reproducibility or interassay variance was evaluated by analyzing the same cDNA sample, representing 1000 SKBR-3 cells on five separate assays performed in five different days (Supplementary Table S2).

3.2. Quantification of CD24, CD44, ALDH1, and TWIST1 mRNA in the EpCAM(+) Fraction in Early Stage BrCa Patients and (HD)

In all EpCAM(+) fractions isolated from 100 early BrCa patient samples and 19 HD *CD24, CD44, ALDH1, HPRT* transcripts were quantified by the developed quadraplex RT-qPCR and *TWIST1* transcripts by the singleplex RT-qPCR assay (Figure 1). Median fold change of *TWIST1* expression in the EpCAM(+) fraction was 0.42 (range: 0–0.95) in HD and 10.06 (range: 2.33–3327) in $TWIST1^{high}$ (Mann-Whitney test, $Z = -1.363$, $p = 0.001$) and 0 (range: 0–0) in $\text{TWIST1}^{\text{low/}-}$ early BrCa patient samples (Mann-Whitney test, $Z = -3.634$, $p < 0.001$) (Figure 1A). Median fold change of *CD24* expression in the EpCAM(+) fraction was 2.00 (range: 1.42–3.81) in HD and 1.91 (range: 0.91–15.14) in $CD24^{\text{high}}$ (Mann-Whitney test, $Z = -0.492$, $p = 0.623$) and 0.62 (range: 0.29–0.88) in $CD24^{\text{low}}$ early BrCa patients (Mann-Whitney test, $Z = -5.577$, $p < 0.001$) (Figure 1B). Median fold change of *CD44* expression in the EpCAM(+) fraction was 0.71 (range: 0.14–1.06) in HD and 2.33 (range: 1.28–202.75) in $CD44^{high}$ (Mann-Whitney test, $Z = -6.084$, $p < 0.001$) and 0.61 (range: 0.01–1.17) in $CD44^{low}$ early BrCa patients (Mann-Whitney test, $Z = -1.084$, $p = 0.278$) (Figure 1C). Median fold change of *ALDH1* expression in the EpCAM(+) fraction was 1.32 (range: 0.69–2.19) in HD and 2.97(range: 2.30–14.72) in $ALDH1^{high}$ (Mann-Whitney test, $Z = -5.119$, $p < 0.001$) and 0.84 (range: 0.06–2.16) in $ALDH1^{low}$ early BrCa patients (Mann-Whitney test, $Z = -2.190$, $p = 0.029$) (Figure 1D).

In 19/100(19%) breast cancer samples tested, *TWIST1* was overexpressed, while in 15/100(15%) samples the $CD24^{-/low}/CD44^{high}$ profile, and in 9/100(9%) the $CD24^{-/low}/ALDH1^{high}$ profile was detected (Figure 2A). There was a positive correlation between *TWIST1* mRNA overexpression and the $CD24^{-/low}/CD44^{high}$ profile (Fisher's Exact Test; $p = 0.008$), while there was no correlation between *TWIST1* mRNA overexpression and the $CD24^{-/low}/ALDH1^{high}$ profile (Fisher's Exact Test; $p = 0.366$) (Table 1). *TWIST1* overexpression and $CD24^{-/low}/CD44^{high}$ and/or $CD24^{-/low}/ALDH1^{high}$ were detected in 7/100(7%) EpCAM(+) samples. The correlation between these characteristics and the clinical variables of the patients revealed an association between *TWIST1* overexpression with lymph node status (chi-square; $p = 0.036$) and HER2 status of the primary tumor (chi-square; $p = 0.006$) (Supplementary Table S1).

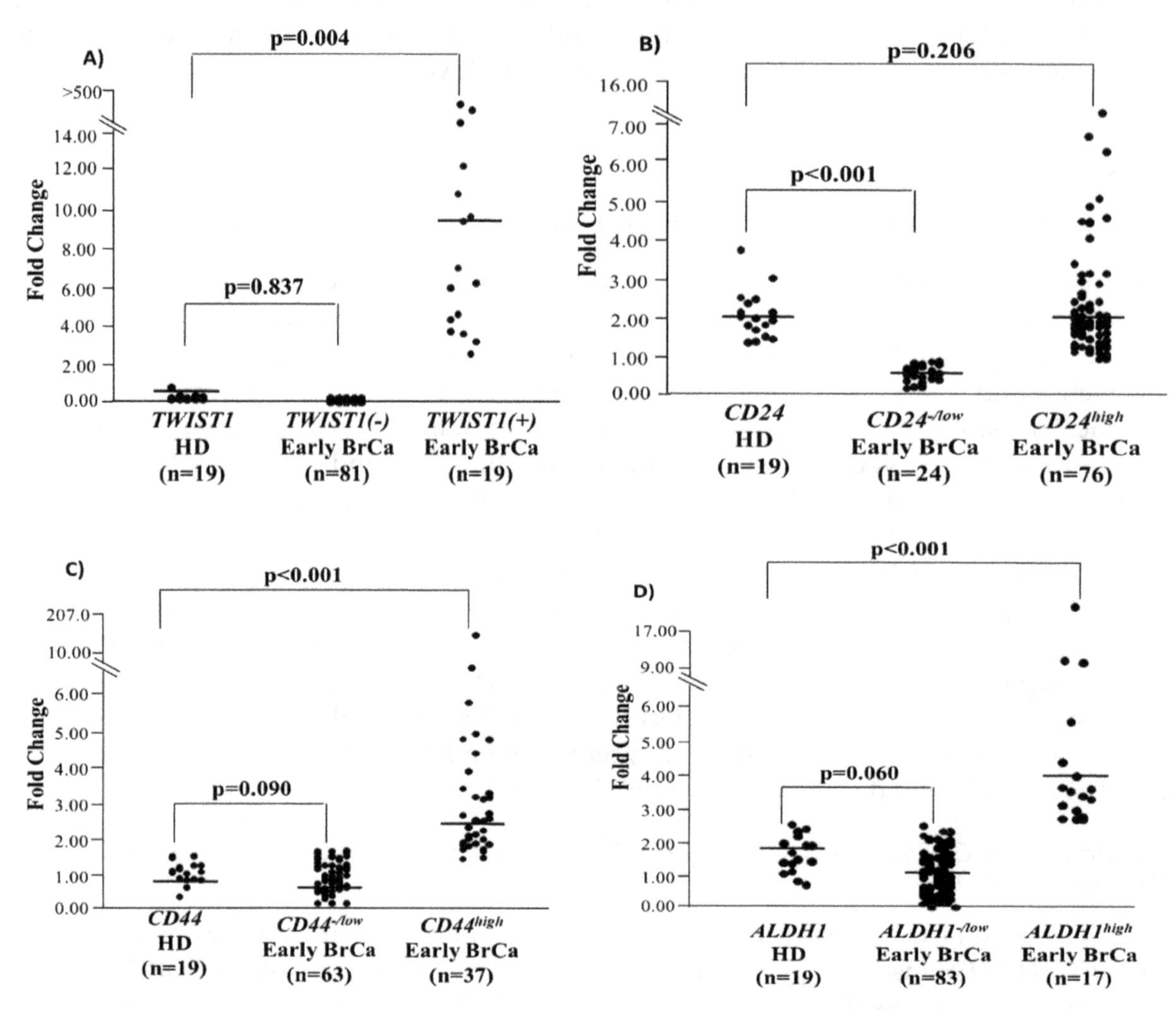

Figure 1. Relative fold change values ($2^{-\Delta\Delta Ct}$) in respect to HPRT expression for: (**A**) *TWIST1* (**B**) *CD24*, (**C**) *CD44*, (**D**) *ALDH1* for early breast cancer patients ($n = 100$) and (HD), ($n = 19$).

3.3. Evaluation of Prognostic Significance

3.3.1. Disease Free Interval

During the follow up period (median: 95 months; range: 4–137 months), 25/100 (25%) patients relapsed and in 9/25 (36%) of them *TWIST1* overexpression was detected in the EpCAM+ CTC fraction (Fisher's Exact Test; $p = 0.019$). Similarly, 6/25 (24%) patients displayed a Stem Cell profile in EpCAM+ CTC fraction (Fisher's Exact Test; $p = 0.194$). In 4/25 (16%) of these patients, both *TWIST1* overexpression and the Stem Cell profile was detected (Fisher's Exact Test; $p = 0.063$) (Supplementary Table S3). The Kaplan–Meier estimates of the cumulative DFI of the patients overexpressing *TWIST1* revealed that these patients had worse survival compared to patients who were negative (83.6mo vs 115.8mo respectively; $p = 0.019$) (Table 2, Figure 3A). However, the stem cell profile alone (86.7mo. vs 113.2mo, respectively in the two groups; log rank test; $p = 0.174$) (Table 2, Supplementary Figure S2A) and both stem cell and mesenchymal characteristics (68.9mo vs 88.8–115.8mo, respectively; $p = 0.087$) (Table 2, Supplementary Figure S2C) failed to show any statistically significant difference even though the mean survival showed a reduced trend. Kaplan–Meier survival analysis of patients with positive axillary lymph nodes and *TWIST1* mRNA overexpression had worst DFI (Table 2, Supplementary Figure S3A) (82.6 mo. vs 88.7–123.3; $p = 0.05$). When all patients were divided into two groups based on the number of positive lymph nodes (1–3, and ≥4 positive nodes) and the overexpression of *TWIST1* [($N_{2\text{-}3}$/*TWIST1*(+), $N_{2\text{-}3}$/*TWIST1*(−), N_1/*TWIST1*(+), and N_1/*TWIST1*(−)], Kaplan–Meier analysis revealed that women harboring more than 3 positive lymph nodes and *TWIST1* that was overexpressed in EpCAM+ CTC fraction had a statistically significant shorter DFI (Table 2, Figure 3C) (mean survival: 68.6mo vs. 103.0–114.3mo.; $p = 0.007$). When patients were dichotomized accordingly to the HR status (ER/PR) in the following groups: a) HR(−)/*TWIST1*(+), b) HR(−)/ *TWIST1*(−), c) HR(+)/*TWIST1*(+) and d) HR(+)/*TWIST1*(−), it was observed that women with HR(−)/*TWIST1*(+) profile were characterized by statistically significant shorter DFI (36mo. vs 102.3–117.9mo.; $p < 0.001$; Figure 3E). A Univariate analysis (Table 3) also revealed the significance of (a) *TWIST1*(+), (b) HR(−)/*TWIST1*(+), (c) *TWIST1*(+) /$N_{2\text{-}3}$, d) SC (+)/ *TWIST1*(+) (Figure 2B) in the risk of disease progression. Multivariate confirmed the prognostic value of HR(−)/*TWIST1*(+) and *TWIST1*(+)/N2-3, in the EpCAM(+) CTC fraction for the prediction of DFI (Table 3) independently from patients' age, tumor T stage, grade, nodal status alone and the HR, and HER2 status of the primary tumor.

Table 1. Correlation between *TWIST1* and *CD44*^{high}/*CD24*$^{-/low}$ and *ALDH1*^{high}/*CD24*$^{-/low}$ expression in early breast cancer EpCAM positive samples ($n = 100$).

	CD44^{high}**/*CD24***$^{-/low}$		p^a	***ALDH1***^{high}**/*CD24***$^{-/low}$		p^a
TWIST1	**Positive**	**Negative**		**Positive**	**Negative**	
Positive	7 (46.7%)	12(14.1%)	**0.008**	3 (33.3)	16(17.6%)	0.366
Negative	8 (53.3%)	73(85.9%)		6 (66.7%)	75 (82.4%)	
Concordance	80/100 (80%)			78/100 (78%)		

[a] Fischer's Exact Test. Bold: highlights the significance of the test.

3.3.2. Overall Survival

Among the 25 patients that relapsed during the follow up period, 14/25 (56.0%) patients died and 11/25 (44.0%) were still alive at the time of the last follow-up. In 6/14 (42.9%) patients that died *TWIST1* overexpression was detected in the EpCAM+ fraction (Fisher's Exact Test; $p = 0.024$). Similarly, 4/14 (28.6%) patients displayed a Stem Cell profile in EpCAM+ CTC fraction (Fisher's Exact Test; $p = 0.217$). In 3/14 (21.4%) of these patients, both *TWIST1* overexpression and *CD24*$^{-/low}$/*CD44*^{high} and/or *CD24*$^{-/low}$/*ALDH1*^{high} profiles (Fisher's Exact Test; $p = 0.055$) were detected (Fisher's Exact Test; $p = 0.055$) (Supplementary Table S3). The Kaplan–Meier estimates of the overall survival (OS) of the patients overexpressing *TWIST1* were significantly different in favor of patients who were negative for *TWIST1* overexpression (106.4 vs 127.2 mo; $p = 0.046$) (Table 2, Figure 3B). Stem Cell profiles (107.3 vs 125.2 mo.; $p = 0.171$) (Table 2, Supplementary Figure S2B) and the co-expression of EMT and SC-associated genes (96.29 vs 109.1–127.3 mo.; $p = 0.118$) (Table 2, Supplementary Figure S2D) failed to show any statistically significant difference. There was no difference in OS in patients with *TWIST1* overexpression according to N0 and N+ lymph node involvement (108.8 mo vs 92–129 mo, respectively; $p = 0.194$; Supplementary Figure S3B). However, when the Kaplan–Meier curves for OS for *TWIST1* overexpression were additionally stratified according to lymph nodes status (Table 2, Figure 3D) and HR status (Table 2, Figure 3F) our data have shown that patients with >3 LN and *TWIST1* overexpression had lower OS (109.8 mo., range: 115–129 mo.; $p = 0.026$); the same was seen for patients that were HR(−) and *TWIST1* was overexpressed (65.7 vs 110.2–131.9 mo.; $p < 0.001$). Univariate analysis showed a significantly higher risk of death in the group of patients positive for *TWIST1* overexpression that had more than 3 lymph nodes affected or co-expressed the stem cell profile (Figure 2B). Multivariate analysis confirmed the prognostic value of *TWIST1* overexpression in combination with $N_{2\text{-}3}$, and in combination with HR(−) status in the EpCAM(+) CTC fraction for the prediction of OS, independently from patients' age, tumor T stage, grade, nodal status, and the status of the receptors ER, PR, HER2 of the primary site (Table 3).

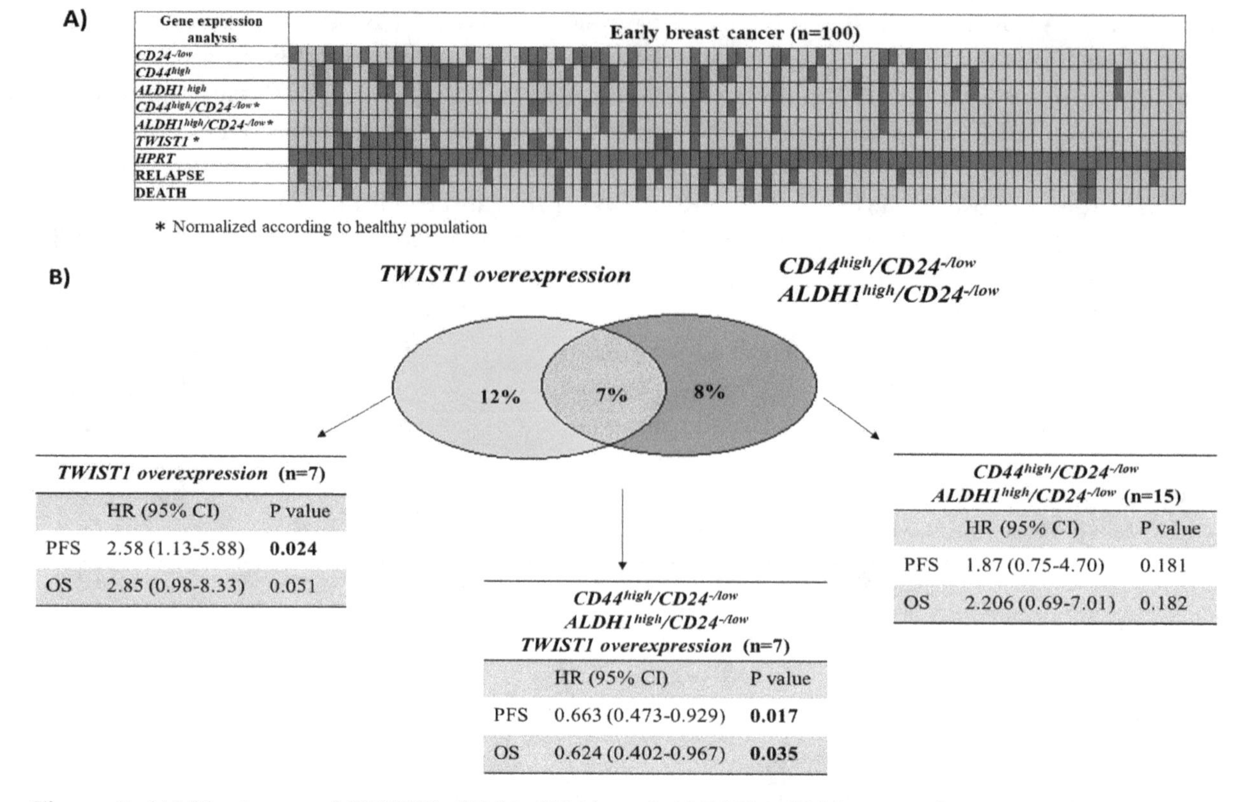

Figure 2. (**A**) Heat map of *TWIST1, CD24, CD44,* and *ALDH1*-mRNA quantification in the EpCAM+ CTC fraction from early stage breast cancer patients (n = 100). Red color represents overexpression, while green color indicates underexpression or lack of expression. Concerning the relapse or death, red color represents the relapse or death, while green color indicates no relapse or alive status. (**B**) Univariate Cox-regression hazard models for TWIST1 overexpression, $CD44^{high}/CD24^{-/low}$, and $ALDH1^{high}/CD24{-}^{/low}$ and the co-expression of the mesenchymal profile, *TWIST1*, and the stem cell profile, $CD44^{high}/CD24^{-/low}$, and $ALDH1^{high}/CD24^{-/low}$.

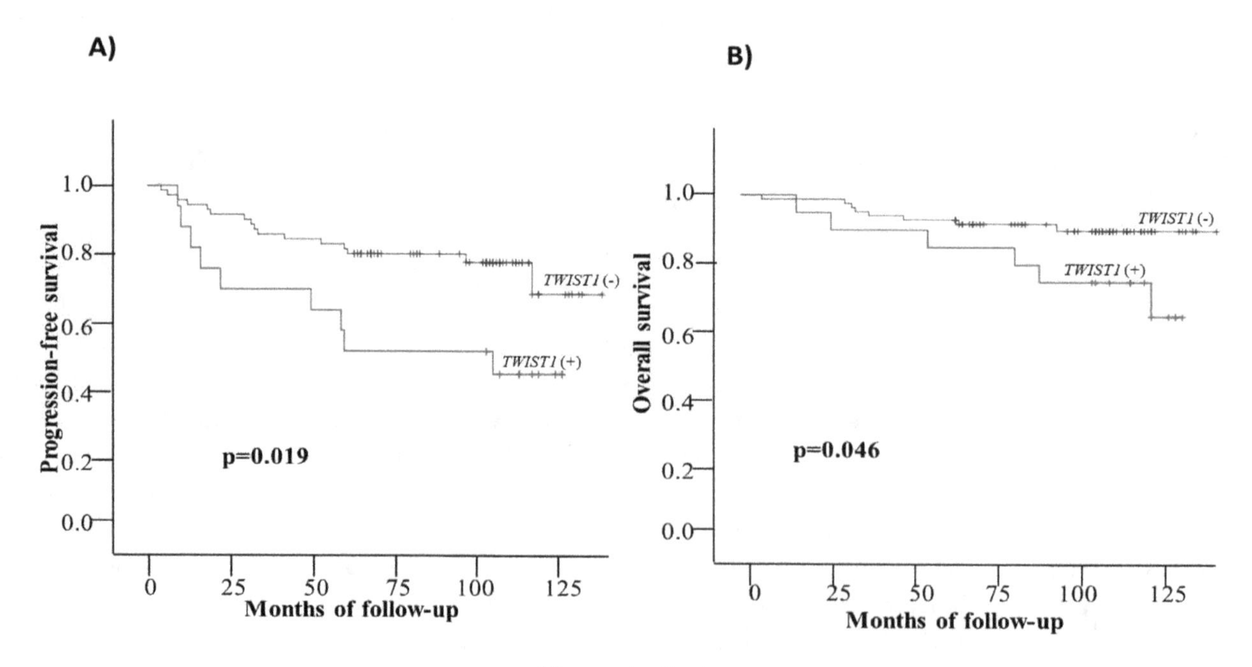

Figure 3. *Cont.*

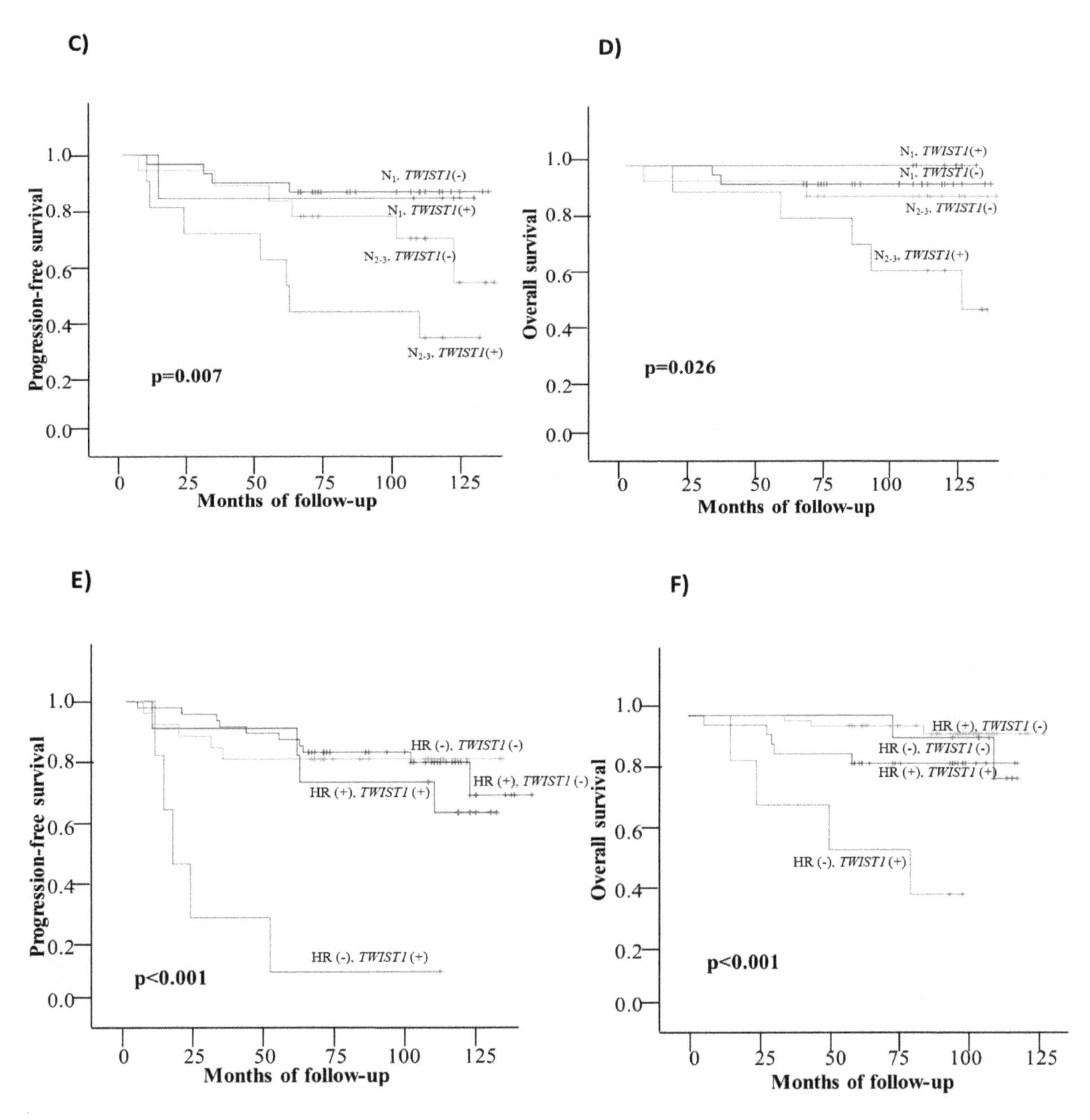

Figure 3. Kaplan–Meier estimates for early BrCa patients: (**A**) DFI: *TWIST1* overexpression, (**B**) OS: *TWIST1* overexpression, (**C**) DFI: *TWIST1* overexpression and number of affected lymph nodes, (**D**) OS: *TWIST1* overexpression and number of affected lymph nodes, (**E**) DFI: *TWIST1* overexpression and HR status, (**F**) OS: *TWIST1* overexpression and HR status.

Table 2. Gene expression in CTCs in respect to DFI and OS.

Gene Expression in CTCs	DFI				OS			
Gene	Mean Survival	95% CI (months)	Range (months)	*p*	Mean Survival	95% CI (Months)	Range (Months)	*p*
TWIST1+	83.6	61.9–105.3	9–125	**0.019**	106.4	90.3–122.3	16–127	**0.046**
TWIST1−	115	106.3–125.2	4–137		127.2	120.7–133.7	6–137	
Stem cell profile positive (SC+)	86.7	66.7–106.8	16–118	0.174	107.3	89.8–124.8	26–127	0.171
Stem cell profile negative (SC−)	113.2	103.5–122.8	4–137		125.2	118.3–132.1	6–137	
TWIST1+/SC+	68.9	39.4–98.31	16–112	0.087	96.2	67.4–125.1	26–127	0.118
TWIST1+/SC−	88.8	61.2–116.4	9–125		111.1	93.3–128	16–125	
TWIST1−/SC+	100.1	78.6–121.7	41–118		109.1	92.8–125.3	47–118	
TWIST1−/SC−	115.8	105.8–126	4–137		127.3	120.4–134.2	6–137	
TWIST1+/LN+	82.6	59.3–105.9	9–125	0.05	108.8	93–124.6	16–127	0.194
TWIST1+/LN−	88.7	30.5–146.8	16–125		92	39.1–144.8	26–125	
TWIST1−/LN+	110.3	99.2–121.4	6–130		121.3	113–129.6	6–130	
TWIST1−/LN−	123.3	110.6–135.9	4–137		128.9	120.2–137.7	37–137	
TWIST1+/N_{2-3}	68.6	40.5–96.7	9–125	**0.007**	98.3	75.5–121.2	16–127	**0.026**
TWIST1+/N_1	104.5	71.8–137.5	13–123		112	94–125	101–118	
TWIST1−/N_{2-3}	103.1	84.1–121.9	6–130		118.8	103.4–134.1	6–130	
TWIST1−/N_1	114.4	101.7–126.9	9–128		121.1	111.9–130.3	30–128	
TWIST1+/HR+	102.3	81.2–123.4	9–125	**<0.001**	121.6	113.9–129.3	79–127	**<0.001**
TWIST1+/HR−	36	8.9–63	10–106		65.7	36.7–94.6	76–106	
TWIST1−/HR+	117.9	107.1–128.7	4–137		131.9	126.2–137.5	37–137	
TWIST1−/HR−	107.7	92.3–123.1	6–127		110.2	96.5–123.84	6–127	

Bold: highlights the significance of the test.

Table 3. Univariate and multivariate analyses for DFI and OS in the early breast cancer patients group (n = 100).

Covariates	Covariate Value	DFI						OS					
		Univariate Cox Regression Analysis			Multivariate Cox Regression Analysis			Univariate Cox Regression Analysis			Multivariate Cox Regression Analysis		
		HR [a]	95% CI [b]	p	HR [a]	95% CI [b]	p	HR [a]	95% CI [b]	p	HR [a]	95% CI [b]	p
Age	≥50 vs <50	0.787	0.357–1.734	0.552	0.432	0.169–1.103	0.079	0.718	0.249–2.070	0.539	0.593	0.169–2.080	0.414
ER	Yes vs No	0.647	0.286–1.463	0.295	4.391	1.040–18.549	**0.044**	0.238	0.079–0.721	**0.011**	0.623	0.092–4.215	0.628
PR	Yes vs No	0.492	0.217–1.114	0.089	0.087	0.021–0.362	**0.001**	0.196	0.054–0.707	**0.013**	0.098	0.011–0.851	**0.035**
HER2	Yes vs No	0.500	0.197–1.269	0.145	0.626	0.232–1.693	0.357	0.381	0.106–1.366	0.139	0.247	0.060–1.023	0.054
Lymph nodes	N_0 vs N_1 vs $N_{2\text{-}3}$	2.207	1.261–3.861	**0.006**	2.433	1.272–4.654	**0.007**	1.659	0.817–3.371	0.162	1.351	0.637–2.862	0.433
Size	≥2cm vs <2cm	3.060	1.049–8.922	**0.041**	4.926	1.225–19.811	**0.025**	7.244	0.947–55.432	0.056	17.450	1.464–208.1	**0.024**
Grade	I/II vs III	1.366	0.570–3.273	0.485	0.753	0.228–2.483	0.641	1.953	0.544–7.008	0.305	0.286	0.040–2.028	0.211
TWIST1	Yes vs No	2.582	1.135–5.875	**0.024**	1.382	0.490–3.899	0.540	2.851	0.975–8.33	0.051	1.481	0.382–5.743	0.570
HR and *TWIST1* status	HR+*TWIST1*+ HR+*TWIST1*- HR-*TWIST1*+ HR-*TWIST1*-	0.486	0.302–0.784	**0.003**	0.597	0.360–0.991	**0.046** [c]	0.576	0.313–1.062	0.077	0.666	0.345–1.284	0.225 [c]
LN and *TWIST1* status	$N_{0\text{-}1}$*TWIST1*+ $N_{0\text{-}1}$*TWIST1*- $N_{2\text{-}3}$*TWIST1*+ $N_{2\text{-}3}$*TWIST1*-	0.540	0.383–0.762	**<0.001**	0.542	0.371–0.792	**0.002** [d]	0.559	0.357–0.875	**0.011**	0.604	0.373–0.976	**0.040** [d]
Stem cell profile	Yes vs No	1.873	0.746–4.703	0.181	1.755	0.526–5.855	0.360	2.206	0.690–7.050	0.182	3.689	0.806–16.884	0.093
Stem cell profile/*TWIST1*+	Yes vs No	0.663	0.473–0.929	**0.017**	0.776	0.521–1.146	0.202 [e]	0.624	0.402–0.967	**0.035**	0.634	0.378–1.065	0.085 [e]

[a] Hazard ratio, estimated from Cox proportional hazard regression mode. [b] Confidence interval of the estimated HR. Results are based on 1000 bootstrap samples and obtained after the Bias corrected and accelerated (BCa) approach. [c] Multivariate model adjusted for age, HER2, LN, Size, Grade, Stem cell. [d] Multivariate model adjusted for age, ER, PR, HER2, Size, Grade, Stem cell. [e] Multivariate model adjusted for age, ER, PR, HER2, LN, Size, Grade. Bold: highlights the significance of the test.

4. Discussion

Molecular characterization of CTCs at the gene expression level has a strong potential to provide novel prognostic and predictive biomarkers. It is now clear through numerous studies that CTCs isolated from breast cancer patients express epithelial markers [38], receptors (ER, PR, HER2, EGFR), stem cell markers [39], and mesenchymal markers [11]. So far, most studies have been performed in the metastatic setting where the number of circulating tumor cells is usually high. However, in the non-metastatic setting of breast cancer, CTCs are not always detected and their numbers are usually very low, thus their molecular characterization is extremely difficult. For this reason, in the early breast cancer setting, a higher volume of peripheral blood used for the analysis of CTCs is very critical. Our group has shown many years ago the prognostic significance of *CK-19* mRNA detection in peripheral blood of early breast cancer patients, using 20 mL of peripheral blood for CTC isolation and further downstream analysis [36,38]. Other groups have also shown that the detection of CTCs in the early breast cancer setting is providing critical prognostic information for these patients [29].

In this study we evaluated for the first time the prognostic significance of *TWIST1, CD24, CD44,* and *ALDH1* transcript quantification in EpCAM-positive circulating tumor cells isolated from peripheral blood of early stage breast cancer patients. We selected *TWIST1* as this is a very established EMT marker; for this reason, we have developed already in 2011 an RT-qPCR assay for the absolute quantification of *TWIST1*-mRNA expression, and we have validated this assay in EpCAM-positive cells isolated by early and metastatic breast cancer patients [33]. Concerning the selection of stem cell markers, this was based on publications by Al-Hajj M et.al. [8] and Ginestier C et.al. [9], who have shown that the breast cancer stem cell phenotypes of (a) $CD44^{+}/CD24^{-/low}$ phenotype and (b) the overexpression of aldehyde dehydrogenase 1 (ALDH1+) are able to form tumors in mice with high tumorigenic capacity.

Multiplex RT-qPCR assays have many benefits due to their wide dynamic range, the low sample volume required, and the reduced time of analysis [35]. Our study was based on an analytically validated novel multiplex assay for the quantification of *CD24, CD44, ALDH1,* and *HPRT* and a single RT-qPCR assay for the quantification of *TWIST1* transcripts. The analytical sensitivity and specificity of the novel quadraplexRT-qPCR assay for the simultaneous detection of *CD24, CD44, ALDH-1,* and *HPRT* transcripts were determined by using calibrators specific for each gene target. Both these assays were validated according to the Minimum Information for Publication of Quantitative Real-Time PCR Experiments (MIQE) guidelines [40].

Relevant prognostic and predictive markers in early breast cancer cohort is of major significance. The SUCCESS A trial has shown that the presence of CTCs, as evaluated in 30 mL of peripheral blood, two years after chemotherapy has been associated with decreased OS and DFS in high-risk early breast cancer patients [41]. Lucci et.al. has also shown that the presence of one or more circulating tumor cells could predict early recurrence and decreased overall survival in chemonaive patients with non-metastatic breast cancer [29]; however, the main limitation of this study is that it was based on CTC enumeration performed in only 7.5 mL of blood. Additionally, molecular characterization of CTC could identify CTC biomarkers that are associated to specific signaling pathways like EMT or CSC. Our findings demonstrate a positive correlation between *TWIST1* overexpression and the $CD24^{-/low}/CD44^{high}$ profile in the EpCAM positive CTC fraction. This is in agreement with previous findings showing that the mesenchymal-like breast cancer stem cells are characterized as $CD24^{-}/CD44^{+}$, while the epithelial-like breast cancer stem cells express high levels of aldehyde dehydrogenase (ALDH) [16]. Univariate analysis revealed a significantly higher risk of relapse and death in the group of patients that expressed both stem cell and mesenchymal characteristics. Mego et al. have shown that patients with *TWIST1*-high tumors had a significantly higher percentage of breast cancer stem cells than patients with *TWIST1*-low tumors [19]. Recently, it was shown that in CTC of NSCLC patients the CD44(+)/CD24(−) population possess epithelial–mesenchymal transition characteristics [42], while another study in metastatic colorectal cancer has shown the prognostic significance of CTC that express both EMT and stem-like genes [22]. At the single CTC-level, Papadaki et.al have shown that CTCs expressing high levels of ALDH1 along with nuclear TWIST expression are more frequently detected in patients with metastatic breast cancer [21] and that these cells represent a chemo-resistant subpopulation with an unfavorable outcome [23]. The main limitation of our study is that we examined the expression of only one EMT marker, *TWIST1* in the EpCAM+ cells of early breast cancer patients. Since there is a high heterogeneity in CTC, it may be possible that we have not detected CTCs that express other mesenchymal markers like Vimentin or Snail. We plan to extend this study by adding more gene expression markers in a biggest sample cohort and correlate our results to the clinical outcome.

According to our results, patients with *TWIST1* overexpression in the EpCAM+ CTCs fraction and more than 3 involved lymph nodes had a significant lower DFI and OS. Similar to our results, recently, Emprou C et al. have shown that in frozen NSCLC tumor samples *TWIST1* is more frequently overexpressed in the N+ group compared to the N0 group showing that partial EMT is involved in lymph node progression in early stages of NSCLC [43], while in primary breast cancer loss of E-cadherin is correlated with more than 3 LN involved in 80% of the patients [44]. Our results also indicate that patients with *TWIST1* overexpression in the EpCAM+ CTCs fraction and a hormone receptor-negative primary tumor had a worse prognosis both for DFI and OS. This is in accordance with previous findings that have shown that the estrogen receptor silencing induces epithelial to mesenchymal transition in breast cancer [45]. It has also been previously shown that in human breast tumors there is an inverse relationship between *TWIST1* and ER expression that may possibly contribute to the generation of hormone-resistant, ER-negative breast cancer [46]. It has also been reported that EMT likely occurs in the basal-like phenotype both in MCF10A cells [47] and in invasive breast cancer carcinomas [48].

5. Conclusions

In conclusion, detection of *TWIST1* overexpression and stem-cell (*CD24, CD44, ALDH1*) transcripts in the $EpCAM^+$ CTC fraction provides prognostic information in early stage breast cancer patients. Overexpression of *TWIST1* in the $EpCAM^+$ CTC fraction in the group of HR negative patients or with >3 positive lymph nodes is associated with worse prognosis.

Supplementary Materials:
Figure S1: Analytical validation of the multiplex RT-qPCR for *CD24, CD44, ALDH1, HPRT* (all measured in triplicate) (A) analytical specificity, (B) RT-qPCR calibration curves (copies/μL). Figure S2: Kaplan–Meier estimates for early BrCa patients with or without the stem cell profile in respect to (A) DFI, (B) OS and with or without co-expression of *TWIST1* and Stem Cell profile in respect to (C) DFI, (D) OS. Figure S3: Kaplan–Meier estimates for early BrCa patients with positive or negative axillary lymph nodes the (A) DFI and (B) OS. Table S1: Clinical characteristics of the patients with early breast cancer ($n = 100$). Table S2: Quadraplex RT-qPCR for *CD24, CD44, ALDH1, HPRT*, evaluation of intra- ($n = 3$) and inter-assay ($n = 5$) precision. Table S3: Correlation of *TWIST1*, $CD44^{high}/CD24^{-/low}$ and/or $ALDH1^{high}/CD24^{-/low}$ and the co-expression of *TWIST1* and $CD44^{high}/CD24^{-/low}$ and/or $ALDH1^{high}/CD24^{-/low}$ with the patients' clinical outcomes.

Author Contributions: Conceptualization, A.S. and E.S.L.; methodology, A.S.; validation, A.S.; formal analysis, A.S.; investigation, A.S. and E.S.L.; resources, M.N. and V.G.; data curation, A.S.; writing—original draft preparation, A.S.; writing—review and editing, E.S.L. and V.G.; supervision, E.S.L.

Abbreviations

EMT	epithelial–mesenchymal transition
MET	mesenchymal-epithelial transition
MICs	metastasis-initiating cells
SC	Stem Cell
HR	Hormone Receptor
BCSC	breast cancer stem cells
CSCs	Cancer stem cells
CTCs	Circulating tumor cells

ER	estrogen receptor
PR	progesterone receptor
HER2	human epidermal growth factor receptor 2
ALDH1	aldehyde dehydrogenase 1
HD	Healthy Donors
RT-qPCR	Quantitative Reverse Transcription Polymerase Chain Reaction
PBMCs	peripheral blood mononuclear cells
PBS	phosphate-buffered saline
LN	lymph nodes

References

1. Lianidou, E.S. Gene expression profiling and DNA methylation analyses of CTCs. *Mol. Oncol.* **2016**, *10*, 431–442. [CrossRef] [PubMed]
2. Bardelli, A.; Pantel, K. Liquid Biopsies, What We Do Not Know (Yet). *Cancer Cell* **2017**, *31*, 172–179. [CrossRef] [PubMed]
3. Boral, D.; Vishnoi, M.; Liu, H.N.; Yin, W.; Sprouse, M.L.; Scamardo, A.; Hong, D.S.; Tan, T.Z.; Thiery, J.P.; Chang, J.C.; et al. Molecular characterization of breast cancer CTCs associated with brain metastasis. *Nat. Commun.* **2017**, *8*, 196. [CrossRef] [PubMed]
4. Jakabova, A.; Bielcikova, Z.; Pospisilova, E.; Matkowski, R.; Szynglarewicz, B.; Staszek-Szewczyk, U.; Zemanova, M.; Petruzelka, L.; Eliasova, P.; Kolostova, K.; et al. Molecular characterization and heterogeneity of circulating tumor cells in breast cancer. *Breast Cancer Res. Treat.* **2017**, *166*, 695–700. [CrossRef] [PubMed]
5. Mansoori, M.; Madjd, Z.; Janani, L.; Rasti, A. Circulating cancer stem cell markers in breast carcinomas: A systematic review protocol. *Syst. Rev.* **2017**, *6*, 262. [CrossRef] [PubMed]
6. Reya, T.; Morrison, S.J.; Clarke, M.F.; Weissman, I.L. Stem cells, cancer, and cancer stem cells. *Nature* **2001**, *414*, 105–111. [CrossRef]
7. Luo, M.; Clouthier, S.G.; Deol, Y.; Liu, S.; Nagrath, S.; Azizi, E.; Wicha, M.S. Breast cancer stem cells: Current advances and clinical implications. *Methods Mol. Biol.* **2015**, *1293*, 1–49.
8. Al-Hajj, M.; Wicha, M.S.; Benito-Hernandez, A.; Morrison, S.J.; Clarke, M.F. Prospective identification of tumorigenic breast cancer cells. *Proc. Natl. Acad. Sci. USA* **2003**, *100*, 3983–3988. [CrossRef]
9. Ginestier, C.; Hur, M.H.; Charafe-Jauffret, E.; Monville, F.; Dutcher, J.; Brown, M.; Jacquemier, J.; Viens, P.; Kleer, C.G.; Liu, S.; et al. ALDH1 is a marker of normal and malignant human mammary stem cells and a predictor of poor clinical outcome. *Cell Stem Cell* **2007**, *1*, 555–567. [CrossRef]
10. Balic, M.; Lin, H.; Young, L.; Hawes, D.; Giuliano, A.; McNamara, G.; Datar, R.H.; Cote, R.J. Most early disseminated cancer cells detected in bone marrow of breast cancer patients have a putative breast cancer stem cell phenotype. *Clin. Cancer Res.* **2006**, *12*, 5615–5621. [CrossRef]
11. Theodoropoulos, P.A.; Polioudaki, H.; Agelaki, S.; Kallergi, G.; Saridaki, Z.; Mavroudis, D.; Georgoulias, V. Circulating tumor cells with a putative stem cell phenotype in peripheral blood of patients with breast cancer. *Cancer Lett.* **2010**, *288*, 99–106. [CrossRef] [PubMed]
12. Bredemeier, M.; Edimiris, P.; Tewes, M.; Mach, P.; Aktas, B.; Schellbach, D.; Wagner, J.; Kimmig, R.; Kasimir-Bauer, S. Establishment of a multimarker qPCR panel for the molecular characterization of circulating tumor cells in blood samples of metastatic breast cancer patients during the course of palliative treatment. *Oncotarget* **2016**, *7*, 41677–41690. [CrossRef] [PubMed]
13. Baccelli, I.; Schneeweiss, A.; Riethdorf, S.; Stenzinger, A.; Schillert, A.; Vogel, V.; Klein, C.; Saini, M.; Bäuerle, T.; Wallwiener, M.; et al. Identification of a population of blood circulating tumor cells from breast cancer patients that initiates metastasis in a xenograft assay. *Nat. Biotechnol.* **2013**, *31*, 539–544. [CrossRef] [PubMed]
14. Schölch, S.; García, S.A.; Iwata, N.; Niemietz, T.; Betzler, A.M.; Nanduri, L.K.; Bork, U.; Kahlert, C.; Thepkaysone, M.-L.; Swiersy, A.; et al. Circulating tumor cells exhibit stem cell characteristics in an orthotopic mouse model of colorectal cancer. *Oncotarget* **2016**, *7*, 27232–27242. [CrossRef] [PubMed]
15. Wang, L.; Li, Y.; Xu, J.; Zhang, A.; Wang, X.; Tang, R.; Zhang, X.; Yin, H.; Liu, M.; Wang, D.D.; et al. Quantified postsurgical small cell size CTCs and EpCAM+ circulating tumor stem cells with cytogenetic abnormalities in hepatocellular carcinoma patients determine cancer relapse. *Cancer Lett.* **2018**, *412*, 99–107. [CrossRef] [PubMed]

16. Liu, S.; Cong, Y.; Wang, D.; Sun, Y.; Deng, L.; Liu, Y.; Martin-Trevino, R.; Shang, L.; McDermott, S.P.; Landis, M.D.; et al. Breast cancer stem cells transition between epithelial and mesenchymal states reflective of their normal counterparts. *Stem Cell Rep.* **2014**, *2*, 78–91. [CrossRef] [PubMed]
17. Aktas, B.; Tewes, M.; Fehm, T.; Hauch, S.; Kimmig, R.; Kasimir-Bauer, S. Stem cell and epithelial-mesenchymal transition markers are frequently overexpressed in circulating tumor cells of metastatic breast cancer patients. *Breast Cancer Res.* **2009**, *11*, R46. [CrossRef]
18. Giordano, A.; Gao, H.; Anfossi, S.; Cohen, E.; Mego, M.; Lee, B.-N.; Tin, S.; De Laurentiis, M.; Parker, C.A.; Alvarez, R.H.; et al. Epithelial-mesenchymal transition and stem cell markers in patients with HER2-positive metastatic breast cancer. *Mol. Cancer* **2012**, *11*, 2526–2534. [CrossRef]
19. Mego, M.; Gao, H.; Lee, B.-N.; Cohen, E.N.; Tin, S.; Giordano, A.; Wu, Q.; Liu, P.; Nieto, Y.; Champlin, R.E.; et al. Prognostic Value of EMT-Circulating Tumor Cells in Metastatic Breast Cancer Patients Undergoing High-Dose Chemotherapy with Autologous Hematopoietic Stem Cell Transplantation. *J. Cancer* **2012**, *3*, 369–380. [CrossRef]
20. Zhang, L.; Ridgway, L.D.; Wetzel, M.D.; Ngo, J.; Yin, W.; Kumar, D.; Goodman, J.C.; Groves, M.D.; Marchetti, D. The identification and characterization of breast cancer CTCs competent for brain metastasis. *Sci. Transl. Med.* **2013**, *5*, 180ra48. [CrossRef]
21. Papadaki, M.A.; Kallergi, G.; Zafeiriou, Z.; Manouras, L.; Theodoropoulos, P.A.; Mavroudis, D.; Georgoulias, V.; Agelaki, S. Co-expression of putative stemness and epithelial-to-mesenchymal transition markers on single circulating tumour cells from patients with early and metastatic breast cancer. *BMC Cancer* **2014**, *14*, 651. [CrossRef] [PubMed]
22. Ning, Y.; Zhang, W.; Hanna, D.L.; Yang, D.; Okazaki, S.; Berger, M.D.; Miyamoto, Y.; Suenaga, M.; Schirripa, M.; El-Khoueiry, A.; et al. Clinical relevance of EMT and stem-like gene expression in circulating tumor cells of metastatic colorectal cancer patients. *Pharm. J.* **2018**, *18*, 29–34. [CrossRef] [PubMed]
23. Papadaki, M.A.; Stoupis, G.; Theodoropoulos, P.A.; Mavroudis, D.; Georgoulias, V.; Agelaki, S. Circulating Tumor Cells with Stemness and Epithelial-to-Mesenchymal Transition Features Are Chemoresistant and Predictive of Poor Outcome in Metastatic Breast Cancer. *Mol. Cancer* **2019**, *18*, 437–447. [CrossRef] [PubMed]
24. Stathopoulou, A.; Vlachonikolis, I.; Mavroudis, D.; Perraki, M.; Kouroussis, C.; Apostolaki, S.; Malamos, N.; Kakolyris, S.; Kotsakis, A.; Xenidis, N.; et al. Molecular Detection of Cytokeratin-19–Positive Cells in the Peripheral Blood of Patients With Operable Breast Cancer: Evaluation of Their Prognostic Significance. *J. Clin. Oncol.* **2002**, *20*, 3404–3412. [CrossRef] [PubMed]
25. Xenidis, N.; Markos, V.; Apostolaki, S.; Perraki, M.; Pallis, A.; Sfakiotaki, G.; Papadatos-Pastos, D.; Kalmanti, L.; Kafousi, M.; Stathopoulos, E.; et al. Clinical relevance of circulating CK-19 mRNA-positive cells detected during the adjuvant tamoxifen treatment in patients with early breast cancer. *Ann. Oncol.* **2007**, *18*, 1623–1631. [CrossRef] [PubMed]
26. Xenidis, N.; Ignatiadis, M.; Apostolaki, S.; Perraki, M.; Kalbakis, K.; Agelaki, S.; Stathopoulos, E.N.; Chlouverakis, G.; Lianidou, E.; Kakolyris, S.; et al. Cytokeratin-19 mRNA-Positive Circulating Tumor Cells After Adjuvant Chemotherapy in Patients With Early Breast Cancer. *J. Clin. Oncol.* **2009**, *27*, 2177–2184. [CrossRef]
27. Xenidis, N.; Perraki, M.; Apostolaki, S.; Agelaki, S.; Kalbakis, K.; Vardakis, N.; Kalykaki, A.; Xyrafas, A.; Kakolyris, S.; Mavroudis, D.; et al. Differential effect of adjuvant taxane-based and taxane-free chemotherapy regimens on the CK-19 mRNA-positive circulating tumour cells in patients with early breast cancer. *Br. J. Cancer* **2013**, *108*, 549–556. [CrossRef]
28. Bidard, F.-C.; Michiels, S.; Riethdorf, S.; Mueller, V.; Esserman, L.J.; Lucci, A.; Naume, B.; Horiguchi, J.; Gisbert-Criado, R.; Sleijfer, S.; et al. Circulating Tumor Cells in Breast Cancer Patients Treated by Neoadjuvant Chemotherapy: A Meta-analysis. *JNCI J. Natl. Cancer Inst.* **2018**, *110*, 560–567. [CrossRef]
29. Lucci, A.; Hall, C.S.; Lodhi, A.K.; Bhattacharyya, A.; Anderson, A.E.; Xiao, L.; Bedrosian, I.; Kuerer, H.M.; Krishnamurthy, S. Circulating tumour cells in non-metastatic breast cancer: A prospective study. *Lancet Oncol.* **2012**, *13*, 688–695. [CrossRef]
30. Georgoulias, V.; Bozionelou, V.; Agelaki, S.; Perraki, M.; Apostolaki, S.; Kallergi, G.; Kalbakis, K.; Xyrafas, A.; Mavroudis, D. Trastuzumab decreases the incidence of clinical relapses in patients with early breast cancer presenting chemotherapy-resistant CK-19mRNA-positive circulating tumor cells: Results of a randomized phase II study. *Ann. Oncol.* **2012**, *23*, 1744–1750. [CrossRef]

31. Ignatiadis, M.; Litière, S.; Rothe, F.; Riethdorf, S.; Proudhon, C.; Fehm, T.; Aalders, K.; Forstbauer, H.; Fasching, P.A.; Brain, E.; et al. Trastuzumab versus observation for HER2 nonamplified early breast cancer with circulating tumor cells (EORTC 90091-10093, BIG 1-12, Treat CTC): A randomized phase II trial. *Ann. Oncol.* **2018**, *29*, 1777–1783. [CrossRef] [PubMed]
32. Schneble, E.J.; Graham, L.J.; Shupe, M.P.; Flynt, F.L.; Banks, K.P.; Kirkpatrick, A.D.; Nissan, A.; Henry, L.; Stojadinovic, A.; Shumway, N.M.; et al. Current approaches and challenges in early detection of breast cancer recurrence. *J. Cancer* **2014**, *5*, 281–290. [CrossRef] [PubMed]
33. Strati, A.; Markou, A.; Parisi, C.; Politaki, E.; Mavroudis, D.; Georgoulias, V.; Lianidou, E. Gene expression profile of circulating tumor cells in breast cancer by RT-qPCR. *BMC Cancer* **2011**, *11*, 422. [CrossRef] [PubMed]
34. Mavroudis, D.; Saloustros, E.; Boukovinas, I.; Papakotoulas, P.; Kakolyris, S.; Ziras, N.; Christophylakis, C.; Kentepozidis, N.; Fountzilas, G.; Rigas, G.; et al. Sequential vs concurrent epirubicin and docetaxel as adjuvant chemotherapy for high-risk, node-negative, early breast cancer: An interim analysis of a randomised phase III study from the Hellenic Oncology Research Group. *Br. J. Cancer* **2017**, *117*, 164–170. [CrossRef] [PubMed]
35. Wittwer, C.T.; Herrmann, M.G.; Gundry, C.N.; Elenitoba-Johnson, K.S.J. Real-Time Multiplex PCR Assays. *Methods* **2001**, *25*, 430–442. [CrossRef]
36. Stathopoulou, A.; Ntoulia, M.; Perraki, M.; Apostolaki, S.; Mavroudis, D.; Malamos, N.; Georgoulias, V.; Lianidou, E.S. A highly specific real-time RT-PCR method for the quantitative determination of CK-19 mRNA positive cells in peripheral blood of patients with operable breast cancer. *Int. J. Cancer* **2006**, *119*, 1654–1659. [CrossRef] [PubMed]
37. Livak, K.J.; Schmittgen, T.D. Analysis of Relative Gene Expression Data Using Real-Time Quantitative PCR and the $2-\Delta\Delta CT$ Method. *Methods* **2001**, *25*, 402–408. [CrossRef] [PubMed]
38. Ignatiadis, M.; Xenidis, N.; Perraki, M.; Apostolaki, S.; Politaki, E.; Kafousi, M.; Stathopoulos, E.N.; Stathopoulou, A.; Lianidou, E.; Chlouverakis, G.; et al. Different prognostic value of cytokeratin-19 mRNA positive circulating tumor cells according to estrogen receptor and HER2 status in early-stage breast cancer. *J. Clin. Oncol.* **2007**, *25*, 5194–5202. [CrossRef]
39. Aktas, B.; Müller, V.; Tewes, M.; Zeitz, J.; Kasimir-Bauer, S.; Loehberg, C.R.; Rack, B.; Schneeweiss, A.; Fehm, T. Comparison of estrogen and progesterone receptor status of circulating tumor cells and the primary tumor in metastatic breast cancer patients. *Gynecol. Oncol.* **2011**, *122*, 356–360. [CrossRef]
40. Bustin, S.A.; Benes, V.; Garson, J.A.; Hellemans, J.; Huggett, J.; Kubista, M.; Mueller, R.; Nolan, T.; Pfaffl, M.W.; Shipley, G.L.; et al. The MIQE guidelines: Minimum information for publication of quantitative real-time PCR experiments. *Clin. Chem.* **2009**, *55*, 611–622. [CrossRef]
41. Trapp, E.; Janni, W.; Schindlbeck, C.; Jückstock, J.; Andergassen, U.; de Gregorio, A.; Alunni-Fabbroni, M.; Tzschaschel, M.; Polasik, A.; Koch, J.G.; et al. Presence of Circulating Tumor Cells in High-Risk Early Breast Cancer During Follow-Up and Prognosis. *Jnci J. Natl. Cancer Inst.* **2019**, *111*, 380–387. [CrossRef] [PubMed]
42. Mirza, S.; Jain, N.; Rawal, R. Evidence for circulating cancer stem-like cells and epithelial-mesenchymal transition phenotype in the pleurospheres derived from lung adenocarcinoma using liquid biopsy. *Tumour Biol.* **2017**, *39*, 1010428317695915. [CrossRef] [PubMed]
43. Emprou, C.; Le Van Quyen, P.; Jégu, J.; Prim, N.; Weingertner, N.; Guérin, E.; Pencreach, E.; Legrain, M.; Voegeli, A.-C.; Leduc, C.; et al. SNAI2 and TWIST1 in lymph node progression in early stages of NSCLC patients. *Cancer Med.* **2018**, *7*, 3278–3291. [CrossRef] [PubMed]
44. Markiewicz, A.; Wełnicka-Jaśkiewicz, M.; Seroczyńska, B.; Skokowski, J.; Majewska, H.; Szade, J.; Żaczek, A.J. Epithelial-mesenchymal transition markers in lymph node metastases and primary breast tumors - relation to dissemination and proliferation. *Am. J. Transl. Res.* **2014**, *6*, 793–808. [PubMed]
45. Voutsadakis, I.A. Epithelial-Mesenchymal Transition (EMT) and Regulation of EMT Factors by Steroid Nuclear Receptors in Breast Cancer: A Review and in Silico Investigation. *J. Clin. Med.* **2016**, *5*, 11. [CrossRef] [PubMed]
46. Vesuna, F.; Lisok, A.; Kimble, B.; Domek, J.; Kato, Y.; van der Groep, P.; Artemov, D.; Kowalski, J.; Carraway, H.; van Diest, P.; et al. Twist contributes to hormone resistance in breast cancer by downregulating estrogen receptor-α. *Oncogene* **2012**, *31*, 3223–3234. [CrossRef]

47. Sarrió, D.; Rodriguez-Pinilla, S.M.; Hardisson, D.; Cano, A.; Moreno-Bueno, G.; Palacios, J. Epithelial-mesenchymal transition in breast cancer relates to the basal-like phenotype. *Cancer Res.* **2008**, *68*, 989–997. [CrossRef] [PubMed]
48. Choi, Y.; Lee, H.J.; Jang, M.H.; Gwak, J.M.; Lee, K.S.; Kim, E.J.; Kim, H.J.; Lee, H.E.; Park, S.Y. Epithelial-mesenchymal transition increases during the progression of in situ to invasive basal-like breast cancer. *Hum. Pathol.* **2013**, *44*, 2581–2589. [CrossRef]

13

The Detection and Morphological Analysis of Circulating Tumor and Host Cells in Breast Cancer Xenograft Models

Loredana Cleris, Maria Grazia Daidone, Emanuela Fina *,† and Vera Cappelletti *,†

Biomarkers Unit, Department of Applied Research and Technological Development, Fondazione IRCCS Istituto Nazionale dei Tumori, 20133 Milan, Italy

* Correspondence: emanuela1.fina@gmail.com (E.F.); vera.cappelletti@istitutotumori.mi.it (V.C.)

† These authors share senior authorship.

Abstract: Hematogenous dissemination may occur early in breast cancer (BC). Experimental models could clarify mechanisms, but in their development, the heterogeneity of this neoplasia must be considered. Here, we describe circulating tumor cells (CTCs) and the metastatic behavior of several BC cell lines in xenografts. MDA-MB-231, BT-474, MDA-MB-453 and MDA-MB-468 cells were injected at the orthotopic level in immunocompromised mice. CTCs were isolated using a size-based method and identified by cytomorphological criteria. Metastases were detected by COX IV immunohistochemistry. CTCs were detected in 90% of animals in each model. In MDA-MB-231, CTCs were observed after 5 weeks from the injection and step wisely increased at later time points. In animals injected with less aggressive cell lines, the load of single CTCs (mean ± SD CTCs/mL: 1.8 ± 1.3 in BT-474, 122.2 ± 278.5 in MDA-MB-453, 3.4 ± 2.5 in MDA-MB-468) and the frequency of CTC clusters (overall 38%) were lower compared to MDA-MB-231 (946.9 ± 2882.1; 73%). All models had lung metastases, MDA-MB-453 and MDA-MB-468 had ovarian foci too, whereas lymph nodal involvement was observed in MDA-MB-231 and MDA-MB-468 only. Interestingly, CTCs showed morphological heterogeneity and were rarely associated to host cells. Orthotopic xenograft of BC cell lines offers valid models of hematogenous dissemination and a possible experimental setting to study CTC-blood microenvironment interactions.

Keywords: circulating tumor cells; metastasis; xenograft models; breast cancer

---

1. Introduction

Metastasis is definitely a hallmark of cancer [1] and represents the main cause of cancer-related deaths [2] due to ineffective therapies. Unraveling the molecular mechanisms of tumor progression would help to anticipate disease outcome and to point the way for selecting personalized treatments. In breast cancer (BC), in particular, the timing of cancer cell dissemination has been largely discussed [3] and has proven to represent an early step in tumor progression [4,5]. In accordance with this, circulating tumor cells (CTCs) can be detected in patients without clinical evidence of secondary lesions [6–8] and, in several studies, the presence of dormant cells has been also reported even in the bone marrow of patients with ductal carcinoma in situ [9–11]. In addition to this grim scenario, BC is, in fact, a group of heterogeneous tumors [12–15], with cancer cells cross-talking with normal cells from the microenvironment [16,17]. More recently, based on copy number and gene expression data from over 2000 tumors, BCs were re-classified into ten clusters associated with distinct clinical outcomes [18,19], with implications for patient management. As the development of drug resistance is often interpreted as an inevitable consequence of tumor heterogeneity [20,21], efforts to address such interrelated themes are urgently needed, especially in non-operable and advanced-stage clinical settings.

At present, the biological events and molecular mechanisms that orchestrate the metastatic process are still not fully understood due to their complexity [22–24]. Functional assays to elucidate the biological meaning of a gene in tumor dissemination or the effect of a compound on metastasis outgrowth have to be necessarily set in organisms. In this field, scientists have largely based their studies on metastasis modeling on laboratory animals, including drosophila, zebrafish, mice, rats and, more rarely, rabbits, companion pets and monkeys with spontaneous onset of cancer [25]. Xenotransplantation of BC cell lines in mice with a compromised immune system is commonly used as a model for metastasis studies. In particular, direct injection into the systemic circulation of the MDA-MB-231 cell line and its derivatives generated several models of metastasis [26–30], either in basal conditions or after selection of organ-specific metastatic variants upon several rounds of transplantation [31,32], providing valuable knowledge on the genetic determinants of metastasis in BC. However, although forced hematogenous dissemination does enable to finely dissect the late steps of the metastatic cascade [31], this strategy is not adequate to recapitulate the initial events of the process as in spontaneous metastasis models, where cells are implanted at the orthotopic level. Moreover, the research mainly focused on a single model type might fail in addressing the heterogeneity issue in BC, thus limiting possible applications to the clinical context [33].

Since the molecular classification of BC has been established, researchers have paid attention to the similarities between cell lines and clinical samples. Studies have shown that the luminal, basal, HER2 and claudin-low clusters identified in BC are mirrored in BC cell lines [34–36]. However, the claudin-low and basal subtypes are over-represented among the BC cell lines used for xenograft models [37]. Indeed, spontaneous metastasis is a rare event when using cell lines belonging to less aggressive subtypes, and only a few models with variable frequencies of metastasis have been described in recent years for MCF7, BT-474 and MDA-MB-453 [38,39]. Not dissimilar from BC cell lines, which however ensure a high tumor take in mice, is the behavior of xenotransplanted BC specimens (PDXs, patient-derived xenografts), whose both development and metastatic organotropism in mice are variable and dependent on the aggressiveness of the tumor of origin. Indeed, despite PDXs representing important preclinical tools since proven to retain over serial passages histopathology, behavior and genomic features of the tumor of origin [40–45], in BC the tumor take efficiency of the luminal subtype in mice is low [46,47], thus generating a bias towards aggressive triple-negative BCs models.

On the basis of these considerations, we have reconsidered the use of xenograft models from BC cell lines for basic metastasis research studies. To this aim, we have (i) transplanted BC cell lines belonging to different molecular subtypes in the mammary fat pad of immunocompromised mice, (ii) set up a method to detect CTCs and small foci of metastatic cells in such xenograft models, and (iii) described the morphological features of BC cell line derived CTCs and host-derived circulating cells.

2. Materials and Methods

2.1. Cell Lines

Cell lines were purchased from the American Type Culture Collection organization and verified for identity via short tandem repeat (STR) profile analysis using the StemElite™ ID System kit (Promega, Madison, WI, USA), which yielded a 100% match on 9 STR loci, and on amelogenin for gender identification, in all cases.

BT-474, MDA-MB-453 and MDA-MB-468 BC cell lines were cultured in Dulbecco's Modified Eagles' Medium (DMEM)/F-12 medium (Lonza, Switzerland) supplemented with 10% South America Fetal Bovine Serum (FBS, Lonza). The MDA-MB-231 BC cell line was cultured in DMEM/F-12 medium supplemented with 5% FBS. Cells were grown at 37 °C in a 95% humidified 5% pCO_2 atmosphere.

All experiments were performed using cells from the second to the eighth in vitro passage from thawing, and showing at least 95% viability by 0.4% Trypan Blue solution exclusion test. Cell culture supernatants were regularly tested for Mycoplasma contamination using the MycoAlert™ Mycoplasma Detection Kit (Lonza) before each injection in mice.

2.2. Animal Models

Animal experiments were performed according to the Italian law D.L. 116/92, and the following additions, which enforced the 2010/63/EU Directive. The study protocols were approved by the Ethical Committee for Animal Experimentation at Fondazione IRCCS Istituto Nazionale dei Tumori (INT), in Milan, (INT_08/2012, and INT_01/2017, which was also approved by Italian Ministry of Health with approval number 452/2017-PR, following the receipt of the D.L. 26/2014). All efforts were deployed to minimize animal suffering [48], following the most recently published version of recommended ARRIVE guidelines [49]. Female NOD.CB17-*Prkdc*^{scid}/J (NOD scid) and NOD.*Cg-Prkdc*^{scid} *Il2rg*^{tm1Wjl}/SzJ (NSG) mice were purchased from Charles River (Wilmington, MA, USA) and The Jackson Laboratory (Sacramento, CA, USA), respectively, and bred by the qualified personnel at INT Animal House Facility in individually ventilated cages, 3 to 5 animals per cage. Animals were anesthetized by intraperitoneal injection of a ketamine (100 mg/kg) and xylazine (5 mg/kg) cocktail before orthotopic injection of cancer cells and before animal sacrifice. Sacrifice procedure was cervical dislocation, performed at a priori set experimental time points or immediately upon signs of moderate suffering (e.g., decrease in activity, hunched appearance, ruffled hair coat, respiratory distress).

The tumor implant was performed under sterile conditions on healthy and normal-weight 7- to 16-week-old anesthetized mice using a 30G needle syringe. Eighty to ninety μL of Dulbecco's Phosphate Buffered Saline (DPBS, Lonza) cell suspensions mixed with 50% ECM Gel from Engelbreth-Holm-Swarm murine sarcoma (Sigma-Aldrich, St. Louis, MO, USA) Matrigel matrix (final concentration 4 mg/mL) were injected in the mammary fat pad (m.f.p.) of the axillary and/or the inguinal mammary gland, according to the scheme reported in Table 1:

Table 1. Scheme of breast cancer cell line xenotransplantation in immunocompromised mice.

Cell Line	Mouse Model	Number of Cells per Injection	Injection Sites (m.f.p.)
BT-474	NSG	5×10^6	4th left
MDA-MB-453	NSG	10^7	4th left
MDA-MB-468	NSG	5×10^6	2nd right and 4th left
MDA-MB-231	NOD scid	5×10^6	2nd right and 4th left

BT-474 cell injection was performed after 24–48 h from subcutaneous implantation of a 0.72 mg 90-day release 17-β-estradiol pellet (Innovative Research of America, Sarasota, FL, USA), performed on the neck lateral side using a trocar. The overall tumor take rate was 100%.

For the time-course experiments with MDA-MB-231 cells, 4 groups of 6 animals each were injected with cells according to the standard scheme. Animals were randomized before sacrifice at the defined time points (day 35, 50, 65 and 80) according to the tumor growth rate and the cage where they had been bred. Tumor take was obtained in 23/24 mice.

Tumor growth was monitored every week using a caliper and the tumor mass (g) was estimated by the $(D \times d^2)/2$ formula, where D and d represent the longest and the shortest diameter, respectively, of the nodule. The tumor load was lower than 10% of the body mass (range: 0.4–9.5%), except for two animals (10.2% and 15.7%) in which tumors had increased rapidly during the latest week.

An intravenous injection was performed using suspensions of 10^6 or 2×10^6 cells in 400 μL of DPBS.

Splenic leukocytes from BALB/c Nude mice were kindly provided by Dr. Claudia Chiodoni from the Molecular Immunology Unit at INT.

2.3. Collection of Tissues and Organs

Blood samples were drawn from anesthetized mice by cardiac puncture, using a 1 mL 26G needle EDTA conditioned syringe (1.8 mg/mL final concentration), stored at 4 °C and processed for CTC isolation within 30 min. Mice were immediately sacrificed and primary tumor nodules and organs (lung, axillary, inguinal subclavian or peritoneal lymph-nodes, ovaries, liver, kidneys, brain, and spleen) were collected and fixed in a 10% neutral buffered formalin solution (Bio-Optica, Milan, Italy) for 18–24 h; samples were then washed with distilled water and stored in 70% ethanol until paraffin embedding.

2.4. Circulating Tumor Cell Isolation and Detection

CTCs were isolated using the ScreenCell®Cyto kit (ScreenCell, Sarcelles, France), according to the manufacturer instructions. Briefly, blood was diluted in DPBS to reach 3 mL and subsequently mixed with 4 mL of the ScreenCell®FC2 proprietary buffer for red blood cell osmotic lysis and cell fixation. When the flux rate decreased due to a microcoagulation phenomenon or the presence of numerous CTCs, the residual blood was filtered on further devices. After filtration, the isolation supports (IS) were stained with Hematoxylin Solution S (Merck, Darmstadt, Germany) for 1 min and a Shandon Eosin Y Aqueous Solution (Thermo Fisher Scientific Inc., Waltham, MA, USA) for 30 s, or with a pure May-Grünwald solution for 2.5 min, followed by a 2.5-min incubation step with a May-Grünwald solution diluted 1:2 with pH 7-adjusted distilled water, and a 10-min incubation step with a Giemsa solution (Merck) 1:10 diluted with pH 7-adjusted distilled water. All samples were analyzed by a referral pathologist at ScreenCell. The cytomorphological analysis and CTC count were performed on the basis of the criteria of malignancy reported by Hofman et al. [50]. Major criteria for CTC identification were a high nucleus-to-cell ratio (i.e., cytoplasm area/whole cell area, ≥0.5) and large nuclear size (≥20 μm diameter), whereas minor criteria included irregular nuclear contours and nuclear hyperchromatism. CTC clusters were defined as groups of two or more CTCs, sometimes mixed with platelets and various leukocytes (i.e., circulating tumor microemboli, CTM), showing criteria of malignancy like those described for single CTCs. The nucleus-to-cell ratios in CTC aggregates are similar to those in single CTCs in [51]. Platelets appear as small, round eosinophilic or grayish particles, and can be found isolated or grouped in plaques, sometimes mixed with deposits of fibrin. Like CTCs, lymphocytes have a high nucleus-to-cell ratio, but they are smaller (7–8 μm diameter). Circulating atypical giant cells were defined as large cells (20–300 μm diameter), with generally voluminous and filamentary cytoplasm, various morphology (e.g., amorphous, round, elongated) and nucleus to cell ratio lower than that of CTCs [52,53]. Samples were defined as CTC-positive (+ve) when at least one single CTC and/or CTC cluster and/or CTM were observed in at least one stained IS.

2.5. Immunofluorescence and Immunohistochemical Staining

Immunofluorescence was performed on unstained ISs upon storage at −20 °C. ISs were incubated in an oven at 37 °C for 1 h, rehydrated in Tris Buffered Saline (TBS) 1× pH 7.4 (Bio-Optica) and blocked for 30 min with 5% bovine serum albumin (BSA, Sigma-Aldrich, St. Louis, MO, USA) in TBS 1X. Tumor cells were stained overnight at 4 °C using a rabbit monoclonal Alexa Fluor®488 conjugated antibody against human cytochrome *c* oxidase subunit IV (COX IV, clone 3E11, isotype IgG; Cell Signaling Technology, Danvers, MA, USA), diluted 1:100 in 5% BSA in TBS 1×. Nuclei were stained with a 5 μg/mL 4′,6-Diamidino-2-phenylindole (DAPI) dilactate solution (Sigma-Aldrich, St. Louis, MO, USA). ISs were mounted on glass slides and covered with a round coverslip using the Fluoroshield Mounting Medium (Abcam, Cambridge, UK). Images were acquired by Nikon Eclipse TE2000-S fluorescence (Nikon, Tokyo, Japan) microscope.

Four-micron thick formalin-fixed paraffin-embedded (FFPE) sections from tumor nodules and organs were deparaffinized by standard protocols and stained using a rabbit monoclonal antibody against human COX IV (clone 3E11, isotype IgG, Cell Signaling Technology, Danvers, MA, USA).

Antigen retrieval was performed at 95 °C for 30 min in a Sodium Citrate Buffer (10 mM Sodium Citrate, 0.05% Tween 20, pH 6.0). Endogenous biotin blocking was performed for liver sections only using the Dako Cytomation Biotin Blocking System (Dako, Troy, MI, USA). Samples were incubated with a 1:1000 diluted (Antibody Diluent, Dako) primary antibody at 4 °C overnight. Antibody visualization was obtained using the EnVision®+ System-HRP Labelled Polymer (Dako). Nuclei were counterstained with a Mayer's Hematoxylin Solution (Bio-Optica). Sections were observed and images acquired by a Nikon Eclipse E600 microscope.

For COX IV specificity verification, 4 consecutive sections from different organs of 3 non-tumor-bearing NOD scid mice were analyzed. For the MDA-MB-231 time-course experiment, 4 sections per lymph-node and 10 sections per lung sample were analyzed. Macroscopic inspection of organs at sacrifice and microscopic analysis by IHC on a series of non-adjacent FFPE sections (series of 4 consecutive stained and 8 consecutive unstained sections), for a total of 24 or 48, according to positivity, were performed for the preliminary assessment of metastasis formation in all kind of models (Experiment 1). For the quantitation of metastasis-positive (+ve) sections (Experiment 2), systematic IHC analysis was focused on a series of 24 or 48 non-adjacent FFPE sections (a series of 8 consecutive stained and 8 consecutive unstained sections) from lung, lymph-nodes and ovary samples. For the artificial metastasis experiment by tail-vein injection, 4 FFPE consecutive sections per lung sample were analyzed.

2.6. Statistical Analysis

Statistical analyses and graph constructions were performed using Graph Pad Prism v5. Differences in tumor mass between axillary and inguinal nodules were assessed using the point by point multiple Student's *t*-test, assuming that all time points were samples from populations with the same standard deviation, and the false discovery rate was set at 1% and determined using the two-stage linear step-up procedure of Benjamini, Krieger and Yekutieli [54].

3. Results

3.1. Technical Protocol for CTC Isolation and Species-Specificity-Based Detection of Tumor Cells in Xenograft Models

For CTC isolation, blood samples were drawn from anesthetized mice by cardiac puncture, which was proven to ensure the highest CTC yield compared to other approaches, according to Eliane et al. [55], and processed with the size-based CTC isolation device provided by ScreenCell®(Figure 1, Panel A; details are reported in Materials and Methods).

CTCs were identified on the basis of the cytomorphological criteria of malignancy already described for cancer patients [50]: in xenograft models, CTCs showed (i) a larger nucleus (generally 13 to 15 μm in diameter) compared to leukocytes, whose nuclei instead appeared slightly larger (about 7–8 μm in diameter) than membrane pores (6.5 ± 0.33 μm), (ii) a high nucleus-to-cell ratio (>0.5 for cell lines, rather than 0.75, cut-off used for clinical samples), (iii) a dense basophilic and irregularly outlined nucleus and (iv) a pale-bluish ring of cytoplasm, which generally appears as a thin rim encircling the nucleus.

Such a blood sampling approach coupled with filtration showed high efficiency in terms of sensitivity, as described in the following paragraphs, and adaptability to murine blood sample processing and cytological analysis for CTC detection, since the sample quality in terms of cellularity was adequate and the cell morphology was well preserved in about 80% (58/71) of samples.

Furthermore, a technical protocol was developed to effectively isolate and unambiguously identify tumor cells in tissues from xenograft mouse models, taking advantage of the human-murine species-specificity.

Given the weak metastatic ability expected in some models, an antibody-based staining protocol was set up in order to facilitate both the quantitation of rare CTCs and to enable screening for metastases in FFPE tissue sections from samples with microscopic and scattered metastatic foci. Immunofluorescence (for CTCs) and immunohistochemistry (IHC) analyses (for tissue sections) were performed using a commercially available antibody specific for the human mitochondrial marker cytochrome *c* oxidase subunit IV (COX IV). The non-cross-reactivity of the antibody with the murine counterpart has been preliminarily verified by immunofluorescence on peripheral blood mononuclear cells from a BALB/c nude mouse used as the control (Figure S1), and by IHC on FFPE sections of several organs from non-tumor-bearing NOD scid mice (Figure 1, Panel C), thus proving to be a reliable method to detect tumor cells in mouse xenografts.

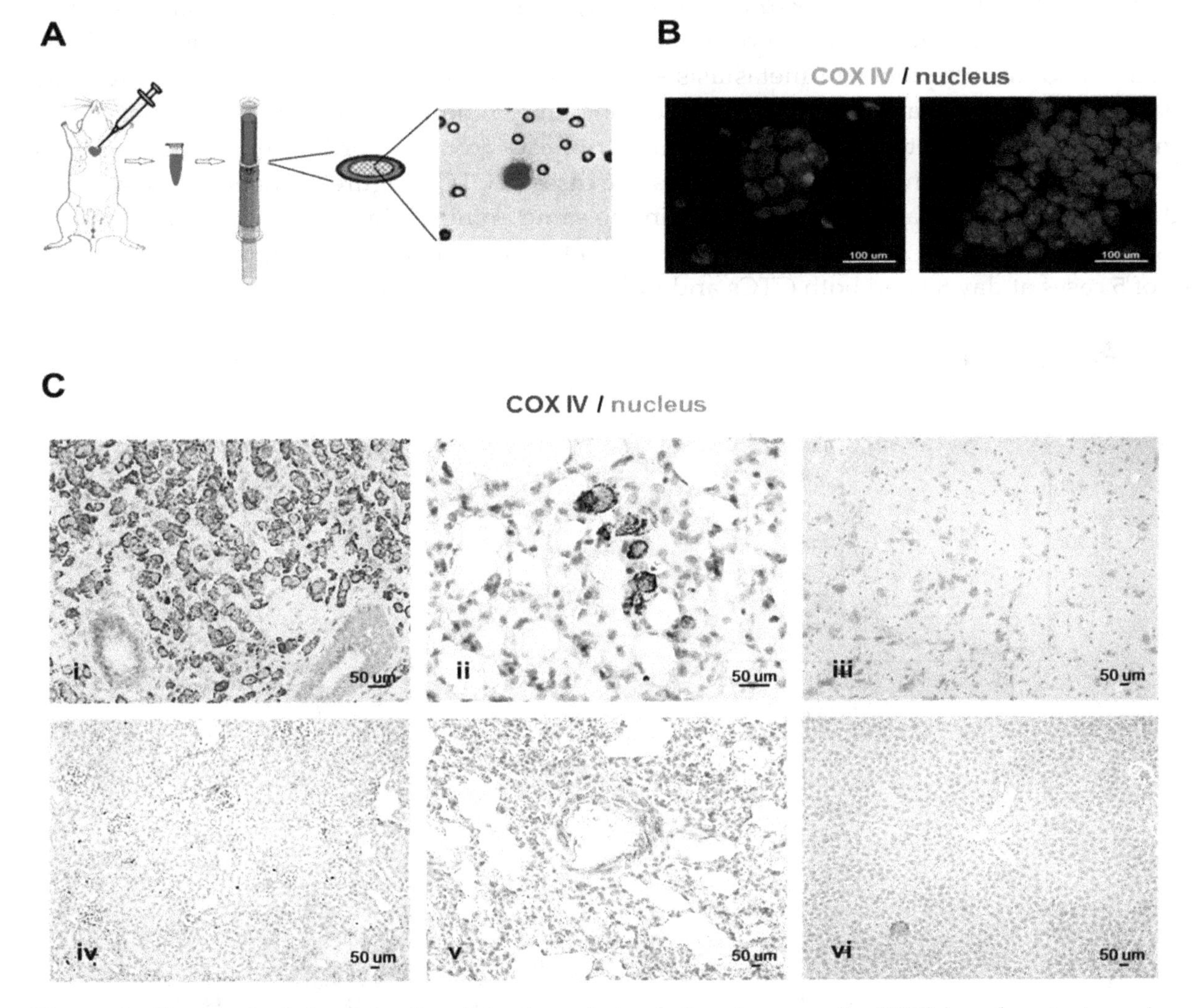

Figure 1. The methodology for the detection of circulating tumor cells (CTCs) and metastases in xenograft models. (**A**) The scheme illustrates a CTC isolation and detection technical approach for application in xenograft mouse models. Briefly, blood was drawn by cardiac puncture and cells were isolated by filtration on a porous membrane and identified by the morphological criteria (or immunostaining). Images represent (**B**) a cluster of CTCs (left) and a cluster of leukocytes (right) isolated from MDA-MB-231 xenografts, acquired by the 4′,6-Diamidino-2-phenylindole (DAPI) and the Fluorescein isothiocyanate (FITC) filters (60× oil immersion objective) and showing COX IV positive and negative staining, respectively; (**C**) COX IV immunohistochemistry stained formalin-fixed paraffin-embedded sections of (**i**) primary tumor nodule (20× objective) and (**ii**) lung metastases (40× objective) from MDA-MB-231 xenograft, and of (**iii**) brain, (**iv**) kidney, (**v**) lung, and (**vi**) liver (10× objective) collected from a non-tumor-bearing mouse.

CTCs were detectable by immunofluorescence and distinguished from leukocytes by the nucleus size and the typical staining pattern, as depicted in Figure 1, Panel B: CTCs organized in clusters have intense cytoplasmic-specific staining and are larger compared to the cluster of leukocytes, which instead have smaller nuclei and show negative staining for COX IV.

3.2. Cancer Cell Dissemination Can Be Monitored from the Early to Late Stages of Tumor Progression in the MDA-MB-231 Xenograft Model

The dynamics of dissemination in the MDA-MB-231 model was investigated in a time-course experiment, where the CTC load and the frequency of lymph-nodal and pulmonary metastases were measured at different time points after tumor cell injection. Overall, the load of single CTCs (mean ± SD: 0.40 ± 0.89; 0.33 ± 0.58; 79.33 ± 181.7; 1,993 ± 4,269; Figure 2, Panel B) and CTC clusters (mean ± SD: 2.33 ± 4.04; 1.75 ± 1.50; 62.00 ± 137.20; 1,229 ± 2,653; Figure 2, Panel C) showed a stepwise increase during progression, which mirrored the primary tumor growth (Figure 2, Panel A). Following a similar trend, the frequency of metastasis +ve cases, assessed in lymph-nodes (axillary, inguinal, subclavian or peritoneal) and lungs, increased during time, although, differently than lungs, metastases at lymph-nodes were detectable since the earliest phases from tumor injection (Figure 2, Panel D). At day 35 CTCs were found in 1 out of 5 assessable cases (2 CTCs, Figure 2, Panel B), consistently with the detection of few metastatic cells at lung in the same animal (Figure 2, Panel D). At day 50 lung metastases were found in 1 out of 3 CTC +ve cases only. On the contrary, 5 out of 6 cases at day 65 and 5 out of 5 cases at day 80 had both CTCs and pulmonary metastases.

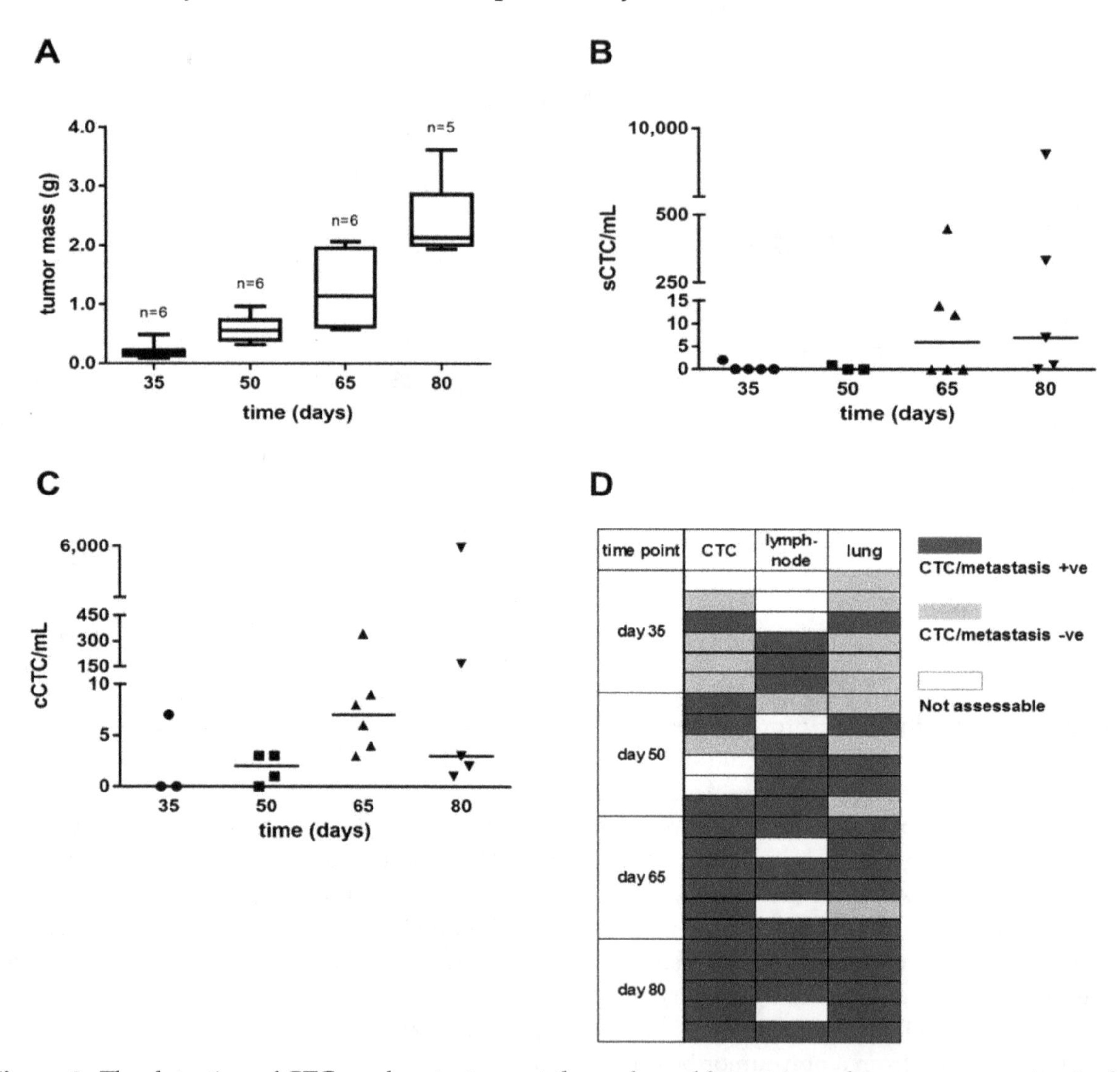

Figure 2. The detection of CTCs and metastases at the early and late stages of tumor progression in the MDA-MB-231 xenograft model. Box and whiskers plots and dot plots represent the distribution of (**A**) the total tumor mass, (**B**) single CTC (sCTC) and (**C**) CTC cluster or tumor microemboli (cCTC) numbers per milliliter of blood (horizontal line representing the median value), at different experimental time points. (**D**) the scheme represents the frequency of CTC-positive (+ve) and of lymph-nodal or pulmonary metastasis-positive (+ve) animals per group, at each experimental time point.

MDA-MB-231 cells were also injected in the tail vein of five animals and their presence in blood was monitored during time. Blood samples collected from two animals, injected with 10^6 or 2×10^6 cells, 1 h after injection contained 1 and 7 sCTCs per milliliter, respectively, thus indicating that the vast majority of cells had reached peripheral districts in short time from forced blood dissemination. The remaining three animals, two injected with 10^6 and one injected with 2×10^6 cells, were sacrificed after 78 days and were all CTC +ve and lung metastasis +ve. cCTCs were detected in all cases and ranged from 1 to 31 per milliliter, while sCTCs (about 280) were found in one animal only, injected with 10^6 cells. Lymph-nodes, ovaries and spleen were all negative for metastases by macroscopic examination and IHC analysis.

3.3. Breast Cancer Cell Lines with Different Subtypes Disseminate in Blood and Show Distinct Organotropism in Xenograft Models

CTC models were obtained by the orthotopic injection of BT-474, MDA-MB-453, MDA-MB-468 and MDA-MB-231 BC cell lines, performed in two independent experiments. The tumor take rate was 100% in all models and the growth rate of nodules was faster in MDA-MB-231 xenografts, where the total mass reached 500 mg after about 50 days from the cell injection, compared to the other models, which reached comparable masses over longer times (Figure S2), thus mirroring the expected level of aggressiveness according to the molecular subtype of each cell line. A significant difference (adjusted p-value <0.01) between the tumor mass of axillary and inguinal nodules was also observed in MDA-MB-468 (axillary versus inguinal mean ± SD tumor mass (g): 0.89 ± 0.22 versus 0.62 ± 0.27 and 1.00 ± 0.31 versus 0.69 ± 0.30, after 92 and 98 days from tumor implant, respectively) and MDA-MB-231 xenografts (axillary versus inguinal mean ± SD tumor mass (g): 1.02 ± 0.38 versus 0.69 ± 0.27, 1.12 ± 0.40 versus 0.77 ± 0.21, and 1.53 ± 0.64 versus 0.83 ± 0.27, after 75, 78, and 83 days from the tumor implant, respectively), with a general trend towards a faster growth rate in the axillary compared to the inguinal mammary fat pad injection site (Figure S2). Interestingly, MDA-MB-231 and the less aggressive BC cell lines were both able to disseminate in blood as sCTCs, (Figure 3, Panel A) found in about 90% of cases, and as cCTCs (both tumor cell clusters and microemboli, Figure 3, Panel B), detected at variable frequency according to the specific xenograft model. Overall, in both experiments, the sCTC and cCTC load per milliliter of blood was higher in the MDA-MB-231 (median(range): 2(0–9625) and 2(0–5973), respectively) compared to the other CTC models (Table S1), in keeping with the aggressiveness and high proliferation rate of these cells.

Among the weakly metastagenic models, MDA-MB-453 showed the highest numbers of sCTC/mL (median(range): 4.5(0–800)), while overall sCTC numbers for BT-474 and MDA-MB-468 ranged from 0 to 8. Moreover, cCTC positivity was approximately 2-fold lower in the less aggressive models (overall 10/26 cases) compared to the highly metastatic MDA-MB-231 xenografts (8/11 cases). Representative images of sCTCs and cCTCs from each model are reported in Figure 3, Panel C.

The metastatic potential of BC cell lines was also assessed in a preliminary exploratory experiment by macroscopic inspection and IHC analysis. Overall, organs presenting with metastasis were lung, lymph-nodes and ovaries, and those without metastasis +ve sections were the liver, brain and spleen. In a second experiment, systematic IHC analysis (Table 2) confirmed that ovarian metastases were detectable in 2 out of 7 MDA-MB-468 and the majority of MDA-MB-453 xenografts, but not in the BT-474, and likely MDA-MB-231 models, since they did not show ovary enlargement at macroscopic inspection. Lymph-nodal involvement was already macroscopically assessable in the MDA-MB-468 and MDA-MB-231 models in 100% of cases, as also confirmed by the IHC analysis, whereas lymph-nodes in BT-474 and MDA-MB-453 models were hardly detectable and collectible, suggesting the absence of massive dissemination via the lymphatic system. Instead, lungs were the metastatic site showing the highest tropism and frequency in all xenograft models. Consistently with CTC numbers, metastatic foci in weakly aggressive models consisted of single scattered cells or small foci of 3–30 cells each compared to the larger clusters, and macroscopic nodules in a few cases, observed in MDA-MB-231 xenografts (Figure S3).

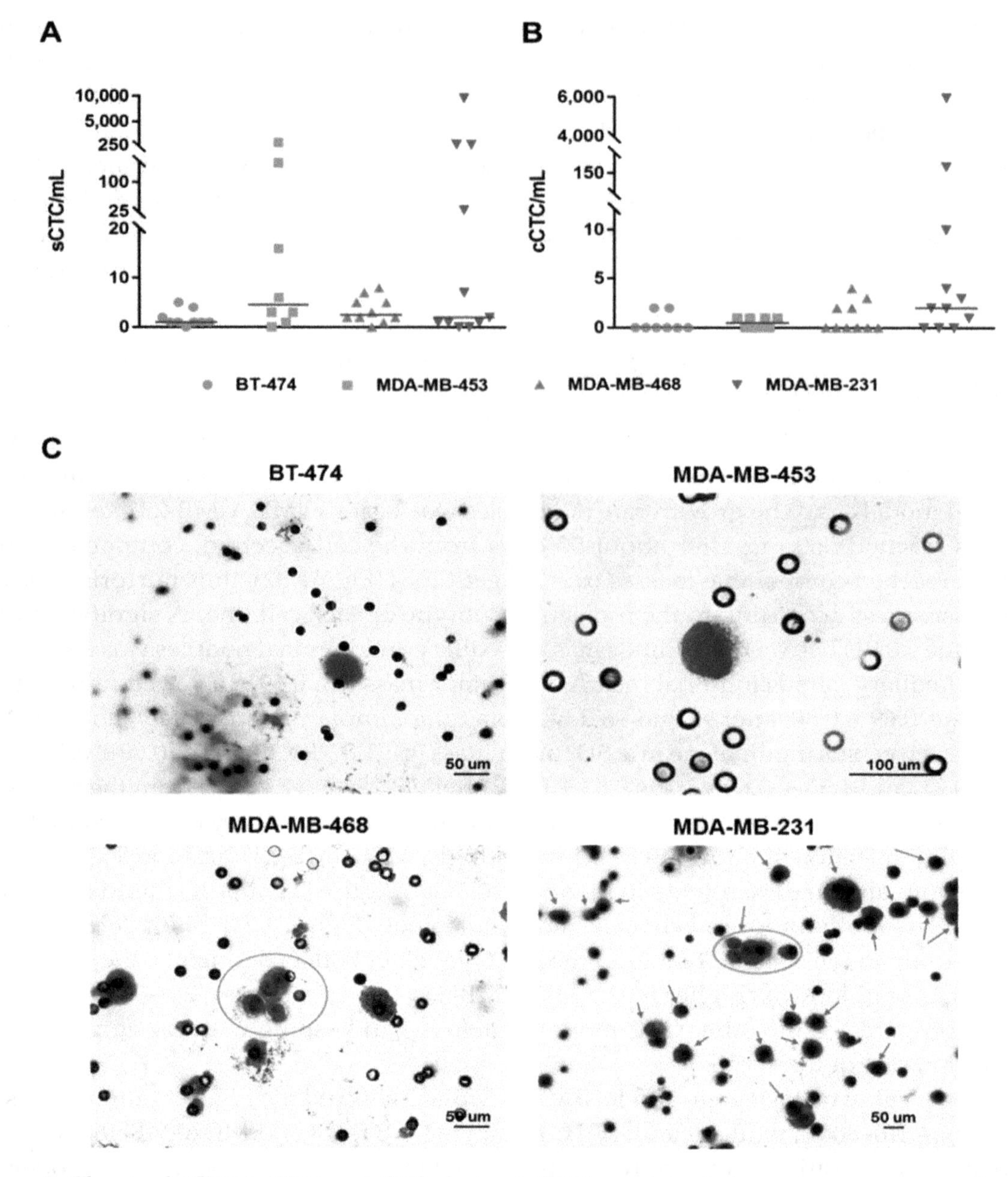

Figure 3. The single CTC (sCTC) and CTC cluster (cCTC) numbers in breast cancer xenograft models. Dot plots represent the distribution (horizontal line corresponding to the median value) of (**A**) sCTCs and (**B**) cCTCs per milliliter of blood. Images (**C**) represent May-Grünwald-Giemsa stained sCTC (40× objective) from BT-474, sCTC (60× oil immersion objective) from MDA-MB-453, cCTC consisting of four cells from MDA-MB-468 (circle, 40×), numerous sCTCs and a cCTC consisting of 4-to-5 cells (arrows, 20× objective) from MDA-MB-231 xenografts.

Table 2. The metastasis sites and frequencies in breast cancer xenograft models.

CTC-Model	Experiment 1									Experiment 2								
	N	Lymph-Node		Lungs		Ovary 1		Ovary 2		N	Lymph-Node		Lungs		Ovary 1		Ovary 2	
		+ve Cases	Positivity Frequency (%) *	+ve Cases	Positivity Frequency (%) *	+ve Cases	Positivity Frequency (%) *	+ve Cases	Positivity Frequency (%) *		+ve Cases	Positivity Frequency (%) *	+ve Cases	Positivity frequency (%) *	+ve Cases	Positivity Frequency (%) *	+ve Cases	Positivity Frequency (%) *
BT-474	3	- §	-	3	25–100 †	0	-	0	-	7	- §	-	7	100	0	-	0	-
MDA-MB-453	3	- §	-	2	80–100 †	0	-	0	-	7	- §	-	7	75–100 †	6	75–100 †	5 ‡	43–100 †
MDA-MB-468	3	- §	-	3	100	1	21	0	-	7	7	100	7	100	2	23–100 †	1	90
MDA-MB-231	5	5	100	5	100	- ¥	-	- ¥	-	6	6	100	6	100	- ¥	-	- ¥	-

* range of positivity frequencies (number of metastasis-positive (+ve) out of 10 to 24 sections in Experiment 1, and out of 24 to 48 sections in Experiment 2). § not detectable or not involved at the macroscopic level. ‡ 1 out of 2 ovaries not assessable in one case. † positivity range. ¥ not assessed.

3.4. Circulating Tumor Cells in Breast Cancer Xenograft Models are Pleomorphic and Circulate with Cells of the Host

Cytological blood samples from different CTC models were analyzed and compared in order to highlight intra-sample and inter-model differences on the basis of morphological criteria (details reported in Materials and Methods). The identified cell subpopulations are hereafter described. As already appreciable in the MDA-MB-231 model (Figure 3, Panel C), single CTCs show a certain degree of morphological heterogeneity (i.e., pleomorphism), each cell with a more or less irregularly outlined nucleus of various sizes and shapes, in addition to the heterogeneity in the whole cell size and morphology (Figure 4, Panel A). The difference in sizes is particularly evident in the two BT-474- and the two MDA-MB-453-derived CTCs depicted in Figure 4, one of them smaller than the other. While in the BT-474 CTC model, the cells display an irregular nucleus, the MDA-MB-453-derived CTCs have a clearly round-shaped nucleus and, besides the larger size, the bigger cell also displays a higher nucleus-to-cell ratio (>0.90) compared to the other, as also indicated by the thinner cytoplasmic rim. sCTCs with low (<0.75) or high (>0.90) nucleus-to-cell ratios were also observed in MDA-MB-468, a few of them also presenting with a multilobulated nucleus. CTCs in the MDA-MB-231 model may present as either round-shaped or polygonal physically interacting cells and may have a widely variable whole size. Interestingly, CTC clusters intermingled with or surrounded by platelets were rarely detected in all models, sporadically also in direct contact with leukocytes, as observed in MDA-MB-453 and MDA-MB-231 (Figure 4, Panel B). Few numbers of cytological figures appearing like atypical giant cells with several shapes (morphological details reported in Materials and Methods), were detected in 13%, 30% and 17% of CTC positive cases from MDA-MB-453, MDA-MB-468 and MDA-MB-231 models, respectively. Despite being present in a minority of CTC +ve cases, atypical giant cells were never found in samples called CTC-negative. Images depicting all cell types described in cytological blood samples from xenograft models is reported in Figure S4. Additionally, a complete list of data describing the presence of circulating cells and metastases in all the analyzed animals for each model is reported in Table S2.

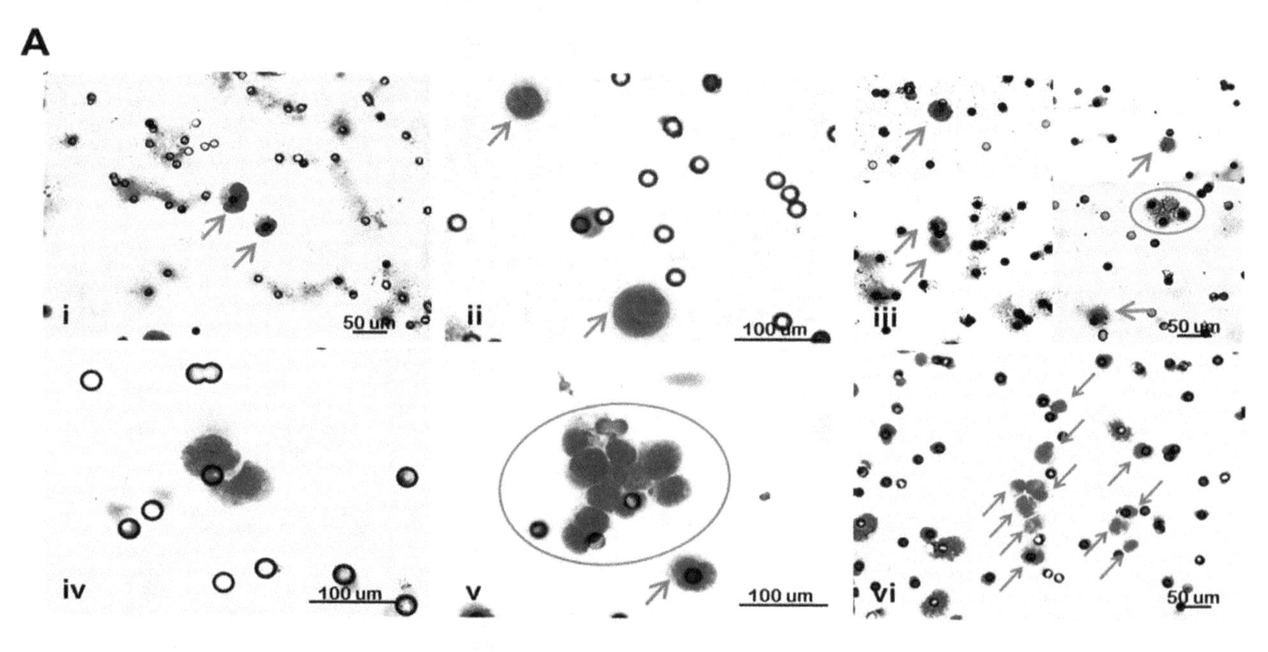

Figure 4. *Cont.*

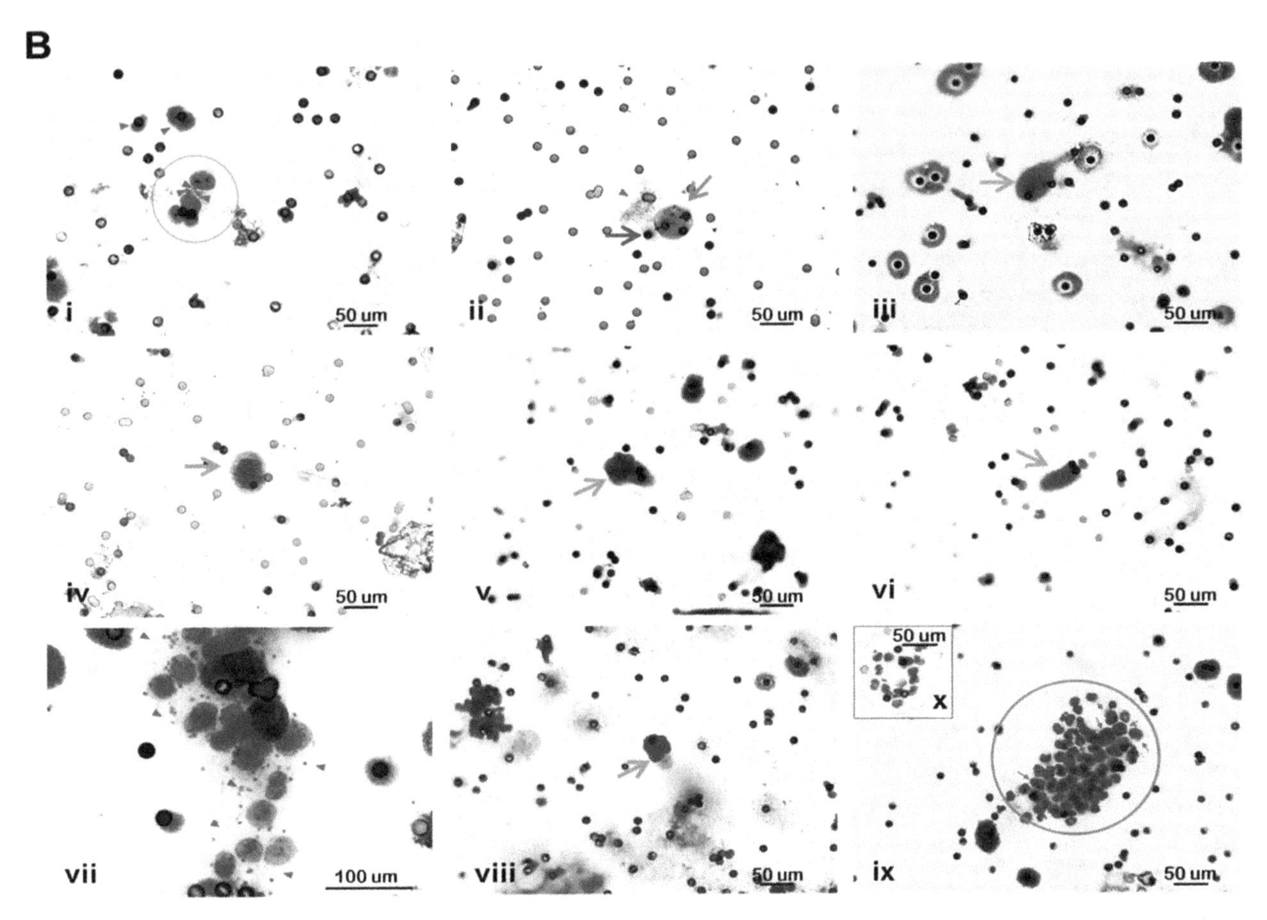

Figure 4. The morphological heterogeneity of CTCs and circulating host cells in BC xenograft models. (**A**) Images are representative of CTC pleomorphism: sCTCs (arrows) with different sizes from (**i**) BT-474 (40× objective) and from (**ii**) MDA-MB-453 (60× oil immersion); (**iii**) sCTCs (arrows) with low (top images) and high (bottom images) nucleus-to-cell ratios, and cCTC (circle) from MDA-MB-468 (40×); (**iv**) the cluster of two CTCs with multilobulated nucleus from MDA-MB-468 (60× oil immersion); (**v**) cCTC from MDA-MB-231 (60× oil immersion); (**vi**) sCTCs and CTCs in clusters (arrows) with different sizes from MDA-MB-231 (40×). (**B**) Images are representative of CTCs and circulating host cells: (**i**) a cluster of four CTCs (circle) and platelets (arrowheads) from BT-474 (40×); (**ii**) sCTC (orange arrow) in cluster with platelets (arrowhead) and one leukocyte (blue arrow) from MDA-MB-453 (40×); (**iii**) atypical giant cells cell from MDA-MB-453 (green arrow, 40×); (**iv,v,vi**) three atypical giant cells from MDA-MB-468 (green arrow, 40×); (**vii**) cCTC and platelets (arrowheads) from MDA-MB-231 (60× oil immersion); (**viii**) an atypical giant cell from MDA-MB-231 (green arrow, 40×); (**ix**) a cluster (circle) of CTCs (arrows showing clearly distinguishable tumor cells) combined with leukocytes (40×).

4. Discussion

Our study demonstrates the reliability of a technical protocol for CTC and metastasis detection in BC xenograft models based on the classical morphological features of malignancy and relying on the advantage of the species-specificity barrier for the identification of cells of human origin in both cytological blood samples and tissues compared to the host counterpart. The described methodology can be, in principle, applied to every kind of xenograft model for CTC and metastasis biology studies, thanks to its sensitivity and specificity in detecting rare circulating cells and also small metastatic foci in weakly metastatic models. Assessment of metastasis formation using a species-specific antibody was proven a reliable method to identify small and rare metastases, especially in weakly metastatic models.

As such, a preliminary validation test with the metastatic MDA-MB-231 BC cell line showed that hematogenous dissemination and metastases may occur even at the earliest stages upon tumor nodule appearances at the orthotopic site, and can be monitored in a time-course experiment. Here, studies were also performed to model BC metastases using BT-474, MDA-MB-453 and MDA-MB-468 cells, providing a comparative analysis, for the first time, of the hematogenous dissemination potential among BC cell lines belonging to different molecular subtypes, in addition to MDA-MB-231, upon injection at the orthotopic level rather than forced metastasis formation assays. Consistently with the growth rate of the primary tumors, xenografts in the murine model, generated using the HER2 positive BT-474 and MDA-MB-453 and the basal A MDA-MB-468 cell lines, according to Neve et al. [34], determined a lower CTC load compared to the numbers of CTCs generated by MDA-MB-231 xenografts. Interestingly, BC cell line xenografts can generate a pleomorphic CTC population, consisting of markedly heterogeneous cells in terms of the whole cellular and/or nuclear size and morphology, as also nucleus to cell ratio, and which also includes clusters of CTCs, released in blood at a reduced frequency compared to the sCTC subset. Intra-clonal size heterogeneity has been already reported in MDA-MB-231-derived clonal subpopulations in vitro [56], suggesting that upon injection in mice cells with different metastatic abilities and morphological features underwent clonal selection and that such clones might be more easily identifiable in the CTC population in xenografts. However, variable sizes and nuclear-cytoplasmic ratios have been also described in CTCs from prostate cancer patients compared to cultured prostate cancer cell lines [57], again indicating that a clonal selection process takes place at the primary or secondary sites during tumor progression. Therefore, CTC morphological heterogeneity might be interpreted as a hallmark of tumor cells which is likely to be more easily assessable among clones of the blood disseminating population.

We have also demonstrated that not only MDA-MB-231 but also other models may represent experimental tools suitable for CTC characterization and metastasis biology studies. Indeed, each cell line was shown to follow preferential dissemination routes, i.e., through blood or lymphatic vessels, as also distinct colonization patterns at distant sites. Lastly, despite the species-barrier, it was surprising to find, for the first time, circulating cells from the murine host in physical contact with tumor cells of human origin, as also atypical cytological figures.

MDA-MB-231 cells were already proven to induce lung metastases when injected in the tail vein of nude mice [58], and the success of transplantation experiments in the m.f.p. ranked them among the most aggressive BC cell lines [59,60]. Since the first reports, studies employing such a cell line have started to proliferate, and even nowadays they represent a large fraction of the literature on BC metastasis biology. On the contrary, ER-positive cell lines such as MCF7, T47D, and BT-474 are able to form tumors only in the presence of an exogenous source of estrogen. However, despite the metastatic origin of these cell lines, they have a limited ability to invade and metastasize [37], unless subjected to a selection of hormone-resistant variants or genetically modified [61]. More recently, severely immunocompromised mice, such as NSG and Rag2$^{-/-}$ $\gamma c^{-/-}$ models, which exhibit T cell, B cell and natural killer cell immunodeficiency, were also explored to generate new metastatic models. This time, MCF7 were able to give rise to metastases at the lymph-node, lung, spleen and, sporadically, even at the renal level when injected in the m.f.p. of NSG mice [38]. BT-474 cells were instead less metastatic in these mice, generating macro-metastases in only a few cases (axillary lymph node in 17% of mice and the spleen in 8%. of mice). The latter result is in contrast with our model where, although no macrometastases were found, small foci were detected in all animals and in all FFPE sections we have analyzed. In another study, bioimaging analysis enabled the detection of multi-organ metastases in Rag2$^{-/-}$ $\gamma c^{-/-}$ mice injected at the orthotopic level with the MDA-MB-453 and BT-474 cell lines [39].

Attempts to explore hematogenous dissemination in BC experimental models were made only recently. In a technical paper published in 2008 [55], different approaches for blood collection were tested to isolate CTCs from tumor-bearing mice, finally demonstrating that the cardiac puncture represents the most suitable approach to reach high yields without interference from contaminating normal murine epithelial cells. The authors also validated a method to enumerate CTCs by applying a modified version of an in vitro diagnostic system for quantifying CTCs in patients, obtaining numbers of CTCs ranging from ~100 to 1000 per milliliter of blood. In line with our results, the reported CTC concentration in the blood of MDA-MB-231 xenograft models was highly variable among different animals. Concerning CTC variability, despite the wide range of cells detected in this model, we have observed a correlation between CTC load and tumor burden, whereas the literature data from experiments with GFP-expressing MDA-MB-231 cells suggested that the primary tumor size is not a strong indicator of CTC load [62]. Hence, such a variability in CTC load in experimental models could also be the result of fluctuations in CTC release, as also suggested by results obtained in a melanoma CTC-model [62]. Here, the authors performed real-time continuous monitoring of CTCs and could estimate a release of 0 to 54 CTCs every 5 min, alternated to CTC-free phases. Differently from data obtained in the MDA-MB-231 CTC model [62], this peak in CTC level mirrored the increase in tumor growth at the same time point, suggesting again that the number of CTCs may indeed correlate with tumor size. In our MDA-MB-231 model, probably due to its pronounced aggressiveness, we had no possibility to define an optimal time point to isolate CTCs before they were able to colonize distant organs since in the time-course experiment CTC release and increase during time mirrored the onset, frequency and extent of pulmonary metastases in matched FFPE sections. On the contrary, dissemination via the lymphatic system was observed at the earliest time points (after 35 days from injection), suggesting that different molecular mechanisms are required to enter the lymphatic system compared to blood vessels. In agreement with this observation, Juratli and colleagues [62] also reported that metastatic foci at lymph nodes can be observed just two weeks after orthotopic inoculation. More recently, two seminal works demonstrated that in experimental models, tumor cells invading lymph nodes are able to enter local blood vessels and exit from them to invade distant organs [63,64]. However, commonalities in clinical tumors need to be demonstrated. The results of our dynamic studies with the MDA-MB-231 model also show that CTCs can be actively released from pulmonary metastases. At the same time, once in the bloodstream, MDA-MB-231 cells were able to rapidly reach peripheral districts and colonize the lung as CTC numbers rapidly dropped out upon intravenous injection and their presence was generally associated to the presence of lymph-node and lung metastases, even at early time points. Different results were observed in a MDA-MB-468 CTC-model already described by Bonnomet and colleagues [65], where CTCs were detectable as early as 8 days after injection and increased 36 days later, after which their levels remained quite constant. Indeed, Bonnomet and colleagues also found lung metastases at later time point only, despite CTC recovery was possible even a few days after cell injection in a time-course experiment [65].

An interesting result emerging from our study is the detection of clusters of CTCs in all the examined BC models. Overall, to our knowledge, the presence of cCTCs has been reported for the triple-negative LM2 MDA-MB-231 [66] and MDA-MB-435 [67] BC cell lines only. We were able to monitor the presence of CTCs using size-based isolation support also in xenograft models obtained from BC cells with different molecular subtypes. Consistently with the demonstration that cCTCs are endowed with higher efficiency in initiating metastasis [66,68], the frequency of cCTCs was 2-fold lower in models with weak metastatic potential compared to MDA-MB-231. Moreover, CTCs generated upon xenograft of BC cell lines were heterogeneous in morphology, even within the same cluster, thus suggesting further heterogeneity at the molecular level, probably as a result of a selection process of cells which are committed to disseminate in blood according to their functionality and ability to cooperate and survive in a foreign microenvironment. Unexpectedly, some clusters of tumor cells also presented with physically interacting platelets and/or leukocytes. CTCs in contact with blood or stromal cells were already described both in experimental models and clinical samples and some cell types, such as platelets, neutrophils and monocytes/macrophages, were also demonstrated to promote epithelial-to-mesenchymal transition and the pre-metastatic niche formation [69–72], even at early BC stages [73], or to assist CTCs during their transit in blood and organ colonization, increasing their metastatic ability [74]. However, to our knowledge, interactions between circulating tumor and host cells in xenografts have not been reported at present. Platelet depletion in a nude mouse model transplanted subcutaneously with SKOV3 ovarian cancer cells provided evidence that platelets are involved in cancer cell growth [75], hence suggesting that even cells from the murine compartment can be actively recruited to cooperate with human tumor cells in xenograft models. With reference to our data, the growth rate of matched axillary and inguinal m.f.p. nodules generally showed a different trend in both MDA-MB-468 and MDA-MB-231 models, with a higher tumor mass at the axillary compared to the inguinal site, even statistically significant during the latest time point measurements. Such data are consistent with the hypothesis that, despite the species specificity barrier, tumor cells can interact and possibly cross-talk with the murine microenvironment in xenograft models. Finally, atypical giant cells, presenting without features of malignancy and, therefore, expected to originate from the host, were also observed in our models. Similarly to clinical samples [52], such a cell type has been generally associated with the presence of CTCs and never found in CTC-negative samples. The origin of circulating atypical giant cells in xenografts has not been investigated here. Our hypothesis is that cells displaying not all the classical features of malignancy and presenting with unusual morphological patterns, which resemble those described for cancer-associated macrophage-like cells by Adams and colleagues [52], might derive from the interaction between tumor and host cells. Fusion hybrids, i.e., hybrid cells derived from fusion events between tumor cells and macrophages, were already described in murine experimental models [76]. If such cells originate in response to the attack from the immune system towards tumor cells, or as a strategy to increase the tumor cell viability and metastatic potential, is yet to be established. Literature data reported that murine peritoneal macrophages can phagocytize apoptotic BC tumor cells from cell lines in vitro, and acquire stem-like features in the following steps [77]. With reference to the xenograft milieu, another study reported the host macrophage invasion and the presence of multinuclear giant cells or foreign body giant cells at the implant site upon the injection of human mesenchymal stem cells with biopolymers in NOD scid mice, thus indicating that severely immunocompromised mice are able to retain a certain level of innate immune responsiveness [78]. On the other hand, cell fusion has been associated with the acquisition of increased metastatic capacity or enhanced drug resistance [79], and the presence of circulating hybrid cells was shown to correlate with the disease stage and patient survival [76]. Overall, we are aware that beside the identification based on morphological criteria although performed by a pathologist experienced in CTC detection in clinical samples, only a molecular characterization of such atypical populations of cells, including leukocytes interacting with CTCs, e.g., with species-specific antibodies and gene expression analyses, would definitely elucidate their nature and confirm the validity of xenograft models for new research lines.

5. Conclusions

In the end, CTCs and metastases can be in vivo modeled from BC cell lines with different subtypes and disseminating potential. In xenografts, several subpopulations of cells circulating in the blood can be identified by applying the classical morphological criteria, thus offering experimental models alternative to MDA-MB-231, as unusual and intriguing tools to investigate tumor-host interactions in the blood microenvironment.

Supplementary Materials:
Figure S1: Species-specificity of anti-human COX IV antibody; Figure S2: Primary tumor growth in breast cancer xenograft models; Figure S3: Lung metastases in breast cancer xenograft models; Figure S4: Cell types observed in cytological blood samples from breast cancer xenograft models; Table S1: Circulating tumor cell load in breast cancer xenograft models; Table S2: Circulating cells and metastases in xenograft models from breast cancer cell lines.

Author Contributions: Conceptualization, E.F.; Methodology and Investigation, E.F. and L.C.; Formal analysis, E.F.; Project administration and Supervision, V.C. and M.G.D.; Funding acquisition, M.G.D. and E.F.; Visualization and Writing—original draft, E.F.; Writing—review & editing, M.G.D., V.C. and L.C.

Acknowledgments: The Authors thank Janine Wechsler, consultant pathologist at ScreenCell, for morphological analysis of cytological samples, and Gloria Morandi, Lucia Gioiosa and Lorena Ventura, technicians at INT IHC Facility, for performing IHC staining. A special thanks to INT Committee for animal welfare (OPBA, Organismo Per il Benessere Animale) for its support in writing the protocol for in vivo experiments.

References

1. Hanahan, D.; Weinberg, R.A. Hallmarks of cancer: The next generation. *Cell* **2011**, *144*, 646–674. [CrossRef] [PubMed]
2. WHO. Cancer. Available online: https://www.who.int/en/news-room/fact-sheets/detail/cancer (accessed on 28 May 2019).
3. Klein, C.A. Parallel progression of primary tumours and metastases. *Nat. Rev. Cancer* **2009**, *9*, 302–312. [CrossRef] [PubMed]
4. Hüsemann, Y.; Geigl, J.B.; Schubert, F.; Musiani, P.; Meyer, M.; Burghart, E.; Forni, G.; Eils, R.; Fehm, T.; Riethmüller, G.; et al. Systemic spread is an early step in breast cancer. *Cancer Cell* **2008**, *13*, 58–68. [CrossRef] [PubMed]
5. Hosseini, H.; Obradović, M.M.S.; Hoffmann, M.; Harper, K.L.; Sosa, M.S.; Werner-Klein, M.; Nanduri, L.K.; Werno, C.; Ehrl, C.; Maneck, M.; et al. Early dissemination seeds metastasis in breast cancer. *Nature* **2016**, *540*, 552–558. [CrossRef] [PubMed]
6. Bidard, F.C.; Mathiot, C.; Delaloge, S.; Brain, E.; Giachetti, S.; de Cremoux, P.; Marty, M.; Pierga, J.Y. Single circulating tumor cell detection and overall survival in nonmetastatic breast cancer. *Ann. Oncol.* **2010**, *21*, 729–733. [CrossRef] [PubMed]
7. Lucci, A.; Hall, C.S.; Lodhi, A.K.; Bhattacharyya, A.; Anderson, A.E.; Xiao, L.; Bedrosian, I.; Kuerer, H.M.; Krishnamurthy, S. Circulating tumour cells in non-metastatic breast cancer: A prospective study. *Lancet Oncol.* **2012**, *13*, 688–695. [CrossRef]
8. Fina, E.; Reduzzi, C.; Motta, R.; Di Cosimo, S.; Bianchi, G.; Martinetti, A.; Wechsler, J.; Cappelletti, V.; Daidone, M.G. Did circulating tumor cells tell us all they could? The missed circulating tumor cell message in breast cancer. *Int. J. Biol. Markers* **2015**, *30*, 429–433. [CrossRef]
9. Sänger, N.; Effenberger, K.E.; Riethdorf, S.; Van Haasteren, V.; Gauwerky, J.; Wiegratz, I.; Strebhardt, K.; Kaufmann, M.; Pantel, K. Disseminated tumor cells in the bone marrow of patients with ductal carcinoma in situ. *Int. J. Cancer* **2011**, *129*, 2522–2526. [CrossRef]
10. Franken, B.; de Groot, M.R.; Mastboom, W.J.; Vermes, I.; van der Palen, J.; Tibbe, A.G.; Terstappen, L.W. Circulating tumor cells, disease recurrence and survival in newly diagnosed breast cancer. *Breast Cancer Res.* **2012**, *14*, R133. [CrossRef]
11. Banys, M.; Hahn, M.; Gruber, I.; Krawczyk, N.; Wallwiener, M.; Hartkopf, A.; Taran, F.A.; Röhm, C.; Kurth, R.; Becker, S.; et al. Detection and clinical relevance of hematogenous tumor cell dissemination in patients with ductal carcinoma in situ. *Breast Cancer Res. Treat.* **2014**, *144*, 531–538. [CrossRef]
12. Bloom, H.J.; Richardson, W.W. Histological grading and prognosis in breast cancer; a study of 1409 cases of which 359 have been followed for 15 years. *Br. J. Cancer* **1957**, *11*, 359–377. [CrossRef] [PubMed]
13. Sotiriou, C.; Pusztai, L. Gene-expression signatures in breast cancer. *N. Engl. J. Med.* **2009**, *360*, 790–800. [CrossRef] [PubMed]
14. Bertos, N.R.; Park, M. Breast cancer—One term, many entities? *J. Clin. Investig.* **2011**, *121*, 3789–3796. [CrossRef] [PubMed]
15. Russnes, H.G.; Navin, N.; Hicks, J.; Borresen-Dale, A.L. Insight into the heterogeneity of breast cancer through next-generation sequencing. *J. Clin. Investig.* **2011**, *121*, 3810–3818. [CrossRef] [PubMed]
16. Korkaya, H.; Liu, S.; Wicha, M.S. Breast cancer stem cells, cytokine networks, and the tumor microenvironment. *J. Clin. Investig.* **2011**, *121*, 3804–3809. [CrossRef] [PubMed]
17. Place, A.E.; Jin Huh, S.; Polyak, K. The microenvironment in breast cancer progression: Biology and implications for treatment. *Breast Cancer Res.* **2011**, *13*, 227. [CrossRef]
18. Curtis, C.; Shah, S.P.; Chin, S.F.; Turashvili, G.; Rueda, O.M.; Dunning, M.J.; Speed, D.; Lynch, A.G.; Samarajiwa, S.; Yuan, Y.; et al. The genomic and transcriptomic architecture of 2,000 breast tumours reveals novel subgroups. *Nature* **2012**, *486*, 346–352. [CrossRef]
19. Dawson, S.J.; Rueda, O.M.; Aparicio, S.; Caldas, C. A new genome-driven integrated classification of breast cancer and its implications. *EMBO J.* **2013**, *32*, 617–628. [CrossRef]
20. Dexter, D.L.; Leith, J.T. Tumor heterogeneity and drug resistance. *J. Clin. Oncol.* **1986**, *4*, 244–257. [CrossRef]
21. Dagogo-Jack, I.; Shaw, A.T. Tumour heterogeneity and resistance to cancer therapies. *Nat. Rev. Clin. Oncol.* **2018**, *15*, 81–94. [CrossRef]

22. Fidler, I.J. The pathogenesis of cancer metastasis: The 'seed and soil' hypothesis revisited. *Nat. Rev. Cancer* **2003**, *3*, 453–458. [CrossRef] [PubMed]
23. Gupta, G.P.; Massagué, J. Cancer metastasis: Building a framework. *Cell* **2006**, *127*, 679–695. [CrossRef] [PubMed]
24. Talmadge, J.E.; Fidler, I.J. AACR centennial series: The biology of cancer metastasis: Historical perspective. *Cancer Res.* **2010**, *70*, 5649–5669. [CrossRef] [PubMed]
25. Saxena, M.; Christofori, G. *Mol. Oncol.* **2013**, *7*, 283–296. [CrossRef] [PubMed]
26. Khanna, C.; Hunter, K. Modeling metastasis in vivo. *Carcinogenesis* **2005**, *26*, 513–523. [CrossRef] [PubMed]
27. Yin, J.J.; Selander, K.; Chirgwin, J.M.; Dallas, M.; Grubbs, B.G.; Wieser, R.; Massagué, J.; Mundy, G.R.; Guise, T.A. TGF-beta signaling blockade inhibits PTHrP secretion by breast cancer cells and bone metastases development. *J. Clin. Investig.* **1999**, *103*, 197–206. [CrossRef] [PubMed]
28. Kang, Y.; Siegel, P.M.; Shu, W.; Drobnjak, M.; Kakonen, S.M.; Cordón-Cardo, C.; Guise, T.A.; Massagué, J. A multigenic program mediating breast cancer metastasis to bone. *Cancer Cell* **2003**, *3*, 537–549. [CrossRef]
29. Harms, J.F.; Welch, D.R. MDA-MB-435 human breast carcinoma metastasis to bone. *Clin. Exp. Metastasis* **2003**, *20*, 327–334. [CrossRef]
30. Minn, A.J.; Gupta, G.P.; Siegel, P.M.; Bos, P.D.; Shu, W.; Giri, D.D.; Viale, A.; Olshen, A.B.; Gerald, W.L.; Massagué, J. Genes that mediate breast cancer metastasis to lung. *Nature* **2005**, *436*, 518–524. [CrossRef]
31. Nguyen, D.X.; Bos, P.D.; Massagué, J. Metastasis: From dissemination to organ-specific colonization. *Nat. Rev. Cancer* **2009**, *9*, 274–284. [CrossRef]
32. Bos, P.D.; Nguyen, D.X.; Massagué, J. Modeling metastasis in the mouse. *Curr. Opin. Pharmacol.* **2010**, *10*, 571–577. [CrossRef] [PubMed]
33. Vargo-Gogola, T.; Rosen, J.M. Modelling breast cancer: One size does not fit all. *Nat. Rev. Cancer* **2007**, *7*, 659–672. [CrossRef] [PubMed]
34. Neve, R.M.; Chin, K.; Fridlyand, J.; Yeh, J.; Baehner, F.L.; Fevr, T.; Clark, L.; Bayani, N.; Coppe, J.P.; Tong, F.; et al. A collection of breast cancer cell lines for the study of functionally distinct cancer subtypes. *Cancer Cell* **2006**, *10*, 515–527. [CrossRef] [PubMed]
35. Mackay, A.; Tamber, N.; Fenwick, K.; Iravani, M.; Grigoriadis, A.; Dexter, T.; Lord, C.J.; Reis-Filho, J.S.; Ashworth, A. A high-resolution integrated analysis of genetic and expression profiles of breast cancer cell lines. *Breast Cancer Res. Treat.* **2009**, *118*, 481–498. [CrossRef] [PubMed]
36. Prat, A.; Parker, J.S.; Karginova, O.; Fan, C.; Livasy, C.; Herschkowitz, J.I.; He, X.; Perou, C.M. Phenotypic and molecular characterization of the claudin-low intrinsic subtype of breast cancer. *Breast Cancer Res.* **2010**, *12*, R68. [CrossRef] [PubMed]
37. Lacroix, M.; Leclercq, G. Relevance of breast cancer cell lines as models for breast tumours: An update. *Breast Cancer Res. Treat.* **2004**, *83*, 249–289. [CrossRef] [PubMed]
38. Iorns, E.; Drews-Elger, K.; Ward, T.M.; Dean, S.; Clarke, J.; Berry, D.; El Ashry, D.; Lippman, M. A new mouse model for the study of human breast cancer metastasis. *PLoS ONE* **2012**, *7*, e47995. [CrossRef] [PubMed]
39. Nanni, P.; Nicoletti, G.; Palladini, A.; Croci, S.; Murgo, A.; Ianzano, M.L.; Grosso, V.; Stivani, V.; Antognoli, A.; Lamolinara, A.; et al. Multiorgan metastasis of human HER-2^{+} breast cancer in Rag2^{-}/$^{-}$;Il2rg^{-}/$^{-}$ mice and treatment with PI3K inhibitor. *PLoS ONE* **2012**, *7*, e39626. [CrossRef] [PubMed]
40. Marangoni, E.; Vincent-Salomon, A.; Auger, N.; Degeorges, A.; Assayag, F.; de Cremoux, P.; de Plater, L.; Guyader, C.; De Pinieux, G.; Judde, J.G.; et al. A new model of patient tumor-derived breast cancer xenografts for preclinical assays. *Clin. Cancer Res.* **2007**, *13*, 3989–3998. [CrossRef]
41. Bergamaschi, A.; Hjortland, G.O.; Triulzi, T.; Sørlie, T.; Johnsen, H.; Ree, A.H.; Russnes, H.G.; Tronnes, S.; Maelandsmo, G.M.; Fodstad, O.; et al. Molecular profiling and characterization of luminal-like and basal-like in vivo breast cancer xenograft models. *Mol. Oncol.* **2009**, *3*, 469–482. [CrossRef]
42. DeRose, Y.S.; Wang, G.; Lin, Y.C.; Bernard, P.S.; Buys, S.S.; Ebbert, M.T.; Factor, R.; Matsen, C.; Milash, B.A.; Nelson, E.; et al. Tumor grafts derived from women with breast cancer authentically reflect tumor pathology, growth, metastasis and disease outcomes. *Nat. Med.* **2011**, *17*, 1514–1520. [CrossRef] [PubMed]
43. Kabos, P.; Finlay-Schultz, J.; Li, C.; Kline, E.; Finlayson, C.; Wisell, J.; Manuel, C.A.; Edgerton, S.M.; Harrell, J.C.; Elias, A.; et al. Patient-derived luminal breast cancer xenografts retain hormone receptor heterogeneity

and help define unique estrogen-dependent gene signatures. *Breast Cancer Res. Treat.* **2012**, *135*, 415–432. [CrossRef] [PubMed]

44. Cassidy, J.W.; Caldas, C.; Bruna, A. Maintaining Tumor Heterogeneity in Patient-Derived Tumor Xenografts. *Cancer Res.* **2015**, *75*, 2963–2968. [CrossRef] [PubMed]
45. Bruna, A.; Rueda, O.M.; Greenwood, W.; Batra, A.S.; Callari, M.; Batra, R.N.; Pogrebniak, K.; Sandoval, J.; Cassidy, J.W.; Tufegdzic-Vidakovic, A.; et al. A Biobank of Breast Cancer Explants with Preserved Intra-tumor Heterogeneity to Screen Anticancer Compounds. *Cell* **2016**, *167*, 260–274. [CrossRef] [PubMed]
46. Zhang, X.; Claerhout, S.; Prat, A.; Dobrolecki, L.E.; Petrovic, I.; Lai, Q.; Landis, M.D.; Wiechmann, L.; Schiff, R.; Giuliano, M.; et al. A renewable tissue resource of phenotypically stable, biologically and ethnically diverse, patient-derived human breast cancer xenograft models. *Cancer Res.* **2013**, *73*, 4885–4897. [CrossRef] [PubMed]
47. Siolas, D.; Hannon, G.J. Patient-derived tumor xenografts: Transforming clinical samples into mouse models. *Cancer Res.* **2013**, *73*, 5315–5319. [CrossRef] [PubMed]
48. Workman, P.; Aboagye, E.O.; Balkwill, F.; Balmain, A.; Bruder, G.; Chaplin, D.J.; Double, J.A.; Everitt, J.; Farningham, D.A.; Glennie, M.J.; et al. Guidelines for the welfare and use of animals in cancer research. *Br. J. Cancer* **2010**, *102*, 1555–1577. [CrossRef]
49. ARRIVE Guidelines. Available online: https://www.nc3rs.org.uk/arrive-guidelines (accessed on 28 May 2019).
50. Hofman, V.J.; Ilie, M.I.; Bonnetaud, C.; Selva, E.; Long, E.; Molina, T.; Vignaud, J.M.; Fléjou, J.F.; Lantuejoul, S.; Piaton, E.; et al. Cytopathologic detection of circulating tumor cells using the isolation by size of epithelial tumor cell method: Promises and pitfalls. *Am. J. Clin. Pathol.* **2011**, *135*, 146–156. [CrossRef]
51. Cho, E.H.; Wendel, M.; Luttgen, M.; Yoshioka, C.; Marrinucci, D.; Lazar, D.; Schram, E.; Nieva, J.; Bazhenova, L.; Morgan, A.; et al. Characterization of circulating tumor cell aggregates identified in patients with epithelial tumors. *Phys. Biol.* **2012**, *9*, 016001. [CrossRef]
52. Adams, D.L.; Martin, S.S.; Alpaugh, R.K.; Charpentier, M.; Tsai, S.; Bergan, R.C.; Ogden, I.M.; Catalona, W.; Chumsri, S.; Tang, C.M.; et al. Circulating giant macrophages as a potential biomarker of solid tumors. *Proc. Natl. Acad. Sci. USA* **2014**, *111*, 3514–3519. [CrossRef]
53. Wechsler, J. *Atlas de Cytologie. Cellules Tumorales Circulantes Des Cancers Solides*; Sauramps Medical: Montpellier, France, 2015; ISBN 979-103030-009-3.
54. Benjamini, Y.; Krieger, A.M.; Yekutieli, D. Adaptive linear step-up procedures that control the false discovery rate. *Biometrika* **2006**, *93*, 491–507. [CrossRef]
55. Eliane, J.P.; Repollet, M.; Luker, K.E.; Brown, M.; Rae, J.M.; Dontu, G.; Schott, A.F.; Wicha, M.; Doyle, G.V.; Hayes, D.F.; et al. Monitoring serial changes in circulating human breast cancer cells in murine xenograft models. *Cancer Res.* **2008**, *68*, 5529–5532. [CrossRef] [PubMed]
56. Nguyen, A.; Yoshida, M.; Goodarzi, H.; Tavazoie, S.F. Highly variable cancer subpopulations that exhibit enhanced transcriptome variability and metastatic fitness. *Nat. Commun.* **2016**, *7*, 11246. [CrossRef] [PubMed]
57. Park, S.; Ang, R.R.; Duffy, S.P.; Bazov, J.; Chi, K.N.; Black, P.C.; Ma, H. Morphological differences between circulating tumor cells from prostate cancer patients and cultured prostate cancer cells. *PLoS ONE* **2014**, *9*, e85264. [CrossRef] [PubMed]
58. Fraker, L.D.; Halter, S.A.; Forbes, J.T. Growth inhibition by retinol of a human breast carcinoma cell line in vitro and in athymic mice. *Cancer Res.* **1984**, *44*, 5757–5763. [PubMed]
59. Price, J.E.; Polyzos, A.; Zhang, R.D.; Daniels, L.M. Tumorigenicity and metastasis of human breast carcinoma cell lines in nude mice. *Cancer Res.* **1990**, *50*, 717–721.
60. Zhang, R.D.; Fidler, I.J.; Price, J.E. Relative malignant potential of human breast carcinoma cell lines established from pleural effusions and a brain metastasis. *Invasion Metastasis* **1991**, *11*, 204–215.
61. Clarke, R. Animal models of breast cancer: Their diversity and role in biomedical research. *Breast Cancer Res. Treat.* **1996**, *39*, 1–6. [CrossRef]
62. Juratli, M.A.; Sarimollaoglu, M.; Nedosekin, D.A.; Melerzanov, A.V.; Zharov, V.P.; Galanzha, E.I. Dynamic Fluctuation of Circulating Tumor Cells during Cancer Progression. *Cancers* **2014**, *6*, 128–142. [CrossRef]

63. Brown, M.; Assen, F.P.; Leithner, A.; Abe, J.; Schachner, H.; Asfour, G.; Bago-Horvath, Z.; Stein, J.V.; Uhrin, P.; Sixt, M.; et al. Lymph node blood vessels provide exit routes for metastatic tumor cell dissemination in mice. *Science* **2018**, *359*, 1408–1411. [CrossRef]
64. Pereira, E.R.; Kedrin, D.; Seano, G.; Gautier, O.; Meijer, E.F.J.; Jones, D.; Chin, S.M.; Kitahara, S.; Bouta, E.M.; Chang, J.; et al. Lymph node metastases can invade local blood vessels, exit the node, and colonize distant organs in mice. *Science* **2018**, *359*, 1403–1407. [CrossRef] [PubMed]
65. Bonnomet, A.; Syne, L.; Brysse, A.; Feyereisen, E.; Thompson, E.W.; Noël, A.; Foidart, J.M.; Birembaut, P.; Polette, M.; Gilles, C. A dynamic in vivo model of epithelial-to-mesenchymal transitions in circulating tumor cells and metastases of breast cancer. *Oncogene* **2012**, *31*, 3741–3753. [CrossRef] [PubMed]
66. Aceto, N.; Bardia, A.; Miyamoto, D.T.; Donaldson, M.C.; Wittner, B.S.; Spencer, J.A.; Yu, M.; Pely, A.; Engstrom, A.; Zhu, H.; et al. Circulating tumor cell clusters are oligoclonal precursors of breast cancer metastasis. *Cell* **2014**, *158*, 1110–1122. [CrossRef] [PubMed]
67. Glinsky, V.V.; Glinsky, G.V.; Glinskii, O.V.; Huxley, V.H.; Turk, J.R.; Mossine, V.V.; Deutscher, S.L.; Pienta, K.J.; Quinn, T.P. Intravascular metastatic cancer cell homotypic aggregation at the sites of primary attachment to the endothelium. *Cancer Res.* **2003**, *63*, 3805–3811. [CrossRef] [PubMed]
68. Cheung, K.J.; Padmanaban, V.; Silvestri, V.; Schipper, K.; Cohen, J.D.; Fairchild, A.N.; Gorin, M.A.; Verdone, J.E.; Pienta, K.J.; Bader, J.S.; et al. Polyclonal breast cancer metastases arise from collective dissemination of keratin 14-expressing tumor cell clusters. *Proc. Natl. Acad. Sci. USA* **2016**, *113*, E854–E863. [CrossRef] [PubMed]
69. Labelle, M.; Begum, S.; Hynes, R.O. Direct signaling between platelets and cancer cells induces an epithelial-mesenchymal-like transition and promotes metastasis. *Cancer Cell* **2011**, *20*, 576–590. [CrossRef] [PubMed]
70. Coffelt, S.B.; Kersten, K.; Doornebal, C.W.; Weiden, J.; Vrijland, K.; Hau, C.S.; Verstegen, N.J.M.; Ciampricotti, M.; Hawinkels, L.J.A.C.; Jonkers, J.; et al. IL-17-producing $\gamma\delta$ T cells and neutrophils conspire to promote breast cancer metastasis. *Nature* **2015**, *522*, 345–348. [CrossRef] [PubMed]
71. Wculek, S.K.; Malanchi, I. Neutrophils support lung colonization of metastasis-initiating breast cancer cells. *Nature* **2015**, *528*, 413–417. [CrossRef]
72. Hanna, R.N.; Cekic, C.; Sag, D.; Tacke, R.; Thomas, G.D.; Nowyhed, H.; Herrley, E.; Rasquinha, N.; McArdle, S.; Wu, R.; et al. Patrolling monocytes control tumor metastasis to the lung. *Science* **2015**, *350*, 985–990. [CrossRef]
73. Linde, N.; Casanova-Acebes, M.; Sosa, M.S.; Mortha, A.; Rahman, A.; Farias, E.; Harper, K.; Tardio, E.; Reyes Torres, I.; Jones, J.; et al. Macrophages orchestrate breast cancer early dissemination and metastasis. *Nat. Commun.* **2018**, *9*, 21. [CrossRef]
74. Szczerba, B.M.; Castro-Giner, F.; Vetter, M.; Krol, I.; Gkountela, S.; Landin, J.; Scheidmann, M.C.; Donato, C.; Scherrer, R.; Singer, J.; et al. Neutrophils escort circulating tumour cells to enable cell cycle progression. *Nature* **2019**, *566*, 553–557. [CrossRef] [PubMed]
75. Yuan, L.; Liu, X. Platelets are associated with xenograft tumor growth and the clinical malignancy of ovarian cancer through an angiogenesis-dependent mechanism. *Mol. Med. Rep.* **2015**, *11*, 2449–2458. [CrossRef] [PubMed]
76. Gast, C.E.; Silk, A.D.; Zarour, L.; Riegler, L.; Burkhart, J.G.; Gustafson, K.T.; Parappilly, M.S.; Roh-Johnson, M.; Goodman, J.R.; Olson, B.; et al. Cell fusion potentiates tumor heterogeneity and reveals circulating hybrid cells that correlate with stage and survival. *Sci. Adv.* **2018**, *4*, eaat7828. [CrossRef] [PubMed]
77. Zhang, Y.; Zhou, N.; Yu, X.; Zhang, X.; Li, S.; Lei, Z.; Hu, R.; Li, H.; Mao, Y.; Wang, X.; et al. Tumacrophage: Macrophages transformed into tumor stem-like cells by virulent genetic material from tumor cells. *Oncotarget* **2017**, *8*, 82326–82343. [CrossRef] [PubMed]
78. Xia, Z.; Ye, H.; Choong, C.; Ferguson, D.J.; Platt, N.; Cui, Z.; Triffitt, J.T. Macrophagic response to human mesenchymal stem cell and poly(epsilon-caprolactone) implantation in nonobese diabetic/severe combined immunodeficient mice. *J. Biomed. Mater. Res. A* **2004**, *71*, 538–548. [CrossRef] [PubMed]
79. Weiler, J.; Dittmar, T. Cell Fusion in Human Cancer: The Dark Matter Hypothesis. *Cells* **2019**, *8*, 132. [CrossRef] [PubMed]

14

CTC-Derived Models: A Window into the Seeding Capacity of Circulating Tumor Cells (CTCs)

Tala Tayoun [1,2,3], Vincent Faugeroux [1,2], Marianne Oulhen [1,2], Agathe Aberlenc [1,2], Patrycja Pawlikowska [2] and Françoise Farace [1,2,*]

1 "Circulating Tumor Cells" Translational Platform, CNRS UMS3655 – INSERM US23AMMICA, Gustave Roussy, Université Paris-Saclay, F-94805 Villejuif, France; tala.tayoun@gustaveroussy.fr (T.T.); vincent.faugeroux@laposte.net (V.F.); marianne.oulhen@gustaveroussy.fr (M.O.); agathe.aberlenc@gustaveroussy.fr (A.A.)
2 INSERM, U981 "Identification of Molecular Predictors and new Targets for Cancer Treatment", F-94805 Villejuif, France; patrycjamarta.pawlikowska@gustaveroussy.fr
3 Faculty of Medicine, Université Paris Sud, Université Paris-Saclay, F-94270 Le Kremlin-Bicetre, France
* Correspondence: francoise.farace@gustaveroussy.fr

Abstract: Metastasis is the main cause of cancer-related death owing to the blood-borne dissemination of circulating tumor cells (CTCs) early in the process. A rare fraction of CTCs harboring a stem cell profile and tumor initiation capacities is thought to possess the clonogenic potential to seed new lesions. The highest plasticity has been generally attributed to CTCs with a partial epithelial-to-mesenchymal transition (EMT) phenotype, demonstrating a large heterogeneity among these cells. Therefore, detection and functional characterization of these subclones may offer insight into mechanisms underlying CTC tumorigenicity and inform on the complex biology behind metastatic spread. Although an in-depth mechanistic investigation is limited by the extremely low CTC count in circulation, significant progress has been made over the past few years to establish relevant systems from patient CTCs. CTC-derived xenograft (CDX) models and CTC-derived ex vivo cultures have emerged as tractable systems to explore tumor-initiating cells (TICs) and uncover new therapeutic targets. Here, we introduce basic knowledge of CTC biology, including CTC clusters and evidence for EMT/cancer stem cell (CSC) hybrid phenotypes. We report and evaluate the CTC-derived models generated to date in different types of cancer and shed a light on challenges and key findings associated with these novel assays.

Keywords: metastasis; tumor-initiating cells (TICs); circulating tumor cells (CTCs); CTC-derived xenografts; CTC-derived ex vivo models

---

1. Introduction

Metastatic spread and its resistance to treatment remain the leading cause of death in cancer patients. This process is fueled by malignant cells that dissociate from the primary tumor and travel through the bloodstream to colonize distant organs. These cells are referred to as "circulating tumor cells" (CTCs) and are able to enter vasculature during the early course of disease. Nonetheless, the majority of the tumor cell population dies during transit as a result of biological and physical constraints such as shear stress and immune surveillance, and only a minor subset of the surviving CTCs (0.01%) acquires the capacity of tumor-initiating cells (TICs) [1–4]. The outcome of tumor dissemination is dependent on a selection process that favors the survival of a small proportion of cancer cells holding the self-renewal ability of stem cells along with TIC properties, which enables them to seed tumors and reconstitute tumor heterogeneity [5–7]. These cells are termed "cancer stem cells"

(CSCs), and CTCs holding a CSC phenotype have been detected and associated with high invasiveness and tumorigenicity in many cancers including breast cancer (BC), colorectal cancer, and glioma [8–11].

An important aspect of CTC research is to study the mechanistic basis behind their TIC properties and explore new CTC-based biomarkers and targeting strategies. The generation of CTC-derived xenografts (CDXs) or CTC-derived cell lines at relevant time points during disease progression is therefore crucial to achieve a longitudinal and functional characterization of these cells, along with in vivo and in vitro pharmacological testing. Although this task remains challenging owing to CTCs scarcity in peripheral blood and technical hurdles related to their enrichment strategies, significant efforts have been made in the establishment of clinically relevant systems to study CTC biology in different cancer types. In this review, we briefly cover basic knowledge of TIC-related properties in CTCs and evaluate the existing CTC-derived models, including both in vivo CDXs and in vitro functional culture assays in different cancers. We also highlight the important findings which have helped unveil new insights into CTC biology and novel therapeutic strategies.

2. Brief Glimpse into TIC-Related Properties of CTCs

CTC profile evolves as the initial events of the metastatic cascade take place. Indeed, CTCs undergo reversible phenotypic alterations to achieve intravasation, survival in vasculature and extravasation, known as epithelial-to-mesenchymal transition (EMT). During EMT—a key phenomenon in embryonic development—cancer cells undergo cytoskeletal changes and typically lose their cell–cell adhesion proteins as well as their polarity to become motile cells and intravasate [12,13]. EMT signatures were detectable in CTCs of BC patients [14–17]. Increasing experimental evidence draws a potential link between EMT and acquisition of stemness [12,13,18,19]. In fact, several EMT-inducing transcription factors have been shown to confer malignancy in neoplastic cells, leading to the emergence of highly aggressive clones with combined EMT/CSC traits [20–23]. Nevertheless, this association is not universal. Indeed, it has been suggested that the loss of the EMT-inducing factor *Prxx1* is required for cancer cells to colonize organs in vivo, which revert to the epithelial state and acquire CSC traits, thus uncoupling EMT and stemness [19,24,25]. Moreover, the requirement of EMT for CTC dissemination has long been subject to debate. Several studies have shown that mesenchymal features in tumor cells may indeed be dispensable for their migratory activity but could contribute molecularly and phenotypically to chemoresistance [26–28]. It is currently hypothesized that CTC subclones displaying an intermediate phenotype between epithelial and mesenchymal have the highest plasticity to adapt to the microenvironment and generate a more aggressive CTC population resistant to conventional chemotherapy and capable of metastatic outgrowth. Our group showed the existence of a hybrid epithelial/mesenchymal (E/M) phenotype in CTCs from patients with non-small cell lung cancer (NSCLC) [29]. Heterogeneous expression of EMT markers within SCLC and NSCLC patient cohorts was described by Hou et al., while Hofman et al. reported the presence of proportions of NSCLC CTCs which expressed the mesenchymal marker vimentin and correlated with shorter disease-free survival [30,31]. Recent data in metastatic BC patients showed the enrichment of CTC subpopulations with a CSC^{+}/partial EMT^{+} signature in patients post-treatment, which correlated with worse clinical outcome [32]. Indeed, the CTC population is described as a highly heterogeneous pool of tumor cells with low numbers of metastasis-initiating cells (MICs) that are sometimes prone to apoptosis [33]. The different factors influencing MIC properties of CTCs and their survival underlie the complexity and inefficiency of organ invasion and macro-metastases formation, relevant both clinically and in experimental mouse models [4,34,35]. Recent advances in single-cell technologies have unraveled CTC-specific genetic mutations and profiling of the CTC population thus points out the emergence of subclones with dynamic phenotypes that contribute to the evolution of the tumor genome during disease progression and treatment [36–39]. CTCs are less frequently found in clusters, also termed "circulating tumor microemboli" (CTM), which travel as 2–50 cells in vasculature and present extremely enhanced metastatic competency [40]. This can be explained by the survival advantage they hold over single CTCs, as CTM were shown to escape anoikis as well as stresses in circulation [30,41]. A recent report showed that these characteristics are due to CSC properties of CTM, notably a CD44-directed cell aggregation mechanism that forms these clusters, promotes their survival and favors polyclonal metastasis [42]. Another group also investigated the factors behind CTM metastatic potential: Gkountela et al. reported that CTC clusters from BC patients and CTC cell lines exhibit a DNA methylation pattern distinct from that of single CTCs and which represents targetable vulnerabilities [43]. Moreover, CTC-neutrophils clusters are occasionally formed in the bloodstream and in vivo evidence shows that this association triggers cell cycle progression and thus drives metastasis formation in BC [44].

3. Brief Introduction to CTC Enrichment and Detection Strategies

A plethora of technologies have been developed over the last decade to respond to specific CTC applications. CTC identification remains a technically challenging task due to the extreme phenotypic heterogeneity and rarity of these cells in the bloodstream and therefore requires methods with high sensitivity and specificity. Enrichment strategies can be based on either biological properties (i.e., cell-surface markers) or physical characteristics (i.e., size, density, electric charge) and are usually combined with detection techniques (e.g., immunofluorescence, immunohistochemistry, FISH) to identify CTCs. CTC capture relies on a positive selection among normal blood cells or a negative selection by leukocyte depletion. Among biologically-based technologies is the CellSearch system (Menarini-Silicon Biosystem, Bologna, Italy). It is the most commonly applied assay for CTC enumeration in which CTCs are captured in whole blood by EpCAM (epithelial cell adhesion molecule)-coated immunomagnetic beads followed by fluorescent detection using anti-cytokeratins (CK 8, CK 18, CK 19), anti-CD45 (leukocyte marker), and a nuclear stain (DAPI). It is the only technology cleared by the US Food and Drug Administration to aid in prognosis for patients with metastatic breast, prostate, and colorectal cancer [45–49]. Although standardized and reproducible, this method has a limited sensitivity most likely due to failure in recognizing cells undergoing EMT and thus inevitably misses an aggressive and clinically relevant CTC subpopulation. Platforms relying on the depletion of leukocytes (negative selection) are being investigated and used to overcome this bias. One example is the widely used RosetteSep technique which enriches CTCs without phenotypic a priori by excluding $CD45^+$ and $CD36^+$ cells in rosettes and eliminating them in a Ficoll-Paque PLUS density-gradient centrifugation. Physical property-based methods including filtration systems have been developed to capture CTCs based on their large size compared to leukocytes, notably the ISET® (*Isolation by Size of Tumor Cells*) (RareCells Diagnostics, Paris, France) and the ScreenCell® (Paris, France) methods, which are able to detect CTCs as well as CTM using microporous polycarbonate filters [50,51]. In line with this notion, we and others have reported an overall higher recovery rate using ISET compared to CellSearch for CTC enumeration in NSCLC and prostate cancer patients [31,52]. Our lab developed a novel CTC detection approach combining ISET filtration with a FISH assay, optimized for the detection of *ALK*- or *ROS1*-rearranged pattern of NSCLC CTCs on filters [53,54]. To ensure a wider coverage of CTC heterogeneity, new devices are being developed (and some commercially) such as the CTC-iChip which relies on both biological and physical properties of CTCs: it applies size-based filtration using microfluidics processing, followed by positive selection of CTCs with EpCAM-conjugated beads or negative selection with $CD45^-$-coated beads to deplete hematopoietic cells [55]. Different technologies have been implemented to isolate live CTCs (without a fixation step) and perform subsequent functional studies. Some strategies have integrated isolation protocols for molecular analysis of single CTCs. One example is the DEPArray™ (Silicon Biosystems S.p.A., Bologna, Italy), a microfluidic system which sorts live single CTCs based on image selection followed by entrapment of CTCs inside dielectrophoretic cages [56–58]. FACS has also been adapted for molecular characterization of CTCs as well as their isolation in the aim of xenograft establishment [59].

At this point, none of the technologies fully respond to the phenotypic heterogeneity of CTCs. Indeed, each method has its own advantages and limitations and researchers have based the development of capture strategies on the specific aim of further CTC characterization studies. New insights in CTC biology should be integrated into current enrichment, detection, and isolation techniques to optimize the process and improve their reliability. As shown in Table 1, RosetteSep and FACS have been used for CDX establishment. Enrichment using RosetteSep may be advantageous owing to the lack of phenotypic a priori on tumorigenic CTCs and a higher recovery rate.

4. CTC-Derived Xenografts

Patient-derived xenograft (PDX) technology has rapidly emerged as a standard translational research platform to improve understanding of cancer biology and test novel therapeutic strategies [60]. PDXs are generated by implantation of surgically-removed tumor tissue (primary or metastasis) into immunodeficient mice. Although these models have proven utility as a preclinical tool in many cancers, their feasibility remains challenged by limited tumor tissue availability, as single-site biopsies may be impossible or detrimental in some malignancies [61]. This limitation can be overcome by the generation of CDX models after enrichment of CTCs collected from a readily accessible blood draw and subsequent injection into immunodeficient mice [62–64]. Nevertheless, it is noteworthy that CDX development still presents an enormous challenge due to low CTC prevalence in several cancers. Until now, CDXs have been established in breast, melanoma, lung and prostate cancer and are discussed in this section (Table 1).

In 2013, Baccelli et al. reported the first experimental proof that primary human luminal BC CTC populations contain MICs in a xenograft assay. Injection of CTCs from 110 patients was performed. Six recipient mice developed bone, lung, and liver metastases within 6–12 months after CTC transplant (~1000 CTCs) from three patients with advanced metastatic BC. Cell sorting analysis of the MIC-containing population shared a common $EpCAM^{+}CD44^{+}MET^{+}CD47^{+}$ phenotype, highlighting a CSC characteristic of CTCs. The authors also showed that the number of CTCs positive for these markers strongly correlated with decreased progression-free survival of metastatic BC patients. This study has therefore revealed a first phenotypic identification of luminal BC CTCs with MIC properties, making them an attractive tool to track and potentially target metastatic development in BC [59]. A second group derived a CDX model from a metastatic triple-negative BC (TNBC) patient for the first time. The patient selected for CDX establishment had advanced TNBC with a very high CTC count obtained with CellSearch analysis (969 CTCs and 74 CTC clusters/7.5 mL). Enriched cells were injected subcutaneously into nude mice and a palpable tumor was observed five months later. The authors carried out a longitudinal study and samples were collected at two different time points (metastasis and progression) during the course of the disease, which allowed real-time assessment of molecular changes between patient tumor, CTCs, and CDXs samples. The obtained CDX phenocopied the patient tumor. Most importantly, RNA sequencing of the CDX tumor disclosed key mechanisms relevant in TNBC biology such as the WNT pathway, which is necessary for the maintenance of CSCs and was shown to correlate with metastasis and poor clinical outcome in TNBC subtypes. CTC analysis also deciphered a panel of potential tumor biomarkers [65]. An additional TNBC CDX model of liver metastasis was established very recently by Vishnoi et al. Similar genomic profiling of metastatic tissue was obtained in four sequential CDX generations, representing the recapitulation of liver metastasis in all the models. Notably, the authors deciphered a first 597-gene CTC signature related to liver metastasis in TNBC which, despite small sample size bias, can provide insight into the mechanistic basis of TNBC disease progression in the liver [66].

In melanoma, Girotti et al. demonstrated the tumorigenicity of advanced-disease CTCs in immunocompromised mice. The authors resorted to CDX development when tumor material was inaccessible for PDX generation. They reported a success rate of 13% with six CDX established, 15 failed attempts, and 26 additional models followed at the time of publication. CDX tumor growth was detectable as of one month after CTC implantation and was sustainable in secondary hosts. Moreover, the CDXs were representative of patient tumors and mirrored therapy response. This proof-of-principle was developed along with PDX technology and circulating tumor DNA analysis as part of a platform to optimize precision medicine for melanoma patients. It explored the TIC properties of melanoma CTCs but did not achieve a biological characterization of these cells [67].

In lung cancer, Hodgkinson et al. showed that CTCs in chemosensitive or chemorefractory SCLC are tumorigenic. CTCs were isolated from six late-stage SCLC patients having never received chemotherapy and were subsequently injected into NSG mice. Each patient presented with more than 400 CTCs and four out of six CTC samples gave rise to CDX tumors detected as of 2.4 months post-implantation. CDXs recapitulated the genomic profile of CellSearch-enriched CTCs and mimicked donor patients response to standard of care chemotherapy (platinum and etoposide), proving the clinical relevance of these models [68]. CDX tumors were subsequently dissociated and expanded into short-term in vitro CDX cultures (Table 2). These cells maintained the genomic landscape of donor tumors as well as their drug sensitivity profiles. CDX-derived cells were also labeled in vitro with the GFP lentivirus and successfully implanted into mice, where they can serve as a tracking tool to study tumor dissemination patterns in vivo [69]. Additional 16 SCLC CDX models were recently generated by Drapkin et al. from CTCs collected at initial diagnosis or at progression, with 38% efficiency. Somatic mutations were maintained between patient tumors and CDX as shown by whole-exome sequencing (WES) and the genomic landscape remained stable throughout early CDX passages showing clonal homogeneity. The authors also developed serial CDX models from one patient at baseline of the combination olaparib and temozolomide and at relapse. Interestingly, the models accurately reflected the evolving drug sensitivity profiles of the patient's malignancy, which highlights the potential utility of serial CDXs to study the evolution of resistance to treatment in SCLC [70]. One CDX model was also described in NSCLC. In this study, CTC samples were retrieved at two different time points: Baseline and post-brain radiotherapy. No CDX was developed at baseline. Notably, no EpCAM$^+$ CTCs were detected during CellSearch analysis at disease progression, yet injection of post-radiotherapy CTCs gave rise to a palpable tumor 95 days after engraft. Phenotypic and molecular characterizations showed no epithelial CTCs, but revealed a sizeable population of phenotypically heterogeneous CTCs mostly expressing the mesenchymal marker vimentin. This study suggests that the absence of EpCAM$^+$ CTCs in NSCLC does not preclude the existence of CTCs with TIC potential in patients and underlines the importance of investigating CTCs undergoing EMT in this malignancy [71].

Our group generated the first CDX model of castration-resistant prostate cancer (CRPC) and derived a permanent ex vivo culture from CDX tumor cells. A total of 22 samples from metastatic CRPC patients were collected, among which seven were obtained from diagnostic leukapheresis (DLA). DLA products were generated as part of the European FP7 program CTCTrap which aimed for an increased CTC yield to perform molecular characterization of the tumor [72]. One patient with a very high CTC count (~20,000 CTCs) obtained by DLA gave rise to a palpable tumor within 5.5 months. Acquisition of key genetic drivers (i.e., *TP53, PTEN*, and *RB1*) that govern the trans-differentiation of CRPC into CRPC-neuroendocrine (CRPC-NE) malignancy was detected in CTCs, highlighting the role of tumorigenic CRPC-NE CTCs in this transformation. Moreover, the obtained in vitro CDX-derived cell line faithfully recapitulated the genetic characteristics and tumorigenicity of the CDX and mimicked patient response to standard of care treatments for CRPC (i.e., enzalutamide and docetaxel) (Table 2) [73,74].

Table 1. Overview of in vivo circulating tumor cell (CTC)-derived models established to date.

CTC-Derived Xenografts

Type of Cancer	Stage	Live CTC Isolation Technique	# of CTCs	Injection Procedure	Take Rate	Passaging	Main Findings	Ref
Breast cancer	Metastatic luminal	FACS isolation (PI$^-$CD45$^-$EpCAM$^+$) or RosetteSep	≥1109 CTCs EpCAM$^+$(CellSearch)	- Dilution in matrigel - Injection in femoral medullar cavity	5%	N/A	- Specific CTC MIC signature EpCAM$^+$CD44$^+$MET$^+$CD47$^+$ - Recapitulation of patient metastases phenotype in CDX metastases - No drug sensitivity study	[59]
	Metastatic triple-negative	Density gradient centrifugation: Histopaque®	969 CTCs EpCAM$^+$ (CellSearch)	- Dilution in matrigel - Subcutaneous inje	3%	Piece of tumor explant or injection of explant culture	- RT-qPCR for genomic profiling of CTC/CDX samples before and after injection - WNT pathway upregulation as a potential therapeutic target in TNBC identified by RNAseq - No drug sensitivity study	[65]
	Metastatic triple-negative	FACS (CD45$^-$/CD34$^-$/CD105$^-$/CD90$^-$ CD73$^-$)	N/A	Intracardiac injection	33%	Minced metastatic liver tissue	- Identification of a TNBC liver metastasis CTC-specific signature (whole-transcriptome)- Survival analyses for signature transcripts	[66]
Melanoma	Stage IV	RosetteSep	N/A	- Dilution in matrigel - Subcutaneous inje	13%	Tumor fragments	- recapitulation of patient response to dabrafetinib in the CDX - concordance in SNV profiles (WES/RNAseq)	[67]
SCLC	Metastatic	RosetteSep	>400 CTCs EpCAM$^+$ (CellSearch)	Dilution in matrigel/subcutaneous	67%	Tumor fragments	- Recapitulation of CTC genomic profile by CDX tumors - CDX mimicked donor's response to chemotherapy	[68]
	Limited or extensive stage	CTC-iChip + RosetteSep Ficoll	N/A	Dilution in matrigel/subcutaneous	38%	Tumor fragments	- Faithful recapitulation of the tumor genome - Reflection of evolving treatment sensitivities of patient tumor	[70]

Table 1. *Cont.*

CTC-Derived Xenografts								
Type of Cancer	**Stage**	**Live CTC Isolation Technique**	**# of CTCs**	**Injection Procedure**	**Take Rate**	**Passaging**	**Main Findings**	**Ref**
NSCLC	Metastatic	RosetteSep	>150 CTCs by FACS (CD45/CD144/ vimentin/CK)	Dilution in matrigel/subcutaneous	100%	Disaggregation of tumor	- Importance of mesenchymal CTCs with tumorigenic capacity	[71]
CRPC	Metastatic	DLA/RosetteSep	~20,000 CTCs $EpCAM^+$ (CellSearch)	Dilution in matrigel/subcutaneous	14%	Tumor fragments	- Recapitulation of genome characteristics in CTC, patient tumor and CDX (WES) - Tumorigenic CTCs with acquired CRPC-NE features	[73]

* N/A: not available; FACS: Fluorescent-activated cell sorting; CDX: CTC-derived xenograft; MIC: Metastasis-initiating cell; TIC: Tumor-initiating cell; TNBC: Triple-negative breast cancer; SCLC: Small-cell lung cancer; SNV: Single nucleotide variant; NSCLC: Non-small cell lung cancer; CRPC: Castration-resistant prostate cancer; NE: Neuroendocrine; WES: Whole-exome sequencing.

Table 2. Overview of CDX-derived ex vivo cultures established to date.

CDX-Derived **Ex Vivo** *Cultures*				
Type of Cancer	**Stage**	**Culturing Conditions**	**Main Findings**	**Ref**
SCLC	Metastatic	HITES medium with ROCK inhibitor—non-adherent cell clusters—short-term	Recapitulate genomic landscape and in vivo drug response Tumorigenic in vivo Lentiviral transduction of one cell line	[69]
CRPC	Metastatic	DMEM/F12 medium—adherent conditions—permanent	Recapitulation of genomic characteristics and standard of care drug response	[73]

5. CTC-Derived Ex Vivo Models

Although CDXs represent classical preclinical mouse models that are relatively easy to handle, they cannot be derived from every patient depending on tumor type and the process could take several months, a time frame that would not provide proper aid for the clinical guidance of donor patients. Expansion of viable CTCs ex vivo may offer an attractive alternative allowing both molecular analysis and high-throughput drug screening in a shorter time, but with CTC scarcity remaining, a fortiori, a significant limitation. In vitro CTC cultures were reported in colon, breast, prostate, and lung cancer and are evaluated in this section (Table 3).

The first long-term colon cancer CTC cell line was derived by Cayrefourcq et al. from a metastatic colon cancer patient who had 302 $EpCAM^+$ CTCs detected by the CellSearch platform. Importantly, the characterized CTC-MCC-41 cell line shared the main genomic features of both the donor patient primary tumor and lymph node metastasis [9]. In a second study, the authors established and characterized eight additional cell lines from the same patient with CTCs collected at different time points during his follow-up. Transcriptomics analyses in the nine cell lines revealed an intermediate epithelial/mesenchymal phenotype promoting their metastatic potential, as well as stem cell-like properties that increased in cell lines isolated at later stages of progression. This may highlight the selection mechanism of treatment-resistant clones with specific phenotypes that drive disease progression. Functional experiments showed that these cells favor angiogenesis in vitro, which was concordant with the secretion of potent angiogenesis inducers such as VEGF and FGF2 as well as the tumorigenicity of these cells in vivo [9,10].

In BC, Zhang et al. presented the characterization of $EpCAM^-$ CTCs and revealed a shared protein signature $HER2^+/EGFR^+/HPSE^+/Notch1^+$ in CTCs competent for brain metastasis. Indeed, the three established CTC lines expressing this signature promoted brain and lung localization after xenotransplantation into nude mice. The authors therefore deciphered a preliminary signature which provides insight into metastatic competency of BC CTCs and pushes towards using CTC research to explore new potential biomarkers [75]. Another study reported the establishment of non-adherent CTC lines under hypoxic conditions (4% O_2) with CTCs issued from six patients with metastatic luminal-subtype BC. Three out of five tested cell lines were tumorigenic in vivo, giving rise to tumors with histological and immunohistochemical similarities with the primary patient tumor. This proof-of-concept study also identified targetable mutations acquired de novo in CTC cell lines, elucidating the importance of monitoring the mutational evolution of the tumor throughout the disease. To explore this, the authors performed sensitivity assays on the CTC lines with large panels of single drug and drug combinations targeting the different mutations identified [76]. In vitro phenotypic analysis of these cell lines and patient CTCs was recently performed. A CTM-specific DNA methylation status was revealed in which binding sites for stemness and proliferation transcription factors were hypomethylated, suggesting potential targets. This pattern correlated with poor prognosis in patients and targeting of clusters with Na+/K+ ATPase inhibitors shed them into single cell and enabled DNA methylation remodeling, leading to suppression of metastasis. These data therefore highlight a key connection between phenotypic properties of CTCs and DNA methylation patterns at specific stemness- and proliferation-related sites [43].

Table 3. Overview of ex vivo CTC-derived models established to date.

CTC-Derived Ex Vivo *Models*							
Type of Cancer	**Stage**	**Live CTC Isolation Technique**	**# of CTCs (CellSearch)**	**Culturing Conditions**	**Success Rate**	**Main Findings**	**Ref**
Colon cancer	Nonresectable metastatic	RosetteSep	≥300	- Hypoxic in medium 1 DMEM/F12 to normoxic conditions in medium 2 RPMI1640 - 2D, sustained for >6months	1%	- Recapitulation of main genomic features - Tumorigenic in vivo - Intermediate EMT + stem cell properties	[9,10]
Breast cancer	Metastatic	FACS	0	- Normoxic stem cell culture medium - 2D	8%	- Tumorigenic in vivo, brain metastasis signature (EpCAM$^-$HER2$^+$/EGFR$^+$/HPSE$^+$/Notch1$^+$)	[75]
	Metastatic luminal	CTC-iChip	3–3000	- Hypoxic, nonadherent - 2D, Sustained for >6 months	83%	- Tumorigenic in vivo - Drug sensitivity panels and CTM-specific methylation profile	[43, 76]
CRPC	Metastatic	RosetteSep-Ficoll	>100	- Growth factors reduced Matrigel/Advanced DMEM/F12 - 3D, sustained for >6 months	6%	- Tumorigenic in vivo	[77]
NSCLC	Early stage	Microfluidic CTC-capture device	1–11	- Matrigel + collagen - 3D, sustained for ~1 month	73%	- Common mutations between cultured CTCs and primary tumor	[78]

Despite successful in vitro expansion of patient CTCs in several cancer types as reported above, important limitations should be noted when handling 2D cultures, including cell morphology alterations due to adherence to plastic and lack of tumor microenvironment. Moreover, cell-cell and spatial interactions in vitro are not fully representative of the setting in the tumor mass in vivo [79]. These constraints can thus interfere with physiological functions and molecular responses of the tumor cells, making them less representative of the actual malignancy. To circumvent this problem, 3D models were proposed in prostate and lung cancer [77,78]. In prostate cancer, Gao et al. generated the first seven fully characterized organoid lines from a CRPC patient including a CTC-derived 3D organoid system from a patient who had more than 100 CTCs in 8 mL of blood. Success rate for the establishment of the CTC-derived organoid was not provided. Whole-exome sequencing (WES) analysis showed that all the 3D models recapitulated the molecular diversity of prostate cancer subtypes and were amenable to pharmacological assays. Engraftment of the CTC-derived organoid in vivo gave rise to tumors with a histological pattern similar to that of the primary cancer. This research, therefore, contributes a patient-derived model of CRPC which, with further optimization, may respond to the pressing need of in vitro models that faithfully recapitulate CRPC [77]. In lung cancer, Zhang et al. developed a novel ex vivo CTC-derived model using a 3D co-culture system which stimulated a microenvironment to sustain tumor development. CTCs were enriched and expanded for a short period of time from 14 to 19 early lung cancer patients. Next-generation sequencing detected several mutations including *TP53* found in both cultured CTCs and matched patient primary tumors [78].

6. Discussion

During the last decade, tremendous technological progress has been made to reliably detect, quantify and characterize CTCs at phenotypic, genomic, and functional levels. The characterization of CTC-derived models has paved the way toward an improved understanding of tumor dissemination by these cells (Figure 1). As depicted in Table 1, procedures for developing CDXs can vary from one study to another. Subcutaneous (SC) injection of cells in mice is the simplest method for tumor engrafts which has been used for decades and was most recently applied for PDX establishment. It facilitates tumor growth monitoring as it does not require fluorescent labeling or imaging. Most CDX models published to date have been developed through SC injection of CTCs. SC tumors do not usually metastasize probably due to the absence of the human microenvironment and the impact of murine angiogenesis, which influence dissemination of primary human tumors. Moreover, as the time-frame needed for tumor growth extends to several months, ethical regulations may not allow waiting for metastatic spread. To this end, these studies were limited to the characterization of the CDX primary tumor. Injection in mouse bone marrow as done by Baccelli et al. may also be an appropriate way to investigate MICs as this microenvironment has been previously described as a reservoir for disseminating tumor cells [35,59]. Conversely, studies aiming to assess metastatic and not only tumorigenic competency of CTCs have resorted to intracardiac injection [66,75]. This method, similarly to tail vein (TV) injection, allows a more rapid spread of the cells as they directly enter the bloodstream and thus mimics CTCs in their original setting. Propagation of CDX models through intracardiac or TV injection is less common or completely lacking, most likely due to potential dissemination bias. Indeed, organ metastasis could be influenced by the injection site of CTCs and defined by the first capillary bed encountered by cells post-injection. TV has been observed to induce lung metastases, thus generating false-positive results [80].

Another important challenge is ensuring the CDX consistently maintains its clinical relevance and serves as a patient surrogate. To this end, stringent validation is required and several aspects must be addressed. Firstly, it is crucial to verify the human origin of the CDX, as spontaneous tumors could grow in immunocompromised mouse models. Secondly, confirming cancer type and comparing the CDX tumor to the donor patient's biopsy through histopathology, followed by genomic studies to assess CDX genomic fidelity with patient tumor. Moreover, in the context of establishing preclinical models for precision medicine, functional drug sensitivity assays are needed to evaluate recapitulation of patient response to therapy in the CDX [68].

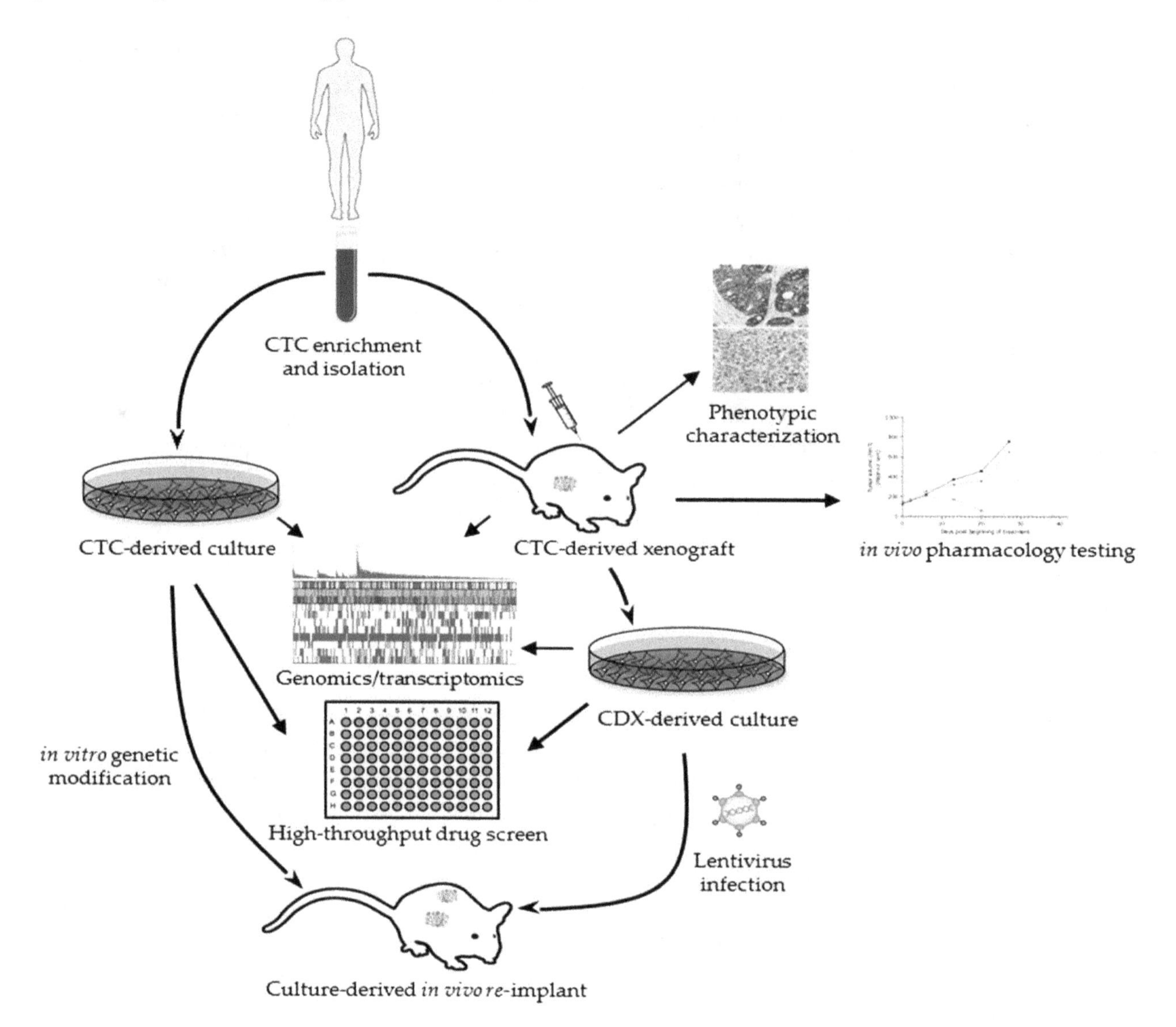

Figure 1. CTC-derived models as tractable systems to explore tumor-initiating cells (TICs) and new therapeutic strategies. CTCs isolated from late-stage cancer patients are used to generate CTC-derived xenografts (CDXs) to perform functional characterizations and pharmacology studies. CDX tumors can be isolated and dissociated into ex vivo cultures for drug screening and genome-wide analyses. CDX-derived cultures are amenable to lentiviral infection and can be re-injected into mice and used as tools to track tumor dissemination. In parallel, CTCs can be expanded in vitro and used as readouts of drug sensitivity. CTC = circulating tumor cell. CDX = CTC-derived xenograft.

Although PDX models serve as reliable tools for tumor modeling, CDXs offer added value for the understanding of tumor biology and metastasis. Detection and characterization of metastasis-competent CTCs using in vivo models offer a more representative molecular snapshot of the disease, as they serve as easily accessible "surrogates" of metastatic tissue, which is otherwise unobtainable in many cancer organs (e.g., bones or lungs) [81]. Indeed, CDX models could help showcase tumor heterogeneity in the metastatic setting in contrast to a localized biopsy in the case of PDX and are attainable at different time points throughout disease progression [63]. Most importantly, CDX models established to date reveal the high tumorigenic capacity of CTCs—even at a low number of cells (as low as 400 CTCs [68]). As reviewed above, CTCs with survival and MIC properties are assumed to be selected for seeding CDX tumors, similar to what has been observed in PDXs [82]. It is expected that the proportion of tumorigenic CTCs may vary between cancer types and patients as well as under selective pressure of treatment, which highlights a potential selective process for the acquisition of minor metastasis-competent CTC subclones [73]. CTC clusters and hybrid E/M CTCs have been described as the most aggressive cells with a high propensity for tumorigenesis. However, it is currently difficult to evaluate the impact each subpopulation could have on CDX tumor take rate. Indeed, as detailed before, Aceto et al. have shown an increased metastatic competency in CTM vs. single CTCs but this remains limited to murine models and is difficult to translate to human subjects [40]. It is worth noting however that, although in vivo models are sustained by the host tissue microenvironment and can faithfully recapitulate the tumor genome, the absence of immune components constitutes an important bias.

On the other hand, CTC expansion ex vivo is promising but is still very far from routine applications as culturing conditions are still under investigation and need further optimization. Therefore, CDX-derived cultures represent an attractive intermediate model to characterize this aggressive population in vitro. In the event of molecular similarities between the two models, CDX and CDX-derived cell lines offer complementary, tractable systems for CTC functional characterization and therapy testing. Re-injection of the CDX-derived cell line in immunodeficient mice could allow the identification of candidate genes in metastasis and chemoresistance mechanisms [64,69]. Additional model systems such as the chick embryo chorioallantoic membrane have also opened up new promising avenues in the in vivo studies of tumor metastasis, as the highly vascularized setting sustains tumor formation and dissemination rapidly after engraft [83]. Moreover, organoids have recently emerged as novel robust 3D models optimized to propagate in vitro and reminiscent of tumoral heterogeneity, with amenability to genetic modifications and drug screening assays [77,84]. One can hypothesize that the establishment of several CTC-derived organoid lines from the same patient could be useful in modeling metastatic disease and acquired CTC mutational profiles to monitor disease progression. However, these models lack in vivo host complexity and recent efforts have been put into the generation of 3D co-cultures in microfluidic devices to model ex vivo tumor microenvironments by the integration of different cell populations (e.g., immune cells, fibroblasts) [78,85].

7. Concluding Remarks

CDX models have shown unprecedented opportunities to provide insight into the complex biology of the metastatic process. However, at the present time, these functional models serve as proof-of-principle tools as their development is limited to late-stage disease settings and high CTC counts. The main goal of functional CTC studies being the identification and characterization of MICs and candidate target genes among CTCs, it is crucial to expand analyses to earlier stages of cancer [63]. Unfortunately, these rare preclinical models are derived from patients in exceptional clinical situations and we are currently unable to predict if the limitation caused by CTC scarcity could be circumvented. Nevertheless, the establishment of CTC-derived models from only a few CTCs is a major achievement today and an invaluable opportunity to decipher new biomarkers, which are urgently needed for novel therapeutic strategies in advanced cancers.

Acknowledgments: We are grateful to the patients and their families.

References

1. Fidler, I.J. Metastasis: Quantitative Analysis of Distribution and Fate of Tumor Emboli Labeled with ^{125}I-5-Iodo-2′ -deoxyuridine23. *JNCI J. Natl. Cancer Inst.* **1970**, *45*, 773–782. [PubMed]
2. Chambers, A.F.; Groom, A.C.; MacDonald, I.C. Dissemination and growth of cancer cells in metastatic sites. *Nat. Rev. Cancer* **2002**, *2*, 563–572. [CrossRef] [PubMed]

3. Luzzi, K.J.; MacDonald, I.C.; Schmidt, E.E.; Kerkvliet, N.; Morris, V.L.; Chambers, A.F.; Groom, A.C. Multistep Nature of Metastatic Inefficiency. *Am. J. Pathol.* **1998**, *153*, 865–873. [CrossRef]
4. Massagué, J.; Obenauf, A.C. Metastatic colonization by circulating tumour cells. *Nature* **2016**, *529*, 298–306. [CrossRef] [PubMed]
5. Fidler, I.J. Tumor heterogeneity and the biology of cancer invasion and metastasis. *Cancer Res.* **1978**, *38*, 2651–2660. [CrossRef]
6. Al-Hajj, M.; Wicha, M.S.; Benito-Hernandez, A.; Morrison, S.J.; Clarke, M.F. Prospective identification of tumorigenic breast cancer cells. *Proc. Natl. Acad. Sci. USA* **2003**, *100*, 3983–3988. [CrossRef] [PubMed]
7. Ricci-Vitiani, L.; Lombardi, D.G.; Pilozzi, E.; Biffoni, M.; Todaro, M.; Peschle, C.; De Maria, R. Identification and expansion of human colon-cancer-initiating cells. *Nature* **2007**, *445*, 111–115. [CrossRef] [PubMed]
8. Agnoletto, C.; Corrà, F.; Minotti, L.; Baldassari, F.; Crudele, F.; Cook, W.; Di Leva, G.; d'Adamo, A.; Gasparini, P.; Volinia, S. Heterogeneity in Circulating Tumor Cells: The Relevance of the Stem-Cell Subset. *Cancers* **2019**, *11*, 483. [CrossRef] [PubMed]
9. Cayrefourcq, L.; Mazard, T.; Joosse, S.; Solassol, J.; Ramos, J.; Assenat, E.; Schumacher, U.; Costes, V.; Maudelonde, T.; Pantel, K.; et al. Establishment and Characterization of a Cell Line from Human Circulating Colon Cancer Cells. *Cancer Res.* **2015**, *75*, 892–901. [CrossRef] [PubMed]
10. Soler, A.; Cayrefourcq, L.; Mazard, T.; Babayan, A.; Lamy, P.-J.; Assou, S.; Assenat, E.; Pantel, K.; Alix-Panabières, C. Autologous cell lines from circulating colon cancer cells captured from sequential liquid biopsies as model to study therapy-driven tumor changes. *Sci. Rep.* **2018**, *8*, 15931. [CrossRef]
11. Liu, T.; Xu, H.; Huang, M.; Ma, W.; Saxena, D.; Lustig, R.A.; Alonso-Basanta, M.; Zhang, Z.; O Rourke, D.M.; Zhang, L.; et al. Circulating glioma cells exhibit stem cell-like properties. *Cancer Res.* **2018**. [CrossRef] [PubMed]
12. Nieto, M.A.; Huang, R.Y.-J.; Jackson, R.A.; Thiery, J.P. EMT: 2016. *Cell* **2016**, *166*, 21–45. [CrossRef] [PubMed]
13. Shibue, T.; Weinberg, R.A. EMT, CSCs, and drug resistance: The mechanistic link and clinical implications. *Nat. Rev. Clin. Oncol.* **2017**, *14*, 611–629. [CrossRef] [PubMed]
14. Kasimir-Bauer, S.; Hoffmann, O.; Wallwiener, D.; Kimmig, R.; Fehm, T. Expression of stem cell and epithelial-mesenchymal transition markers in primary breast cancer patients with circulating tumor cells. *Breast Cancer Res.* **2012**, *14*. [CrossRef] [PubMed]
15. Yu, M.; Bardia, A.; Wittner, B.S.; Stott, S.L.; Smas, M.E.; Ting, D.T.; Isakoff, S.J.; Ciciliano, J.C.; Wells, M.N.; Shah, A.M.; et al. Circulating Breast Tumor Cells Exhibit Dynamic Changes in Epithelial and Mesenchymal Composition. *Science* **2013**, *339*, 580–584. [CrossRef] [PubMed]
16. Kallergi, G.; Papadaki, M.A.; Politaki, E.; Mavroudis, D.; Georgoulias, V.; Agelaki, S. Epithelial to mesenchymal transition markers expressed in circulating tumour cells of early and metastatic breast cancer patients. *Breast Cancer Res.* **2011**, *13*. [CrossRef] [PubMed]
17. Giordano, A.; Gao, H.; Anfossi, S.; Cohen, E.; Mego, M.; Lee, B.-N.; Tin, S.; Laurentiis, M.D.; Parker, C.A.; Alvarez, R.H.; et al. Epithelial–Mesenchymal Transition and Stem Cell Markers in Patients with HER2-Positive Metastatic Breast Cancer. *Mol. Cancer Ther.* **2012**, *11*, 2526–2534. [CrossRef]
18. Tam, W.L.; Weinberg, R.A. The epigenetics of epithelial-mesenchymal plasticity in cancer. *Nat. Med.* **2013**, *19*, 1438–1449. [CrossRef]
19. Puisieux, A.; Brabletz, T.; Caramel, J. Oncogenic roles of EMT-inducing transcription factors. *Nat. Cell Biol.* **2014**, *16*, 488–494. [CrossRef]
20. Mani, S.A.; Guo, W.; Liao, M.-J.; Eaton, E.N.; Ayyanan, A.; Zhou, A.Y.; Brooks, M.; Reinhard, F.; Zhang, C.C.; Shipitsin, M.; et al. The Epithelial-Mesenchymal Transition Generates Cells with Properties of Stem Cells. *Cell* **2008**, *133*, 704–715. [CrossRef]
21. Morel, A.-P.; Lièvre, M.; Thomas, C.; Hinkal, G.; Ansieau, S.; Puisieux, A. Generation of Breast Cancer Stem Cells through Epithelial-Mesenchymal Transition. *PLoS ONE* **2008**, *3*, e2888. [CrossRef] [PubMed]
22. Polyak, K.; Weinberg, R.A. Transitions between epithelial and mesenchymal states: Acquisition of malignant and stem cell traits. *Nat. Rev. Cancer* **2009**, *9*, 265–273. [CrossRef] [PubMed]
23. Creighton, C.J.; Li, X.; Landis, M.; Dixon, J.M.; Neumeister, V.M.; Sjolund, A.; Rimm, D.L.; Wong, H.; Rodriguez, A.; Herschkowitz, J.I.; et al. Residual breast cancers after conventional therapy display mesenchymal as well as tumor-initiating features. *Proc. Natl. Acad. Sci. USA* **2009**, *106*, 13820–13825. [CrossRef] [PubMed]

24. Ocaña, O.H.; Córcoles, R.; Fabra, Á.; Moreno-Bueno, G.; Acloque, H.; Vega, S.; Barrallo-Gimeno, A.; Cano, A.; Nieto, M.A. Metastatic Colonization Requires the Repression of the Epithelial-Mesenchymal Transition Inducer Prrx1. *Cancer Cell* **2012**, *22*, 709–724. [CrossRef] [PubMed]
25. Beerling, E.; Seinstra, D.; de Wit, E.; Kester, L.; van der Velden, D.; Maynard, C.; Schäfer, R.; van Diest, P.; Voest, E.; van Oudenaarden, A.; et al. Plasticity between Epithelial and Mesenchymal States Unlinks EMT from Metastasis-Enhancing Stem Cell Capacity. *Cell Rep.* **2016**, *14*, 2281–2288. [CrossRef] [PubMed]
26. Fischer, K.R.; Durrans, A.; Lee, S.; Sheng, J.; Li, F.; Wong, S.T.C.; Choi, H.; El Rayes, T.; Ryu, S.; Troeger, J.; et al. Epithelial-to-mesenchymal transition is not required for lung metastasis but contributes to chemoresistance. *Nature* **2015**, *527*, 472–476. [CrossRef] [PubMed]
27. Zheng, X.; Carstens, J.L.; Kim, J.; Scheible, M.; Kaye, J.; Sugimoto, H.; Wu, C.-C.; LeBleu, V.S.; Kalluri, R. Epithelial-to-mesenchymal transition is dispensable for metastasis but induces chemoresistance in pancreatic cancer. *Nature* **2015**, *527*, 525–530. [CrossRef]
28. Bailey; Martin Insights on CTC Biology and Clinical Impact Emerging from Advances in Capture Technology. *Cells* **2019**, *8*, 553. [CrossRef]
29. Lecharpentier, A.; Vielh, P.; Perez-Moreno, P.; Planchard, D.; Soria, J.C.; Farace, F. Detection of circulating tumour cells with a hybrid (epithelial/mesenchymal) phenotype in patients with metastatic non-small cell lung cancer. *Br. J. Cancer* **2011**, *105*, 1338–1341. [CrossRef]
30. Hou, J.-M.; Krebs, M.; Ward, T.; Sloane, R.; Priest, L.; Hughes, A.; Clack, G.; Ranson, M.; Blackhall, F.; Dive, C. Circulating tumor cells as a window on metastasis biology in lung cancer. *Am. J. Pathol.* **2011**, *178*, 989–996. [CrossRef]
31. Hofman, V.; Ilie, M.I.; Long, E.; Selva, E.; Bonnetaud, C.; Molina, T.; Vénissac, N.; Mouroux, J.; Vielh, P.; Hofman, P. Detection of circulating tumor cells as a prognostic factor in patients undergoing radical surgery for non-small-cell lung carcinoma: Comparison of the efficacy of the CellSearch Assay^{TM} and the isolation by size of epithelial tumor cell method. *Int. J. Cancer* **2011**, *129*, 1651–1660. [CrossRef] [PubMed]
32. Papadaki, M.A.; Stoupis, G.; Theodoropoulos, P.A.; Mavroudis, D.; Georgoulias, V.; Agelaki, S. Circulating Tumor Cells with Stemness and Epithelial-to-Mesenchymal Transition Features Are Chemoresistant and Predictive of Poor Outcome in Metastatic Breast Cancer. *Mol. Cancer Ther.* **2019**, *18*, 437–447. [CrossRef] [PubMed]
33. Wong, C.W.; Lee, A.; Shientag, L.; Yu, J.; Dong, Y.; Kao, G.; Al-Mehdi, A.B.; Bernhard, E.J.; Muschel, R.J. Apoptosis: An early event in metastatic inefficiency. *Cancer Res.* **2001**, *61*, 333–338. [PubMed]
34. Bednarz-Knoll, N.; Alix-Panabières, C.; Pantel, K. Clinical relevance and biology of circulating tumor cells. *Breast Cancer Res. BCR* **2011**, *13*, 228. [CrossRef] [PubMed]
35. Braun, S.; Vogl, F.D.; Naume, B.; Janni, W.; Osborne, M.P.; Coombes, R.C.; Schlimok, G.; Diel, I.J.; Gerber, B.; Gebauer, G.; et al. A Pooled Analysis of Bone Marrow Micrometastasis in Breast Cancer. *N. Engl. J. Med.* **2005**, *353*, 793–802. [CrossRef]
36. Faugeroux, V.; Lefebvre, C.; Pailler, E.; Pierron, V.; Marcaillou, C.; Tourlet, S.; Billiot, F.; Dogan, S.; Oulhen, M.; Vielh, P.; et al. An Accessible and Unique Insight into Metastasis Mutational Content Through Whole-exome Sequencing of Circulating Tumor Cells in Metastatic Prostate Cancer. *Eur. Urol. Oncol.* **2019**. [CrossRef] [PubMed]
37. Fernandez, S.V.; Bingham, C.; Fittipaldi, P.; Austin, L.; Palazzo, J.; Palmer, G.; Alpaugh, K.; Cristofanilli, M. TP53 mutations detected in circulating tumor cells present in the blood of metastatic triple negative breast cancer patients. *Breast Cancer Res. BCR* **2014**, *16*, 445. [CrossRef]
38. Jordan, N.V.; Bardia, A.; Wittner, B.S.; Benes, C.; Ligorio, M.; Zheng, Y.; Yu, M.; Sundaresan, T.K.; Licausi, J.A.; Desai, R.; et al. HER2 expression identifies dynamic functional states within circulating breast cancer cells. *Nature* **2016**, *537*, 102–106. [CrossRef]
39. Heitzer, E.; Auer, M.; Gasch, C.; Pichler, M.; Ulz, P.; Hoffmann, E.M.; Lax, S.; Waldispuehl-Geigl, J.; Mauermann, O.; Lackner, C.; et al. Complex Tumor Genomes Inferred from Single Circulating Tumor Cells by Array-CGH and Next-Generation Sequencing. *Cancer Res.* **2013**, *73*, 2965–2975. [CrossRef]
40. Aceto, N.; Bardia, A.; Miyamoto, D.T.; Donaldson, M.C.; Wittner, B.S.; Spencer, J.A.; Yu, M.; Pely, A.; Engstrom, A.; Zhu, H.; et al. Circulating Tumor Cell Clusters Are Oligoclonal Precursors of Breast Cancer Metastasis. *Cell* **2014**, *158*, 1110–1122. [CrossRef]
41. Aceto, N.; Toner, M.; Maheswaran, S.; Haber, D.A. En Route to Metastasis: Circulating Tumor Cell Clusters and Epithelial-to-Mesenchymal Transition. *Trends Cancer* **2015**, *1*, 44–52. [CrossRef] [PubMed]

42. Liu, X.; Taftaf, R.; Kawaguchi, M.; Chang, Y.-F.; Chen, W.; Entenberg, D.; Zhang, Y.; Gerratana, L.; Huang, S.; Patel, D.B.; et al. Homophilic CD44 Interactions Mediate Tumor Cell Aggregation and Polyclonal Metastasis in Patient-Derived Breast Cancer Models. *Cancer Discov.* **2019**, *9*, 96–113. [CrossRef] [PubMed]
43. Gkountela, S.; Castro-Giner, F.; Szczerba, B.M.; Vetter, M.; Landin, J.; Scherrer, R.; Krol, I.; Scheidmann, M.C.; Beisel, C.; Stirnimann, C.U.; et al. Circulating Tumor Cell Clustering Shapes DNA Methylation to Enable Metastasis Seeding. *Cell* **2019**, *176*, 98–112.e14. [CrossRef] [PubMed]
44. Szczerba, B.M.; Castro-Giner, F.; Vetter, M.; Krol, I.; Gkountela, S.; Landin, J.; Scheidmann, M.C.; Donato, C.; Scherrer, R.; Singer, J.; et al. Neutrophils escort circulating tumour cells to enable cell cycle progression. *Nature* **2019**, *566*, 553–557. [CrossRef] [PubMed]
45. Allard, W.J.; Matera, J.; Miller, M.C.; Repollet, M.; Connelly, M.C.; Rao, C.; Tibbe, A.G.J.; Uhr, J.W. Tumor Cells Circulate in the Peripheral Blood of All Major Carcinomas but not in Healthy Subjects or Patients with Nonmalignant Diseases. *Clin. Cancer Res* **2004**, *10*, 6897–6904. [CrossRef] [PubMed]
46. Cristofanilli, M.; Budd, G.T.; Ellis, M.J.; Stopeck, A.; Matera, J.; Miller, M.C.; Reuben, J.M.; Doyle, G.V.; Allard, W.J.; Terstappen, L.W.M.M.; et al. Circulating Tumor Cells, Disease Progression, and Survival in Metastatic Breast Cancer. *N. Engl. J. Med.* **2004**, *351*, 781–791. [CrossRef] [PubMed]
47. Cristofanilli, M.; Hayes, D.F.; Budd, G.T.; Ellis, M.J.; Stopeck, A.; Reuben, J.M.; Doyle, G.V.; Matera, J.; Allard, W.J.; Miller, M.C.; et al. Circulating Tumor Cells: A Novel Prognostic Factor for Newly Diagnosed Metastatic Breast Cancer. *J. Clin. Oncol.* **2005**, *23*, 1420–1430. [CrossRef]
48. De Bono, J.S.; Scher, H.I.; Montgomery, R.B.; Parker, C.; Miller, M.C.; Tissing, H.; Doyle, G.V.; Terstappen, L.W.W.M.; Pienta, K.J.; Raghavan, D. Circulating Tumor Cells Predict Survival Benefit from Treatment in Metastatic Castration-Resistant Prostate Cancer. *Clin. Cancer Res.* **2008**, *14*, 6302–6309. [CrossRef]
49. Cohen, S.J.; Punt, C.J.A.; Iannotti, N.; Saidman, B.H.; Sabbath, K.D.; Gabrail, N.Y.; Picus, J.; Morse, M.; Mitchell, E.; Miller, M.C.; et al. Relationship of Circulating Tumor Cells to Tumor Response, Progression-Free Survival, and Overall Survival in Patients With Metastatic Colorectal Cancer. *J. Clin. Oncol.* **2008**, *26*, 3213–3221. [CrossRef]
50. Vona, G.; Sabile, A.; Louha, M.; Sitruk, V.; Romana, S.; Schütze, K.; Capron, F.; Franco, D.; Pazzagli, M.; Vekemans, M.; et al. Isolation by Size of Epithelial Tumor Cells. *Am. J. Pathol.* **2000**, *156*, 57–63. [CrossRef]
51. Desitter, I.; Guerrouahen, B.S.; Benali-Furet, N.; Wechsler, J.; Jänne, P.A.; Kuang, Y.; Yanagita, M.; Wang, L.; Berkowitz, J.A.; Distel, R.J.; et al. A new device for rapid isolation by size and characterization of rare circulating tumor cells. *Anticancer Res.* **2011**, *31*, 427–441. [PubMed]
52. Farace, F.; Massard, C.; Vimond, N.; Drusch, F.; Jacques, N.; Billiot, F.; Laplanche, A.; Chauchereau, A.; Lacroix, L.; Planchard, D.; et al. A direct comparison of CellSearch and ISET for circulating tumour-cell detection in patients with metastatic carcinomas. *Br. J. Cancer* **2011**, *105*, 847–853. [CrossRef] [PubMed]
53. Pailler, E.; Adam, J.; Barthélémy, A.; Oulhen, M.; Auger, N.; Valent, A.; Borget, I.; Planchard, D.; Taylor, M.; André, F.; et al. Detection of Circulating Tumor Cells Harboring a Unique *ALK* Rearrangement in *ALK* -Positive Non–Small-Cell Lung Cancer. *J. Clin. Oncol.* **2013**, *31*, 2273–2281. [CrossRef] [PubMed]
54. Pailler, E.; Oulhen, M.; Borget, I.; Remon, J.; Ross, K.; Auger, N.; Billiot, F.; Ngo Camus, M.; Commo, F.; Lindsay, C.R.; et al. Circulating Tumor Cells with Aberrant *ALK* Copy Number Predict Progression-Free Survival during Crizotinib Treatment in *ALK* -Rearranged Non–Small Cell Lung Cancer Patients. *Cancer Res.* **2017**, *77*, 2222–2230. [CrossRef] [PubMed]
55. Ozkumur, E.; Shah, A.M.; Ciciliano, J.C.; Emmink, B.L.; Miyamoto, D.T.; Brachtel, E.; Yu, M.; Chen, P.-i.; Morgan, B.; Trautwein, J.; et al. Inertial Focusing for Tumor Antigen-Dependent and -Independent Sorting of Rare Circulating Tumor Cells. *Sci. Transl. Med.* **2013**, *5*, 179ra47. [CrossRef] [PubMed]
56. Fabbri, F.; Carloni, S.; Zoli, W.; Ulivi, P.; Gallerani, G.; Fici, P.; Chiadini, E.; Passardi, A.; Frassineti, G.L.; Ragazzini, A.; et al. Detection and recovery of circulating colon cancer cells using a dielectrophoresis-based device: KRAS mutation status in pure CTCs. *Cancer Lett.* **2013**, *335*, 225–231. [CrossRef] [PubMed]
57. Bulfoni, M.; Gerratana, L.; Del Ben, F.; Marzinotto, S.; Sorrentino, M.; Turetta, M.; Scoles, G.; Toffoletto, B.; Isola, M.; Beltrami, C.A.; et al. In patients with metastatic breast cancer the identification of circulating tumor cells in epithelial-to-mesenchymal transition is associated with a poor prognosis. *Breast Cancer Res.* **2016**, *18*. [CrossRef]

58. Ross, K.; Pailler, E.; Faugeroux, V.; Taylor, M.; Oulhen, M.; Auger, N.; Planchard, D.; Soria, J.-C.; Lindsay, C.R.; Besse, B.; et al. The potential diagnostic power of circulating tumor cell analysis for non-small-cell lung cancer. *Expert Rev. Mol. Diagn.* **2015**, *15*, 1605–1629. [CrossRef] [PubMed]
59. Baccelli, I.; Schneeweiss, A.; Riethdorf, S.; Stenzinger, A.; Schillert, A.; Vogel, V.; Klein, C.; Saini, M.; Bäuerle, T.; Wallwiener, M.; et al. Identification of a population of blood circulating tumor cells from breast cancer patients that initiates metastasis in a xenograft assay. *Nat. Biotechnol.* **2013**, *31*, 539–544. [CrossRef]
60. Hidalgo, M.; Amant, F.; Biankin, A.V.; Budinská, E.; Byrne, A.T.; Caldas, C.; Clarke, R.B.; de Jong, S.; Jonkers, J.; Mælandsmo, G.M.; et al. Patient Derived Xenograft Models: An Emerging Platform for Translational Cancer Research. *Cancer Discov.* **2014**, *4*, 998–1013. [CrossRef]
61. Byrne, A.T.; Alférez, D.G.; Amant, F.; Annibali, D.; Arribas, J.; Biankin, A.V.; Bruna, A.; Budinská, E.; Caldas, C.; Chang, D.K.; et al. Interrogating open issues in cancer precision medicine with patient-derived xenografts. *Nat. Rev. Cancer* **2017**, *17*, 254–268. [CrossRef] [PubMed]
62. Blackhall, F.; Frese, K.K.; Simpson, K.; Kilgour, E.; Brady, G.; Dive, C. Will liquid biopsies improve outcomes for patients with small-cell lung cancer? *Lancet Oncol.* **2018**, *19*, e470–e481. [CrossRef]
63. Pantel, K.; Alix-Panabieres, C. Functional Studies on Viable Circulating Tumor Cells. *Clin. Chem.* **2016**, *62*, 328–334. [CrossRef] [PubMed]
64. Lallo, A.; Schenk, M.W.; Frese, K.K.; Blackhall, F.; Dive, C. Circulating tumor cells and CDX models as a tool for preclinical drug development. *Transl. Lung Cancer Res.* **2017**, *6*, 397–408. [CrossRef] [PubMed]
65. Pereira-Veiga, T.; Abreu, M.; Robledo, D.; Matias-Guiu, X.; Santacana, M.; Sánchez, L.; Cueva, J.; Palacios, P.; Abdulkader, I.; López-López, R.; et al. CTCs-derived xenograft development in a triple negative breast cancer case. *Int. J. Cancer* **2018**, *144*, 2254–2265. [CrossRef]
66. Vishnoi, M.; Haowen Liu, N.; Yin, W.; Boral, D.; Scamardo, A.; Hong, D.; Marchetti, D. The identification of a TNBC liver metastasis gene signature by sequential CTC-xenograft modelling. *Mol. Oncol.* **2019**. [CrossRef]
67. Girotti, M.R.; Gremel, G.; Lee, R.; Galvani, E.; Rothwell, D.; Viros, A.; Mandal, A.K.; Lim, K.H.J.; Saturno, G.; Furney, S.J.; et al. Application of Sequencing, Liquid Biopsies, and Patient-Derived Xenografts for Personalized Medicine in Melanoma. *Cancer Discov.* **2016**, *6*, 286–299. [CrossRef]
68. Hodgkinson, C.L.; Morrow, C.J.; Li, Y.; Metcalf, R.L.; Rothwell, D.G.; Trapani, F.; Polanski, R.; Burt, D.J.; Simpson, K.L.; Morris, K.; et al. Tumorigenicity and genetic profiling of circulating tumor cells in small-cell lung cancer. *Nat. Med.* **2014**, *20*, 897–903. [CrossRef]
69. Lallo, A.; Gulati, S.; Schenk, M.W.; Khandelwal, G.; Berglund, U.W.; Pateras, I.S.; Chester, C.P.E.; Pham, T.M.; Kalderen, C.; Frese, K.K.; et al. Ex vivo culture of cells derived from circulating tumour cell xenograft to support small cell lung cancer research and experimental therapeutics. *Br. J. Pharmacol.* **2019**, *176*, 436–450. [CrossRef]
70. Drapkin, B.J.; George, J.; Christensen, C.L.; Mino-Kenudson, M.; Dries, R.; Sundaresan, T.; Phat, S.; Myers, D.T.; Zhong, J.; Igo, P.; et al. Genomic and Functional Fidelity of Small Cell Lung Cancer Patient-Derived Xenografts. *Cancer Discov.* **2018**, *8*, 600–615. [CrossRef]
71. Morrow, C.J.; Trapani, F.; Metcalf, R.L.; Bertolini, G.; Hodgkinson, C.L.; Khandelwal, G.; Kelly, P.; Galvin, M.; Carter, L.; Simpson, K.L.; et al. Tumourigenic non-small-cell lung cancer mesenchymal circulating tumour cells: A clinical case study. *Ann. Oncol.* **2016**, *27*, 1155–1160. [CrossRef] [PubMed]
72. Andree, K.C.; Mentink, A.; Zeune, L.L.; Terstappen, L.W.M.M.; Stoecklein, N.H.; Neves, R.P.; Driemel, C.; Lampignano, R.; Yang, L.; Neubauer, H.; et al. Toward a real liquid biopsy in metastatic breast and prostate cancer: Diagnostic LeukApheresis increases CTC yields in a European prospective multicenter study (CTCTrap). *Int. J. Cancer* **2018**, *143*, 2584–2591. [CrossRef] [PubMed]
73. Faugeroux, V.; Pailler, E.; Deas, O.; Brulle-Soumare, L.; Hervieu, C.; Marty, V.; Alexandrova, K.; Andree, K.C.; Stoecklein, N.H.; Tramalloni, D.; et al. Genetic characterization of a Unique Neuroendocrine Transdifferentiation Prostate Circulating Tumor Cell - Derived eXplant (CDX) Model. *Nat. Commun.* **2019**. under review.
74. Faugeroux, V.; Pailler, E.; Deas, O.; Michels, J.; Mezquita, L.; Brulle-Soumare, L.; Cairo, S.; Scoazec, J.-Y.; Marty, V.; Queffelec, P.; et al. Development and characterization of novel non-small cell lung cancer (NSCLC) Circulating Tumor Cells (CTCs)-derived xenograft (CDX) models. In Proceedings of the AACR Annual Meeting 2018, Chicago, IL, USA, 14–18 April 2018.

75. Zhang, L.; Ridgway, L.D.; Wetzel, M.D.; Ngo, J.; Yin, W.; Kumar, D.; Goodman, J.C.; Groves, M.D.; Marchetti, D. The Identification and Characterization of Breast Cancer CTCs Competent for Brain Metastasis. *Sci. Transl. Med.* **2013**, *5*, 180ra48. [CrossRef] [PubMed]
76. Yu, M.; Bardia, A.; Aceto, N.; Bersani, F.; Madden, M.W.; Donaldson, M.C.; Desai, R.; Zhu, H.; Comaills, V.; Zheng, Z.; et al. Ex vivo culture of circulating breast tumor cells for individualized testing of drug susceptibility. *Science* **2014**, *345*, 216–220. [CrossRef] [PubMed]
77. Gao, D.; Vela, I.; Sboner, A.; Iaquinta, P.J.; Karthaus, W.R.; Gopalan, A.; Dowling, C.; Wanjala, J.N.; Undvall, E.A.; Arora, V.K.; et al. Organoid Cultures Derived from Patients with Advanced Prostate Cancer. *Cell* **2014**, *159*, 176–187. [CrossRef]
78. Zhang, Z.; Shiratsuchi, H.; Lin, J.; Chen, G.; Reddy, R.M.; Azizi, E.; Fouladdel, S.; Chang, A.C.; Lin, L.; Jiang, H.; et al. Expansion of CTCs from early stage lung cancer patients using a microfluidic co-culture model. *Oncotarget* **2014**, *5*, 12383–12397. [CrossRef]
79. Tellez-Gabriel, M.; Cochonneau, D.; Cadé, M.; Jubelin, C.; Heymann, M.-F.; Heymann, D. Circulating Tumor Cell-Derived Pre-Clinical Models for Personalized Medicine. *Cancers* **2018**, *11*, 19. [CrossRef]
80. Khanna, C. Modeling metastasis in vivo. *Carcinogenesis* **2004**, *26*, 513–523. [CrossRef]
81. Alix-Panabières, C.; Pantel, K. Challenges in circulating tumour cell research. *Nat. Rev. Cancer* **2014**, *14*, 623–631. [CrossRef]
82. Eirew, P.; Steif, A.; Khattra, J.; Ha, G.; Yap, D.; Farahani, H.; Gelmon, K.; Chia, S.; Mar, C.; Wan, A.; et al. Dynamics of genomic clones in breast cancer patient xenografts at single-cell resolution. *Nature* **2015**, *518*, 422–426. [CrossRef] [PubMed]
83. Stoletov, K.; Willetts, L.; Paproski, R.J.; Bond, D.J.; Raha, S.; Jovel, J.; Adam, B.; Robertson, A.E.; Wong, F.; Woolner, E.; et al. Quantitative in vivo whole genome motility screen reveals novel therapeutic targets to block cancer metastasis. *Nat. Commun.* **2018**, *9*. [CrossRef] [PubMed]
84. Praharaj, P.P.; Bhutia, S.K.; Nagrath, S.; Bitting, R.L.; Deep, G. Circulating tumor cell-derived organoids: Current challenges and promises in medical research and precision medicine. *Biochim. Biophys. Acta BBA - Rev. Cancer* **2018**, *1869*, 117–127. [CrossRef] [PubMed]
85. Nguyen, M.; De Ninno, A.; Mencattini, A.; Mermet-Meillon, F.; Fornabaio, G.; Evans, S.S.; Cossutta, M.; Khira, Y.; Han, W.; Sirven, P.; et al. Dissecting Effects of Anti-cancer Drugs and Cancer-Associated Fibroblasts by On-Chip Reconstitution of Immunocompetent Tumor Microenvironments. *Cell Rep.* **2018**, *25*, 3884–3893.e3. [CrossRef] [PubMed]

15

Insights on CTC Biology and Clinical Impact Emerging from Advances in Capture Technology

Patrick C. Bailey [1] and Stuart S. Martin [1,2,*]

1 Marlene and Stewart Greenebaum Comprehensive Cancer Center, School of Medicine (UMGCCC), University of Maryland, Baltimore, MD 21201, USA; pcbailey@umaryland.edu
2 Department of Physiology, School of Medicine, University of Maryland, Baltimore, MD 21201, USA
* Correspondence: ssmartin@som.umaryland.edu.

Abstract: Circulating tumor cells (CTCs) and circulating tumor microemboli (CTM) have been shown to correlate negatively with patient survival. Actual CTC counts before and after treatment can be used to aid in the prognosis of patient outcomes. The presence of circulating tumor materials (CTMat) can advertise the presence of metastasis before clinical presentation, enabling the early detection of relapse. Importantly, emerging evidence is indicating that cancer treatments can actually increase the incidence of CTCs and metastasis in pre-clinical models. Subsequently, the study of CTCs, their biology and function are of vital importance. Emerging technologies for the capture of CTC/CTMs and CTMat are elucidating vitally important biological and functional information that can lead to important alterations in how therapies are administered. This paves the way for the development of a "liquid biopsy" where treatment decisions can be informed by information gleaned from tumor cells and tumor cell debris in the blood.

Keywords: circulating tumor cells; CTC; liquid biopsy; CTM; CTMat; CTC biology; CTC capture technology

---

1. Introduction

Cancer remains a leading cause of death in all areas of the world [1]. The primary cause of death however, is not the primary tumor but metastases. The complete biology of metastasis remains unclear, but several general processes are recognized. The initial steps are understood to include the local invasion of the tumor into neighboring tissues followed by intravasation into the circulation, involving either the epithelial to mesenchymal transition (EMT) or the physical shedding of tumor cells into leaky, poorly formed vessels. Both EMT and shedding lead to the dissemination of tumor cells into the lymphatic and hematogenic systems [2]. Of these two methods, hematogenous spread is the most lethal.

Integral to the process of dissemination is circulation in the vasculature. Detached cells are termed circulating tumor cells (CTCs) or, in the case of cell clusters, circulating tumor microemboli (CTM). These cells circulate until they either attach to the vessel endothelium or become lodged in small capillaries. From this point, there can either be migration through the tissue or, in the case of CTMs, possible vascular rupture [3]. Cells which have survived these processes can serve as the seeds of eventual metastatic recurrence.

It has been estimated that tumor cells shed from the primary tumor at a rate of 3.2×10^6 cells per gram of tumor tissue per day, but over half quickly perish [4]. What remains is one cell per 10^{6-7} leukocytes [5]. The rarity and importance of these CTCs has led to the development of many technologies designed to enrich for this small population. Among the challenges inherent in isolating CTCs are the methodologies used for characterizing them. The two main methods that have been employed involve cell surface markers and the physical characteristics of the cell [6], both of which have advantages and pitfalls. The intent of this review is not to exhaustively catalog technologies, but to discuss the principles behind several stand-outs, the importance of CTC isolation in general, possible applications in functional studies and the clinical importance of CTCs in view of biology and new ideas in dissemination modality.

2. Diagnostic Importance of CTCs

The presence of CTCs in the blood has been proportionally correlated with poor prognosis, and CTMs are even more strongly correlated with patient outcome [7,8]. For a widespread use of CTC/CTM detection as a diagnostic tool, clinical acceptance is critical. The American Society of Clinical Oncology (ASCO), the National Academy of Clinical Biochemistry, the American Association for Clinical Chemistry, and the American Joint Committee on Cancer have all declined to recommend CTC/CTMat assays in the detection, monitoring or staging of cancer until the benefits of the technique are clarified [9–11].

The CellSearch system was approved by the FDA in 2004 for the clinical detection of CTCs but there are numerous challenges inherent in the platform. Problems of physics, statistics, translation, preparation time, and the constraint of fixed cells stained for limited biomarkers have led to inconsistent results [12]. These challenges impact results in detection rate, patient positivity, and correlation with prognosis [6,13–15]. Discounting phenotypic heterogeneity between CTCs, there are also numerous technical factors involved in these discrepancies, including differences in technique and bias between operators, sample size and lack of a common reference standard, among many others.

Toward a standard protocol that minimizes these issues, two new trends have a great deal of potential. These are the detection of circulating tumor materials (CTMat) and telomerase activity. As previously mentioned, half of the cells shed from the primary tumor die in circulation. Due to many factors, the membranes of these cells are perforated and cellular contents leak into the blood stream [16]. The physical forces in drawing blood are also a contributing factor to the destruction of viable cells, leading to the accumulation of cellular debris. CTMat is usually captured by the same methods outlined below, but where standard capture technologies would overlook these cell fragments as negative, CTMat capture technology can visualize and enumerate them. Using the CellSpotter technology, which can differentiate between intact tumor cells, damaged tumor cells and tumor cell fragments, CTMat was found to comprise the largest subpopulation in 18 blood samples from prostate cancer patients [16]. CTMat has not only been found to correlate well with viable CTC detection in prognostic capacity, but could also potentially provide an avenue for standardization, insofar as CTMat detection can be more easily quantified. It is also less restrictive in the identification of targets and the process of imaging can be automated [17].

In contrast to the release of cell fragments through apoptosis in the blood stream, another component of CTMat, circulating tumor DNA (ctDNA), is believed to stem mainly from cellular death in the solid tumor [18]. Levels of ctDNA have been found to correlate well with primary tumor resection, chemotherapy and metastasis [19,20]. Although the difficulty in producing primers for PCR of ctDNA fragments is not trivial, this process has been shown to discover relapse well before other conventional methods [21,22] Indeed, ctDNA is already being used for treatment response monitoring, the early detection of relapse [23,24] and even therapy decision (e.g., therapies related to the presence of mutant Epidermal Growth Factor Receptor [EGFR]) [25]. ctDNA from viral associated cancer has also been employed to monitor treatment response [26]. To this end, the analysis of ctDNA can be used to monitor therapeutic success. Increases in mutant alleles as a result of therapy resistance have been shown in patients monitored over a period of two years [25,27]. Finally, the FDA has approved the Cobas EGFR Mutation Test v2 as a companion diagnostic for non-small cell lung cancer therapy with Erlotinib. Standard clinical imaging detection involves the visualizing of a tumor mass, which is a process requiring millions of cells. ctDNA can be monitored and relapse discovered well before this timepoint.

Many of the most utilized platforms for the detection of CTCs utilize epithelial markers for identification, such as cytokeratin and EpCAM (epithelial-cell-adhesion-molecule). This can provide information as to cellular origin but neglects biological behavior. It has also been reported that tumor cells can downregulate or completely lose expression of these epithelial markers during the process of migration and/or dissemination [28]. This creates difficulty for epithelial-based isolations due to their reliance on the EpCAM surface marker for their capture technology. Telomerase, however, has been found to be re-activated in most cancers including prostate, ovarian, breast, lung, colon and bladder [29–32]. Telomerase activity is also associated with malignancy, is often detected in stage IV cancers and is a marker of stem cell activity [33]. Despite the requirement of lysing the sample for assay preparation, the above factors make this enzymatic activity an attractive choice to detect circulating tumor cells for diagnosis. Especially appealing is the possible application of this assay in the detection of relapse. Basal telomerase activity levels due to T-cell activity and other factors could be established and significant variations from this (apart from infections) could indicate possible tumor relapse.

Subsequent increases in activity could also reduce the occurrence of false positives. A possible second step to this process that would circumvent the establishment of basal activity would be to negatively select (as outlined below) leukocytes from the sample. If used in combination with monitoring ctDNA, this could be a powerful tool for treating relapse much earlier than currently possible (Figure 1).

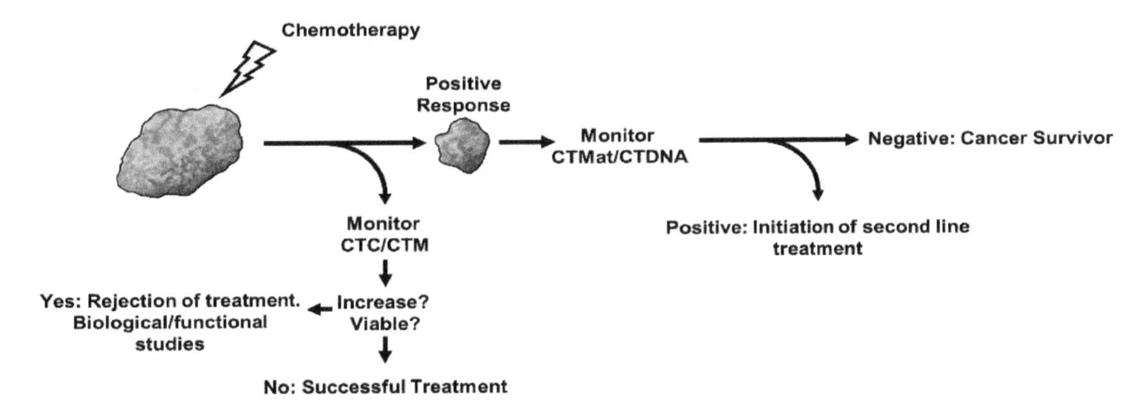

Figure 1. Workflow concept for the analysis of therapy and the early detection of relapse. After chemotherapy, patient CTCs can be analyzed for viability. An increase in viable CTCs can indicate increased mobilization and possible increased risk of relapse. After successful treatment, monitoring patient blood for telomerase activity or ctDNA can give a clinician a much earlier indication of relapse.

3. Clinical Relevance

The mobilization of tumor cells into the circulation is integral to distal metastasis. Current thought is that treatment failure due to metastasis is caused by micrometastasis present at the time of treatment or residual local disease [34]. However, there is mounting evidence that treatment methods themselves could cause an increased dissemination of cells into the vasculature or even the activation of dormant metastatic sites [35–41]. As outlined below, surgery, radiotherapy and systemic chemotherapy can alter tumor biology and possibly influence the risk of metastasis in unforeseen ways. The increase in CTCs as a side effect of treatment is a consideration that deserves careful study.

The effect of radiotherapy on metastasis has long been studied. Early studies indicated that lower doses of radiation resulted in higher rates of metastasis. Breast cancers transplanted into mice and subjected to non-curative doses of radiation had a 43.5% rate of metastasis compared to 9.6% in the control [42]. Metastasis rates were also 10% higher in transplanted mammary tumors given radiation in addition to resection compared to surgery alone [43]. In experiments with lung cancer and fibrosarcoma, it was shown that irradiated mice had higher rates of distal recurrence compared to control. This was initially explained by the activation of dormant micrometastasis and the modification of local tumor cells into a more aggressive and invasive phenotype [44].

Typical regimens of radiotherapy involve fractionated low doses over the course of many days. After longer periods, tumor cells have typically lost reproductive capacity with successful treatment. However, during the early course of the therapy, tumor cells are much more likely to repair therapy-induced DNA damage [45]. These cells have a higher probability of survival if disseminated into the blood stream.

This can be the result of surrounding tissue damage as well as the increased plasticity and genomic instability of irradiated cells [46]. Radiation-induced hypoxia was reported to upregulate the expression of surface markers that increased invasiveness [47]. An increased expression of Vascular Endothelial Growth Factor (VEGF) has also been observed following treatment [48].

The importance of radiation as a therapy cannot be understated. Its clinical value has been demonstrated in many settings. Nevertheless, it has been recently reported that radiation therapy on Non-Small Cell Lung Carcinoma (NSCLC) can mobilize CTCs into the blood stream early in therapy [49]. CTC counts were highest after the first doses of radiation and were shown to originate from the primary tumor. These cells were shown to have increased growth capacity in culture compared to CTCs collected pre-treatment. They also had increased mesenchymal characteristics and were more often found in clusters [8].

Not only radiation, but surgical procedures and chemotherapy have been linked to increased CTCs. Both needle and incisional biopsies have been correlated with increased CTC counts [50,51]. Tumors have also been reported to have formed along the track left by the biopsy needle [52]. Survival rates and local dissemination have been found to be worse with pre-operative biopsies in colorectal cancer, and increased CTCs compared to baseline have also been found both during and after surgery as well [53]. Karigiannis and colleagues have recently reported that neoadjuvant paclitaxel increases both CTCs and metastasis in an MMTV-PyMT (mouse mammary tumor virus-polyoma middle tumor-antigen) murine model [38]. After harvesting the lungs of mice treated with neoadjuvant paclitaxel, they found an increase in both the number and incidence of micrometastasis as well as the presence of single metastatic cells. There was also a twofold increase in CTCs in all experimental models examined, which included xenotransplanted cell lines, the spontaneous PyMT transgenic model and patient-derived xenografts (PDX) [38]. The interrelation between therapy, CTCs and metastasis underscores the vital need to understand the biology of rare circulating cells with the goal of developing targeted treatments. If conventional therapies can potentially increase CTC count and conversely metastasis in some cases, then combination treatments targeting CTCs can potentially improve outcomes.

4. Isolation of Cells

The importance of CTCs in diagnosis, prognosis and therapy outcome seems to be clear. Several technologies have been developed for their capture and enumeration. The assays involving ctDNA and CTMat are exciting prospects in the monitoring of recurrence, but neither involve the capture of CTCs for further analysis. Problematically, even with whole cell capture, many techniques kill the cell along the way. Even the FDA-approved gold standard of CTC detection, the CellSearch system, involves chemical fixation. This process is lethal to cells and does not allow for further characterization of viable cells or expansion in culture. Many of the technologies reported in table 1 involve chemical fixation. This does not preclude the modification of the platform's protocol such that live cells may be captured, but what is commonly reported is outlined in Table 1. In contrast to this, there are many established and developing technologies that have proven to be more sensitive than the CellSearch system and are also designed to capture viable cells, allowing for further biological study [6].

Table 1. Circulating tumor cell (CTC) technologies. CTC isolation technologies grouped by category and isolation criteria. Modified from Ferreira et al. 2016 [54]. *refers to the reference in question.

Subcategory	Platform	Enrichment Principle	Live Cell Analysis Reported *	Company
		Label-Based		
		Positive Enrichment Immunoaffinity		
Micropost Arrays	CTC-Chip [55]	EpCAM	Yes	
	GEDI Chip [56]	PSMA/HER2, Size	No	
	OncoCEE [57]	Antibody Cocktail	No	Biocept Inc. San Diego, CA, USA
Microfluidic Surface Capture	Biofluidica CTC system [58]	EpCAM	Yes	Biofluidica Inc.San Diego, CA, USA
	CytoTrapNano [59]	EpCAM	No	Cytolumina. Los Angeles, CA, USA
	GEM Chip [60]	EpCAM	Yes	
	HTMSU [61]	EpCAM	No	
	Graphene Oxide Chip [62]	EpCAM	No	
	Herringbone Chip [63]	EpCAM	No	
Microfluidic Magnetic	Ephesia [64]	EpCAM	Yes	
	Magnetic Sifter [60]	EpCAM	No	
	LiquidBiopsy [65]	Antibody Cocktail	No	Thermo Fisher, Waltham, MA, USA
	Isoflux [66]	EpCAM	No	Fluxion Biosciences, Alameda, CA, USA
Magnetic	CellSearch [67]	EpCAM	No	Silicon Biosystems, Huntington Valley, PA, USA
	AdnaTest [68]	Antibody Cocktail	No	Qiagen, Hilden, Germany
	MACS [69]	EpCAM	No	Miltenyi Biotec, Bergisch Gladbach, North Rhine-Westphalia, Germany
	MagSweeper [70]	EpCAM	No	
Magnetic in vivo	CellCollector [71]	EpCAM	Yes	GILUPI, Potsdam, Germany
		Negative Enrichment Immunoaffinity		
Magnetic	EasySep [72]	CD45	No	STEMCELL, Vancouver, BC, Canada
	QMS [73]		Yes	
	MACS [74]		Yes	Miltenyi Biotec, Bergisch Gladbach, North Rhine-Westphalia, Germany
Microfluidic/Magnetic	CTC-iChip [75]	CD45, CD66b, Size	Yes	

Table 1. *Cont.*

Subcategory	Platform	Enrichment Principle	Live Cell Analysis Reported *	Company
		Label-Free		
		Density		
	Ficoll-Paque [76]	Density	Yes	GE Healthcare Bio-Sciences, Pittsburg, PA, USA
	OncoQuick [77]	Density, Size	Yes	Greiner Bio-One, Kremsmünster, Austria
	RosetteSep [78]	Density, Antibody Cocktail	Yes	STEMCELL, Vancouver, BC, Canada
	Accucyte and CyteSealer [79]	Density	Yes	Rarecyte, Seattle, WA, USA
		Size		
Filtration	Parsortix [80]	Size, Deformability	Yes	Angle, King of Prussia, PA, USA
	Microwall Chip [81]		Yes	
	ScreenCell [82]		Yes	ScreenCell, Westford, MA, USA
	Resettable Cell Trap [83]		Yes	
	Flexible Micro Spring Array (FMSA) [84]		Yes	
	FaCTchecker [85]		Yes	Circulogix, Hallandale Beach, FL, USA
	Crescent Chip [86]		Yes	
	ISET [87]		Yes	RareCells Diagnostics, Paris Cedes, France
	CellSieve [88]		Yes	Creatv Microtech, Potomac, MD, USA
	Cluster Chip [89]		Yes	
Fluid Dynamics	Vortex [90]	Size	Yes	Vortex Biosciences, Pleasanton, CA, USA
	Double Spiral Chip [91]		Yes	
	Micropinching Chip [92]		Yes	
	ClearCell FX [93]		Yes	Genomax Technologies, Singapore
		Electric		
	ApoStream [94]	Electrical Signature	Yes	Apocell, Houston, TX, USA
	DEPArray [95]		Yes	Silicon Biosystems, Huntington Valley, PA, USA

There are several competing modalities in CTC capture methodology, but all of them fall under two conceptual umbrellas: label-based and label-free. Label-based (or affinity-based) capture is the most widely used strategy, with CellSearch as the only technology approved by the US Food and Drug Administration. The prevailing idea behind this methodology is that tumor cells display different surface markers than blood cells and can therefore be separated from the rest of the circulatory cells on this basis. The three most commonly employed biomarkers utilized for tumor cell selection and identification are the epithelial-cell-adhesion-molecule (EpCAM), cytokeratins, and the antigen CD45 [96]. EpCAM is used to positively select for CTCs, while CD45 negatively depletes white blood cells and cytokeratins are used to positively identify CTCs post-enrichment. These three biomarkers have been expanded upon in some technologies in the use of antibody cocktails including, for example, the human epidermal growth factor 2 (HER2) for breast cancer and the prostate-specific membrane antigen (PSMA) for prostate cancer. In most cases, magnetic beads are conjugated to the antibodies allowing for a magnetic field to capture the cell after the antibody binds to its target. Capture strategies also include microfluidic devices with surface-coated antibodies. Cells of interest bind to these antibodies as the sample flows over the surface. Unfortunately, due to the complexity of CTC biomarker expression, there is no single antigen which allows for 100% error-free capture. This makes effective capture a continuing challenge. Table 1 outlines a variety of capture technologies that fall under the umbrellas of "label-based" and "label-free". Platforms are further characterized by their enrichment principle and their reported capture of live cells.

The CellSearch and Adnatest platforms both make use of magnetic beads attached to antibodies to EpCAM, but Adnatest employs additional cancer-specific antibodies depending on the requirement. CellSearch uses downstream immunostaining to identify CTCs. Positive ID is dependent on the expression of cytokeratins, negative expression of CD45 and positive DAPI nuclear stain. The Adnatest further differs from CellSearch in that it does not rely on downstream immunostaining. Instead, it employs cell lysis and RT-PCR to measure tumor-associated gene expression. A limitation of these technologies is a reliance on EpCAM. EpCAM expression has been shown to vary widely, and cells with low or negative expression can be missed by these platforms [96–100]. Cytokeratin expression can also be lost following EMT [101]. A further drawback is that neither of these technologies allows for further live-cell phenotypic analysis as the captured cells are either fixed or lysed.

Several technologies have been formulated that bypass the requirement for fixation or lysis. Recent advances in microfabrication have allowed the creation of devices with features smaller than a cell. With controlled use of the properties of fluid, cellular contact with these microstructures can be directed. The first among these devices to be developed utilized arrays of antibody-coated microposts [55]. In these devices, sample blood is passed over the chip allowing for the capture of marker-expressing cells. Although some require the pre-lysis of red cells, many enable the use of whole blood with no pre-preparation. The accompanying drawback is that flow rates are most often quite slow at @1–2 mL/h [55,56,102]. The most commonly employed antibody is EpCAM, but several devices employ a cocktail of antibodies that can be specialized for the particular cancer being studied. Today, there are many devices available including the CTC chip, nanopillar chip, micropillar chip, GEDI (geometrically enhanced differential immunocapture) chip, and the OncoCEE among others. These devices have all shown higher capture efficiency than the CellSearch [6], and have the advantage of smaller size and lower cost than the magnetic benchtop devices.

The CTC-chip's first iteration (preceding the herringbone chip) captured a median of 155 cells/mL in each of 55 samples tested from 68 patients with non-small cell lung cancer, while the CellSearch only captured cells in 20% of patient samples and had a mean of <6 cells/mL [103]. The GEDI chip employs hydrodynamic chromatography by offsetting the microposts in such a way as to separate cells by size and minimize non-specific leucocyte adhesion [56]. The OncoCEE employs a customizable cocktail that can include antibodies for both cancer and mesenchymal specific markers. It also allows for in situ fluorescent staining of the captured cells by staining the capture antibodies [57].

To increase imaging and production efficiency, the field has begun to explore the idea of surface-capture devices that eschew the concept of posts altogether. Microchannels and surface patterns are designed to maximize mixing and surface contact with cells. The simpler design allows for larger scale production and with opaque posts and three-dimensional structure removed, imaging is enhanced. Another welcome enhancement is the allowance of higher flow rates, leading to more rapid throughput [60,62,63]. Devices which use this technology include the microvortex herringbone chip, sinusoidal chip, GEM chip, and the graphene oxide chip.

Biomarkers may also be used to negatively enrich samples containing CTCs. Blood cell markers such as CD45 and/or CD66 can be used to deplete white blood cells from the larger population enriching for CTCs in the remainder. Technologies utilizing this method include EasySep and RosetteSep. RosetteSep incorporates the additional step of density centrifugation, while EasySep uses a magnetic field. A pitfall inherent in this technique is the fact that not all cells in the blood express these markers, resulting in a much poorer purity than with positive selection [74,104,105]. Another downfall is possible CTC loss being caught up in the large movement of concentrated blood cells during depletion. For these reasons, this technique is often used as a preparatory step for other enrichment methods [106].

Despite the utility and many benefits of cellular biomarkers, there are drawbacks as well. It is becoming established that tumor cells express EpCAM at varying levels. In fact, expression can be ablated entirely in some sub-populations, including those which have undergone EMT [107]. Tumor cells have also been reported to express the white blood cell marker CD45 [108]. With these problems in mind, alternative assays which employ only the biophysical properties of the cell have been developed.

These label-free physical detection methods include cell size, deformability, density and electric charge. The most widely employed biophysical selection criterion is cellular size [12]. Tumor cells are larger on average than blood cells [109], and this morphological difference is employed to differentially capture CTCs and CTMs. There are multiple platforms which use these properties such as the micro double spiral chip, the Parsortix and Vortex systems, the micro crescent chip, the Cellsee system, micro column wall chip, ISET, Clear Cell FX, cluster chip, micro pinching chip and the CellSieve among others. Each of these assays have proven to be more selective than the CellSearch system in isolating tumor cells [6].

There are different ways of using size in the process of selection, however. Two-dimensional microfiltration involves a single membrane with variable pore size used to filter out smaller cells while leaving the larger CTCs trapped on the membrane. Cell pore sizes come in a variety of sizes ranging from 6 to 9 um. CellSieve filtration has not only been used to detect cancer-associated macrophages and cancer-associated macrophage-like cells, [110,111] but, using 7.5 mL patient samples, it detected CTCs in 100% of metastatic breast cancer patients tested [88]. CellSieve, ISET and ScreenCell use this methodology, but require pre-processing of the patient sample. FMSA (Flexible Microspring Array) can use whole blood and has been validated in the detection of CTCs in 76% of samples tested in various cancers [112].

Three-dimensional filtration systems exploit the larger size of tumor cells, but use multiple layers of filter to capture them. The FaCTChecker, Parsortix system, and cluster chip fall into this category. The FaCTChecker takes advantage of multiple vertical layers with different sized pores [113], while the Parsortix has developed a horizontal stair-type scheme that reduces the channel width stepwise [80]. Viable CTCs can be harvested using either platform. Our lab has employed the Parsortix system to isolate CTCs from breast cancer patients. We subsequently tethered these live cells on a proprietary PEM+Lipid technology [114] and imaged them for Microtentacles (Figure 2). The Cluster Chip is unique in size selection technologies, as its sole target are CTMs. Many technologies have reported on the capture of CTMs, but this novel approach enriches for them specifically while allowing single CTCs to pass through [89]. The design involves staggered rows of triangular pillars. The repeating unit of the design is the cluster trap. This three-triangle arrangement is reminiscent of a biohazard sign insofar as two triangles side by side to create a tunnel that is bifurcated by the third triangle beneath them. This simple design can capture CTMs as small as two cells. The utility of the device was shown in breast, melanoma and prostate cancers, isolating clusters in 41%, 30% and 31% of patients, respectively [89]. Large downsides to filtration systems exist, however. Despite the capture of viable cells without labels that are difficult to remove, the systems are prone to clogging and parallel processing is needed for large volumes. Purity is also an issue as it can range below 10%.

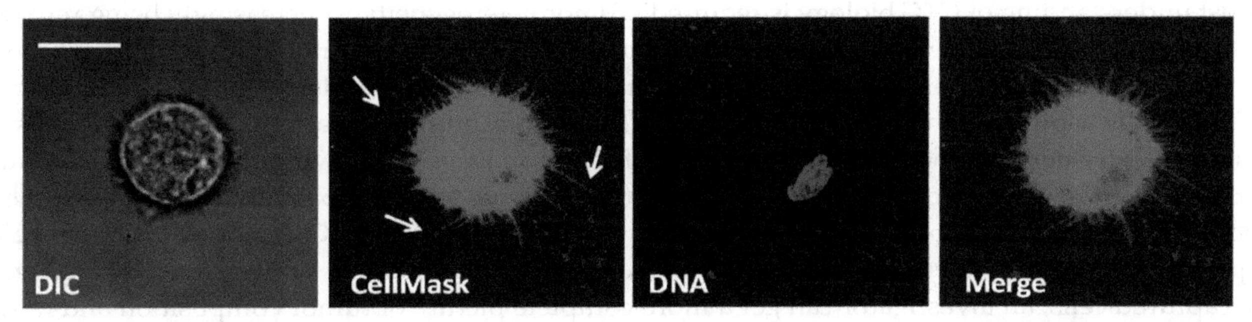

Figure 2. Live CTCs isolated with Parsortix technology. Whole blood was taken from a stage IV metastatic breast cancer patient. The Angle Parsortix was used to isolate CTCs from the blood (15 CTCs in 10 mL). CTCs were tethered to proprietary PEM+Lipid slides and stained with CellMask membrane dye (red). Cells are $CD45^-$ and contain a nucleus (blue). Arrows indicate microtubule-based structures termed Microtentacles (McTN).

Two exciting new technologies to recently emerge involve the use of inertial fluid forces to passively separate CTCs from the rest of the blood population based on cell size. A combination of shear gradient and wall lift forces interact to stably trap the CTCs. The Vortex platform capitalizes on these forces to inertially focus trapped CTCs in micro vortices created in reservoirs apart from the main fluid channel. Smaller blood cells simply flow by in the main stream. CTCs remain in the device until a slower flow rate flushes them out of the reservoirs. The Vortex Chip processes the standard 7.5 mL sample size in 20 minutes using whole unprocessed blood. Confirmation has come in breast and lung cancers with a purity of 57–94%, much higher than that normally attained with size-based techniques [90]. The ClearCell FX uses inertial forces in combination with secondary flow arising from curved channels [115]. When a channel is curved, there is a difference in the flow rates between the center of the channel and the walls. This difference in flow rates is termed a "Dean's" flow and, when combined with inertial forces, can be calculated to precisely position cells. The trapezoidal channel results in larger cells on the shorter wall and smaller cells on the larger wall. This channel then splits into two collection outlets, where CTCs are isolated and captured. This technology requires red cell lysis prior to flow but has an impressive 8-minute run time. It has been confirmed in breast and lung cancers with a higher capture rate than the Vortex [116]. Both processes involve minimal stress on cells without the use of labels and are much simpler to fabricate than those previously mentioned.

Dielectrophoresis (DEP) exploits the electrical characteristics of tumor cells. These characteristics depend on phenotype, composition and morphology. DEP polarizes cells by using a nonuniform electric field. This results in the ability to physically manipulate the cells by exerting attractive or repulsive forces (positive pDEP or negative nDEP). ApoStream employs a strategy wherein the electrical field separates tumor cells and leukocytes, using differences in their conductivity. The field attracts CTCs and repels leukocytes. After pre-processing by centrifugation, the ApoStream can process captured CTCs from 10mL of whole blood in less than an hour [117].

The DEPArray applies the second DEP strategy, retention, by trapping single cells in DEP cages generated via an array of individually controllable electrodes [118]. DEPArray as a platform is not designed for the bulk enrichment of cells, however. It is intended for single cell capture. Multiple studies have shown the utility of the technology in this capacity [95,119,120], but an unfortunate drawback is large cell loss during sample preparation [121].

5. CTC Biology

The prognostic importance of CTC counts is well established, but counts have not yet been widely employed to affect clinical decisions, due to unclear relevance to treatment. CTC counts have therefore not been recommended clinically to affect treatment decisions, as of yet [122]. Consequently, a more robust understanding of CTC biology is required. Tumor heterogeneity is increasingly being reported in the literature, not only between primary and secondary tumors, but intratumor as well. There can be as many as six different clonal cell lines within just one tumor [123]. Standard biopsy techniques such as fine needle aspiration and core biopsy are insufficient to capture this variety. These techniques, by design, take tissue from one area of the tumor for further analysis. Even with multiple samples, such as those taken in prostate cancers, there is not sufficient tissue to encompass all of the heterogeneity. "Liquid biopsy" is a term being increasingly used to describe analysis of CTC populations. The CTC population is thought to encompass more of the clonal populations in a tumor [122]. By analyzing the captured cells, an investigator can get a more complete picture of tumor composition and how it changes over time.

Studies of the composition of CTCs can further shed light into the process of metastasis. The complete process of metastasis is unclear, but conventional wisdom describes a process where tumor cells undergo the epithelial mesenchymal transition (EMT). This process involves cells detaching from the main tumor body, migrating through the extracellular matrix and extravasating into the circulation (Figure 3). During this process, the cell downregulates the expression of its epithelial markers, such as E-cadherin, and upregulates EMT markers, such as N-cadherin, snail, twist, vimentin and detyrosinated tubulin [123].) CTC/CTMs have been shown to upregulate vimentin and detyrosinated tubulin as well [124]. After extravasation, the cell then undergoes the reverse process of mesenchymal-to-epithelial transition (MET). This has been widely held to be the main mode of metastatic dissemination, but new reports have begun to challenge this.

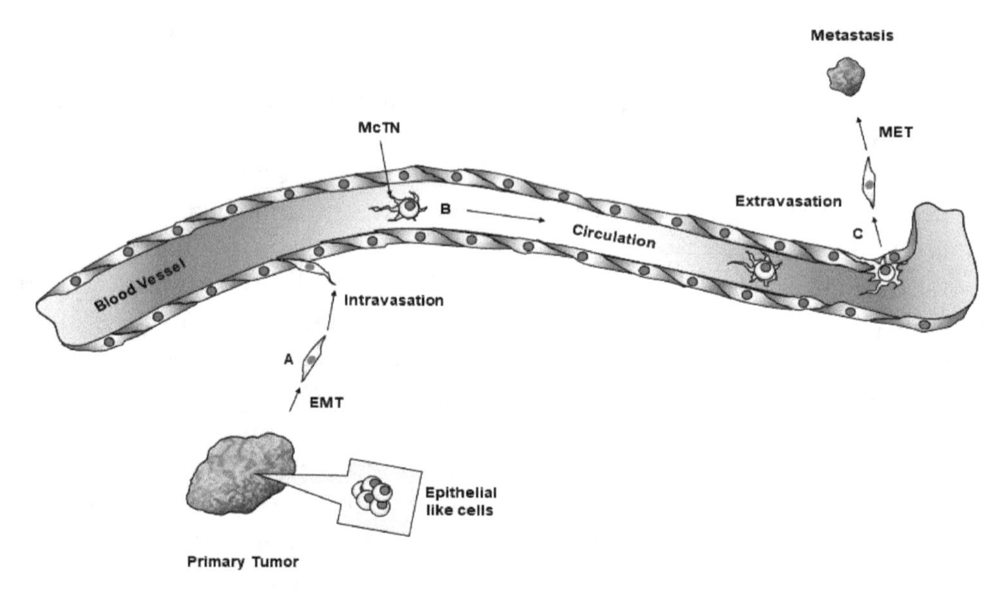

Figure 3. Epithelial to Mesenchymal Transition (EMT) and metastasis. (**A**) Epithelial-like cells in the primary tumor undergo a transition to a mesenchymal phenotype and migrate towards the vasculature. (**B**) Detached tumor cells in the circulatory vessels display microtubule-based structures, termed Microtentacles (McTN). (**C**) McTN aid in reattachment and extravasation. Extravasated cells undergo a mesenchymal to epithelial transition, and seed tumors at distal sites.

Fischer and colleagues described an experiment with a triple transgenic mouse that tracked mesenchymal lineage in breast cancer tissue. The system utilized an irreversible color switch that was activated by the expression of fsp1, a crucial protein in EMT initiation. With the expression of fsp1, cells experiencing EMT would undergo an irreversible color change from red to green, allowing for the tracking of any metastatic cell that had gone through the process. What was observed was that the vast majority of metastatic tumor tissue was red and had not undergone EMT. This was confirmed using multiple oncogenes and EMT tracing proteins. Interestingly, the following chemotherapy tumor recurrence was mostly green [125]. Similar findings were reported independently in the same issue of *Nature*, from a lab using twist and snail in pancreatic tumor lines [126].

The ramifications of these findings are manifold and beyond the scope of this review to cover. It is however important to note that this is a proof-of-principle that the process of EMT can be dispensable for initial metastasis, in some cases. This underscores the importance of understanding the biology in circulating cells. Which proteins CTCs express, and the resulting phenotypes, are crucial to understanding how cancer spreads to distal sites. It is indeed possible that the bulk of tumor spread results from simple CTC shedding into the vasculature. This does not reduce the importance of EMT in cancer, however. Cancer cells displaying the mesenchymal phenotype have been shown to be more aggressive, stem-like, and resistant to treatment [127]. Both Zheng and Fischer also observed EMT cells persisting after treatment despite original metastasis composition. What this highlights is that there can be multiple modes of metastasis, and the study of cells in transition can give us insights into the process.

Aceto et al. have recently shown that CTMs are 23–50 times more metastatic than CTCs [8]. Their use of fluorescently labeled cells also highlighted that clusters arise from oligoclonal groupings of cells that differentially express the cell junction protein plakoglobin. These studies, along with the results of Zheng and Fischer, further emphasize the importance of circulating cell study. They give us insight into the probable mechanism of metastasis. In the 323 lung foci that Aceto observed, 171 were CTM-derived, although CTMs only comprised 2–5% of the total population of tumor cells in the circulation.

Previous thought was that CTMs were likely to break up in the physical pressures of the blood stream, or to become lodged very quickly in smaller capillaries, negating their capability of seeding distant metastasis [128]. Recent work has shown this is not the case. Au et al. demonstrated with microscopy and capillary tubing that tumor clusters migrated in a single file fashion without dissociation. Moreover, the clusters were viable upon capillary exit [129]. Taken together with the evidence that clusters have a much higher metastatic potential, the benefit of elucidating biological differences between CTCs and CTMs is clear. In fact, very recent evidence has indicated that the disruption of CTMs leads to the suppression of metastasis [130].

It has been hypothesized that CTMs could arise either by passive shedding or through collective migration [101,131]. Collective migration has been observed in multiple tumor types, but it has only been directly correlated to local invasion [101]. Metastasis, through collective migration, has merely been inferred by the presence of clusters in the blood. Tumor vasculature is improperly formed, tortuous, leaky, and possessive of blind shunts [132]. It has been reported that tumor cells can actually replace vascular endothelium in places, a process known as vasculogenic mimicry [133]. With these factors in mind, it is quite feasible that CTCs and CTMs mainly arise through the passive sloughing of cells. This would correlate well with the data showing that breast cancers arising from *neu* and PyMT transgenes undergo very little EMT.

Interstitial fluid pressure (IFP) could contribute to CTC shedding as well. IFP is the fluid pressure measured within tumors and is the direct result of hyperpermeable blood vessels. Fluid and plasma proteins extravasate into the tumor tissue and elevate the pressure in the interstitium [47]. Not only could this increased pressure disrupt cell–cell junctions, but it could cause physical pressures that assist in cells detaching from the tumor bulk. High IFP is correlated strongly with poor prognosis [134]. As higher interstitial pressure is a direct result of improperly formed vessels, and stronger pressure could result in cell detachment, it follows that cells could break off at a higher rate as capillaries become leakier.

6. Functional CTC Studies

Translating lab research into clinical practice involves the study of how cells function, both in vitro and in vivo. As outlined above, it has been clearly shown that higher CTC counts in peripheral blood correlates with poor prognosis. Functional studies can broaden the spectrum of applications to CTC analyses. The challenges in obtaining stable cultures are significant but advances in CTC expansion from patient samples have been achieved. The subsequent functional studies can give clues into the identity of metastasis-initiating cells and can point the way to new avenues of therapy. A workflow, as outlined in Figure 4, illustrates the concept of CTC study, beginning with isolation and ending with the functional study of cultured CTCs. The first step in a workflow of this kind would be sample preparation and isolation using one of the methods outlined above. This would result in the capture of differing circulating materials, depending on the capture technology. These captured materials could eventually be used for prognosis and relapse decisions.

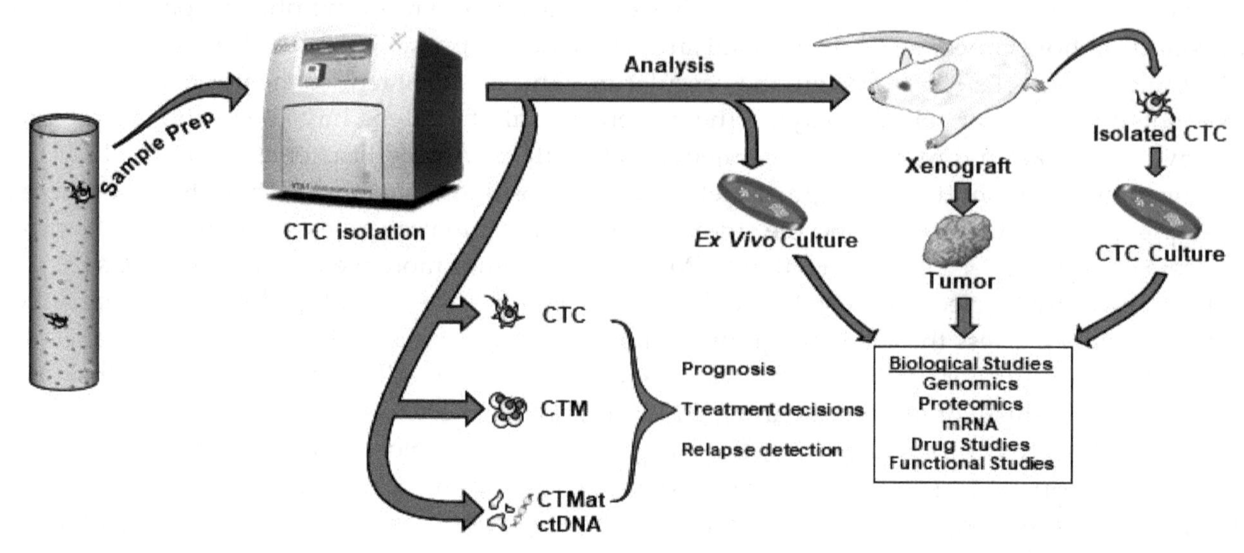

Figure 4. Workflow concept for the isolation of CTCs and subsequent analysis. Patient blood is passed through a capture device which enriches for tumor cells. Captured cells are then identified, enumerated and characterized. Cells can then be cultured and subjected to further biological and functional analysis.

Functional analysis of CTCs has been performed in multiple studies. Zhang et al. reported a protocol for the primary culture of breast cancer CTCs from patients with advanced stage and brain metastases [135]. The cultures survived for several weeks. This study allowed the elucidation of several biomarkers, including HER2 and EGFR, as brain metastasis selected markers (BMSM). Cells which expressed this BMSM signature exhibited significant invasiveness and resulted in brain metastases in murine xenografts. Oligoclonal breast cancer CTC cell lines were cultured for >6 months in 2014 [136]. Of five tested lines, three proved to be tumorigenic. The culture allowed for the discovery of new mutations in the estrogen receptor gene, fibroblast growth factor and PIK3CA. A long term culture of a CTC line from prostate cancer was also established using a novel 3D organoid system [137]. This included TRMPRSS2-ERG fusion proteins, overexpression of SPINK1 and SPOP and CHD1 mutations and loss, respectively. Lung cancer CTCs were successfully expanded ex vivo using a 3D co-culture which used a simulated tumor microenvironment. CTCs expanded from 14/19 patient samples and had matched mutations with their respective primary tumors, including tp53 [138].

Captured breast cancer CTCs were injected into murine tibia bone resulting in lung, liver and bone metastases [104]. The study of protein expression in the metastasis revealed universal expression of EpCAM, MET, CD44 and CD47. This could reveal important information on necessary proteins in the process of engraftment and metastatic outgrowth. Further study in an additional cohort revealed that metastases increased with the number of CD44/CD47/MET/EpCAM-positive cells. Importantly, these cells were obtained from advanced stage patients with high numbers of CTCs. This underscores the need to obtain and expand tumor cells from early stage patients to confirm this protein expression profile as metastasis-initiating in all stages.

Migratory capabilities of isolated metastatic prostate CTCs were shown in NOD/SCID mice [139]. Tumor cells were found in the spleen and the bone marrow after xenografting. Hodgkinson et al. showed that CTC xenografts of small cell lung cancer (SCLC) are not only tumorigenic in murine models but respond similarly to chemotherapy as in the original donor patient. SCLC patients have been reported to have the highest CTC counts of all solid tumors [140]. Notably, these tumors are often inoperable and difficult to biopsy. Expanding tumors which mirror patient response is an important step in furthering treatment less invasively.

7. Conclusions

Metastasis remains the number one cause of death in cancer patients. This is the result of the migration of cells from the primary tumor to distal sites. Indispensable to this process is the migration/shedding of CTCs into the vasculature. These circulating tumor cells can be analyzed for a breadth of beneficial information. Currently, prognostic indications can be made based on the enumeration of CTCs in the blood. With further technological development, the presence of metastasis could be detected before clinical manifestation, by monitoring tumor materials in the blood. It is also feasible that patients with known genetic risk factors could be monitored for ctDNA, using primers for known tumor mutations. This could possibly advance diagnosis by years, and increase survival rates significantly.

Even after disease control is accomplished with surgery and/or therapy, metastasis can remain a problem. This can be partially due to cancer cell mobilization caused by therapy itself. Radiation has been shown to select for and to convert tumor cells to phenotypes that are more mobile and aggressive, allowing for the generation of metastases. Tissue disruption and the leakage of blood containing tumor cells during surgery can also promote tumor spread. This includes procedures such as routine biopsy.

These problems underscore the need for the capture and study of viable tumor cells. Many technologies exist, but many involve the fixation of cells and their subsequent death. Emerging platforms have developed ways to isolate live CTCs which allow for downstream biological analysis. These studies have led to valuable insights into the mechanisms of metastasis and cellular survival in the harsh environment of the circulation. Functional studies with cultured CTCs and xenografts have revealed important information on protein expression and genetic composition. With the standardization of capture techniques, inconsistencies in efficiency can be greatly reduced, allowing for more robust information to be attained.

All these principles could support the goal of improving drug discovery to reduce metastasis. The current cancer detection and drug treatment paradigm involves tumor growth and visualization. Current technological parameters limit the tumors we can visualize to upwards of ten million cells. A shift of focus to the detection of ctDNA/CTMat/CTC/CTMs can improve detection sensitivity and improve treatment strategies. If surgery and radiation can promote cellular dissemination, then therapies that specifically target circulating cells could increase survival outcomes and reduce distal recurrence. Overall, developing therapies that target cancer's ability to ever survive in circulation can prevent metastasis before it occurs.

Author Contributions: P.C.B. and S.S.M. wrote the manuscript.

References

1. Jemal, A.; Bray, F.; Center, M.M.; Ferlay, J.; Ward, E.; Forman, D. Global cancer statistics. *CA Cancer J. Clin.* **2011**, *61*, 69–90. [CrossRef] [PubMed]
2. Talmadge, J.E.; Fidler, I.J. AACR centennial series: The biology of cancer metastasis: Historical perspective. *Cancer Res.* **2010**, *70*, 5649–5669. [CrossRef] [PubMed]
3. Zhang, X.; Nie, D.; Chakrabarty, S. Growth factors in tumor microenvironment. *Front. Biosci. Landmark Ed.* **2010**, *15*, 151–165. [CrossRef] [PubMed]

4. Butler, T.P.; Gullino, P.M. Quantitation of cell shedding into efferent blood of mammary adenocarcinoma. *Cancer Res.* **1975**, *35*, 512–516. [PubMed]
5. Ross, A.A.; Cooper, B.W.; Lazarus, H.M.; Mackay, W.; Moss, T.J.; Ciobanu, N.; Tallman, M.S.; Kennedy, M.J.; Davidson, N.E.; Sweet, D.; et al. Detection and viability of tumor cells in peripheral blood stem cell collections from breast cancer patients using immunocytochemical and clonogenic assay techniques. *Blood* **1993**, *82*, 2605–2610. [PubMed]
6. Hong, B.; Zu, Y. Detecting circulating tumor cells: Current challenges and new trends. *Theranostics* **2013**, *3*, 377–394. [CrossRef] [PubMed]
7. Hou, J.M.; Krebs, M.G.; Lancashire, L.; Sloane, R.; Backen, A.; Swain, R.K.; Priest, L.J.; Greystoke, A.; Zhou, C.; Morris, K.; et al. Clinical significance and molecular characteristics of circulating tumor cells and circulating tumor microemboli in patients with small-cell lung cancer. *J. Clin. Oncol.* **2012**, *30*, 525–532. [CrossRef] [PubMed]
8. Aceto, N.; Bardia, A.; Miyamoto, D.T.; Donaldson, M.C.; Wittner, B.S.; Spencer, J.A.; Yu, M.; Pely, A.; Engstrom, A.; Zhu, H.; et al. Circulating tumor cell clusters are oligoclonal precursors of breast cancer metastasis. *Cell* **2014**, *158*, 1110–1122. [CrossRef] [PubMed]
9. Amin, M.B.; Edge, S.; Greene, F.; Byrd, D.R.; Brookland, R.K.; Washington, M.K.; Gershenwald, J.E.; Compton, C.C.; Hess, K.R.; Sullivan, D.C.; et al. *American Joint Committee on Cancer; American Cancer Society. AJCC Cancer Staging Manual*, 8th ed.; American Joint Committee on Cancer; Springer: Chicago, IL, USA, 2017; p. xvii. 1024p.
10. Mittendorf, E.A.; Bartlett, J.M.S.; Lichtensztajn, D.L.; Chandarlapaty, S. Incorporating Biology Into Breast Cancer Staging: American Joint Committee on Cancer, Eighth Edition, Revisions and Beyond. *Am. Soc. Clin. Oncol. Educ. Book* **2018**, 38–46. [CrossRef] [PubMed]
11. Sturgeon, C.M.; Duffy, M.J.; Stenman, U.H.; Lilja, H.; Brunner, N.; Chan, D.W.; Babaian, R.; Bast, R.C., Jr.; Dowell, B.; Esteva, F.J.; et al. National Academy of Clinical Biochemistry laboratory medicine practice guidelines for use of tumor markers in testicular, prostate, colorectal, breast, and ovarian cancers. *Clin. Chem.* **2008**, *54*, e11–e79. [CrossRef]
12. Van der Toom, E.E.; Verdone, J.E.; Gorin, M.A.; Pienta, K.J. Technical challenges in the isolation and analysis of circulating tumor cells. *Oncotarget* **2016**, *7*, 62754–62766. [CrossRef] [PubMed]
13. Grover, P.K.; Cummins, A.G.; Price, T.J.; Roberts-Thomson, I.C.; Hardingham, J.E. Circulating tumour cells: the evolving concept and the inadequacy of their enrichment by EpCAM-based methodology for basic and clinical cancer research. *Ann. Oncol.* **2014**, *25*, 1506–1516. [CrossRef] [PubMed]
14. Wang, L.; Balasubramanian, P.; Chen, A.P.; Kummar, S.; Evrard, Y.A.; Kinders, R.J. Promise and limits of the CellSearch platform for evaluating pharmacodynamics in circulating tumor cells. *Semin. Oncol.* **2016**, *43*, 464–475. [CrossRef] [PubMed]
15. Flores, L.M.; Kindelberger, D.W.; Ligon, A.H.; Capelletti, M.; Fiorentino, M.; Loda, M.; Cibas, E.S.; Janne, P.A.; Krop, I.E. Improving the yield of circulating tumour cells facilitates molecular characterisation and recognition of discordant HER2 amplification in breast cancer. *Br. J. Cancer* **2010**, *102*, 1495–1502. [CrossRef] [PubMed]
16. Rao, G.C.; Larson, C.; Repollet, M.; Rutner, H.; Terstappen, L.W.; O'hara, S.M.; Gross, S. Analysis of Circulating Tumor Cells, Fragments, and Debris. U.S. Patent 8,329,422, 11 December 2012.
17. Coumans, F.A.; Doggen, C.J.; Attard, G.; de Bono, J.S.; Terstappen, L.W. All circulating EpCAM+CK+CD45- objects predict overall survival in castration-resistant prostate cancer. *Ann. Oncol.* **2010**, *21*, 1851–1857. [CrossRef] [PubMed]
18. Schwarzenbach, H.; Alix-Panabieres, C.; Muller, I.; Letang, N.; Vendrell, J.P.; Rebillard, X.; Pantel, K. Cell-free tumor DNA in blood plasma as a marker for circulating tumor cells in prostate cancer. *Clin. Cancer Res.* **2009**, *15*, 1032–1038. [CrossRef]
19. Leary, R.J.; Kinde, I.; Diehl, F.; Schmidt, K.; Clouser, C.; Duncan, C.; Antipova, A.; Lee, C.; McKernan, K.; De La Vega, F.M.; et al. Development of personalized tumor biomarkers using massively parallel sequencing. *Sci. Transl. Med.* **2010**, *2*, 20ra14. [CrossRef] [PubMed]
20. Kaiser, J. Medicine. Keeping tabs on tumor DNA. *Science* **2010**, *327*, 1074. [CrossRef]
21. Garcia-Murillas, I.; Schiavon, G.; Weigelt, B.; Ng, C.; Hrebien, S.; Cutts, R.J.; Cheang, M.; Osin, P.; Nerurkar, A.; Kozarewa, I.; et al. Mutation tracking in circulating tumor DNA predicts relapse in early breast cancer. *Sci. Transl. Med.* **2015**, *7*, 302ra133. [CrossRef]

22. Tie, J.; Wang, Y.; Tomasetti, C.; Li, L.; Springer, S.; Kinde, I.; Silliman, N.; Tacey, M.; Wong, H.L.; Christie, M.; et al. Circulating tumor DNA analysis detects minimal residual disease and predicts recurrence in patients with stage II colon cancer. *Sci. Transl. Med.* **2016**, *8*, 346ra392. [CrossRef]
23. Schmiegel, W.; Scott, R.J.; Dooley, S.; Lewis, W.; Meldrum, C.J.; Pockney, P.; Draganic, B.; Smith, S.; Hewitt, C.; Philimore, H.; et al. Blood-based detection of RAS mutations to guide anti-EGFR therapy in colorectal cancer patients: Concordance of results from circulating tumor DNA and tissue-based RAS testing. *Mol. Oncol.* **2017**, *11*, 208–219. [CrossRef] [PubMed]
24. Siravegna, G.; Bardelli, A. Genotyping cell-free tumor DNA in the blood to detect residual disease and drug resistance. *Genome Biol.* **2014**, *15*, 449. [CrossRef] [PubMed]
25. Ulz, P.; Heitzer, E.; Geigl, J.B.; Speicher, M.R. Patient monitoring through liquid biopsies using circulating tumor DNA. *Int. J. Cancer* **2017**, *141*, 887–896. [CrossRef] [PubMed]
26. Chera, B.S.; Kumar, S.; Beaty, B.T.; Marron, D.; Jefferys, S.R.; Green, R.L.; Goldman, E.C.; Amdur, R.; Sheets, N.; Dagan, R.; et al. Rapid Clearance Profile of Plasma Circulating Tumor HPV Type 16 DNA during Chemoradiotherapy Correlates with Disease Control in HPV-Associated Oropharyngeal Cancer. *Clin Cancer Res.* **2019**. [CrossRef] [PubMed]
27. Murtaza, M.; Dawson, S.J.; Tsui, D.W.; Gale, D.; Forshew, T.; Piskorz, A.M.; Parkinson, C.; Chin, S.F.; Kingsbury, Z.; Wong, A.S.; et al. Non-invasive analysis of acquired resistance to cancer therapy by sequencing of plasma DNA. *Nature* **2013**, *497*, 108–112. [CrossRef] [PubMed]
28. Hou, J.M.; Krebs, M.; Ward, T.; Morris, K.; Sloane, R.; Blackhall, F.; Dive, C. Circulating tumor cells, enumeration and beyond. *Cancers* **2010**, *2*, 1236–1250. [CrossRef] [PubMed]
29. Fizazi, K.; Morat, L.; Chauveinc, L.; Prapotnich, D.; De Crevoisier, R.; Escudier, B.; Cathelineau, X.; Rozet, F.; Vallancien, G.; Sabatier, L.; et al. High detection rate of circulating tumor cells in blood of patients with prostate cancer using telomerase activity. *Ann. Oncol.* **2007**, *18*, 518–521. [CrossRef] [PubMed]
30. Sapi, E.; Okpokwasili, N.I.; Rutherford, T. Detection of telomerase-positive circulating epithelial cells in ovarian cancer patients. *Cancer Detect. Prev.* **2002**, *26*, 158–167. [CrossRef]
31. Soria, J.C.; Gauthier, L.R.; Raymond, E.; Granotier, C.; Morat, L.; Armand, J.P.; Boussin, F.D.; Sabatier, L. Molecular detection of telomerase-positive circulating epithelial cells in metastatic breast cancer patients. *Clin. Cancer Res.* **1999**, *5*, 971–975.
32. Gauthier, L.R.; Granotier, C.; Soria, J.C.; Faivre, S.; Boige, V.; Raymond, E.; Boussin, F.D. Detection of circulating carcinoma cells by telomerase activity. *Br. J. Cancer* **2001**, *84*, 631–635. [CrossRef]
33. Maurelli, R.; Zambruno, G.; Guerra, L.; Abbruzzese, C.; Dimri, G.; Gellini, M.; Bondanza, S.; Dellambra, E. Inactivation of p16INK4a (inhibitor of cyclin-dependent kinase 4A) immortalizes primary human keratinocytes by maintaining cells in the stem cell compartment. *FASEB J.* **2006**, *20*, 1516–1518. [CrossRef] [PubMed]
34. Pantel, K.; Deneve, E.; Nocca, D.; Coffy, A.; Vendrell, J.P.; Maudelonde, T.; Riethdorf, S.; Alix-Panabieres, C. Circulating epithelial cells in patients with benign colon diseases. *Clin. Chem.* **2012**, *58*, 936–940. [CrossRef] [PubMed]
35. Martin, O.A.; Anderson, R.L.; Narayan, K.; MacManus, M.P. Does the mobilization of circulating tumour cells during cancer therapy cause metastasis? *Nat. Rev. Clin. Oncol.* **2017**, *14*, 32–44. [CrossRef] [PubMed]
36. Mego, M.; Gao, H.; Lee, B.N.; Cohen, E.N.; Tin, S.; Giordano, A.; Wu, Q.; Liu, P.; Nieto, Y.; Champlin, R.E.; et al. Prognostic Value of EMT-Circulating Tumor Cells in Metastatic Breast Cancer Patients Undergoing High-Dose Chemotherapy with Autologous Hematopoietic Stem Cell Transplantation. *J. Cancer* **2012**, *3*, 369–380. [CrossRef] [PubMed]
37. Inhestern, J.; Oertel, K.; Stemmann, V.; Schmalenberg, H.; Dietz, A.; Rotter, N.; Veit, J.; Gorner, M.; Sudhoff, H.; Junghanss, C.; et al. Prognostic Role of Circulating Tumor Cells during Induction Chemotherapy Followed by Curative Surgery Combined with Postoperative Radiotherapy in Patients with Locally Advanced Oral and Oropharyngeal Squamous Cell Cancer. *PLoS ONE* **2015**, *10*, e0132901. [CrossRef] [PubMed]
38. Karagiannis, G.S.; Pastoriza, J.M.; Wang, Y.; Harney, A.S.; Entenberg, D.; Pignatelli, J.; Sharma, V.P.; Xue, E.A.; Cheng, E.; D'Alfonso, T.M.; et al. Neoadjuvant chemotherapy induces breast cancer metastasis through a TMEM-mediated mechanism. *Sci. Transl. Med.* **2017**, *9*. [CrossRef] [PubMed]
39. Wang, Y.; Li, W.; Patel, S.S.; Cong, J.; Zhang, N.; Sabbatino, F.; Liu, X.; Qi, Y.; Huang, P.; Lee, H.; et al. Blocking the formation of radiation-induced breast cancer stem cells. *Oncotarget* **2014**, *5*, 3743–3755. [CrossRef]

40. Hu, X.; Ghisolfi, L.; Keates, A.C.; Zhang, J.; Xiang, S.; Lee, D.K.; Li, C.J. Induction of cancer cell stemness by chemotherapy. *Cell Cycle* **2012**, *11*, 2691–2698. [CrossRef]
41. Xu, Z.Y.; Tang, J.N.; Xie, H.X.; Du, Y.A.; Huang, L.; Yu, P.F.; Cheng, X.D. 5-Fluorouracil chemotherapy of gastric cancer generates residual cells with properties of cancer stem cells. *Int. J. Biol. Sci.* **2015**, *11*, 284–294. [CrossRef]
42. Kaplan, H.S.; Murphy, E.D. The effect of local roentgen irradiation on the biological behavior of a transplantable mouse carcinoma; increased frequency of pulmonary metastasis. *J. Natl. Cancer Inst.* **1949**, *9*, 407–413.
43. Sheldon, P.W.; Fowler, J.F. The effect of low-dose pre-operative X-irradiation of implanted mouse mammary carcinomas on local recurrence and metastasis. *Br. J. Cancer* **1976**, *34*, 401–407. [CrossRef] [PubMed]
44. Camphausen, K.; Moses, M.A.; Beecken, W.D.; Khan, M.K.; Folkman, J.; O'Reilly, M.S. Radiation therapy to a primary tumor accelerates metastatic growth in mice. *Cancer Res.* **2001**, *61*, 2207–2211. [PubMed]
45. Eriksson, D.; Stigbrand, T. Radiation-induced cell death mechanisms. *Tumour Biol.* **2010**, *31*, 363–372. [CrossRef] [PubMed]
46. Butof, R.; Dubrovska, A.; Baumann, M. Clinical perspectives of cancer stem cell research in radiation oncology. *Radiother. Oncol.* **2013**, *108*, 388–396. [CrossRef] [PubMed]
47. Rofstad, E.K.; Galappathi, K.; Mathiesen, B.S. Tumor interstitial fluid pressure-a link between tumor hypoxia, microvascular density, and lymph node metastasis. *Neoplasia* **2014**, *16*, 586–594. [CrossRef] [PubMed]
48. Gorski, D.H.; Beckett, M.A.; Jaskowiak, N.T.; Calvin, D.P.; Mauceri, H.J.; Salloum, R.M.; Seetharam, S.; Koons, A.; Hari, D.M.; Kufe, D.W.; et al. Blockage of the vascular endothelial growth factor stress response increases the antitumor effects of ionizing radiation. *Cancer Res.* **1999**, *59*, 3374–3378. [PubMed]
49. Martin, O.A.; Anderson, R.L.; Russell, P.A.; Cox, R.A.; Ivashkevich, A.; Swierczak, A.; Doherty, J.P.; Jacobs, D.H.; Smith, J.; Siva, S.; et al. Mobilization of viable tumor cells into the circulation during radiation therapy. *Int. J. Radiat. Oncol. Biol. Phys.* **2014**, *88*, 395–403. [CrossRef] [PubMed]
50. Polascik, T.J.; Wang, Z.P.; Shue, M.; Di, S.; Gurganus, R.T.; Hortopan, S.C.; Ts'o, P.O.; Partin, A.W. Influence of sextant prostate needle biopsy or surgery on the detection and harvest of intact circulating prostate cancer cells. *J. Urol.* **1999**, *162*, 749–752. [CrossRef]
51. Kusukawa, J.; Suefuji, Y.; Ryu, F.; Noguchi, R.; Iwamoto, O.; Kameyama, T. Dissemination of cancer cells into circulation occurs by incisional biopsy of oral squamous cell carcinoma. *J. Oral Pathol. Med.* **2000**, *29*, 303–307. [CrossRef] [PubMed]
52. Jones, O.M.; Rees, M.; John, T.G.; Bygrave, S.; Plant, G. Biopsy of resectable colorectal liver metastases causes tumour dissemination and adversely affects survival after liver resection. *Br. J. Surg.* **2005**, *92*, 1165–1168. [CrossRef]
53. Weitz, J.; Kienle, P.; Lacroix, J.; Willeke, F.; Benner, A.; Lehnert, T.; Herfarth, C.; von Knebel Doeberitz, M. Dissemination of tumor cells in patients undergoing surgery for colorectal cancer. *Clin. Cancer Res.* **1998**, *4*, 343–348. [PubMed]
54. Ferreira, M.M.; Ramani, V.C.; Jeffrey, S.S. Circulating tumor cell technologies. *Mol. Oncol.* **2016**, *10*, 374–394. [CrossRef] [PubMed]
55. Nagrath, S.; Sequist, L.V.; Maheswaran, S.; Bell, D.W.; Irimia, D.; Ulkus, L.; Smith, M.R.; Kwak, E.L.; Digumarthy, S.; Muzikansky, A.; et al. Isolation of rare circulating tumour cells in cancer patients by microchip technology. *Nature* **2007**, *450*, 1235–1239. [CrossRef] [PubMed]
56. Galletti, G.; Sung, M.S.; Vahdat, L.T.; Shah, M.A.; Santana, S.M.; Altavilla, G.; Kirby, B.J.; Giannakakou, P. Isolation of breast cancer and gastric cancer circulating tumor cells by use of an anti HER2-based microfluidic device. *Lab Chip* **2014**, *14*, 147–156. [CrossRef] [PubMed]
57. Mikolajczyk, S.D.; Millar, L.S.; Tsinberg, P.; Coutts, S.M.; Zomorrodi, M.; Pham, T.; Bischoff, F.Z.; Pircher, T.J. Detection of EpCAM-Negative and Cytokeratin-Negative Circulating Tumor Cells in Peripheral Blood. *J. Oncol.* **2011**, *2011*, 252361. [CrossRef] [PubMed]
58. Kamande, J.W.; Hupert, M.L.; Witek, M.A.; Wang, H.; Torphy, R.J.; Dharmasiri, U.; Njoroge, S.K.; Jackson, J.M.; Aufforth, R.D.; Snavely, A.; et al. Modular microsystem for the isolation, enumeration, and phenotyping of circulating tumor cells in patients with pancreatic cancer. *Anal. Chem.* **2013**, *85*, 9092–9100. [CrossRef] [PubMed]
59. Wang, S.; Liu, K.; Liu, J.; Yu, Z.T.; Xu, X.; Zhao, L.; Lee, T.; Lee, E.K.; Reiss, J.; Lee, Y.K.; et al. Highly efficient capture of circulating tumor cells by using nanostructured silicon substrates with integrated chaotic micromixers. *Angew. Chem. Int. Ed. Engl.* **2011**, *50*, 3084–3088. [CrossRef] [PubMed]

60. Sheng, W.; Ogunwobi, O.O.; Chen, T.; Zhang, J.; George, T.J.; Liu, C.; Fan, Z.H. Capture, release and culture of circulating tumor cells from pancreatic cancer patients using an enhanced mixing chip. *Lab Chip* **2014**, *14*, 89–98. [CrossRef] [PubMed]
61. Adams, A.A.; Okagbare, P.I.; Feng, J.; Hupert, M.L.; Patterson, D.; Gottert, J.; McCarley, R.L.; Nikitopoulos, D.; Murphy, M.C.; Soper, S.A. Highly efficient circulating tumor cell isolation from whole blood and label-free enumeration using polymer-based microfluidics with an integrated conductivity sensor. *J. Am. Chem. Soc.* **2008**, *130*, 8633–8641. [CrossRef] [PubMed]
62. Yoon, H.J.; Kim, T.H.; Zhang, Z.; Azizi, E.; Pham, T.M.; Paoletti, C.; Lin, J.; Ramnath, N.; Wicha, M.S.; Hayes, D.F.; et al. Sensitive capture of circulating tumour cells by functionalized graphene oxide nanosheets. *Nat. Nanotechnol.* **2013**, *8*, 735–741. [CrossRef]
63. Stott, S.L.; Hsu, C.H.; Tsukrov, D.I.; Yu, M.; Miyamoto, D.T.; Waltman, B.A.; Rothenberg, S.M.; Shah, A.M.; Smas, M.E.; Korir, G.K.; et al. Isolation of circulating tumor cells using a microvortex-generating herringbone-chip. *Proc. Natl. Acad. Sci. USA* **2010**, *107*, 18392–18397. [CrossRef] [PubMed]
64. Saliba, A.E.; Saias, L.; Psychari, E.; Minc, N.; Simon, D.; Bidard, F.C.; Mathiot, C.; Pierga, J.Y.; Fraisier, V.; Salamero, J.; et al. Microfluidic sorting and multimodal typing of cancer cells in self-assembled magnetic arrays. *Proc. Natl. Acad. Sci. USA* **2010**, *107*, 14524–14529. [CrossRef] [PubMed]
65. Winer-Jones, J.P.; Vahidi, B.; Arquilevich, N.; Fang, C.; Ferguson, S.; Harkins, D.; Hill, C.; Klem, E.; Pagano, P.C.; Peasley, C.; et al. Circulating tumor cells: Clinically relevant molecular access based on a novel CTC flow cell. *PLoS ONE* **2014**, *9*, e86717. [CrossRef] [PubMed]
66. Harb, W.; Fan, A.; Tran, T.; Danila, D.C.; Keys, D.; Schwartz, M.; Ionescu-Zanetti, C. Mutational Analysis of Circulating Tumor Cells Using a Novel Microfluidic Collection Device and qPCR Assay. *Transl. Oncol.* **2013**, *6*, 528–538. [CrossRef] [PubMed]
67. Cristofanilli, M.; Budd, G.T.; Ellis, M.J.; Stopeck, A.; Matera, J.; Miller, M.C.; Reuben, J.M.; Doyle, G.V.; Allard, W.J.; Terstappen, L.W.; et al. Circulating tumor cells, disease progression, and survival in metastatic breast cancer. *N. Engl. J. Med.* **2004**, *351*, 781–791. [CrossRef] [PubMed]
68. Musella, V.; Pietrantonio, F.; Di Buduo, E.; Iacovelli, R.; Martinetti, A.; Sottotetti, E.; Bossi, I.; Maggi, C.; Di Bartolomeo, M.; de Braud, F.; et al. Circulating tumor cells as a longitudinal biomarker in patients with advanced chemorefractory, RAS-BRAF wild-type colorectal cancer receiving cetuximab or panitumumab. *Int. J. Cancer* **2015**, *137*, 1467–1474. [CrossRef] [PubMed]
69. Pluim, D.; Devriese, L.A.; Beijnen, J.H.; Schellens, J.H. Validation of a multiparameter flow cytometry method for the determination of phosphorylated extracellular-signal-regulated kinase and DNA in circulating tumor cells. *Cytometry A* **2012**, *81*, 664–671. [CrossRef]
70. Deng, Y.; Zhang, Y.; Sun, S.; Wang, Z.; Wang, M.; Yu, B.; Czajkowsky, D.M.; Liu, B.; Li, Y.; Wei, W.; et al. An integrated microfluidic chip system for single-cell secretion profiling of rare circulating tumor cells. *Sci. Rep.* **2014**, *4*, 7499. [CrossRef]
71. Saucedo-Zeni, N.; Mewes, S.; Niestroj, R.; Gasiorowski, L.; Murawa, D.; Nowaczyk, P.; Tomasi, T.; Weber, E.; Dworacki, G.; Morgenthaler, N.G.; et al. A novel method for the in vivo isolation of circulating tumor cells from peripheral blood of cancer patients using a functionalized and structured medical wire. *Int. J. Oncol.* **2012**, *41*, 1241–1250. [CrossRef]
72. Liu, Z.; Fusi, A.; Klopocki, E.; Schmittel, A.; Tinhofer, I.; Nonnenmacher, A.; Keilholz, U. Negative enrichment by immunomagnetic nanobeads for unbiased characterization of circulating tumor cells from peripheral blood of cancer patients. *J. Transl. Med.* **2011**, *9*, 70. [CrossRef]
73. Lara, O.; Tong, X.; Zborowski, M.; Chalmers, J.J. Enrichment of rare cancer cells through depletion of normal cells using density and flow-through, immunomagnetic cell separation. *Exp. Hematol.* **2004**, *32*, 891–904. [CrossRef] [PubMed]
74. Lara, O.; Tong, X.; Zborowski, M.; Farag, S.S.; Chalmers, J.J. Comparison of two immunomagnetic separation technologies to deplete T cells from human blood samples. *Biotechnol. Bioeng.* **2006**, *94*, 66–80. [CrossRef] [PubMed]
75. Ozkumur, E.; Shah, A.M.; Ciciliano, J.C.; Emmink, B.L.; Miyamoto, D.T.; Brachtel, E.; Yu, M.; Chen, P.I.; Morgan, B.; Trautwein, J.; et al. Inertial focusing for tumor antigen-dependent and -independent sorting of rare circulating tumor cells. *Sci. Transl. Med.* **2013**, *5*, 179ra147. [CrossRef] [PubMed]
76. Harouaka, R.A.; Nisic, M.; Zheng, S.Y. Circulating tumor cell enrichment based on physical properties. *J. Lab. Autom.* **2013**, *18*, 455–468. [CrossRef] [PubMed]

77. Rosenberg, R.; Gertler, R.; Friederichs, J.; Fuehrer, K.; Dahm, M.; Phelps, R.; Thorban, S.; Nekarda, H.; Siewert, J.R. Comparison of two density gradient centrifugation systems for the enrichment of disseminated tumor cells in blood. *Cytometry* **2002**, *49*, 150–158. [CrossRef] [PubMed]
78. He, W.; Kularatne, S.A.; Kalli, K.R.; Prendergast, F.G.; Amato, R.J.; Klee, G.G.; Hartmann, L.C.; Low, P.S. Quantitation of circulating tumor cells in blood samples from ovarian and prostate cancer patients using tumor-specific fluorescent ligands. *Int. J. Cancer* **2008**, *123*, 1968–1973. [CrossRef] [PubMed]
79. Campton, D.E.; Ramirez, A.B.; Nordberg, J.J.; Drovetto, N.; Clein, A.C.; Varshavskaya, P.; Friemel, B.H.; Quarre, S.; Breman, A.; Dorschner, M.; et al. High-recovery visual identification and single-cell retrieval of circulating tumor cells for genomic analysis using a dual-technology platform integrated with automated immunofluorescence staining. *BMC Cancer* **2015**, *15*, 360. [CrossRef]
80. Xu, L.; Mao, X.; Imrali, A.; Syed, F.; Mutsvangwa, K.; Berney, D.; Cathcart, P.; Hines, J.; Shamash, J.; Lu, Y.J. Optimization and Evaluation of a Novel Size Based Circulating Tumor Cell Isolation System. *PLoS ONE* **2015**, *10*, e0138032. [CrossRef]
81. Mohamed, H.; Murray, M.; Turner, J.N.; Caggana, M. Isolation of tumor cells using size and deformation. *J. Chromatogr. A* **2009**, *1216*, 8289–8295. [CrossRef]
82. Yanagita, M.; Luke, J.J.; Hodi, F.S.; Janne, P.A.; Paweletz, C.P. Isolation and characterization of circulating melanoma cells by size filtration and fluorescent in-situ hybridization. *Melanoma Res.* **2018**, *28*, 89–95. [CrossRef]
83. Qin, X.; Park, S.; Duffy, S.P.; Matthews, K.; Ang, R.R.; Todenhofer, T.; Abdi, H.; Azad, A.; Bazov, J.; Chi, K.N.; et al. Size and deformability based separation of circulating tumor cells from castrate resistant prostate cancer patients using resettable cell traps. *Lab Chip* **2015**, *15*, 2278–2286. [CrossRef] [PubMed]
84. Harouaka, R.A.; Zhou, M.D.; Yeh, Y.T.; Khan, W.J.; Das, A.; Liu, X.; Christ, C.C.; Dicker, D.T.; Baney, T.S.; Kaifi, J.T.; et al. Flexible micro spring array device for high-throughput enrichment of viable circulating tumor cells. *Clin. Chem.* **2014**, *60*, 323–333. [CrossRef] [PubMed]
85. Zhou, M.D.; Hao, S.; Williams, A.J.; Harouaka, R.A.; Schrand, B.; Rawal, S.; Ao, Z.; Brenneman, R.; Gilboa, E.; Lu, B.; et al. Separable bilayer microfiltration device for viable label-free enrichment of circulating tumour cells. *Sci. Rep.* **2014**, *4*, 7392. [CrossRef] [PubMed]
86. Tan, S.J.; Lakshmi, R.L.; Chen, P.; Lim, W.T.; Yobas, L.; Lim, C.T. Versatile label free biochip for the detection of circulating tumor cells from peripheral blood in cancer patients. *Biosens. Bioelectron.* **2010**, *26*, 1701–1705. [CrossRef]
87. Vona, G.; Sabile, A.; Louha, M.; Sitruk, V.; Romana, S.; Schutze, K.; Capron, F.; Franco, D.; Pazzagli, M.; Vekemans, M.; et al. Isolation by size of epithelial tumor cells: A new method for the immunomorphological and molecular characterization of circulatingtumor cells. *Am. J. Pathol.* **2000**, *156*, 57–63. [CrossRef]
88. Adams, D.L.; Zhu, P.; Makarova, O.V.; Martin, S.S.; Charpentier, M.; Chumsri, S.; Li, S.; Amstutz, P.; Tang, C.M. The systematic study of circulating tumor cell isolation using lithographic microfilters. *RSC Adv.* **2014**, *9*, 4334–4342. [CrossRef]
89. Sarioglu, A.F.; Aceto, N.; Kojic, N.; Donaldson, M.C.; Zeinali, M.; Hamza, B.; Engstrom, A.; Zhu, H.; Sundaresan, T.K.; Miyamoto, D.T.; et al. A microfluidic device for label-free, physical capture of circulating tumor cell clusters. *Nat. Methods* **2015**, *12*, 685–691. [CrossRef] [PubMed]
90. Sollier, E.; Go, D.E.; Che, J.; Gossett, D.R.; O'Byrne, S.; Weaver, W.M.; Kummer, N.; Rettig, M.; Goldman, J.; Nickols, N.; et al. Size-selective collection of circulating tumor cells using Vortex technology. *Lab Chip* **2014**, *14*, 63–77. [CrossRef]
91. Sun, J.; Li, M.; Liu, C.; Zhang, Y.; Liu, D.; Liu, W.; Hu, G.; Jiang, X. Double spiral microchannel for label-free tumor cell separation and enrichment. *Lab Chip* **2012**, *12*, 3952–3960. [CrossRef]
92. Bhagat, A.A.; Hou, H.W.; Li, L.D.; Lim, C.T.; Han, J. Pinched flow coupled shear-modulated inertial microfluidics for high-throughput rare blood cell separation. *Lab Chip* **2011**, *11*, 1870–1878. [CrossRef]
93. Warkiani, M.E.; Guan, G.; Luan, K.B.; Lee, W.C.; Bhagat, A.A.; Chaudhuri, P.K.; Tan, D.S.; Lim, W.T.; Lee, S.C.; Chen, P.C.; et al. Slanted spiral microfluidics for the ultra-fast, label-free isolation of circulating tumor cells. *Lab Chip* **2014**, *14*, 128–137. [CrossRef] [PubMed]
94. Gupta, V.; Jafferji, I.; Garza, M.; Melnikova, V.O.; Hasegawa, D.K.; Pethig, R.; Davis, D.W. ApoStream(), a new dielectrophoretic device for antibody independent isolation and recovery of viable cancer cells from blood. *Biomicrofluidics* **2012**, *6*, 24133. [CrossRef] [PubMed]

95. Polzer, B.; Medoro, G.; Pasch, S.; Fontana, F.; Zorzino, L.; Pestka, A.; Andergassen, U.; Meier-Stiegen, F.; Czyz, Z.T.; Alberter, B.; et al. Molecular profiling of single circulating tumor cells with diagnostic intention. *EMBO Mol. Med.* **2014**, *6*, 1371–1386. [CrossRef] [PubMed]
96. Hayes, D.F.; Cristofanilli, M.; Budd, G.T.; Ellis, M.J.; Stopeck, A.; Miller, M.C.; Matera, J.; Allard, W.J.; Doyle, G.V.; Terstappen, L.W. Circulating tumor cells at each follow-up time point during therapy of metastatic breast cancer patients predict progression-free and overall survival. *Clin. Cancer Res.* **2006**, *12*, 4218–4224. [CrossRef] [PubMed]
97. Riethdorf, S.; Fritsche, H.; Muller, V.; Rau, T.; Schindlbeck, C.; Rack, B.; Janni, W.; Coith, C.; Beck, K.; Janicke, F.; et al. Detection of circulating tumor cells in peripheral blood of patients with metastatic breast cancer: A validation study of the CellSearch system. *Clin. Cancer Res.* **2007**, *13*, 920–928. [CrossRef]
98. Hofman, P.; Popper, H.H. Pathologists and liquid biopsies: To be or not to be? *Virchows Arch.* **2016**, *469*, 601–609. [CrossRef]
99. De Wit, S.; van Dalum, G.; Lenferink, A.T.; Tibbe, A.G.; Hiltermann, T.J.; Groen, H.J.; van Rijn, C.J.; Terstappen, L.W. The detection of EpCAM(+) and EpCAM(-) circulating tumor cells. *Sci. Rep.* **2015**, *5*, 12270. [CrossRef]
100. Connelly, M.; Wang, Y.; Doyle, G.V.; Terstappen, L.; McCormack, R. Re: Anti-epithelial cell adhesion molecule antibodies and the detection of circulating normal-like breast tumor cells. *J. Natl. Cancer Inst.* **2009**, *101*, 895. [CrossRef]
101. Friedl, P.; Wolf, K. Tumour-cell invasion and migration: diversity and escape mechanisms. *Nat. Rev. Cancer* **2003**, *3*, 362–374. [CrossRef]
102. Kirby, B.J.; Jodari, M.; Loftus, M.S.; Gakhar, G.; Pratt, E.D.; Chanel-Vos, C.; Gleghorn, J.P.; Santana, S.M.; Liu, H.; Smith, J.P.; et al. Functional characterization of circulating tumor cells with a prostate-cancer-specific microfluidic device. *PLoS ONE* **2012**, *7*, e35976. [CrossRef]
103. Sequist, L.V.; Nagrath, S.; Toner, M.; Haber, D.A.; Lynch, T.J. The CTC-chip: An exciting new tool to detect circulating tumor cells in lung cancer patients. *J. Thorac. Oncol.* **2009**, *4*, 281–283. [CrossRef] [PubMed]
104. Baccelli, I.; Schneeweiss, A.; Riethdorf, S.; Stenzinger, A.; Schillert, A.; Vogel, V.; Klein, C.; Saini, M.; Bauerle, T.; Wallwiener, M.; et al. Identification of a population of blood circulating tumor cells from breast cancer patients that initiates metastasis in a xenograft assay. *Nat. Biotechnol.* **2013**, *31*, 539–544. [CrossRef] [PubMed]
105. Yang, L.; Lang, J.C.; Balasubramanian, P.; Jatana, K.R.; Schuller, D.; Agrawal, A.; Zborowski, M.; Chalmers, J.J. Optimization of an enrichment process for circulating tumor cells from the blood of head and neck cancer patients through depletion of normal cells. *Biotechnol. Bioeng.* **2009**, *102*, 521–534. [CrossRef] [PubMed]
106. Lustberg, M.; Jatana, K.R.; Zborowski, M.; Chalmers, J.J. Emerging technologies for CTC detection based on depletion of normal cells. *Recent Results Cancer Res.* **2012**, *195*, 97–110. [CrossRef] [PubMed]
107. Yu, M.; Bardia, A.; Wittner, B.S.; Stott, S.L.; Smas, M.E.; Ting, D.T.; Isakoff, S.J.; Ciciliano, J.C.; Wells, M.N.; Shah, A.M.; et al. Circulating breast tumor cells exhibit dynamic changes in epithelial and mesenchymal composition. *Science* **2013**, *339*, 580–584. [CrossRef] [PubMed]
108. Ramakrishnan, M.; Mathur, S.R.; Mukhopadhyay, A. Fusion-derived epithelial cancer cells express hematopoietic markers and contribute to stem cell and migratory phenotype in ovarian carcinoma. *Cancer Res.* **2013**, *73*, 5360–5370. [CrossRef]
109. Dolfi, S.C.; Chan, L.L.; Qiu, J.; Tedeschi, P.M.; Bertino, J.R.; Hirshfield, K.M.; Oltvai, Z.N.; Vazquez, A. The metabolic demands of cancer cells are coupled to their size and protein synthesis rates. *Cancer Metab.* **2013**, *1*, 20. [CrossRef] [PubMed]
110. Adams, D.L.; Martin, S.S.; Alpaugh, R.K.; Charpentier, M.; Tsai, S.; Bergan, R.C.; Ogden, I.M.; Catalona, W.; Chumsri, S.; Tang, C.M.; et al. Circulating giant macrophages as a potential biomarker of solid tumors. *Proc. Natl. Acad. Sci. USA* **2014**, *111*, 3514–3519. [CrossRef]
111. Adams, D.L.; Adams, D.K.; Alpaugh, R.K.; Cristofanilli, M.; Martin, S.S.; Chumsri, S.; Tang, C.M.; Marks, J.R. Circulating Cancer-Associated Macrophage-Like Cells Differentiate Malignant Breast Cancer and Benign Breast Conditions. *Cancer Epidemiol. Biomarkers Prev.* **2016**, *25*, 1037–1042. [CrossRef]
112. Kaifi, J.T.; Kunkel, M.; Das, A.; Harouaka, R.A.; Dicker, D.T.; Li, G.; Zhu, J.; Clawson, G.A.; Yang, Z.; Reed, M.F.; et al. Circulating tumor cell isolation during resection of colorectal cancer lung and liver metastases: A prospective trial with different detection techniques. *Cancer Biol. Ther.* **2015**, *16*, 699–708. [CrossRef]
113. Hao, S.; Nisic, M.; He, H.; Tai, Y.C.; Zheng, S.Y. Separable Bilayer Microfiltration Device for Label-Free Enrichment of Viable Circulating Tumor Cells. *Methods Mol. Biol.* **2017**, *1634*, 81–91. [CrossRef] [PubMed]

114. Chakrabarti, K.R.; Andorko, J.I.; Whipple, R.A.; Zhang, P.; Sooklal, E.L.; Martin, S.S.; Jewell, C.M. Lipid tethering of breast tumor cells enables real-time imaging of free-floating cell dynamics and drug response. *Oncotarget* **2016**, *7*, 10486–10497. [CrossRef] [PubMed]
115. Hou, H.W.; Warkiani, M.E.; Khoo, B.L.; Li, Z.R.; Soo, R.A.; Tan, D.S.; Lim, W.T.; Han, J.; Bhagat, A.A.; Lim, C.T. Isolation and retrieval of circulating tumor cells using centrifugal forces. *Sci. Rep.* **2013**, *3*, 1259. [CrossRef] [PubMed]
116. Khoo, B.L.; Lee, S.C.; Kumar, P.; Tan, T.Z.; Warkiani, M.E.; Ow, S.G.; Nandi, S.; Lim, C.T.; Thiery, J.P. Short-term expansion of breast circulating cancer cells predicts response to anti-cancer therapy. *Oncotarget* **2015**, *6*, 15578–15593. [CrossRef] [PubMed]
117. Shim, S.; Stemke-Hale, K.; Noshari, J.; Becker, F.F.; Gascoyne, P.R. Dielectrophoresis has broad applicability to marker-free isolation of tumor cells from blood by microfluidic systems. *Biomicrofluidics* **2013**, *7*, 11808. [CrossRef] [PubMed]
118. Manaresi, N.; Romani, A.; Medoro, G.; Altomare, L.; Leonardi, A.; Tartagni, M.; Guerrieri, R. A CMOS chip for individual cell manipulation and detection. *IEEE J. Solid State Circuits* **2003**, *38*, 2297–2305. [CrossRef]
119. Carpenter, E.L.; Rader, J.; Ruden, J.; Rappaport, E.F.; Hunter, K.N.; Hallberg, P.L.; Krytska, K.; O'Dwyer, P.J.; Mosse, Y.P. Dielectrophoretic capture and genetic analysis of single neuroblastoma tumor cells. *Front. Oncol.* **2014**, *4*, 201. [CrossRef] [PubMed]
120. Fernandez, S.V.; Bingham, C.; Fittipaldi, P.; Austin, L.; Palazzo, J.; Palmer, G.; Alpaugh, K.; Cristofanilli, M. TP53 mutations detected in circulating tumor cells present in the blood of metastatic triple negative breast cancer patients. *Breast Cancer Res.* **2014**, *16*, 445. [CrossRef] [PubMed]
121. Peeters, D.J.; De Laere, B.; Van den Eynden, G.G.; Van Laere, S.J.; Rothe, F.; Ignatiadis, M.; Sieuwerts, A.M.; Lambrechts, D.; Rutten, A.; van Dam, P.A.; et al. Semiautomated isolation and molecular characterisation of single or highly purified tumour cells from CellSearch enriched blood samples using dielectrophoretic cell sorting. *Br. J. Cancer* **2013**, *108*, 1358–1367. [CrossRef]
122. Zill, A.; Mortimer, S.; Banks, K.; Nagy, R.; Chudova, D.; Jackson, C.; Baca, A.; Ye, J.Z.; Lanman, B.; Talasaz, A.; et al. Somatic genomic landscape of over 15,000 patients with advanced-stage cancer from clinical next-generation sequencing analysis of circulating tumor DNA. Proceedings of ASCO Annual Meeting, Chicago, IL, USA, 31 May–4 June 2019.
123. Zeisberg, M.; Neilson, E.G. Biomarkers for epithelial-mesenchymal transitions. *J. Clin. Investig.* **2009**, *119*, 1429–1437. [CrossRef]
124. Kallergi, G.; Aggouraki, D.; Zacharopoulou, N.; Stournaras, C.; Georgoulias, V.; Martin, S.S. Evaluation of alpha-tubulin, detyrosinated alpha-tubulin, and vimentin in CTCs: Identification of the interaction between CTCs and blood cells through cytoskeletal elements. *Breast Cancer Res.* **2018**, *20*, 67. [CrossRef] [PubMed]
125. Fischer, K.R.; Durrans, A.; Lee, S.; Sheng, J.; Li, F.; Wong, S.T.; Choi, H.; El Rayes, T.; Ryu, S.; Troeger, J.; et al. Epithelial-to-mesenchymal transition is not required for lung metastasis but contributes to chemoresistance. *Nature* **2015**, *527*, 472–476. [CrossRef] [PubMed]
126. Zheng, X.; Carstens, J.L.; Kim, J.; Scheible, M.; Kaye, J.; Sugimoto, H.; Wu, C.C.; LeBleu, V.S.; Kalluri, R. Epithelial-to-mesenchymal transition is dispensable for metastasis but induces chemoresistance in pancreatic cancer. *Nature* **2015**, *527*, 525–530. [CrossRef] [PubMed]
127. Gurzu, S.; Turdean, S.; Kovecsi, A.; Contac, A.O.; Jung, I. Epithelial-mesenchymal, mesenchymal-epithelial, and endothelial-mesenchymal transitions in malignant tumors: An update. *World J. Clin. Cases* **2015**, *3*, 393–404. [CrossRef] [PubMed]
128. Paterlini-Brechot, P.; Benali, N.L. Circulating tumor cells (CTC) detection: Clinical impact and future directions. *Cancer Lett.* **2007**, *253*, 180–204. [CrossRef] [PubMed]
129. Au, S.H.; Storey, B.D.; Moore, J.C.; Tang, Q.; Chen, Y.L.; Javaid, S.; Sarioglu, A.F.; Sullivan, R.; Madden, M.W.; O'Keefe, R.; et al. Clusters of circulating tumor cells traverse capillary-sized vessels. *Proc. Natl. Acad. Sci. USA* **2016**, *113*, 4947–4952. [CrossRef] [PubMed]
130. Gkountela, S.; Castro-Giner, F.; Szczerba, B.M.; Vetter, M.; Landin, J.; Scherrer, R.; Krol, I.; Scheidmann, M.C.; Beisel, C.; Stirnimann, C.U.; et al. Circulating Tumor Cell Clustering Shapes DNA Methylation to Enable Metastasis Seeding. *Cell* **2019**, *176*, 98–112 e114. [CrossRef]
131. Christiansen, J.J.; Rajasekaran, A.K. Reassessing epithelial to mesenchymal transition as a prerequisite for carcinoma invasion and metastasis. *Cancer Res.* **2006**, *66*, 8319–8326. [CrossRef]

132. Nagy, J.A.; Chang, S.H.; Dvorak, A.M.; Dvorak, H.F. Why are tumour blood vessels abnormal and why is it important to know? *Br. J. Cancer* **2009**, *100*, 865–869. [CrossRef]
133. Sun, B.; Zhang, D.; Zhao, N.; Zhao, X. Epithelial-to-endothelial transition and cancer stem cells: Two cornerstones of vasculogenic mimicry in malignant tumors. *Oncotarget* **2017**, *8*, 30502–30510. [CrossRef]
134. Heldin, C.H.; Rubin, K.; Pietras, K.; Ostman, A. High interstitial fluid pressure—An obstacle in cancer therapy. *Nat. Rev. Cancer* **2004**, *4*, 806–813. [CrossRef] [PubMed]
135. Zhang, L.; Ridgway, L.D.; Wetzel, M.D.; Ngo, J.; Yin, W.; Kumar, D.; Goodman, J.C.; Groves, M.D.; Marchetti, D. The identification and characterization of breast cancer CTCs competent for brain metastasis. *Sci. Transl. Med.* **2013**, *5*, 180ra148. [CrossRef] [PubMed]
136. Yu, M.; Bardia, A.; Aceto, N.; Bersani, F.; Madden, M.W.; Donaldson, M.C.; Desai, R.; Zhu, H.; Comaills, V.; Zheng, Z.; et al. Cancer therapy. Ex vivo culture of circulating breast tumor cells for individualized testing of drug susceptibility. *Science* **2014**, *345*, 216–220. [CrossRef] [PubMed]
137. Gao, D.; Vela, I.; Sboner, A.; Iaquinta, P.J.; Karthaus, W.R.; Gopalan, A.; Dowling, C.; Wanjala, J.N.; Undvall, E.A.; Arora, V.K.; et al. Organoid cultures derived from patients with advanced prostate cancer. *Cell* **2014**, *159*, 176–187. [CrossRef] [PubMed]
138. Zhang, Z.; Shiratsuchi, H.; Lin, J.; Chen, G.; Reddy, R.M.; Azizi, E.; Fouladdel, S.; Chang, A.C.; Lin, L.; Jiang, H.; et al. Expansion of CTCs from early stage lung cancer patients using a microfluidic co-culture model. *Oncotarget* **2014**, *5*, 12383–12397. [CrossRef]
139. Rossi, E.; Rugge, M.; Facchinetti, A.; Pizzi, M.; Nardo, G.; Barbieri, V.; Manicone, M.; De Faveri, S.; Chiara Scaini, M.; Basso, U.; et al. Retaining the long-survive capacity of Circulating Tumor Cells (CTCs) followed by xeno-transplantation: not only from metastatic cancer of the breast but also of prostate cancer patients. *Oncoscience* **2014**, *1*, 49–56. [CrossRef] [PubMed]
140. Hodgkinson, C.L.; Morrow, C.J.; Li, Y.; Metcalf, R.L.; Rothwell, D.G.; Trapani, F.; Polanski, R.; Burt, D.J.; Simpson, K.L.; Morris, K.; et al. Tumorigenicity and genetic profiling of circulating tumor cells in small-cell lung cancer. *Nat. Med.* **2014**, *20*, 897–903. [CrossRef]

Molecular Characterization of Circulating Tumor Cells Enriched by a Microfluidic Platform in Patients with Small-Cell Lung Cancer

Eva Obermayr [1,*], Christiane Agreiter [1], Eva Schuster [1], Hannah Fabikan [2], Christoph Weinlinger [2], Katarina Baluchova [3], Gerhard Hamilton [4], Maximilian Hochmair [2] and Robert Zeillinger [1]

1 Molecular Oncology Group, Department of Obstetrics and Gynecology, Comprehensive Cancer Center, Medical University of Vienna, Waehringer Guertel 18-20, 1090 Vienna, Austria
2 Department of Respiratory and Critical Care Medicine, Sozialmedizinisches Zentrum Baumgartner Höhe, Sanatoriumstrasse 2, 1140 Vienna, Austria
3 Division of Oncology, Biomedical Center Martin, Jessenius Faculty of Medicine in Martin, Comenius University in Bratislava, Malá Hora 4C, 036 01 Martin, Slovakia
4 Department of Surgery, Medical University of Vienna, Waehringer Guertel 18-20, 1090 Vienna, Austria
* Correspondence: eva.obermayr@meduniwien.ac.at

Abstract: At initial diagnosis, most patients with small-cell lung cancer (SCLC) present with metastatic disease with a high number of tumor cells (CTCs) circulating in the blood. We analyzed RNA transcripts specific for neuroendocrine and for epithelial cell lineages, and Notch pathway delta-like 3 ligand (*DLL3*), the actionable target of rovalpituzumab tesirine (Rova-T) in CTC samples. Peripheral blood samples from 48 SCLC patients were processed using the microfluidic Parsortix™ technology to enrich the CTCs. Blood samples from 26 healthy donors processed in the same way served as negative controls. The isolated cells were analyzed for the presence of above-mentioned transcripts using quantitative PCR. In total, 16/51 (31.4%) samples were CTC-positive as determined by the expression of epithelial cell adhesion molecule 1 (*EpCAM)*, cytokeratin 19 (*CK19)*, chromogranin A (*CHGA)*, and/or synaptophysis (*SYP)*. The epithelial cell lineage-specific *EpCAM* and/or *CK19* gene expression was observed in 11 (21.6%) samples, and positivity was not associated with impaired survival. The neuroendocrine cell lineage-specific *CHGA* and/or *SYP* were positive in 13 (25.5%) samples, and positivity was associated with poor overall survival. *DLL3* transcripts were observed in four (7.8%) SCLC blood samples and *DLL3*-positivity was similarly associated with poor overall survival (OS). CTCs in SCLC patients can be assessed using epithelial and neuroendocrine cell lineage markers at the molecular level. Thus, the implementation of liquid biopsy may improve the management of lung cancer patients, in terms of a faster diagnosis, patient stratification, and on-treatment therapy monitoring.

Keywords: small-cell lung carcinoma; circulating tumor cells; microfluidics; gene expression analysis; synaptophysin; chromogranin A; rovalpituzumab tesirine

---

1. Introduction

Lung cancer is the most common cancer worldwide. In 2018, a total of 2.1 million new cases were estimated, accounting for 11.6% of all new cancer diagnoses [1,2]. In general, two major types of lung cancer exist: non-small-cell lung cancer (NSCLC), which accounts for about 85% of all lung cancer cases, and small-cell lung cancer (SCLC), which is diagnosed in approximately 15% of all lung cancers. For patients with early-stage NSCLC, a surgical resection offers the best opportunity for cure, while in advanced cases a systemic therapy is the standard of care. SCLC, however, is usually diagnosed rather late when the cancer has already disseminated. In this case a multimodal therapy which includes chemotherapy and radiotherapy is considered the gold standard [3]. Due to these different therapeutic approaches, it is of utmost importance to have a reliable diagnostic platform to differentiate between SCLC and NSCLC.

SCLC belongs to the group of neuroendocrine tumors of the lung. It is diagnosed using hematoxylin and eosin stained sections of the biopsied tissue. However, the histopathological diagnosis of SCLC based on its distinctive morphology may be difficult due to limited material supply from biopsied tissue or aspirated cytological specimens [4]. In some cases the diagnosis of SCLC may be further confirmed by immunohistochemistry using the neuroendocrine markers chromogranin (CHGA), synaptophysin (SYP), and neural cell adhesion molecule 1 (NCAM1) [5,6]. In recent years, the Notch pathway delta-like 3 ligand (DLL3) has gradually gained more interest since it is frequently and selectively expressed on tumorous tissue in SCLC patients and hence it has been associated with neuroendocrine tumorigenesis. Most importantly it is the therapeutic target of the antibody-drug conjugate rovalpituzumab tesirine (Rova-T) [7].

In contrast to conventional tissue biopsies or cytological preparations, liquid biopsies that contain circulating tumor cells (CTCs) and/or circulating tumor DNA, represent a novel approach that illuminates the whole molecular profile of a tumor at the time of sampling [8,9]. Liquid biopsies are taken by a simple blood draw and, thus, are less stressful for the patient, more conventionally used and less expensive than tissue biopsies. For this reason, liquid biopsies can be taken several times to monitor the temporal heterogeneity of the tumor. Especially in lung cancer, liquid biopsies may outperform tissue biopsies with respect to the tumor's accessibility at resection. In addition, small tissue samples are often already exhausted after routine histological staining and hence no longer available for advanced analysis. Furthermore, longitudinal sampling for monitoring of any development of therapy resistance is almost impossible with tissue biopsies [10].

The presence and clinical significance of CTCs has already been shown in many types of malignancies, among them e.g., breast, colorectal, prostate, and lung cancer. In contrast to most other cancer types, SCLC is characterized by a large number of CTCs in the circulation [11]. Several studies have shown the prognostic value of CTC counts in SCLC, most of them using the US Food and Drug Administration (FDA) cleared CellSearch test [11–17]. In addition to the number of CTCs found, their molecular characterization may be a part of a more comprehensive approach providing further information on e.g., downregulation of epithelial markers or presence of druggable targets. Recently, we have demonstrated that processing blood samples using the microfluidic Parsortix™ technology considerably improved the molecular analysis of the enriched CTCs [18].

Considering the abundance of CTCs and the ease of obtaining/performing liquid biopsies extends the possibilities for differential diagnosis and patient stratification. For these reasons we believe that the molecular characterization of CTCs in SCLC may be of uppermost importance for this type of lung cancer. In the present study we applied a recently developed workflow which combines a microfluidic enrichment of CTCs and a qPCR-based analysis for evaluating the gene expression levels of markers of the epithelial (epithelial cell adhesion molecule 1, *EpCAM* and cytokeratin 19, *CK19*) and neuroendocrine (*CHGA*, *SYP*, *NCAM1* and enolase 2 *ENO2*) cell lineage origin, in addition to the druggable target *DLL3*.

2. Materials and Methods

Blood samples were taken from patients with SCLC at the Department of Respiratory and Critical Care Medicine at Sozialmedizinisches Zentrum Baumgartner Höhe, Vienna, Austria. Control blood samples came from healthy donors without a history of cancer. All donors signed an informed consent. The study was approved by the Ethics Committee of the Medical University of Vienna, Austria (EK366/2003 and EK2266/2018).

The blood was collected in Vacuette EDTA tubes (Greiner Bio-One) and processed on the same day in accordance with a recently published protocol employing the label-free microfluidic Parsortix™ technology (Angle plc., UK) [18]. The key component of the device is a microscope slide sized disposable separation cassette, which contains a series of steps with a precisely defined height. Rare cells (e.g., CTCs) are captured within the separation cassette based on their less deformable nature and usually larger size compared to blood cells. Before separation, the blood was diluted with an equal volume of phosphate buffered saline (PBS) and directly processed using a Parsortix™ technology. In this study a separation cassette with a critical step size of 6.5 μm was used, and the separation was performed at 99 mbar pressure. After the separation was completed the captured cells were recovered using a back-flush cycle and immediately lysed by adding 350 μl RLT lysis buffer (Qiagen). The lysates were stored at −80 ° C until RNA extraction.

Total RNA was extracted from the cell lysates using the RNeasy Micro Kit (Qiagen) without DNase treatment. The total amount of RNA was converted into cDNA using the SuperScript VILO Mastermix (Invitrogen). qPCR was done in duplicates in a 10 μL total reaction volume on the ViiA7 Real-Time PCR System using the TaqMan® Universal Mastermix II and exon spanning TaqMan® assays specific for *EpCAM, NCAM1, CHGA, SYP, DLL3, ENO2*, and *CDKN1B* (Life Technologies) with thermal cycling parameters (50 °C for 2 min; 95 °C for 10 min followed by 40 cycles at 95 °C for 15 s and 60 °C for 1 min). A qPCR specific for *CK19* was performed at 65 °C annealing/extension with forward and reverse primers that correspond to published primer sequences and with a FAM™ labeled hydrolysis probe (5′-TgTCCTgCAgATCgACAACgCCC-3′) [19]. Raw data were analyzed using the ViiA7 Software v1.1 with automatic threshold setting and baseline correction. If the fluorescent signal did not reach the threshold in both duplicate reactions, the sample was regarded as negative.

The SCLC CTC lines used for the spiking experiments were derived from patients' blood samples [20]. They were trypsinized at about 70% confluence and stained with CellTrace Violet (Invitrogen) according to the manufacturer's protocol. Subsequently, 100 stained cells were added manually to a 10 mL control blood sample, which was then processed using the Parsortix™ technology as described above. The tumor cells captured in the separation cassette were counted using a fluorescence microscope (Olympus BX50).

The Pearson's chi-square and Fisher's exact test were used to assess the relationship between clinicopathological characteristics of the patients and the presence or absence of the respective gene transcripts. Overall survival (OS) was defined as the period of time in months between blood draw and either death or the last date the patient was seen alive. Kaplan–Meier survival analyses and log-rank testing were used to compare survival outcomes [21]. Cox proportional-hazards regression was used to determine univariate and multiple hazards ratios (HR) for OS [7]. The included covariates were the stage of disease at blood draw (primary vs. progressive disease) and the presence vs. absence of the respective transcripts. The model was built using a forward stepwise method by entering all variables at a p value of less than 0.05 and removing them at a p value of greater than 0.10. The statistical analysis was performed with SPSS version 19.0 (SPSS Inc., Chicago, IL). The level of significance was set at $p < 0.05$. Graphs were done using GraphPad Prism version 6.01.

3. Results

3.1. Patients and Samples

The characteristics of 48 patients with a histopathologically confirmed diagnosis of SCLC are shown in Table 1. The SCLC patients were 51 to 78 years old (mean/median age at 64.6/63.5 years), and all patients but one were former or current smokers, with a median of 60 pack years (range 20 to 150). Thirty-four patients died within the observation period, with a median overall survival of 7 months (range 0 to 14 months). The 14 patients who were still alive at study completion were surveyed over a median period of 14 months (range 0 to 19 months). All blood samples were taken before treatment, either at primary diagnosis (n = 27), or when progression or recurrence of the disease was observed (n

= 24). In total, 51 blood samples were available, as blood samples from three patients with progressive disease were taken at two serial time points. The volume of blood was 18 mL in 58.8% of the samples, 17 mL to 14 mL in 33.3%, and 10 mL to 8 mL in 7.8% of the samples. In the control group, 18 mL of blood was taken from 26 healthy donors.

Table 1. Characteristics of 48 small-cell lung cancer (SCLC) patients included in the study.

Characteristics	*n* (%)
Age	
Mean (median)	63.5 y (64.6 y)
Range	51.0–78.0 y
Gender	
Male	30 (62.5%)
Female	18 (37.5%)
Tobacco abuse	
Current smokers	13 (27.1%)
Former smokers	26 (54.2%)
Never smokers	1 (2.1%)
Unknown	8 (16.7%)
UICC 8th edition TNM stage at diagnosis [1]	
III	4 (11.4%)
IV	31 (88.6%)
Unknown	13 (27.1%)
Outcome at study completion	
Dead	34 (70.8%)
Alive	14 (29.2%)
Blood draw for CTCs	
At primary diagnosis	27 (56.3%)
At progression/recurrence	21 (43.8%)

[1] UICC, International Union for Cancer Control.

3.2. Spiking Experiments

The efficiency of the microfluidic Parsortix™ system for capturing cultivated SCLC cells derived from four CTC lines [20] in a separation cassette with a critical gap size of 6.5 μm is shown in Figure 1. The overall mean capture efficiency of all four cell lines was 27.8% (SD 16.4%). The gene expression levels of the epithelial (*EpCAM* and *CK19*) and neuroendocrine (*CHGA* and *SYP*) cell lineage origins were assessed in the same four CTC lines using qPCR, showing a wide-ranging pattern of gene expression.

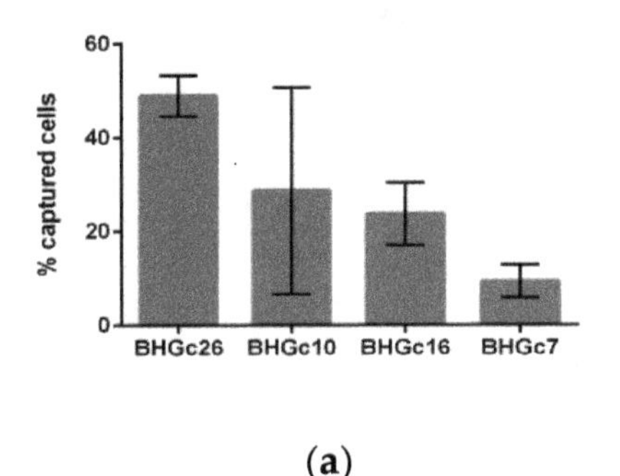

(a)

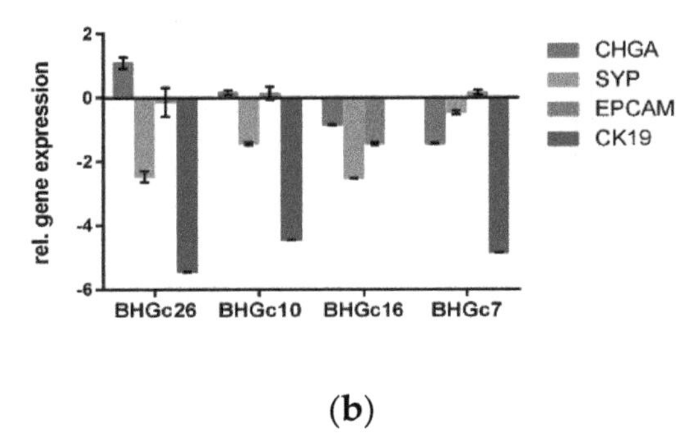

(b)

Figure 1. Characteristics of the microfluidic enrichment procedure and of the tumor cell lines used for the spiking experiments are illustrated. (**a**) Four SCLC tumor cell lines (BHGc26, BHGc10, BHGc16, and BHGc7) were fluorescently labeled and spiked into blood (100 cells per 10 mL) in triplicates. The graph depicts the mean percentage and the standard deviation of tumor cells captured in the Parsortix™ microfluidic cassette. (**b**) The gene expression levels of the epithelial and neuroendocrine cell lineage specific markers of the same cell lines are shown relative to the expression level of *cyclin dependent kinase inhibitor 1B* as reference gene. The graphs depict the means and the standard deviation from duplicate qPCRs amplifications.

3.3. Epithelial and Neuroendocrine Markers in Controls and SCLC Blood Samples

EpCAM, *CK19*, and *CHGA* transcripts were not detected in any of the control blood samples (Figure 2a–c). In contrast, *SYP* levels above the detection limit of qPCR were observed in 1/26 (3.8%), and *ENO2* and *NCAM1* transcripts in 24/26 (92.3%) and 19/26 (73.1%) controls, respectively (Figure 2d–f). Due to the high number of *ENO2*- and *NCAM1*-positive healthy donor samples, these markers were considered as less appropriate for CTC detection and thus excluded from further analyses.

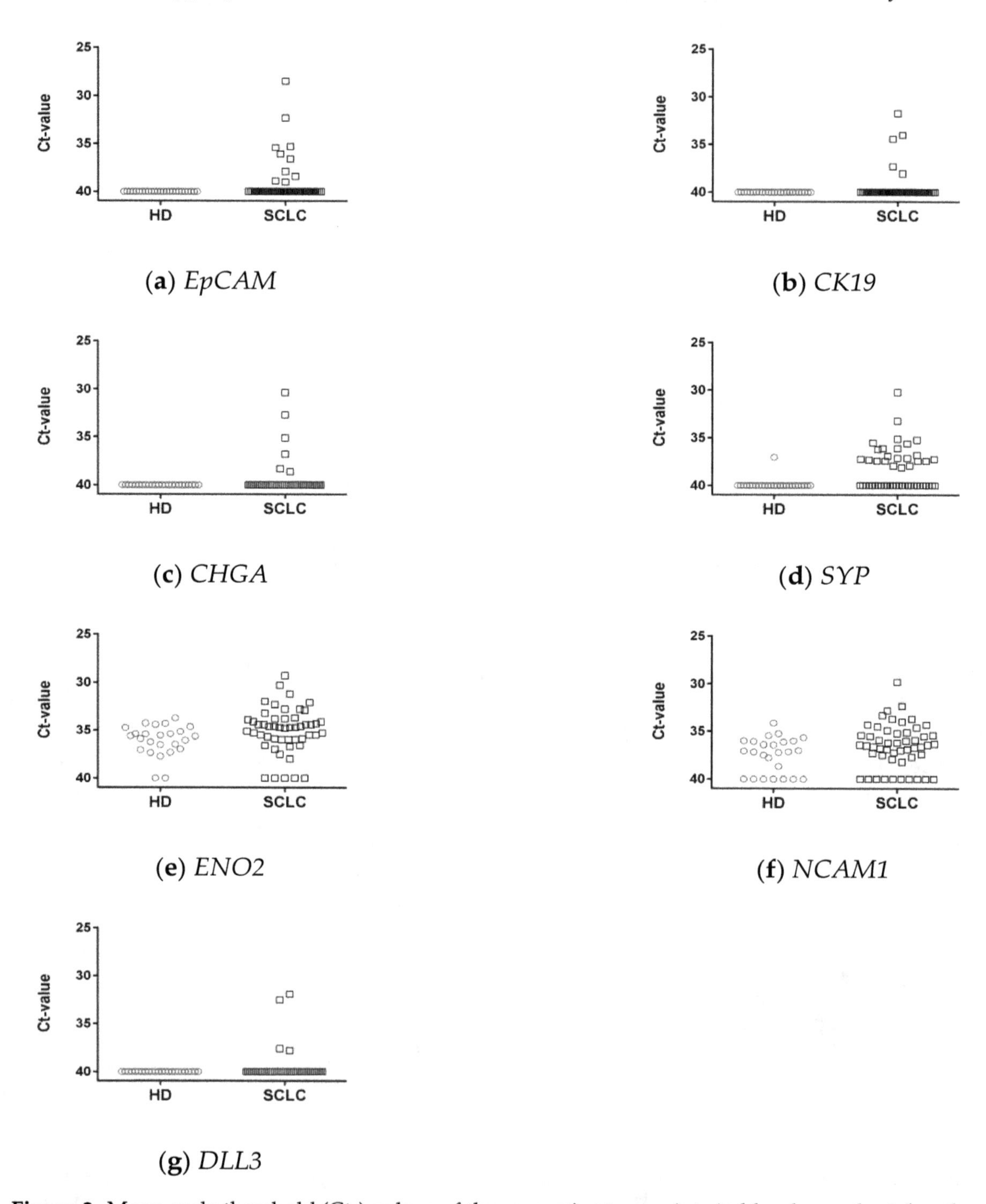

Figure 2. Mean cycle threshold (Ct-) values of the respective transcripts in blood samples taken from 26 healthy donors (HD) and 48 patients with small-cell lung cancer (SCLC). (**a**) *EpCAM*, epithelial cell adhesion molecule; (**b**) *CK19*, cytokeratin 19; (**c**) *CHGA*, chromogranin A; (**d**) *SYP*, synaptophysin; (**e**) *ENO2*, enolase 2; (**f**) *NCAM1*, neural cell adhesion molecule 1; (**g**) *DLL3*, Notch pathway delta-like 3 ligand.

In contrast, *EpCAM*-, *CK19*-, or *CHGA*-positivity above the detection limit of qPCR was observed in 10 (19.6%), 4 (7.8%), and 6 (11.8%) of the 51 samples obtained from SCLC patients. Due to the observed *SYP* gene expression in a single control blood sample, the threshold for *SYP*-positivity in the patients' samples was set at Ct = 37.0. Thus, *SYP* transcript levels below that threshold were observed in 40 (78.4%), and above that threshold in 11 (21.6%) of the 51 SCLC samples. These 11 samples were assigned as *SYP*-positive. In none of the gene transcripts did (*EpCAM, CK19, CHGA,* and *SYP*)-positivity differ significantly between the blood samples taken at diagnosis and disease progression.

In total, 16/51 (31.4%) samples were CTC-positive due to the expression of at least one of *EpCAM, CK19, CHGA,* and *SYP* markers (Figure 3). The expression of epithelial markers (*EpCAM* and/or *CK19*) was observed in 11 (21.6%), and of neuroendocrine markers (*CHGA* and/or *SYP*) in 13 (25.5%) samples. Among the 16 CTC-positive blood samples, three (18.8%) and five (31.3%) were characterized by the presence of just epithelial or neuroendocrine markers, respectively, and eight samples (50.0%) by both types.

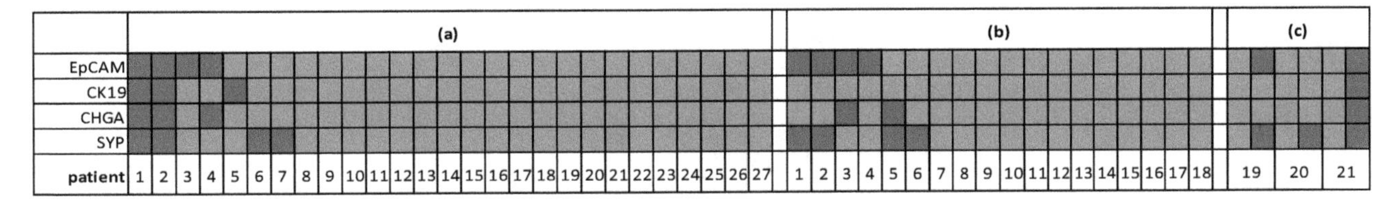

Figure 3. Heat map for *EpCAM, CK19, CHGA,* and *SYP* in the 51 microfluidic enriched blood samples of patients with small-cell lung cancer. (**a**) Twenty-seven samples were taken at diagnosis, (**b**) 18 samples were taken at progression/recurrence, and (**c**) displays serial blood draws taken from three patients during disease progression. Red and green squares indicate positive and negative gene expression per tested sample, respectively.

3.4. *Alterations of Transcript Levels during Disease Progression*

From three patients with progressive disease two serial blood samples were taken at the start of the consecutive lines of treatment. In two cases the second blood was taken two months after the first blood draw (patients 19 and 20 in Figure 3c), and in one case after three months (patient 21 in Figure 3c). At the first blood draw all patients were negative in all markers tested; however, at the second blood draw all patients were PCR-positive for at least *SYP* (see Figure 3c). All patients died within 1.5 months of the second blood draw.

3.5. *Epithelial and Neuroendocrine Specific Markers and Patient Outcome*

The blood samples were stratified on the basis of the epithelial cell lineage-specific gene transcripts *EpCAM* and *CK19* into the epi-positive (n = 11) and the epi-negative group (n = 40), and on the basis of neuroendocrine-specific transcripts *SYP* and *CHGA* into the nec-positive (n = 13) and nec-negative (n = 38) group. The presence of *EpCAM* and/or *CK19* transcripts in the epi-positive group at primary diagnosis may be associated with a shorter OS of the patients (Figure 4a); future studies with larger sample sizes may prove whether or not this difference is statistically significant. Similarly, the presence of *EpCAM* and/or *CK19* transcripts at disease progression was not related to OS (Figure 4b). In contrast, nec-positive patients had a significantly shorter OS than nec-negative patients, both at primary diagnosis and at disease progression. That association of *SYP* and/or *CHGA* transcripts with OS was observed with the presence of these neuroendocrine markers both at primary diagnosis (median OS 4 vs. 11 months, log-rank p = 0.007; Figure 4c), as well as at progression (median OS 1 vs. 5 months, log-rank p = 0.014; Figure 4d). Irrespective of whether the sample was taken at primary diagnosis or at disease progression, nec-positive patients had a high-risk of an early death (HR 3.475, 95% CI 1.685–7.164; p = 0.001).

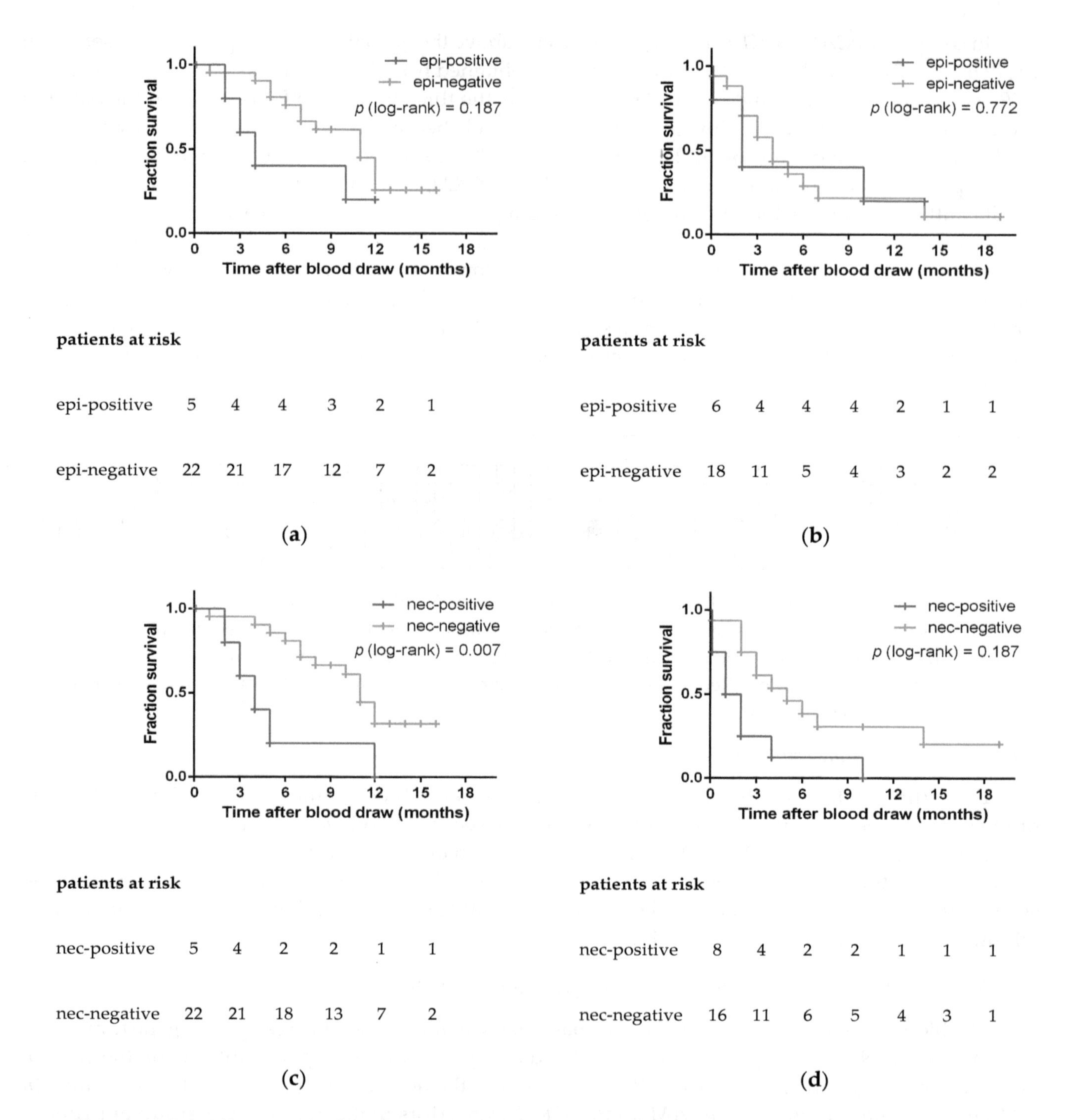

Figure 4. Overall survival of small-cell lung cancer patients according to presence (red) or absence (green) of epithelial (*EpCAM, CK19*) and neuroendocrine markers (*SYP, CHGA*). The figures (**a**) and (**c**) display samples taken at primary diagnosis, whereas the figures (**b**) and (**d**) display samples taken at progression. Log-rank testing was used to compare survival outcomes. epi, epithelial; nec, neuroendocrine.

3.6. DLL3 in Controls and SCLC Blood Samples

DLL3 transcripts were observed in 4/51 (7.8%) of the SCLC blood samples and in none of the 26 control blood samples (Figure 2g). Three *DLL3*-positive blood samples were taken at primary diagnosis, and one was taken from patient 19 at the second blood draw (see Figure 3c). Due to the small number of DLL3-positive patients, we did not stratify the patients into two groups depending on the time-point of blood draw. All four *DLL3*-positive patients had a significantly shorter OS than *DLL3*-negative patients (median OS 2 vs. 7 months, log-rank $p = 0.003$; Figure 5). The risk of dying earlier was 3.793 (95% CI 2.803–115.6) higher in the *DLL3*-positive group compared to the *DLL3*-negative group.

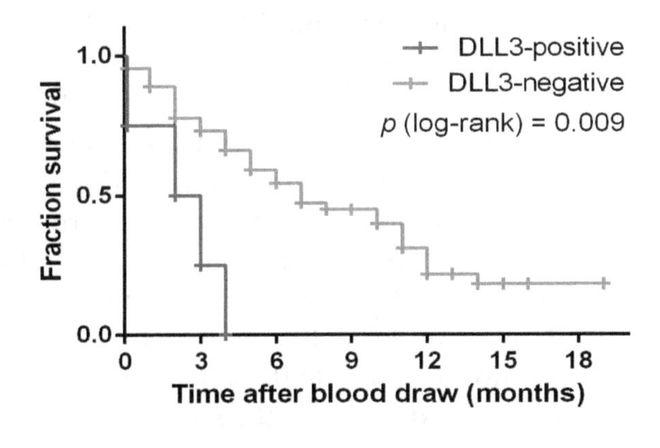

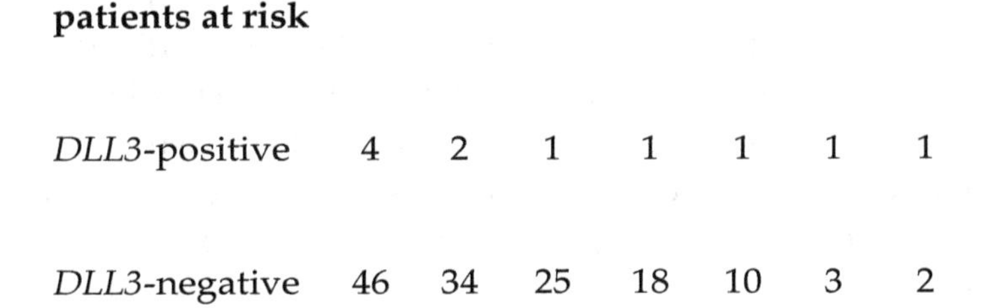

patients at risk

DLL3-positive	4	2	1	1	1	1	1
DLL3-negative	46	34	25	18	10	3	2

Figure 5. Overall survival according to presence (red) or absence (green) of *DLL3*. Log-rank testing was used to compare survival outcomes.

4. Discussion

We have applied a recently established workflow for molecular detection of CTCs [18] in blood samples taken from patients with SCLC, which is a highly aggressive neuroendocrine tumor of the lung. The enrichment of the CTCs was achieved with the microfluidic Parsortix™ technology, and the molecular analysis of the harvested cells was performed using markers that are specific to epithelial (*EpCAM* and *CK19*) and to neuroendocrine cell lineages (*SYP*, *CHGA*, *ENO2*, *NCAM1*), and to *DLL3*, an actionable target of antibody-drug conjugate rovalpituzumab tesirine (Rova-T). To the best of our knowledge, this is the first study that investigates neuroendocrine markers and *DLL3* in CTCs of SCLC patients at a molecular level.

We detected *EpCAM* and/or *CK19* transcripts in 21.6%, and neuroendocrine *CHGA* and/or *SYP* transcripts in 25.5% of the 51 SCLC blood samples. Interestingly, five of the 16 (31.3%) qPCR-positive samples were identified by the presence of neuroendocrine-specific transcripts alone. Similarly, three (18.8%) CTC-positive samples expressed the epithelial markers alone.

The percentage of CTC-positive samples due to the expression of epithelial markers of 21.6% in our cohort is smaller than reported by others in SCLC [22]. Applying the FDA approved CellSearch-based approach for the detection and enumeration of CTCs, positive findings were observed in 50% to 86% of the patients by other investigators [12,23,24]. The reason for the low detection rate of CTCs in our study may be the low overall sensitivity of our approach reflecting the need to split the sample into aliquots to analyze the expression of multiple genes individually. Improved sensitivity could be achieved by multi-plexing the gene expression analysis to avoid splitting the sample. A further improvement of the overall approach may also be achieved by employing a gene-specific pre-amplification of the respective targets prior to qPCR. In a recent study we have demonstrated that targeted pre-amplification in Parsortix™-enriched blood samples is feasible [18].

Another clear limitation of our study is the possibly low efficiency of the enrichment procedure to isolate CTCs from SCLC blood samples. The spiking experiments showed only moderate capture rates of SCLC CTC lines (mean capture efficiency 27.8%, SD 16.4%). These capture rates are lower than reported for the breast cancer cell line MCF-7 using the same type of separation cassette (average 63% [25]). In line with our observations, the capture rates varied depending on the type of cell line used from 30% to 87% in renal carcinoma [26]. The four cell lines used in the present study had been established from patients' CTCs [20]. Their diverse gene expression pattern (see Figure 1b) may reflect the initial heterogeneity of their provenance and might contribute to varying capture efficiencies. In the present study we did not check the number of tumor cells after harvesting; however, results from a recent study imply that the recovery rate may vary depending on the type of cell line from 62%–84% [27]. The number of harvested tumor cells can be increased by intensifying the back-flush cycle; however a higher recovery will only be achieved at the cost of a lower purity of the tumor cells.

Using larger volumes of blood may be a further attempt to increase the sensitivity of the assay in future studies. In our study all four samples with a volume of 10 mL blood or less were negative for all gene markers investigated. We did not exclude these few blood samples from the survival analyses shown in Figures 4 and 5, as that would not alter the significance of the analyses.

There is a single study applying the Parsortix™ technology for the enrichment of blood samples from SCLC patients. In that study, Chudziak et al. found CTCs in all 12 patients, as assessed by immune-fluorescent cytokeratin-specific staining of the enriched cells [16]. In contrast to our approach that group used blood collection tubes containing a preservative which is known to increase the rigidity of the cell, and thereby increasing the number of cells captured in the microfluidic cassette. This fact, along with the more advanced stage of disease in their study population may be the reason for the divergent CTC-positivity rate obtained in that study as compared to ours.

Another weakness of our study may be the limited sample size. Because of that and the low overall sensitivity we had very few positive samples. Thus we were not able to investigate association of the patients' prognosis and the respective gene expression levels, and the interplay of epithelial and neuroendocrine markers in a more detailed way.

High numbers of CTCs at diagnosis, as assessed by CellSearch, were associated with a poor prognosis (reviewed by [10]), yet the investigators reported divergent CTC numbers as a threshold for defining a group of patients with poor prognosis. [14,23]. However, we did not observe any significant impact of the expression of the epithelial markers on the OS. Nonetheless, patients who were epi-positive at primary diagnosis died earlier than epi-negative patients. A statistical significance may be reached by increasing the number of patients in future studies.

Studies investigating CTCs in other neuroendocrine tumors, such as those originating from the prostate, thyroid gland, or the intestine, mainly applied epithelial cell lineage-specific markers and protein-based technologies for the enrichment and analysis of CTCs [28]. However, CTCs may be missed when epithelial markers, such as *EpCAM*, are downregulated in the tumor tissue, as was shown in neuroendocrine tumors of the lung [29]. In addition, tumor cells can undergo epithelial-to-mesenchymal transition and lose their epithelial phenotype [24]. In the present study we detected *CHGA* and/or *SYP* transcripts in 13 samples; this absolute number corresponds to 25.5% of all 51 SCLC blood samples, and to 81.3% of all 16 CTC-positive samples. That the percentage is still not 100% may be because of low numbers of CTCs in some samples. Furthermore, Guinee et al. demonstrated the absence of neuroendocrine markers in just about 20% of the specimen by immunohistochemical staining [6]. The fact that one third of the qPCR-positive samples was identified by the presence of these neuroendocrine transcripts already indicates that epithelial markers alone may not be sufficient to detect CTCs in neuroendocrine tumors such as SCLC. One observation in this respect is of particular interest. The presence of selected neuroendocrine markers was associated with a worse outcome and not the presence of used epithelial markers. This even applies irrespective of the time the markers were detected—be it prior to treatment at initial diagnosis or when the disease has already progressed.

To the best of our knowledge there is just a single study investigating the clinical relevance of neuroendocrine markers in CTCs: Recently, Pal and colleagues quantified the percentage of

SYP-positive CTCs in blood samples taken from castration-resistant prostate cancer patients using the open fluorescent channel of the CellSearch platform [30]. They observed an increasing number of *SYP*-positive CTCs with the onset of resistance to androgen-receptor targeting drugs, which are assumed to stimulate the transition to the neuroendocrine phenotype [31].

5. Conclusions

Besides the neuroendocrine markers *SYP* and *CHGA*, our study also clearly shows that *DLL3* can be detected in CTC-enriched blood samples. Traditional patient stratification for personalized treatment options, such as Rova-T, is based the analysis of tissue samples that were taken long before the disease progression occurred. In contrast, liquid biopsy samples can be taken right before the start of treatment, and may thus provide a snapshot analysis of promising targets for personalized treatments, such as *DLL3*. Apart from treatment stratification, liquid biopsies can be taken at several consecutive points in time to assess the response to treatment. Despite the promising results of our study, the findings need to be validated in larger studies of SCLC patients. In conclusion, the molecular analysis of CTCs may add relevant information to traditional tissue biopsies or cytological specimens in small-cell lung cancer patients, especially in treatment selection and patient monitoring.

Author Contributions: Conceptualization, E.O. and R.Z.; data curation, K.B.; funding acquisition, R.Z.; investigation, E.O.; methodology, E.O., C.A., E.S., and C.W.; project administration, R.Z.; Resources, H.F. and M.H.; supervision, G.H., M.H., and R.Z.; writing—original draft, E.O.; writing—review and editing, K.B. and R.Z.

Acknowledgments: This study received support from ANGLE plc in the form of an in-kind contribution of Parsortix™ devices and microfluidic separation cassettes. The authors thank Gabriele Klaming for language editing, and the team at ANGLE plc for their technical support. Last but not least, the authors would like to thank all patients and voluntary donors for providing blood samples.

References

1. Wong, M.C.S.; Lao, X.Q.; Ho, K.F.; Goggins, W.B.; Tse, S.L.A. Incidence and mortality of lung cancer: Global trends and association with socioeconomic status. *Sci. Rep.* **2017**, *7*, 14300. [CrossRef] [PubMed]
2. Ferlay, J.; Colombet, M.; Soerjomataram, I.; Dyba, T.; Randi, G.; Bettio, M.; Gavin, A.; Visser, O.; Bray, F. Cancer incidence and mortality patterns in Europe: Estimates for 40 countries and 25 major cancers in 2018. *Eur. J. Cancer* **2018**. [CrossRef] [PubMed]
3. Fruh, M.; De Ruysscher, D.; Popat, S.; Crino, L.; Peters, S.; Felip, E. Small-cell lung cancer (SCLC): ESMO Clinical Practice Guidelines for diagnosis, treatment and follow-up. *Ann. Oncol.* **2013**, *24* (Suppl. 6), vi99–vi105. [CrossRef]
4. Travis, W.D. Update on small cell carcinoma and its differentiation from squamous cell carcinoma and other non-small cell carcinomas. *Mod. Pathol.* **2012**, *25* (Suppl. 1), S18–S30. [CrossRef]
5. Nicholson, S.A.; Beasley, M.B.; Brambilla, E.; Hasleton, P.S.; Colby, T.V.; Sheppard, M.N.; Falk, R.; Travis, W.D. Small cell lung carcinoma (SCLC): A clinicopathologic study of 100 cases with surgical specimens. *Am. J. Surg. Pathol.* **2002**, *26*, 1184–1197. [CrossRef] [PubMed]
6. Guinee, D.G., Jr.; Fishback, N.F.; Koss, M.N.; Abbondanzo, S.L.; Travis, W.D. The spectrum of immunohistochemical staining of small-cell lung carcinoma in specimens from transbronchial and open-lung biopsies. *Am. J. Clin. Pathol.* **1994**, *102*, 406–414. [CrossRef] [PubMed]
7. Saunders, L.R.; Bankovich, A.J.; Anderson, W.C.; Aujay, M.A.; Bheddah, S.; Black, K.; Desai, R.; Escarpe, P.A.; Hampl, J.; Laysang, A.; et al. A DLL3-targeted antibody-drug conjugate eradicates high-grade pulmonary neuroendocrine tumor-initiating cells in vivo. *Sci. Transl. Med.* **2015**, *7*, 302ra136. [CrossRef] [PubMed]
8. Kuhn, P.; Bethel, K. A fluid biopsy as investigating technology for the fluid phase of solid tumors. *Phys. Biol.* **2012**, *9*, 010301. [CrossRef] [PubMed]

9. Alberter, B.; Klein, C.A.; Polzer, B. Single-cell analysis of CTCs with diagnostic precision: Opportunities and challenges for personalized medicine. *Expert Rev. Mol. Diagn.* **2016**, *16*, 25–38. [CrossRef] [PubMed]
10. Kapeleris, J.; Kulasinghe, A.; Warkiani, M.E.; Vela, I.; Kenny, L.; O'Byrne, K.; Punyadeera, C. The Prognostic Role of Circulating Tumor Cells (CTCs) in Lung Cancer. *Front. Oncol.* **2018**, *8*, 311. [CrossRef]
11. Hou, J.M.; Krebs, M.G.; Lancashire, L.; Sloane, R.; Backen, A.; Swain, R.K.; Priest, L.J.; Greystoke, A.; Zhou, C.; Morris, K.; et al. Clinical significance and molecular characteristics of circulating tumor cells and circulating tumor microemboli in patients with small-cell lung cancer. *J. Clin. Oncol.* **2012**, *30*, 525–532. [CrossRef] [PubMed]
12. Normanno, N.; Rossi, A.; Morabito, A.; Signoriello, S.; Bevilacqua, S.; Di Maio, M.; Costanzo, R.; De Luca, A.; Montanino, A.; Gridelli, C.; et al. Prognostic value of circulating tumor cells' reduction in patients with extensive small-cell lung cancer. *Lung Cancer* **2014**, *85*, 314–319. [CrossRef] [PubMed]
13. Hiltermann, T.J.; Pore, M.M.; van den Berg, A.; Timens, W.; Boezen, H.M.; Liesker, J.J.; Schouwink, J.H.; Wijnands, W.J.; Kerner, G.S.; Kruyt, F.A.; et al. Circulating tumor cells in small-cell lung cancer: A predictive and prognostic factor. *Ann. Oncol* **2012**, *23*, 2937–2942. [CrossRef] [PubMed]
14. Naito, T.; Tanaka, F.; Ono, A.; Yoneda, K.; Takahashi, T.; Murakami, H.; Nakamura, Y.; Tsuya, A.; Kenmotsu, H.; Shukuya, T.; et al. Prognostic impact of circulating tumor cells in patients with small cell lung cancer. *J. Thorac. Oncol.* **2012**, *7*, 512–519. [CrossRef] [PubMed]
15. Cheng, Y.; Liu, X.Q.; Fan, Y.; Liu, Y.P.; Liu, Y.; Ma, L.X.; Liu, X.H.; Li, H.; Bao, H.Z.; Liu, J.J.; et al. Circulating tumor cell counts/change for outcome prediction in patients with extensive-stage small-cell lung cancer. *Future Oncol.* **2016**, *12*, 789–799. [CrossRef] [PubMed]
16. Chudziak, J.; Burt, D.J.; Mohan, S.; Rothwell, D.G.; Mesquita, B.; Antonello, J.; Dalby, S.; Ayub, M.; Priest, L.; Carter, L.; et al. Clinical evaluation of a novel microfluidic device for epitope-independent enrichment of circulating tumour cells in patients with small cell lung cancer. *Analyst* **2016**, *141*, 669–678. [CrossRef] [PubMed]
17. Huang, C.H.; Wick, J.A.; Sittampalam, G.S.; Nirmalanandhan, V.S.; Ganti, A.K.; Neupane, P.C.; Williamson, S.K.; Godwin, A.K.; Schmitt, S.; Smart, N.J.; et al. A multicenter pilot study examining the role of circulating tumor cells as a blood-based tumor marker in patients with extensive small-cell lung cancer. *Front. Oncol.* **2014**, *4*, 271. [CrossRef] [PubMed]
18. Obermayr, E.; Maritschnegg, E.; Agreiter, C.; Pecha, N.; Speiser, P.; Helmy-Bader, S.; Danzinger, S.; Krainer, M.; Singer, C.; Zeillinger, R. Efficient leukocyte depletion by a novel microfluidic platform enables the molecular detection and characterization of circulating tumor cells. *Oncotarget* **2018**, *9*, 812–823. [CrossRef] [PubMed]
19. Stathopoulou, A.; Ntoulia, M.; Perraki, M.; Apostolaki, S.; Mavroudis, D.; Malamos, N.; Georgoulias, V.; Lianidou, E.S. A highly specific real-time RT-PCR method for the quantitative determination of CK-19 mRNA positive cells in peripheral blood of patients with operable breast cancer. *Int. J. Cancer* **2006**, *119*, 1654–1659. [CrossRef]
20. Klameth, L.; Rath, B.; Hochmaier, M.; Moser, D.; Redl, M.; Mungenast, F.; Gelles, K.; Ulsperger, E.; Zeillinger, R.; Hamilton, G. Small cell lung cancer: Model of circulating tumor cell tumorospheres in chemoresistance. *Sci. Rep.* **2017**, *7*, 5337. [CrossRef]
21. Kaplan, E.L.; Meier, P. Nonparametric estimation from incomplete observations. *J. Am. Stat. Assoc.* **1958**, *53*, 457–481. [CrossRef]
22. Foy, V.; Fernandez-Gutierrez, F.; Faivre-Finn, C.; Dive, C.; Blackhall, F. The clinical utility of circulating tumour cells in patients with small cell lung cancer. *Transl. Lung Cancer Res.* **2017**, *6*, 409–417. [CrossRef] [PubMed]
23. Hou, J.M.; Greystoke, A.; Lancashire, L.; Cummings, J.; Ward, T.; Board, R.; Amir, E.; Hughes, S.; Krebs, M.; Hughes, A.; et al. Evaluation of circulating tumor cells and serological cell death biomarkers in small cell lung cancer patients undergoing chemotherapy. *Am. J. Pathol.* **2009**, *175*, 808–816. [CrossRef] [PubMed]
24. Messaritakis, I.; Politaki, E.; Kotsakis, A.; Dermitzaki, E.K.; Koinis, F.; Lagoudaki, E.; Koutsopoulos, A.; Kallergi, G.; Souglakos, J.; Georgoulias, V. Phenotypic characterization of circulating tumor cells in the peripheral blood of patients with small cell lung cancer. *PLoS ONE* **2017**, *12*, e0181211. [CrossRef] [PubMed]
25. Lampignano, R.; Yang, L.; Neumann, M.H.D.; Franken, A.; Fehm, T.; Niederacher, D.; Neubauer, H. A Novel Workflow to Enrich and Isolate Patient-Matched EpCAM(high) and EpCAM(low/negative) CTCs Enables the Comparative Characterization of the PIK3CA Status in Metastatic Breast Cancer. *Int. J. Mol. Sci.* **2017**, *18*, 1885. [CrossRef] [PubMed]

26. Maertens, Y.; Humberg, V.; Erlmeier, F.; Steffens, S.; Steinestel, J.; Bogemann, M.; Schrader, A.J.; Bernemann, C. Comparison of isolation platforms for detection of circulating renal cell carcinoma cells. *Oncotarget* **2017**, *8*, 87710–87717. [CrossRef] [PubMed]
27. Miller, M.C.; Robinson, P.S.; Wagner, C.; O'Shannessy, D.J. The Parsortix Cell Separation System-A versatile liquid biopsy platform. *Cytom. A* **2018**. [CrossRef]
28. Rizzo, F.M.; Meyer, T. Liquid Biopsies for Neuroendocrine Tumors: Circulating Tumor Cells, DNA, and MicroRNAs. *Endocrinol. Metab. Clin. N. Am.* **2018**, *47*, 471–483. [CrossRef]
29. Khan, M.S.; Tsigani, T.; Rashid, M.; Rabouhans, J.S.; Yu, D.; Luong, T.V.; Caplin, M.; Meyer, T. Circulating tumor cells and EpCAM expression in neuroendocrine tumors. *Clin. Cancer Res.* **2011**, *17*, 337–345. [CrossRef]
30. Pal, S.K.; He, M.; Chen, L.; Yang, L.; Pillai, R.; Twardowski, P.; Hsu, J.; Kortylewski, M.; Jones, J.O. Synaptophysin expression on circulating tumor cells in patients with castration resistant prostate cancer undergoing treatment with abiraterone acetate or enzalutamide. *Urol. Oncol.* **2018**, *36*, 162.e1–162.e6. [CrossRef]
31. Small, E.J.; Huang, J.; Youngren, J.; Sokolov, A.; Aggarwal, R.R.; Thomas, G.; True, L.D.; Zhang, L.; Foye, A.; Alumkal, J.J.; et al. Characterization of neuroendocrine prostate cancer (NEPC) in patients with metastatic castration resistant prostate cancer (mCRPC) resistant to abiraterone (Abi) or enzalutamide (Enz): Preliminary results from the SU2C/PCF/AACR West Coast Prostate Cancer Dream Team (WCDT). *J. Clin. Oncol.* **2015**, *33*, 5003. [CrossRef]

Permissions

All chapters in this book were first published by MDPI; hereby published with permission under the Creative Commons Attribution License or equivalent. Every chapter published in this book has been scrutinized by our experts. Their significance has been extensively debated. The topics covered herein carry significant findings which will fuel the growth of the discipline. They may even be implemented as practical applications or may be referred to as a beginning point for another development.

The contributors of this book come from diverse backgrounds, making this book a truly international effort. This book will bring forth new frontiers with its revolutionizing research information and detailed analysis of the nascent developments around the world.

We would like to thank all the contributing authors for lending their expertise to make the book truly unique. They have played a crucial role in the development of this book. Without their invaluable contributions this book wouldn't have been possible. They have made vital efforts to compile up to date information on the varied aspects of this subject to make this book a valuable addition to the collection of many professionals and students.

This book was conceptualized with the vision of imparting up-to-date information and advanced data in this field. To ensure the same, a matchless editorial board was set up. Every individual on the board went through rigorous rounds of assessment to prove their worth. After which they invested a large part of their time researching and compiling the most relevant data for our readers.

The editorial board has been involved in producing this book since its inception. They have spent rigorous hours researching and exploring the diverse topics which have resulted in the successful publishing of this book. They have passed on their knowledge of decades through this book. To expedite this challenging task, the publisher supported the team at every step. A small team of assistant editors was also appointed to further simplify the editing procedure and attain best results for the readers.

Apart from the editorial board, the designing team has also invested a significant amount of their time in understanding the subject and creating the most relevant covers. They scrutinized every image to scout for the most suitable representation of the subject and create an appropriate cover for the book.

The publishing team has been an ardent support to the editorial, designing and production team. Their endless efforts to recruit the best for this project, has resulted in the accomplishment of this book. They are a veteran in the field of academics and their pool of knowledge is as vast as their experience in printing. Their expertise and guidance has proved useful at every step. Their uncompromising quality standards have made this book an exceptional effort. Their encouragement from time to time has been an inspiration for everyone.

The publisher and the editorial board hope that this book will prove to be a valuable piece of knowledge for researchers, students, practitioners and scholars across the globe.

List of Contributors

Joanna Budna-Tukan, Monika Świerczewska, Agnieszka Jankowiak and Michał Nowicki
Department of Histology and Embryology, Poznan University of Medical Sciences, 60-781 Poznan, Poland

Martine Mazel and Catherine Alix-Panabières
Laboratory of Rare Human Circulating Cells (LCCRH), University Medical Centre of Montpellier, 34093 Montpellier, France

Wojciech A. Cieślikowski, Agnieszka Ida and Andrzej Antczak
Department of Urology, Poznan University of Medical Sciences, 61-285 Poznan, Poland

Klaus Pantel
Department of Tumor Biology, University Medical Centre Hamburg-Eppendorf, 20246 Hamburg, Germany

David Azria
Radiation Oncology Department, Montpellier Cancer Institute, 34298 Montpellier, France

Maciej Zabel
Division of Histology and Embryology, Department of Human Morphology and Embryology, Wroclaw Medical University, 50-368 Wroclaw, Poland
Division of Anatomy and Histology, University of Zielona Góra, 65-046 Zielona Góra, Poland

Kartik Anand
Houston Methodist Cancer Center, Houston, TX 77030, USA

Jason Roszik and Sapna Patel
Department of Melanoma Medical Oncology, UT MD Anderson Cancer Center, Houston, TX 77030, USA

Dan Gombos
Department of Head and Neck Surgery, Section of Ophthalmology, UT MD Anderson Cancer Center, Houston, TX 77030, USA

Joshua Upshaw, Vanessa Sarli, Salyna Meas, Anthony Lucci and Carolyn Hall
Department of Surgical Oncology, UT MD Anderson Cancer Center, Houston, TX 77030, USA

Chiara Nicolazzo, Angela Gradilone, Francesca Belardinilli, Flavia Loreni and Paola Gazzaniga
Department of Molecular Medicine, Circulating tumor cells Unit, Sapienza University of Rome, 00161 Rome, Italy

Alessandra Emiliani, Cristina Raimondi, Patrizia Seminara and Silverio Tomao
Department of Radiological, Oncological and Pathological Sciences, Division of Medical Oncology, Sapienza University of Rome, 00161 Rome, Italy

Ann Zeuner and Federica Francescangeli
Department of Hematology, Oncology and Molecular Medicine, Istituto Superiore di Sanità, 00161 Rome, Italy

Valentina Magri
Department of Surgical Sciences, Sapienza University of Rome, 00161 Rome, Italy

Ivana Bratic Hench, Luigi Costa, Jürgen Hench, Heike Püschel, Christian Ruiz, Markus Tolnay, Lukas Bubendorf and Tatjana Vlajnic
Institute of Pathology, University Hospital Basel, Schönbeinstrasse 40, 4031 Basel, Switzerland

Richard Cathomas
Department of Oncology/Hematology, Cantonal Hospital Graubünden, 7000 Chur, Switzerland

Natalie Fischer
Department of Oncology, Cantonal Hospital Winterthur, 8401 Winterthur, Switzerland

Christian Rothermundt
Department of Oncology/Hematology, Cantonal Hospital St. Gallen, 9007 St. Gallen, Switzerland

Silke Gillessen
Department of Oncology/Hematology, Cantonal Hospital St. Gallen, 9007 St. Gallen, Switzerland
University of Bern, 3012 Bern, Switzerland

Thomas Hermanns
Department of Urology, University Hospital Zurich, University of Zurich, 8091 Zurich, Switzerland

Eloïse Kremer
Swiss Group for Clinical Cancer Research (SAKK) Coordinating Center, 3008 Bern, Switzerland

Walter Mingrone
Department of Medical Oncology, Cantonal Hospital Olten, 4600 Olten, Switzerland

Ricardo Pereira Mestre
Clinic of Medical Oncology, Oncology Institute of Southern Switzerland, 6500 Bellinzona, Switzerland

Philippe Von Burg
Department of Oncology/Hematology, Hospital of Solothurn, 4500 Solothurn, Switzerland

Landon Wark, Aline Rangel-Pozzo and Sabine Mai
Cell Biology, Research Institute of Oncology and Hematology, University of Manitoba, Cancer Care Manitoba, Winnipeg, MB R3E 0V9, Canada

Harvey Quon, Aldrich Ong and Darrel Drachenberg
Manitoba Prostate Center, Cancer Care Manitoba, Section of Urology, Department of Surgery, University of Manitoba, Winnipeg, MB R3E 0V9, Canada

François-Clément Bidard and Luc Cabel
Department of Medical Oncology, Institut Curie, PSL Research University, 75005 Paris, France
Circulating Tumor Biomarkers Laboratory, Institut Curie, PSL Research University, 75005 Paris, France
UVSQ, Paris Saclay University, 92210 Saint Cloud, France

Nicolas Kiavue
Department of Medical Oncology, Institut Curie, PSL Research University, 75005 Paris, France

Jordan Madic, Adrien Saliou, Aurore Rampanou and Charlotte Proudhon
Circulating Tumor Biomarkers Laboratory, Institut Curie, PSL Research University, 75005 Paris, France

Thibault Mazard
Department of Digestive Oncology, ICM Regional Cancer Institute of Montpellier, 34298 Montpellier, France

Marc Ychou
Department of Digestive Oncology, ICM Regional Cancer Institute of Montpellier, 34298 Montpellier, France
Department of Oncology, Montpellier University, 34000 Montpellier, France

Marc-Henri Stern
INSERM U830, Institut Curie, PSL Research University, 75005 Paris, France

Charles Decraene
Circulating Tumor Biomarkers Laboratory, Institut Curie, PSL Research University, 75005 Paris, France
CNRS UMR144, Institut Curie, PSL Research University, 75005 Paris, France

Olivier Bouché
Department of Medical Oncology, Hôpital Robert Debré, Reims University Hospital, 51100 Reims, France

Michel Rivoire
Department of Digestive Oncology, Centre Léon Bérard, 69008 Lyon, France

François Ghiringhelli
INSERM U866, Centre Georges-François Leclerc, 21000 Dijon, France

Eric Francois
Department of Medical Oncology, Centre Antoine Lacassagne, 06189 Nice, France

Rosine Guimbaud
Department of Digestive Oncology, CHU de Toulouse, 31059 Toulouse, France

Laurent Mineur
Department of Digestive Oncology, Institut Sainte Catherine, 84000 Avignon, France

Faiza Khemissa-Akouz
Department of Gastroenterology, Hôpital Saint Jean, 66000 Perpignan, France

Driffa Moussata
Department of Gastroenterology, CHRU de Tours, 37044 Tours, France

Jean-Yves Pierga
Department of Medical Oncology, Institut Curie, PSL Research University, 75005 Paris, France
Circulating Tumor Biomarkers Laboratory, Institut Curie, PSL Research University, 75005 Paris, France
Université Paris Descartes, 75270 Paris, France

Trevor Stanbury
UCGI Group, R&D UNICANCER, 75654 Paris, France

Simon Thézenas
Biometrics Unit, ICM Regional Cancer Institute of Montpellier, 34298 Montpellier, France

Pascale Mariani
Department of Surgical Oncology, Institut Curie, PSL Research University, 75005 Paris, France

Simon Heeke and Baharia Mograbi
Université Côte d'Azur, CHU Nice, FHU OncoAge, 06000 Nice, France
Université Côte d'Azur, CNRS UMR7284, Inserm U1081, Institute for Research on Cancer and Aging, Nice (IRCAN), FHU OncoAge, 06000 Nice, France

Paul Hofman
Université Côte d'Azur, CHU Nice, FHU OncoAge, 06000 Nice, France
Université Côte d'Azur, CNRS UMR7284, Inserm U1081, Institute for Research on Cancer and Aging, Nice (IRCAN), FHU OncoAge, 06000 Nice, France
Laboratory of Clinical and Experimental Pathology and Biobank BB-0033-00025, Pasteur Hospital, FHU OncoAge, 06000 Nice, France

Sara R. Bang-Christensen
Centre for Medical Parasitology at Department for Immunology and Microbiology, Faculty of Health and Medical Sciences, University of Copenhagen and Department of Infectious Disease, Copenhagen University Hospital, 2200 Copenhagen, Denmark
VarCT Diagnostics, 2200 Copenhagen, Denmark

Rasmus S. Pedersen, Marina A. Pereira, Thomas M. Clausen, Caroline Løppke, Nicolai T. Sand, Theresa D. Ahrens, Amalie M. Jørgensen, Louise Goksøyr, Swati Choudhary, Tobias Gustavsson, Robert Dagil, Adam F. Sander, Thor G. Theander, Ali Salanti and Mette Ø. Agerbæk
Centre for Medical Parasitology at Department for Immunology and Microbiology, Faculty of Health and Medical Sciences, University of Copenhagen and Department of Infectious Disease, Copenhagen University Hospital, 2200 Copenhagen, Denmark

Yi Chieh Lim and Petra Hamerlik
Danish Cancer Society Research Center, 2100 Copenhagen, Denmark

Mads Daugaard
Department of Urologic Sciences, University of British Columbia, and Vancouver Prostate Centre, Vancouver, BC V6H 3Z6, Canada

Mathias H. Torp and Olga Østrup
Centre for Genomic Medicine, Copenhagen University Hospital, 2100 Copenhagen, Denmark

Max Søgaard
ExpreS2ion Biotechnologies, SCION-DTU Science Park, 2970 Hørsholm, Denmark

Ulrik Lassen
Department of Oncology, Copenhagen University Hospital, 2100 Copenhagen, Denmark

Olga Chernysheva, Irina Markina, Lev Demidov, Natalia Kupryshina, Svetlana Chulkova, Alexandra Palladina, Alina Antipova and Nikolai Tupitsyn
FSBI "N.N. Blokhin Russian Cancer Research Center" of Ministry of Health of the Russian Federation, 115478 Moscow, Russia

Carmen Garrido-Navas, Alba Ortigosa and Jose Luis Garcia Puche
GENYO, Centre for Genomics and Oncological Research (Pfizer/University of Granada/Andalusian Regional Government), PTS Granada Av. de la Ilustración, 114, 18016 Granada, Spain;

Diego de Miguel-Pérez and Jose Antonio Lorente
GENYO, Centre for Genomics and Oncological Research (Pfizer/University of Granada/Andalusian Regional Government), PTS Granada Av. de la Ilustración, 114, 18016 Granada, Spain
Laboratory of Genetic Identification, Department of Legal Medicine, University of Granada, Av. de la Investigación, 11, 18071 Granada, Spain

Maria José Serrano
GENYO, Centre for Genomics and Oncological Research (Pfizer/University of Granada/Andalusian Regional Government), PTS Granada Av. de la Ilustración, 114, 18016 Granada, Spain
Integral Oncology Division, Virgen de las Nieves University Hospital, Av. Dr. Olóriz 16, 18012 Granada, Spain

Jose Exposito-Hernandez, Victor Amezcua, Javier Valdivia and Rosa Guerrero
Integral Oncology Division, Virgen de las Nieves University Hospital, Av. Dr. Olóriz 16, 18012 Granada, Spain

Clara Bayarri
GENYO, Centre for Genomics and Oncological Research (Pfizer/University of Granada/Andalusian Regional Government), PTS Granada Av. de la Ilustración, 114, 18016 Granada, Spain
Department of Thoracic Surgery, Virgen de las Nieves University Hospital, Av. de las Fuerzas Armadas, 2, 18014 Granada, Spain

James W. T. Toh
Medical Oncology, Ingham Institute of Applied Research, School of Medicine, Western Sydney University and SWS Clinical School, UNSW Sydney 2170, NSW, Australia
Division of Colorectal Surgery, Department of Surgery, Westmead Hospital, Sydney 2145, Australia
Department of Colorectal Surgery, Concord Hospital and Discipline of Surgery, Sydney Medical School, University of Sydney, Sydney 2137, Australia

Scott MacKenzie and Les Bokey
Liverpool Clinical School, Western Sydney University, Sydney 2170, Australia

Stephanie H. Lim
Medical Oncology, Ingham Institute of Applied Research, School of Medicine, Western Sydney University and SWS Clinical School, UNSW Sydney 2170, NSW, Australia

Pierre Chapuis
Department of Colorectal Surgery, Concord Hospital and Discipline of Surgery, Sydney Medical School, University of Sydney, Sydney 2137, Australia

Paul de Souza and Kevin J. Spring
Medical Oncology, Ingham Institute of Applied Research, School of Medicine, Western Sydney University and SWS Clinical School, UNSW Sydney 2170, NSW, Australia
Liverpool Clinical School, Western Sydney University, Sydney 2170, Australia

Areti Strati and Evi S. Lianidou
Analysis of Circulating Tumor Cells Lab, Department of Chemistry, University of Athens, 15771 Athens, Greece

Michail Nikolaou
Medical Oncology Unit, "Elena Venizelou" Hospital, 11521 Athens, Greece

Vassilis Georgoulias
Metropolitan General Hospital, 15562 Athens, Greece

Loredana Cleris, Maria Grazia Daidone, Emanuela Fina and Vera Cappelletti
Biomarkers Unit, Department of Applied Research and Technological Development, Fondazione IRCCS Istituto Nazionale dei Tumori, 20133 Milan, Italy

Tala Tayoun
"Circulating Tumor Cells" Translational Platform, CNRS UMS3655 - INSERM US23AMMICA, Gustave Roussy, Université Paris-Saclay, F-94805 Villejuif, France
INSERM, U981 "Identification of Molecular Predictors and new Targets for Cancer Treatment", F-94805 Villejuif, France
Faculty of Medicine, Université Paris Sud, Université Paris-Saclay, F-94270 Le Kremlin-Bicetre, France

Vincent Faugeroux, Marianne Oulhen, Agathe Aberlenc and Françoise Farace
"Circulating Tumor Cells" Translational Platform, CNRS UMS3655 - INSERM US23AMMICA, Gustave Roussy, Université Paris-Saclay, F-94805 Villejuif, France
INSERM, U981 "Identification of Molecular Predictors and new Targets for Cancer Treatment", F-94805 Villejuif, France

Patrycja Pawlikowska
INSERM, U981 "Identification of Molecular Predictors and new Targets for Cancer Treatment", F-94805 Villejuif, France

Patrick C. Bailey
Marlene and Stewart Greenebaum Comprehensive Cancer Center, School of Medicine (UMGCCC), University of Maryland, Baltimore, MD 21201, USA

Stuart S. Martin
Marlene and Stewart Greenebaum Comprehensive Cancer Center, School of Medicine (UMGCCC), University of Maryland, Baltimore, MD 21201, USA
Department of Physiology, School of Medicine, University of Maryland, Baltimore, MD 21201, USA

Eva Obermayr, Christiane Agreiter, Eva Schuster and Robert Zeillinger
Molecular Oncology Group, Department of Obstetrics and Gynecology, Comprehensive Cancer Center, Medical University of Vienna, Waehringer Guertel 18-20, 1090 Vienna, Austria

Hannah Fabikan, Christoph Weinlinger and Maximilian Hochmair
Department of Respiratory and Critical Care Medicine, Sozialmedizinisches Zentrum Baumgartner Höhe, Sanatoriumstrasse 2, 1140 Vienna, Austria

Katarina Baluchova
Division of Oncology, Biomedical Center Martin, Jessenius Faculty of Medicine in Martin, Comenius University in Bratislava, Malá Hora 4C, 036 01 Martin, Slovakia

Gerhard Hamilton
Department of Surgery, Medical University of Vienna, Waehringer Guertel 18-20, 1090 Vienna, Austria

Index

Printed in the USA
CPSIA information can be obtained
at www.ICGtesting.com
JSHW051410091023
49903JS00006B/356